Progress in Mathematics
Volume 89

G. van der Geer F. Oort J. Steenbrink
Editors

Arithmetic Algebraic Geometry

1991

Birkhäuser
Boston · Basel · Berlin

G. van der Geer
Mathematisch Instituut
Universiteit van Amsterdam
Plantage Muidergracht 24
1018 TV Amsterdam
The Netherlands

F. Oort
Mathematisch Instituut
Rijksuniversiteit Utrecht
Budapestlaan 6
3508 TA Utrecht
The Netherlands

J. Steenbrink
Mathematisch Instituut
Katholieke Universiteit Nijmegen
Toernooiveld
6525 ED Nijmegen
The Netherlands

Printed on acid-free paper.

Camera-ready copy prepared by the authors.
Printed and bound by Edwards Brothers, Incorporated, Ann Arbor, Michigan.
Printed in the United States of America.

9 8 7 6 5 4 3 2 1

ISBN 0-8176-3513-0
ISBN 3-7643-3513-0

Contents

Participants

The following is the list of participants of the conference *Arithmetic Algebraic Geometry, Texel '89*, which was held on Texel Island during the last week of April 1989.

J. van Beele (Leiden), F. Beukers (Utrecht), S. Bloch (Chicago), J. Brinkhuis (Rotterdam), H. Carayol (Strasbourg), T. Chinburg (Philadelphia), B. Edixhoven (Utrecht), T. Ekedahl (Stockholm), J. Franke (Berlin), G. Frey (Saarbrücken), J. van Geel (Gent), B. van Geemen (Utrecht), G. van der Geer (Amsterdam), R. de Jeu (Chicago), A.J. de Jong (Nijmegen), W. van der Kallen (Utrecht), T. Katsura (Tokyo), J. van der Lingen (Amsterdam), R. Livné (Tel Aviv), J.-F. Mestre (Paris), J.P. Murre (Leiden), R. Noot (Utrecht), F. Oort (Utrecht), M. van der Put (Groningen), K. Ribet (Berkeley), K. Rubin (New York), N. Schappacher (Bonn), C.G. Schmidt (Groningen), R. Schoof (Utrecht), J.-P. Serre (Paris), B. de Smit (Amsterdam), J. Steenbrink (Nijmegen), P. Stevenhagen (Amsterdam), J. Stienstra (Utrecht), J. Top (Utrecht), R. Versseput (Amsterdam), P. Vojta (Berkeley), J. Wildeshaus (Cambridge), D. Zagier (Maryland, Bonn), Y.G. Zarhin (Moscow).

Contributors

T. Chinburg Department of Mathematics, University of Pennsylvania, Philadelphia, PA 19104, USA

J. de Jong Mathematisch Instituut, Katholieke Universiteit Nijmegen, Toernooiveld, 6525 ED Nijmegen, The Netherlands

B. Edixhoven Department of Mathematics, University of California, Berkeley, CA 94720, USA

T. Ekedahl Matematiska Institutionen, Stockholms Universitet, Box 6071, S-11385 Stockholm, Sweden

J. Franke Karl-Weierstrass-Institut für Mathematik, Mohrenstrasse 39, DDR-1080 Berlin, GDR

G. Frey Institut für Experimentelle Mathematik, Universität Essen, D-4300 Essen, FRG

E. Kani Department of Mathematics and Statistics, Jeffery Hall, Queen's University, Kingston, Ontario, K7L 3N6, Canada

J.-F. Mestre Département de Mathématiques et d'Informatique, École Normale Supérieure, Mathématiques, 45 rue d'Ulm, F-75230 Paris Cedex 05, France

R. Noot Mathematisch Instituut, Rijksuniversiteit Utrecht, Budapestlaan 6, 3508 TA Utrecht, The Netherlands

F. Oort Mathematisch Instituut, Rijksuniversiteit Utrecht, Budapestlaan 6, 3508 TA Utrecht, The Netherlands

A.N. Parshin Steklov Mathematical Institute, Ul. Vavilova 42, Moscow 117966, GSP-1, USSR

K. Ribet Mathematics Department, University of California, Berkeley, CA 94720, USA

K. Rubin Department of Mathematics, Ohio State University, Columbus, OH 43210, USA

R. Rumely Department of Mathematics, University of Georgia, Athens, GA 30602, USA

N. Schappacher Max-Planck-Institut für Mathematik, Gottfried-Clarenstrasse 26, D-5300 Bonn 3, FRG

R. Schoof Mathematisch Instituut, Rijksuniversiteit Utrecht, Budapestlaan 6, 3508 TA Utrecht, The Netherlands

J. Stienstra Mathematisch Instituut, Rijksuniversiteit Utrecht, Budapestlaan 6, 3508 TA Utrecht, The Netherlands

B. van Geemen Mathematisch Instituut, Rijksuniversiteit Utrecht, Budapestlaan 6, 3508 TA Utecht, The Netherlands

P. Vojta Department of Mathematics, University of California, Berkeley, CA 94720, USA

D. Zagier Max-Planck-Institut für Mathematik, Gottfried-Clarenstrasse 26, D-5300 Bonn 3, FRG

Y.G. Zarhin Research Computing Centre, USSR Academy of Sciences, Puschino, Moscow Region 142292, USSR

Introduction

Es zeigt sich hier einmal mehr, dass die Zahlentheorie zwar mit recht die Königin der Mathematik genannt wird, sie aber ihren Glanz, wie auch Königinnen selbst, nicht so sehr aus sich selbst als vielmehr aus den Kräften ihrer Untertanen zieht.

G. Faltings (1984)

At the moment fascinating developments are taking place in arithmetic algebraic geometry. Very prominent among these is that of Arakelov geometry. This is a way of "completing" a variety over the ring of integers of a number field by adding fibres over the archimedean places. In this way the analogy between algebraic number fields and function fields of algebraic curves is extended to a more precise analogy between arithmetic varieties and varieties fibered over a complete curve. Thus a completely new tool for attacking arithmetic problems has emerged from the Russian school of arithmetic algebraic geometry. The importance of this development lies not only in its direct results (like the proof of the Mordell Conjecture), but also in the link it establishes between number theory and complex analytic geometry.

Another fascinating development is the appearance of the relation between arithmetic geometry and Nevanlinna theory, or more precisely between diophantine approximation theory and the value distribution theory of holomorphic maps. Vojta has formulated conjectures generalizing Mordell's Conjecture by using the unexplained similarity between non-degenerate holomorphic maps from $\mathbb{C}^n$ to a complex variety V and sets of infinitely many K-rational points of a variety V defined over a number field K which are non-degenerate (in the sense that they do not lie in a proper Zariski-closed subset of V).

Besides these developments there is a lot of other activity in the field : algebraic geometry is applied in many ways to solve arithmetic problems. It was against this background that we decided to organize a conference on arithmetic algebraic geometry on Texel Island in 1989. The present volume appears on the occasion of this conference, though the contributions do not always correspond to lectures given at the conference.

We would like to take the opportunity to thank the participants of the conference and especially the speakers, who made the conference into a success. We would also like to express our gratitude to the institutions that financed the conference : de Universiteit van Amsterdam, de Rijksuniversiteit Utrecht, de Katholieke Universiteit Nijmegen, de Koninklijke Nederlandse Akademie van Wetenschappen, de Nederlandse Organisatie voor Wetenschappelijk Onderzoek, de Vertrouwenscommissie and IBM Nederland.

Gerard van der Geer
Frans Oort
Jozef Steenbrink

April 1990

Well-Adjusted Models for Curves over Dedekind Rings

T. Chinburg[1] R. Rumely[1]

§0. Introduction.

Let $\mathcal{O}$ be an excellent Dedekind ring with perfect residue fields, and let $Y = Spec(\mathcal{O})$. Let $\mathcal{C}$ be a curve over Y. (For precise definitions, see §1; we assume $\mathcal{C}$ has a smooth geometrically irreducible general fibre, but we do not assume $\mathcal{C}$ is regular or complete). In this paper we will prove a relative minimal models Theorem, and a variant due to M. Artin of the Deligne-Mumford stable reduction Theorem. To state these results, let $M(\mathcal{C})$ be the set of regular curves $\mathcal{C}'$ for which there is a proper birational Y-morphism $\mathcal{C}' \to \mathcal{C}$. Let $\geq$ be the partial order on $M(\mathcal{C})$ defined by $\mathcal{C}' \geq \mathcal{C}''$ if there is a proper morphism $\mathcal{C}' \to \mathcal{C}''$ over $\mathcal{C}$.

Theorem 0.1. *Up to a unique isomorphism over $\mathcal{C}$, there is a unique minimal element $\check{\mathcal{C}}$ of $M(\mathcal{C})$.*

The curve $\check{\mathcal{C}}$ will be called the minimal regular curve over $\mathcal{C}$.

If $\mathcal{C}$ is complete, Theorem 0.1 is related to the classical minimal models theorem of [8] and [13] in the following way. A complete curve $\mathcal{C}$ has a unique minimal model $\mathcal{C}^m$ only if $\mathcal{C}$ has positive genus. In this case, $\mathcal{C}^m$ is up to isomorphism over Y the unique complete regular Y-curve such that for every complete regular Y-curve $\mathcal{C}'$ which is birational to $\mathcal{C}$ there is a proper birational Y-morphism $\mathcal{C}' \to \mathcal{C}^m$. If $\mathcal{C}$ is complete then $\check{\mathcal{C}}$ is as well, but $\mathcal{C}^m$ need not be in $M(\mathcal{C})$ and $\check{\mathcal{C}}$ is not equal to $\mathcal{C}^m$ in general.

Let K be the fraction field of $\mathcal{O}$ and let L be a finite extension of K. Define $\mathcal{O}_L$ be the integral closure of $\mathcal{O}$ in L. Let $Y_L = Spec(\mathcal{O}_L)$ and $\mathcal{C}_L = \mathcal{C} \times_Y Y_L$. The Deligne-Mumford stable reduction Theorem (cf. [5]) states that if $\mathcal{C}$ is complete and has genus at least two, there is a finite extension L_1 of K such that if L contains L_1, the fibers over Y_L of the minimal model $(\mathcal{C}_L)^m$ of $\mathcal{C}_L$ are reduced with at most ordinary double points.

Using this result, Artin proved the following Theorem without any assumption of the genus of $\mathcal{C}$.

Theorem 0.2 (Artin). *Let $\mathcal{C}$ be a complete curve over Y. There is a finite extension L_0 of K so that if L contains L_0, the minimal regular Y_L-curve $\check{\mathcal{C}}_L$ over $\mathcal{C}_L$ is well-adjusted, in the following sense. Let $(\check{\mathcal{C}}_L)_s$ be the*

[1]Both authors are partially supported by MSRI, NSF and Sloan Foundation grants

fibre of $\check{C}_L$ over the closed point s of Y_L. Then for all s, $(\check{C}_L)_s$ is reduced and has irreducible components which are non-singular and geometrically irreducible. A pair of such irreducible components intersects in at most one point, and this intersection is transverse if it exists.

We note that Theorem 0.2 implies that the intersection point and the (distinct) tangent directions of any two intersecting irreducible components of $(\check{C}_L)_s$ are defined over the residue field $k(s)$ of s. Since $k(s)$ is perfect, the irreducible components of $(\check{C}_L)_s$ are smooth.

Artin's original proof of Theorem 0.2 involved constructing certain two-dimensional algebraic spaces, which he then argued must be schemes. In the proof we give in §2 and §3 we construct the required algebraic spaces directly within the category of schemes.

We end this introduction by mentioning an application of Theorems 0.1 and 0.2 which is developed in [4]. Suppose that $\mathcal{O}$ is the ring of integers of a non-archimedean local field and that $\mathcal{C}$ is a complete regular curve over $Spec(\mathcal{O})$. Let $\mathbf{C}$ be the completion of an algebraic closure of K. In [4] we use Theorems 0.1 and 0.2 to define a canonical Green's function $((x, y))$ on pairs of points x and y of $\mathcal{C}(\mathbf{C})$. This function has non-negative real values, and is a non-archimedean counterpart of the Green's function $G(x, y)$ of the Arakelov metric on a Riemann surface of positive genus. As in the archimedean case, the function $((x, y))$ arises when one constructs extensions of Neron's local height pairing to divisors of arbitrary degree which have desirable properties from the point of view of intersection theory. For a precise statement of these properties and the relation of $((x, y))$ to intersection theory, see [4].

Acknowledgements: The authors would like to thank M. Artin for notes of his proof of Theorem 0.2. Thanks also go to the referee for suggesting an improvement in Theorem 0.1, and to H. Gillet, D. Harbater, W. McCallum, L. Moret-Bailly and L. Szpiro for useful conversations.

§1. Minimal regular curves over Dedekind rings.

Let $\mathcal{O}$ be an excellent Dedekind ring having perfect residue fields and quotient field K. Let $Y = Spec(\mathcal{O})$.

Definition 1.1. *A curve over Y is a scheme $\mathcal{C}$ together with a morphism $\pi: \mathcal{C} \to Y$ such that*

(i) π is flat and of finite type,

(ii) the irreducible components of the fibers of π are of dimension 1, and

(iii) the general fiber $\mathcal{C}_{/K} = K \otimes_{\mathcal{O}} \mathcal{C}$ is a smooth, geometrically irreducible algebraic curve over K.

If π is proper, then C will be called a complete curve.

Let C be a regular curve over Y. An exceptional curve E on C is a prime divisor of C with the following property: There is a proper birational morphism $f : C \to C'$ from C to another regular curve C' over Y which is an isomorphism off E, and for which $f(E)$ is a closed point of C'. By [3, Cor. 2.1], C' is determined up to isomorphism over Y by C and E. We will call f the blowing down of E on C, or alternately the contraction of E on C.

Proposition 1.2. *Let C be a curve over Y. Define $M(C)$ to be the set of regular curves C' over Y for which there is a proper birational Y-morphism $C' \to C$. Let $\geq$ be the partial order on $M(C)$ defined by letting $C' \geq C''$ if there is a proper morphism $C' \to C''$ over C. Then $M(C)$ is non-empty. Let C' be an element of $M(C)$. Blowing down successively all exceptional curves which do not dominate an irreducible component of a fiber of C over Y results after finitely many steps in a minimal element of $M(C)$. If C' and C'' are two minimal elements of $M(C)$, then C' and C'' are uniquely isomorphic over C.*

We first prove a Lemma.

Lemma 1.3. *Let C be a curve over Y. Then C is integral, and K is algebraically closed in the function field $K(C)$ of C.*

Proof. Since $C_{/K}$ is smooth, $C_{/K}$ is geometrically regular by [7, Theorem III.10.2]. Since $C_{/K}$ has been assumed geometrically irreducible, $C_{/K}$ must be geometrically integral. Therefore $C_{/K}$ is integral. Suppose the local ring $\mathcal{O}_p$ of C at some point p is not reduced. Let $y = \pi(p)$, and let $\mathcal{O}_y$ be the local ring of y on Y. Let $\pi_p^\sharp$: $\mathcal{O}_y \to \mathcal{O}_p$ be the flat local homomorphism induced by π ; $\pi_p^\sharp$ is faithfully flat by [11, Prop. I.2.7(d)]. We may tensor the inclusion $\mathcal{O}_y \to K$ with $\mathcal{O}_p$ over $\mathcal{O}_y$ to arrive at an injection $\mathcal{O}_p \otimes_{\mathcal{O}} \mathcal{O}_y \to \mathcal{O}_p \otimes_{\mathcal{O}} K$. Now $\pi_p^\sharp$ is injective by [11, Prop. I.2.18], so $\mathcal{O}_p \otimes_{\mathcal{O}_y} \mathcal{O}_y = \mathcal{O}_p$. Thus because $\mathcal{O}_p$ has nilpotents, $\mathcal{O}_p \otimes_{\mathcal{O}_y} K$ does as well. Hence $C_{/K}$ is not reduced, but this contradicts the fact that $C_{/K}$ is integral. Therefore C is reduced. Since π is an open morphism by [11, Theorem I.2.12], C must be irreducible because $C_{/K}$ is irreducible. Thus C is integral. Since $C_{/K}$ is geometrically integral, K must be algebraically closed in $K(C)$.

Corollary 1.4. *For all curves C, the normalization $C^\sharp$ of C in the function field $K(C)$ of C is well-defined. The natural morphism $C^\sharp \to C$ is finite and birational.*

Proof. We have assumed that $\mathcal{O}$ is an excellent Dedekind ring. By [6, EGA IV, Prop. 7.8.6], a scheme which is locally of finite type over an excellent scheme is excellent. Hence C is excellent, so by [6, EGA IV, Scholie 7.8.3(vi)], $C^\sharp \to C$ is finite and birational. Proof of Proposition 1.2. By resolution of singularities for two dimensional schemes [9,1], we can find a regular curve C' over Y and a proper birtional morphism $C' \to C^\sharp$ over Y. Since finite morphisms are proper, Corollary 1.4 implies that the resulting Y-morphism $C' \to C$ is proper and birational. Thus $M(C)$ is non-empty.

Suppose now that C' is an arbitrary element of $M(C)$. An exceptional curve E on C' is an irreducible component of reducible fiber of C' over a closed point of Y. Because the general fiber of C' is irreducible, [6, EGA IV, 9.7.8] implies that the set of points of Y over which the fiber of C' is reducible is a proper closed set. Hence C' has finitely many reducible fibers because $\mathcal{O}$ is a Dedekind ring. Blowing down E on C' reduces by 1 the number of irreducible components of reducible fibers of C'. Hence if one successively blows down exceptional curves which do not dominate a component of a fiber of C, one arrives after finitely many steps at a curve C_0 having no exceptional curves of this kind. By the Factorization Theorem ([8, Theorem 1.15], [13, Chapter I, section 2]), C_0 is a minimal element of $M(C)$.

Suppose now that C' and C'' are two minimal elements of $M(C)$ which are not isomorphic over C. We will show that this leads to a contradiction using a variation of the argument in [13, pp. 136-139].

By the Factorization Theorem and [8, Prop. 4.2], we have a diagram

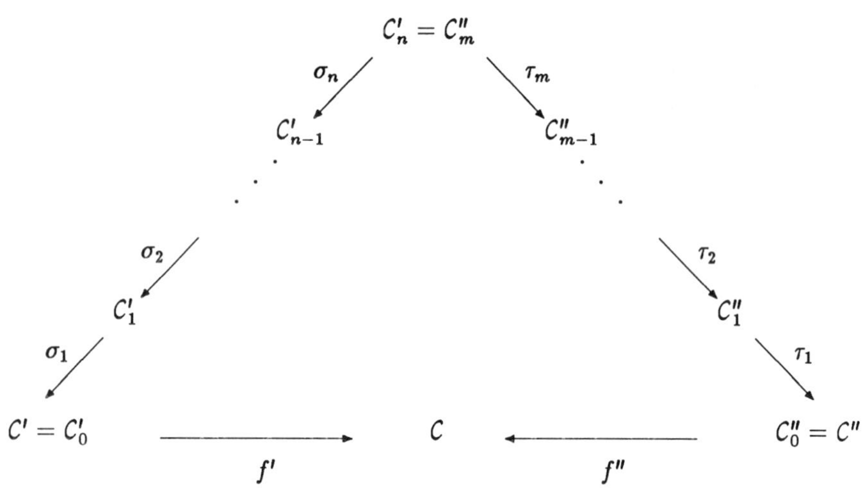

in which the σ_i and τ_j are morphisms which contract a single exceptional curve, the C'_i and C''_j are elements of $M(C)$, and f' and f'' are proper birational Y- morphisms. We may assume that a diagram of this kind has been chosen so that $m+n$ is minimized. Then $m+n \geq 0$, since we have supposed C' and C'' are not isomorphic over C. If $m=0$, then σ_n would contract an exceptional curve on $C'_n = C''$ which does not dominate a component of a fiber of C over Y. But there are no such curves on C'' since C'' is a minimal element of $M(C)$. Hence $m > 0$, and by symmetry, $n > 0$.

Let L be the exceptional curve on $C'_n = C''_m$ which is contracted by τ_m. We claim that $\sigma_1 \circ \ldots \circ \sigma_n(L)$ must be a curve on $C' = C'_0$. Otherwise $\sigma_1 \circ \ldots \circ \sigma_n(L)$ is a closed point of C'. Then there would be a proper birational Y-morphism $\mu \colon C''_{m-1} \to C'$ such that $\sigma_1 \circ \ldots \circ \sigma_n = \mu \circ \tau_m$. Use the Factorization Theorem to write μ as the composition $\mu_1 \circ \ldots \circ \mu_r$ of morphisms μ_i which each contract a single exceptional curve. The irreducible components of the fibers of $C'_n = C''_m$ which are contracted by the morphism $\sigma_1 \circ \ldots \circ \sigma_n = \mu_1 \circ \ldots \circ \mu_r \circ \tau_m$ are the proper transforms of the exceptional curves blown down by the σ_i (resp. by the μ_j and by τ_m). Thus there are exactly $n = r+1$ contracted components of fibers of C'_n. But we now have a diagram

$$\begin{array}{ccccc} & & C''_{m-1} & & \\ & \mu = \mu_1 \circ \ldots \circ \mu_r \swarrow & & \searrow \tau_1 \circ \ldots \circ \tau_{m-1} & \\ C' & \xrightarrow{\ f'\ } & C & \xleftarrow{\ f''\ } & C'' \end{array}$$

with $r + m - 1 = m + n - 2$, and this contradicts the minimality of $m+n$. Thus $\sigma_1 \circ \ldots \circ \sigma_n(L)$ is a curve on $C' = C'_0$.

For $0 \leq i \leq n-1$ let L_i be the curve on C'_i which is the image of L under $\sigma_{i+1} \circ \ldots \circ \sigma_n$, and define $L_n = L$. Then L_{i+1} is the proper transform of L_i with respect to the morphism $\sigma_{i+1} \colon C'_{i+1} \to C'_i$. Since L is contracted by τ_m, $w = f'(L_0)$ is a closed point on C. Let $k(w)$ be the residue field of w. By the proof of [8, Prop. 2.9], L is isomorphic to $P^1(N)$ for some finite extension N of $k(w)$. Let x_i be the point of C'_i which is blown up by $\sigma_{i+1} \colon C'_{i+1} \to C'_i$. Let s_i be the multiplicity of L_i at x_i, where $s_i = 0$ if L_i does not contain x_i. As in [13, p. 138-139], we have

$$\begin{aligned} -[N: k(w)] &= (L, L)_n \\ &= (L_0, L_0)_0 - \sum_{i=0}^{n-1} s_i^2 [k(x_i): k(w))] \end{aligned}$$

(1.1)

where $(\ ,\ )_j$ denotes the intersection pairing on the fiber of C'_j over w, and where $k(x_i)$ is the residue field of x_i.

Suppose that all the s_i are 0. Then there is a neighborhood of L which is mapped isomorphically onto a neighborhood of L_0 by $\sigma_1 \circ \ldots \circ \sigma_n$. Therefore L_0 would be an exceptional curve on C'_0. Since L_0 is mapped to a point on C, this contradicts the fact that $C'_0 = C'$ is a minimal element of $M(C)$. Therefore some s_i is positive.

Let $i \leq n-1$ be the largest integer for which $s_i > 0$. Since L_{i+1} is the proper transform of L_i under $\sigma_{i+1}: C'_{i+1} \to C'_i$, we have

(1.2) $$(L_{i+1}, \sigma_{i+1}^{-1}(x_i))_{i+1} = s_i[k(x_i): k(w)] > 0$$

By our definition of s_i, $\sigma_{i+1} \circ \ldots \circ \sigma_n$ maps L isomorphically to L_{i+1}. Hence L_{i+1} is isomorphic to $P^1(N)$, so the residue field of every closed point of L_{i+1} contains N. Thus the left hand side of (1.2) is at least $[N: k(w)]$. It now follows from (1.1) that $(L_0, L_0)_0 \geq 0$. But $(\ ,\ )_0$ is negative definite (cf. [13, pp. 95-96]), so the contradiction shows C' and C'' must have been isomorphic over C.

To show that C' and C'' are uniquely isomorphic over C, we must show that the only automorphism ϕ of C' over C is the identity. Since $C' \to C$ is birational, ϕ is the identity on a dense open subset of C'. Because C' is reduced and separated over Y, ϕ is the identity by [7, Ex. 4.2, p. 105].

Definition 1.5. *Let C be a curve over Y. The elements of $M(C)$ will be called regular curves over C. Let $\check{C}$ denote a minimal element of $M(C)$. We will identify all of the minimal elements of $M(C)$ with each other, and we will call $\check{C}$ the minimal regular curve over C.*

§2. Artin's well-adjusted models theorem.

Definition 2.1. *A curve C over Y is well-adjusted if C is regular, and if for all closed points s of Y, the fiber C_s of C over Y has the following properties:*

(i) C_s *is reduced;*

(ii) *the irreducible components of* C_s *are non-singular, geometrically irreducible, and intersect transversely; and*

(iii) *each pair of irreducible components of* C_s *intersect in at most one point.*

Definition 2.2. *Suppose that* L *is an algebraic extension of* K. *Let* O_L *be the integral closure of* O *in* L, *and let* $Y_L = Spec(O_L)$. *If* C *is a curve over* Y, *let* C_L *be the curve* $O_L \otimes_O C$ *over* Y_L.

Theorem 2.3. (Artin) *Suppose* C *is a complete curve over* Y. *Then there is a finite extension* L *of* K *for which the minimal regular curve* $\check{C}_L$ *over* C_L *is well-adjusted.*

In the remainder of this section we reduce the proof of Artin's Theorem to the proof of the assertion in Corollary 2.11 below.

Proposition 2.4. *Suppose* C *is a complete well-adjusted curve over* Y. *The contraction of an exceptional curve on* C *results in a complete well-adjusted curve over* Y.

Proof. Let $f: C \to C'$ be the blow down of an exceptional curve E on C. Let s be the closed point of Y for which the fiber C_s contains E. The curve C' is regular and complete, and f is surjective and an isomorphism off E. Hence the fibers of C' satisfy conditions (i) and (ii) of Definition 2.1. To check condition (iii), it will suffice to show that E intersects at most one other component of C_s. This is so because E has self intersection -1 ([8, Thm. 3.9]), and E has trivial intersection with C_s.

In view of Proposition 1.2, we have the following Corollary on contracting all exceptional curves which do not dominate a component of a closed fiber of C_L.

Corollary 2.5. *With the hypotheses of Theorem 2.3, suppose* L *is a finite extension of* K *for which there is a well-adjusted regular curve in the set* $M(C_L)$ *of Proposition 1.2. Then* $\check{C}_L$ *is well-adjusted.*

Proposition 2.6. *Suppose* C *is a regular curve over* Y, *whose fibers are reduced with at most ordinary double points. Let* L *be a finite extension of* K. *Suppose* x_L *is a singular point of* $C_L = O_L \otimes_O C$. *Let* x *be the point of* C *under* C_L. *Let* y_L *and* y *be the points of* Y_L *and* Y *under* x_L *and* x, *respectively. Then* x *is a singular point on the fiber* C_y. *The completion of* O_{C_L,x_L} *is isomorphic to* $O_{Y_L,y_L}[[z,w]]/(zw - \pi_y)$, *where* π_y *is a uniformizer in* $O_{Y,y}$. *Let* e *be the ramification degree of* y_L *over* y. *Performing* $[\frac{e}{2}]$

successive blow-ups resolves the singularity at x_L on C_L. When this is done for all the singular points of C_L, one arrives at the minimal regular curve $\check{C}_L$ over C_L. The closed fibers of $\check{C}_L$ are reduced with at most ordinary double points.

Proof. This follows readily from Proposition 1.2 and the argument in [5, pp. 84-85], the latter applying to curves over arbitrary excellent Dedekind rings. Suppose now that C in Proposition 2.6 is well adjusted. Let F be an irreducible component of the fiber T of $\check{C}_L$ over y_L. Then F is either one of a chain of rational curves introduced by the blow ups in Proposition 2.6, or F is the unique component of T over the (geometrically irreducible) image of F in C. This leads to the following Corollary.

Corollary 2.7. *If C is a well-adjusted curve over Y, and L is a finite extension of K, then the minimal regular curve $\check{C}_L$ over C_L is well adjusted.*

Proposition 2.8. *If Theorem 2.3 is true when $\mathcal{O}$ is an excellent* DVR *with perfect residue field, then Theorem 2.3 is true for all excellent Dedekind rings with perfect residue fields.*

Proof. The general fiber C_K of C is smooth and geometrically irreducible. Let S be the set of closed points s of Y where the fiber C_s is not smooth and geometrically irreducible. Because $\mathcal{O}$ is a Dedekind ring, it follows from [6, EGA IV, 9.7.8] and [11, Prop. I.3.24] that S is finite. By [10, Def. 28.D and Thm. 61], C_s is non-singular if s is not in S.

If L is a finite extension of K and s is in S, let $(\mathcal{O}_L)_s$ be the integral closure of $\mathcal{O}_{Y,s}$ in L. We apply Theorem 2.3 and Corollary 2.7 to the curves $\mathcal{O}_{Y,s} \otimes_{\mathcal{O}} C$ over $Spec(\mathcal{O}_{Y,s})$ for s in S. Since S is finite, we conclude that there is a single finite extension L of K such that for $s \in S$, the minimal regular curve over $(\mathcal{O}_{Y,s} \otimes_{\mathcal{O}} C)_L$ is well-adjusted. The characterization of minimal regular curves in Proposition 1.2 shows that $(\mathcal{O}_L)_s \otimes \check{C}_L$ is the minimal regular curve over $(\mathcal{O}_{Y,s} \otimes_{\mathcal{O}} C)_L$. Hence the fibers of $\check{C}_L$ at all points of Y_L over S satisfy conditions (i)-(iii) of Definition 2.1. These conditions hold automatically for fibers above points not over S, since such fibers are non-singular by Proposition 2.6, and they are also geometrically irreducible.

Proposition 2.9. *Suppose Theorem 2.3 is true when $\mathcal{O}$ is an excellent* DVR *with algebraically closed residue field. Then Theorem 2.3 is true whenever $\mathcal{O}$ is an excellent* DVR *with perfect residue field.*

Proof. Let $\mathcal{O}$ be an excellent DVR with perfect residue field. Let $\mathcal{O}^{sh}$ be the strict Henselization of $\mathcal{O}$ ([11, p. 38]). The residue field of $\mathcal{O}^{sh}$ is the separable closure of that of $\mathcal{O}$, which is thus algebraically closed. Define $C^{sh} = \mathcal{O}^{sh} \otimes_{\mathcal{O}} C$. Apply Theorem 2.3 to C^{sh}. One finds that there is a finite extension L_0 of the fraction field K^{sh} of $\mathcal{O}^{sh}$ such that $\check{C}_{L_0}$ is well-adjusted.

Our objective now is to show that if L is a sufficiently large finite extension of K inside L_0, then there is a well-adjusted regular Y_L-curve $\mathcal{C}_0$ such that $\check{\mathcal{C}}_{L_0} = \mathcal{O}_{L_0} \otimes_{\mathcal{O}_L} \mathcal{C}_0$, and for which there is a proper birational Y_L-morphism $\mathcal{C}_0 \to \mathcal{C}_L$. Corollary 2.5 will then show $\check{\mathcal{C}}_L$ is well-adjusted, and Proposition 2.9 will be proved.

The curve $\check{\mathcal{C}}_{L_0}$ is of finite type over $Y_{L_0} = Spec(\mathcal{O}_{L_0})$. We may hence describe $\check{\mathcal{C}}_{L_0}$ by patching data involving finitely many ring homomorphisms between finitely many finitely generated $\mathcal{O}_{L_0}$-algebras. Thus we may find a finite extension L of K inside L_0 and a scheme $\mathcal{C}_0$ over Y_L for which there is a Cartesian square

$$\begin{array}{ccc} \check{\mathcal{C}}_{L_0} & \xrightarrow{\ f\ } & \mathcal{C}_0 \\ \downarrow & & \downarrow \\ Y_{L_0} & \xrightarrow{\ h\ } & Y_L \end{array}$$

(2.1)

in which h is induced by the inclusion of $\mathcal{O}_L$ into $\mathcal{O}_{L_0}$, and where f and h are faithfully flat. Let $t: \check{\mathcal{C}}_{L_0} \to \mathcal{C}$ by the Y-morphism which is the composition of $\check{\mathcal{C}}_{L_0} \to \mathcal{C}_{L_0}$ and $\mathcal{C}_{L_0} \to \mathcal{C}$. We may further require that $t = t_L \circ f$ for some Y-morphism $t_L: \mathcal{C}_0 \to \mathcal{C}$.

By faithfully flat descent (cf. [11, p. 20], [6, EGA IV 2.6, 2.7], [6, EGA O_{IV}, 17.3.3]), $\mathcal{C}_0$ must be a complete regular curve over Y_L because $\check{\mathcal{C}}_{L_0}$ has these properties over Y_{L_0}. Enlarge L so that the radical ideal of $\mathcal{O}_L$ generates that of $\mathcal{O}_{L_0}$. Then the fibers of $\mathcal{C}_0$ over Y_L must be reduced because those of $\check{\mathcal{C}}_{L_0}$ are reduced. As in the proof of Proposition 2.8, there are only finitely many singular fibers of $\mathcal{C}_0$ over Y_L, and there are only finitely many singular points on these fibers. By enlarging L by a further finite extension, we may make $\mathcal{C}_0$ well-adjusted because $\check{\mathcal{C}}_{L_0}$ is well-adjusted. Since $t_L: \mathcal{C}_0 \to \mathcal{C}$ is a Y-morphism and $\mathcal{C}_0$ is a curve over $\mathcal{O}_L$, we have a Y_L-morphism $1 \otimes t_L: \mathcal{C}_0 \to \mathcal{O}_L \otimes_{\mathcal{O}} \mathcal{C} = \mathcal{C}_L$. We assumed $\mathcal{C}$ is complete, so $\mathcal{C}_L$ is proper over Y_L. Since $\mathcal{C}_0$ has also been shown to be proper over Y_L, it follows from [7, Cor. II. 4.8(e)] that $1 \otimes t_L: \mathcal{C}_0 \to \mathcal{C}_L$ is proper.

It remains only to show that $1 \otimes t_L$ is birational.

Because (2.1) is Cartesian, the homomorphism of function fields $(1 \otimes t_L)_*: K(\mathcal{C}_L) = L \otimes_K K(\mathcal{C}) \to K(\mathcal{C}_0)$ induced by $1 \otimes t_L$ induces an isomorphism $L_0 \otimes_L (L \otimes_K K(C)) = K(\check{C}_{L_0}) \to L_0 \otimes_L K(C_0)$. Since L_0 is faithfully flat over L, $(1 \otimes t_L)_*$ must be an isomorphism, so we are done.

Proposition 2.10. *Suppose $\mathcal{O}$ is an excellent DVR with algebraically closed residue field. Suppose C is a regular complete curve over Y for which the special fiber C_s is reduced and has at most ordinary double points. Let π be a uniformizer for $\mathcal{O}$, and let $L = K(\pi^{\frac{1}{4}})$. Then $\check{C}_L$ is well-adjusted.*

<u>Proof</u>. Let $L' = K(\pi^{\frac{1}{2}})$. Let $C^* = \check{C}_{L'}$, be the $Y_{L'}$-curve which results as in Proposition 2.6 from blowing up the singular points of $C_{L'} = \mathcal{O}_{L'} \otimes_{\mathcal{O}} C$. Let $C^*_{s'}$ be the fiber of C^* over the unique closed point s' of $Y_{L'}$. An ordinary double point p on C_s is either (i) a singular point of an irreducible component F_p of C_s, with no other irreducible component of C_s containing p, or (ii) the intersection of two irreducible components F_p and F'_p of C_s, with p a non-singular point of F_p and F'_p. We may identify the unique closed fiber $(C_{L'})_{s'}$ of $C_{L'}$ with C_s. In case (i), blowing up p on $C_{L'}$ resolves the singularity of F_p at p, and introduces a reduced rational curve E_p which intersects the proper transform F^*_p of F_p in two ordinary double points, with E_p intersecting no other components of $C^*_{s'}$:

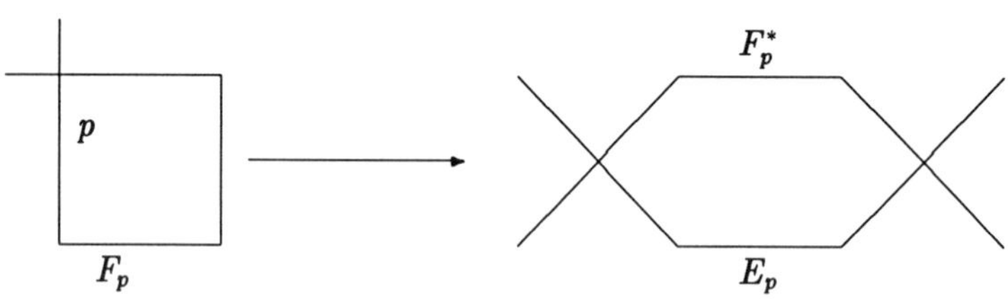

In case (ii), blowing up p on $C_{L'}$, introduces a reduced rational curve E_p which intersects each of the proper transforms F^*_p and F'^*_p at ordinary double points, with E_p intersecting no other components of $C^{*\prime}_{s}$. Furthermore, each of E_p, F^*_p and F'^*_p are non-singular over p, and F^*_p and F'^*_p do not intersect above p:

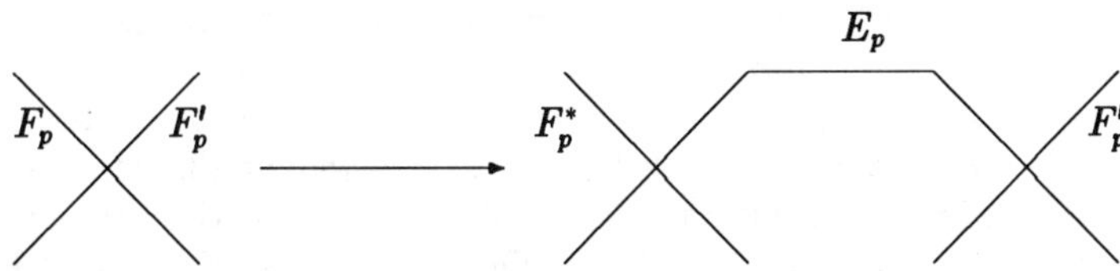

Thus the special fiber $C^*_{s'}$ is reduced, and each of the irreducible components of $C^*_{s'}$ are non-singular, though pairs of components of $C^*_{s'}$ may intersect more than once. However, from case (ii) above we see that when one blows up the singular points of $(\mathcal{O}_L \otimes_{\mathcal{O}_{L'}} C^*)$, one arrives at a well adjusted curve C^{**} over Y_L, where $C^{**} = \check{C}_L$ by Proposition 2.6.

Corollary 2.11. *To show Theorem 2.3, it will be enough to prove the following. Let $\mathcal{O}$ be an excellent* DVR *with algebraically closed residue field. Let C be a complete normal curve over Y. Then there is a finite extension L of K and regular curve C^* over C_L such that the special fiber of C^* is reduced with at most ordinary double points.*

Proof. Let $\mathcal{O}$ be as in Corollary 2.11, and suppose C_1 is a complete curve over Y. By Corollary 1.4, $C = C_1^\natural$ is a complete normal curve over Y which is finite over C_1 and birational to C_1. Apply the hypotheses of Corollary 2.11 to C, and then use Proposition 2.10. We conclude that Theorem 2.3 holds for C_1 , and thus that Theorem 2.3 is true when $\mathcal{O}$ is as in Corollary 2.11. Now Propositions 2.9 and 2.8 show that Theorem 2.3 holds in general.

§3. Completion of the proof of Artin's Theorem.

Throughout this section, $\mathcal{O}$ will be an excellent DVR with algebraically closed residue field k, and C will be a complete normal curve over Y. Our objective is to show that there is a finite extension L of K and a curve C^* over Y_L satisfying the conditions of Corollary 2.11.

Theorem 3.1 (Deligne and Mumford [5]). *Suppose that the generic fiber $C_{/K}$ of C has genus at least 2. Then there is a finite extension L of K such that the closed fibers of the minimal model C_L^m of $L \otimes_K C_{/K}$ are reduced with at most ordinary double points.*

Recall that C_L^m is a regular curve over Y_L, and there is a birational map from C_L^m to C_L over Y_L. This map may not be a morphism, and our given curve $\mathcal{C}$ may not have generic fiber of genus at least 2. To get around these difficulties, we construct etale covers of open subsets of $\mathcal{C}$ such that the Deligne-Mumford Theorem can be applied to the normalizations of suitable projective closures of these covers. In this way one finds for each closed point of $\mathcal{C}$ a local solution in the etale topology to the problem of finding L and $\mathcal{C}^*$ as in Corollary 2.11. The last step is to descend this solution to a global solution in the Zariski topology.

Proposition 3.2. *Let p be a closed point of $\mathcal{C}$. There is an open neighborhood $\mathcal{V}$ of p in $\mathcal{C}$, a complete normal curve $\mathcal{W}$ over Y, and an open subset $\mathcal{U}$ of $\mathcal{W}$ with the following properties:*

(i) *There exists a finite étale Y-morphism $g: \mathcal{U} \to \mathcal{V}$.*

(ii) *There are no rational curves in the special fiber of $\mathcal{W}$.*

(iii) *The general fiber of $\mathcal{W}$ has genus at least 2.*

Proof. Let $\mathcal{V}_1 = Spec(B)$ be an affine open neighborhood of p in $\mathcal{C}$. Since B is a finitely generated $\mathcal{O}$-algebra, there is a closed immersion of $\mathcal{V}_1$ into affine n-space $A^n_{\mathcal{O}}$ over $\mathcal{O}$ for some integer n. Let $\mathcal{C}_1$ be the closure of $\mathcal{V}_1$ in $P^n_{\mathcal{O}}$, and give $\mathcal{C}_1$ the reduced induced scheme structure. By Corollary 1.4, $\mathcal{C}_1^\sharp \to \mathcal{C}_1$ is finite and surjective. Since finite morphisms are projective, and the composition of projective morphisms is projective, $\mathcal{C}_1^\sharp$ is projective over $\mathcal{O}$.

Let $\mathcal{C}_1^\sharp \to P^t_{\mathcal{O}}$ be a projective embedding.

Let F be the reduction of the special fiber $(\mathcal{C}_1^\sharp)_s$ of $\mathcal{C}_1^\sharp$. Then F is the union of finitely many (possibly singular) integral complete curves over the algebraically closed residue field k of $\mathcal{O}$. Let $m \geq 5$ be an odd integer prime to the residue characteristic of k. Let $P^t_{\mathcal{O}} \to P^{t'}_{\mathcal{O}}$ be the m-tuple embedding, where $t' = \binom{t+m}{t} - 1$. The intersection of the image of $P^t_{\mathcal{O}}$ with a hyperplane section of $P^{t'}_{\mathcal{O}}$ is the zero locus of a homogeneous form of degree m on $P^t_{\mathcal{O}}$.

On applying Bertini's Theorem to $P^{t'}_k$, we find a homogeneous form $\overline{G}$ of degree m on P^t_k so that the intersection of the hypersurface $\overline{H}$ defined by $\overline{G}$ with each component of F is a regular divisor on F which does not contain p.

Let G be a homogeneous form of degree m on $P^t_{\mathcal{O}}$ which reduces to $\overline{G}$. Let H be the hypersurface defined by G. Let I be the intersection of H with $\mathcal{C}_1^\sharp$. Let $I_{/K} = K \otimes_{\mathcal{O}} I$. Since I is closed and does not contain the special fiber of $\mathcal{C}_1^\sharp$, $I_{/K}$ is the union of a finite number of closed points on $(\mathcal{C}_1^\sharp)_{/K}$. By construction, $I \cap F$ is a regular divisor on F. Suppose that $I_{/K}$ is not a regular divisor on $(C_1^\sharp)_{/K}$. Let $P_{/K}$ be a singular point of $I_{/K}$ and let P be the closure of $P_{/K}$ in $C_1^\sharp$. Since $C_1^\sharp$ is proper over $Y_L = Spec(\mathcal{O}_L)$, the image of P in Y_L is closed. Since this image contains the generic point of Y_L , it contains all of Y_L, so there is a point P_s on $P \cap (C_1^\sharp)_s$. Since I is closed, I contains P. A local equation g for H at P_s defines H in some open affine neighborhood $Spec(B)$ of P_s on $C_1^\sharp$. Then $P_{/K}$ lies in $Spec(B)$, and we have $D_{/K} \subseteq D \subseteq B$ if $D_{/K}$ (resp. D) is the prime ideal of B which defines $P_{/K}$ (resp. P_s). If I is not regular at $P_{/K}$, then g lies in $D^2_{/K}$. But then g lies in D^2, so P_s is not a regular point of $I \cap F$, which is a contradiction. Hence $I_{/K}$ is a regular divisor on $(C_1^\sharp)_{/K}$. Suppose $x_0, \ldots, x_{t+1}$ are homogeneous coordinates in $P_{\mathcal{O}}^{t+1}$. Let Z be the closed subscheme of $P_{\mathcal{O}}^{t+1}$ defined by $x_{t+1}^m - G = 0$. Regard $P_{\mathcal{O}}^t$ has having homogeneous coordinates $x_0, \ldots, x_t$, so that $P_{\mathcal{O}}^t$ is in this way a closed subscheme of $P_{\mathcal{O}}^{t+1}$. Define $\mathcal{W}_1 = C_1^\sharp \times_{P_{\mathcal{O}}^{t+1}} Z$, where $C_1^\sharp \subseteq P_{\mathcal{O}}^t \subseteq P_{\mathcal{O}}^{t+1}$.

Since $I \cap F$ and $I \cap (C_1^\sharp)_{/K}$ are regular, and we chose m to be odd and relatively prime to the residue characteristic of k, one can check the following facts using Kummer theory. The scheme $\mathcal{W}_1$ is a projective curve over Y. The morphism $\mathcal{W}_1 \to C_1^\sharp$ is etale over every point of $C_1^\sharp$ not in I. In particular, this is so for the point p. Suppose D is a reduced irreducible component of the fiber of $\mathcal{W}_1$ over the point y of Y. Let D' be the image of D in $C_1^\sharp$. Then D and D' are complete algebraic curves over the residue field $k(y)$ of y (where $k(y) = k$ or K). The morphism $D \to D'$ is separable of degree m, and is tamely and totally ramified above the divisor $I \cap D'$ of D'. Here $I \cap D'$ is a non-singular effective divisor of degree m supported on nonsingular points of D'. When $I \cap D'$ is viewed as a divisor on the normalization $D'^\sharp$ of D', the above properties also hold for the induced morphism $D^\sharp \to D'^\sharp$. Thus Hurwitz's Theorem and $m \geq 5$ imply $D^\sharp$ has genus at least 2. In particular, D cannot be a rational curve.

An etalé extension of a normal ring is normal by [11, Prop. I. 3.17]. Hence the above implies that there is an open neighborhood $\mathcal{V}$ of p in $\mathcal{V}_1$ such that the inverse image $\mathcal{U}_1$ of $\mathcal{V}$ in $\mathcal{W}_1$ is normal, and the induced morphism $\mathcal{U}_1 \to \mathcal{V}$ is finite and étale. Let $\mathcal{W} = \mathcal{W}_1^\sharp$, and let $\mathcal{U}$ be the inverse image of $\mathcal{U}_1$ in $\mathcal{W}$. Since $\mathcal{U}_1$ is normal, $\mathcal{U}$ is isomorphic to $\mathcal{U}_1$. We have shown that the fibers of $\mathcal{W}_1$ contain no rational curves, and that the normalizations of the reduced irreducible components of these fibers have genus at least 2. Hence $\mathcal{U}$ and $\mathcal{W}$ satisfy all the conditions of Proposition 3.2.

Now Theorem 3.1 and Proposition 2.6 give the following Corollary.

Corollary 3.3. *Let p, $\mathcal{V}$, $\mathcal{W}$ and $\mathcal{U}$ be in Proposition 3.2. There is a finite extension L_p of K with the following property. For each finite extension L of L_p, there is a complete regular Y_L-curve $\mathcal{X}$ birational to $\mathcal{W}_L^\sharp$ such that*

$\mathcal{X}_s$ *is reduced with at most ordinary double points for all closed points* s *of* Y_L.

Lemma 3.4. *Let* p, $\mathcal{V}$, $\mathcal{W}$, $\mathcal{U}$ *be as in Proposition 3.2. Let* L *be a finite extension of* K. *Define* $\mathcal{U}' = (\mathcal{O}_L \otimes_{\mathcal{O}} \mathcal{U})^\sharp$, $\mathcal{V}' = (\mathcal{O}_L \otimes_{\mathcal{O}} \mathcal{V})^\sharp$ *and let* $g_L : \mathcal{U}' \to \mathcal{V}'$ *be the morphism induced by* $g: \mathcal{U} \to \mathcal{V}$. *Then* g_L *is finite, surjective and étale, and there are no rational curves in a closed fiber* $(\mathcal{W}_L^\sharp)_{s'}$ *of* $\mathcal{W}_L^\sharp$.

Proof. Since $g: \mathcal{U} \to \mathcal{V}$ is finite and étale, the induced morphism $\mathcal{U}'' = \mathcal{O}_L \otimes_{\mathcal{O}} \mathcal{U} \to \mathcal{V}'' = \mathcal{O}_L \otimes_{\mathcal{O}} \mathcal{V}$ is as well. We have a Cartesian square

$$\begin{array}{ccc}
\mathcal{U}'' \times_{\mathcal{V}''} (\mathcal{V}''^{\#}) & \xrightarrow{\quad g'' \quad} & \mathcal{V}''^{\#} \\
\downarrow & & \downarrow \\
\mathcal{U}'' & \xrightarrow{\qquad\quad} & \mathcal{V}''
\end{array}$$

Thus g'' is finite and étale. Now $\mathcal{V}''^\sharp$ is normal and g'' is étale, so by [11, Prop. I.3.17], $\mathcal{U}'' \times_{\mathcal{V}''} (\mathcal{V}''^\sharp)$ is normal. Therefore $\mathcal{U}' = (\mathcal{O}_L \otimes_{\mathcal{O}} \mathcal{U})^\sharp$ is equal to $\mathcal{U}'' \otimes_{\mathcal{V}''} (\mathcal{V}''^\sharp)$, so $g_L : \mathcal{U}' \to \mathcal{V}' = \mathcal{V}''^\sharp$ is finite and étale. Since $\mathcal{V}'$ is connected, g_L is surjective by [11, Remark I.2.14]. Suppose s' is a closed point of Y_L such that $(\mathcal{W}_L^\sharp)_{s'}$ contains a rational component. Since $\mathcal{W}_L^\sharp \to \mathcal{O}_L \otimes_{\mathcal{O}} \mathcal{W}$ is finite and birational, $(\mathcal{O}_L \otimes_{\mathcal{O}} \mathcal{W})_{s'}$ would contain a rational component. Since $Y_L \to Y$ is flat and finite, $\mathcal{O}_L \otimes_{\mathcal{O}} \mathcal{W} \to \mathcal{W}$ is flat and finite. Hence a rational component of $(\mathcal{O}_L \otimes_{\mathcal{O}} \mathcal{W})_{s'}$ would have image equal to a rational component of the closed fibre $\mathcal{W}_s$ of $\mathcal{W}$. But this is impossible because by Proposition 3.2, $\mathcal{W}_s$ contains no rational components.

We now let p, $\mathcal{V}$, $\mathcal{U}$ and $\mathcal{W}$ be as in Proposition 3.2. Let L be a finite extension of K containing the field L_p of Corollary 3.3. Let $h: \mathcal{X} \to \mathcal{W}_L^\sharp$ be a birational map of the kind in Corollary 3.3.

Proposition 3.5. *The map* $h: \mathcal{X} \to \mathcal{W}_L^\sharp$ *is a proper, birational and surjective morphism.*

Before proving this Proposition, let us review what has been constructed thus far. Let $\mathcal{V}'$ and $\mathcal{U}'$ be as in Lemma 3.4. Let p' be a point of $\mathcal{V}'$ over p.

Corollary 3.6. *Let* $\overline{\mathcal{U}}' = h^{-1}(\mathcal{U}')$ *and let* $H = h|_{\overline{\mathcal{U}'}}$. *Then we have a diagram*

$$
\begin{array}{l}
\overline{\mathcal{U}'} \subset \mathcal{X} = \text{regular curve over } Y_L \text{ with reduced closed fibers} \\
\qquad\qquad\qquad \text{having at most ordinary double points.} \\
\Big\downarrow \quad H = \text{proper, surjective, birational } Y_L\text{-morphism} \\
\mathcal{U}' = (\mathcal{O}_L \otimes_{\mathcal{O}} \mathcal{U})^{\#} \subset \mathcal{W}_L^{\#} \\
\Big\downarrow \quad g_L = \text{finite, étale, surjective } Y_L\text{-morphism} \\
p' \in \mathcal{V}' = (\mathcal{O}_L \otimes_{\mathcal{O}} \mathcal{V})^{\#} \subset \mathcal{C}_L^{\#} \\
\Big\downarrow \quad Y\text{-morphism} \\
p \in \mathcal{V} \subset \mathcal{C}
\end{array}
$$

where $\mathcal{V}$ is an open neighborhood of p in $\mathcal{C}$, p' is a point over p in $\mathcal{C}_L^{\sharp}$, and $\overline{\mathcal{U}'}$ is a dense open subset of $\mathcal{X}$.

Proof of Proposition 3.5. By [8, Prop. 4.2] and [3, Prop. 2.2], we can factor h in the following way:

$$
\begin{array}{ccc}
 & \mathcal{W}_{max} & \\
{\scriptstyle t_1} \swarrow & & \searrow {\scriptstyle t_2} \\
 & & \mathcal{W}_{min} \\
 & & \Big\downarrow {\scriptstyle m} \\
\mathcal{X} & \xrightarrow{\quad h \quad} & \mathcal{W}_L^{\#}
\end{array}
$$

(3.1)

Here $m\colon \mathcal{W}_{min} \to (\mathcal{W}_L)^{\sharp}$ is the minimal desingularization constructed by

Lipman's process [9, 1]. $\mathcal{W}_{max}$ is a regular complete curve over Y_L, and t_1 and t_2 are proper birational Y_L-morphisms.

Let z be a point of $\mathcal{W}_L^\sharp$. Define $E = (m \circ t_2)^{-1}(z)$. By Zariski's Main Theorem, either (i) E is a point where $m \circ t_2$ is an isomorphism, or (ii) z is a closed point of $\mathcal{W}_L^\sharp$, and E is the connected union of finite set of complete curves over the residue field $k(z) = k$ of z. We now show that $t_1(E)$ is in the domain of definition D of the birational map $h: \mathcal{X} \to \mathcal{W}_L^\sharp$.

Suppose first that E is a point. The morphism $t_1: \mathcal{W}_{max} \to \mathcal{X}$ is an isomorphism off of the union of a set of rational curves over k, these rational curves being the proper transforms of the exceptional curves which appear in the factorization of t_1 into a series of blow downs. If E lies on one of these rational curves, then since $m \circ t_2$ is an isomorphism in a neighborhood of E, there would be a rational curve in a closed fiber of $\mathcal{W}_L^\sharp$ which contains $m \circ t_2(E) = z$. By Lemma 3.4, this is impossible. Thus t_1 is an isomorphism in a neighborhood of E, and $t_1(E) \subseteq D$.

Suppose now that z is a closed point of $\mathcal{W}_L^\sharp$ and that E is the connected union of a set of complete curves over $k(z) = k$. Let x be a point of $t_1(E)$. If x is not blown up by t_1, then t_1^{-1} is defined in a neighborhood of x, so x is in D. Suppose now that x is blown up by t_1. Then x is closed, $t_1^{-1}(x) = \cup_i F_i$ is the connected union of complete rational curves over $k(x) = k$, and $t_1^{-1}(x)$ intersects E. Suppose that $F_i \cap E \neq \phi$. If $m \circ t_2(F_i)$ is not equal to z, then $m \circ t_2(F_i)$ is a rational curve in the fiber of $\mathcal{W}_L^\sharp$ containing z, contradicting the fact that there are no such curves. Hence $m \circ t_2(F_i) = z$, so $F_i \subseteq E$. Since $t_1^{-1}(x)$ is connected, it follows that $t_1^{-1}(x) \subseteq E$.

Since $\mathcal{X}$ is normal, $\mathcal{O}_{\mathcal{X},x}$ is the intersection of the valuation rings of the function field $K(\mathcal{X})$ of $\mathcal{X}$ which contain $\mathcal{O}_{\mathcal{X},x}$ by [2, Ch. VI, sect. 1, no. 3, Thm. 3, p. 92]. Now $t_1: \mathcal{W}_{max} \to \mathcal{X}$ is proper and birational. Hence the valuative criterion of properness implies that each valuation ring of $K(\mathcal{X}) = K(\mathcal{W}_{max})$ which contains $\mathcal{O}_{\mathcal{X},x}$ dominates the valuation ring of some point of $t_1^{-1}(x)$. Thus

$$\mathcal{O}_{\mathcal{X},x} = \cap_{w \in t_1^{-1}(x)} \mathcal{O}_{\mathcal{W}_{max},w}.$$

Now $m \circ t_2: \mathcal{W}_{max} \to \mathcal{W}_L^\sharp$ is birational, so the local ring $\mathcal{O}_z$ of z on $\mathcal{W}_L^\sharp$ is contained in $\mathcal{O}_{\mathcal{W}_{max},w}$ for all $w \in E = (m \circ t_2)^{-1}(z)$. Since $t_1^{-1}(x) \subseteq E$, it follows that $\mathcal{O}_z$ is a subring of $\mathcal{O}_{\mathcal{X},x}$. Because x and z are closed points, this shows $\mathcal{O}_{\mathcal{X},x}$ dominates $\mathcal{O}_z$. Hence by [8, Cor. 1.13], the birational map $h: \mathcal{X} \to \mathcal{W}_L^\sharp$ is defined by x. This completes the proof that $t_1(E) \subseteq D$.

Since z was an arbitrary point of $\mathcal{W}_L^\sharp$, we conclude that $\mathcal{X} = t_1((m \circ t_2)^{-1}(\mathcal{W}_L^\sharp))$ is contained in D. Hence $h: \mathcal{X} \to \mathcal{W}_L^\sharp$ is a birational surjective morphism. To show h is proper, we must show by the valuative criterion that for each commutative diagram

$$
\begin{array}{ccc}
\mathrm{Spec}(F) & \xrightarrow{\quad f \quad} & \mathcal{X} \\
\Big\downarrow & & \Big\downarrow h \\
\mathrm{Spec}(R_F) & \xrightarrow{\quad g \quad} & \mathcal{W}_L^{\#}
\end{array}
$$

(3.2)

in which R_F is a valuation ring with quotient field F, there is a unique morphism $Spec(R_F) \rightarrow \mathcal{X}$ compatible with the diagram. Let s_1 be the unique point of $Spec(F)$, and let $x_1 = f(s_1)$. Then f determines an injection of the residue field $k(x_1)$ of x_1 into F. The fiber $t_1^{-1}(x_1)$ of $t_1 \colon \mathcal{W}_{max} \rightarrow \mathcal{X}$ over x_1 is a single point unless x_1 is one of the points blown up by t_1, in which case x_1 is closed and $t_1^{-1}(x_1)$ is a connected union of complete curves over $k(x_1) = k$. In either case, we can find a point x_1' in $t_1^{-1}(x_1)$ which has the same residue field as x_1. Define $f' \colon Spec(F) \rightarrow \mathcal{W}_{max}$ to be the morphism induced by the inclusion $k(x_1') = k(x_1) \rightarrow F$. Since the diagram (3.1) commutes, we must have a commutative diagram

$$
\begin{array}{ccc}
\mathrm{Spec}(F) & \xrightarrow{\quad f' \quad} & \mathcal{W}_{max} \\
\Big\downarrow & & \Big\downarrow m \circ t_2 \\
\mathrm{Spec}(R_F) & \xrightarrow{\quad g \quad} & (\mathcal{W}_L)^{\#}
\end{array}
$$

(3.3)

Since $\mathcal{W}_{max}$ is proper over $\mathcal{W}_L^\sharp$, the morphism $m \circ t_2$ in this diagram is proper. Hence the valuative critierion says there is a unique morphism $\tau: Spec(R_F) \to \mathcal{W}_{max}$ compatible with the diagram. It follows that

$$\tau_1 = t_1 \circ \tau: Spec(R_F) \to \mathcal{X}$$

is compatible with (3.2). Suppose now that there is a second distinct morphism $\tau_1': Spec(R_F) \to \mathcal{X}$ compatible with (3.2). Since $t_1: \mathcal{W}_{max} \to \mathcal{X}$ is proper, we can lift τ_1' to a morphism $\tau': Spec(R_F) \to \mathcal{W}_{max}$. Since $\tau_1' \neq \tau_1$, τ is not equal to τ'. On the other hand, both τ and τ' are compatible with the diagram in (3.3), so $\mathcal{W}_{max}$ is not separated over $\mathcal{W}_L^\sharp$, which is impossible. This contradiction proves $h: \mathcal{X} \to \mathcal{W}_L^\sharp$ is proper.

Remark 3.7. The inverse image $\mathcal{U}^*$ of $\mathcal{U}'$ in $\mathcal{W}_{\max}$ may contain exceptional curves which are blown down by $t_1: \mathcal{W}_{max} \to \mathcal{X}$ to ordinary double points on a closed fiber of $\mathcal{X}$. Such exceptional curves will have multiplicity two on a closed fiber of $\mathcal{W}_{max}^*$, so $\mathcal{U}^*$ need not have reduced fibers. This is why we work with $\overline{\mathcal{U}}'$ rather than $\mathcal{U}^*$.

Corollary 3.8. *For each closed point p of $\mathcal{C}$, let $\mathcal{V} = \mathcal{V}_p$ and $\mathcal{U} = \mathcal{U}_p$ be as in Proposition 3.2. There will be a finite set B of p for which the corresponding $\mathcal{V}_p$ cover $\mathcal{C}$. Let L be a finite extension of K containing the fields L_p of Corollary 3.3 for all p in B. Then the closed fibers of $\mathcal{C}_L^\sharp$ are reduced.*

Proof. The $(\mathcal{O}_L \otimes_{\mathcal{O}} \mathcal{V}_p)^\sharp$ of $p \in B$ cover $\mathcal{C}_L^\sharp$. By Lemma 3.4, for each such p we have the finite étale cover $(\mathcal{O}_L \otimes_{\mathcal{O}} \mathcal{U}_p)^\sharp \to (\mathcal{O}_L \otimes_{\mathcal{O}} \mathcal{V}_p)^\sharp$. By Corollary 3.6, each of these $(\mathcal{O}_L \otimes_{\mathcal{O}} \mathcal{U}_p)^\sharp$ is covered by an open subset of a Y_L curve whose closed fibers are reduced. Hence the closed fibers of $(\mathcal{O}_L \otimes_{\mathcal{O}} \mathcal{U}_p)^\sharp$ are reduced, so closed fibers of $\mathcal{C}_L^\sharp$ must be reduced.

We record the following consequence of Corollary 3.8.

Corollary 3.9. *From now on let L be a fixed finite extension of K chosen as in Corollary 3.8. Let S be the set of points of $\mathcal{C}_L^\sharp$ which are singular on $\mathcal{C}_L^\sharp$ or on some closed fiber $(\mathcal{C}_L^\sharp)_s$ of $\mathcal{C}_L^\sharp$. Then S is a finite set of closed points.*

Definition 3.10 ([1, pp. 267-268]). *A surface is a noetherian, normal, connected, excellent scheme Z of dimension 2. A modification of Z is a surface Z' together with a projective birational map $f: Z' \to Z$.*

As in [1], a modification $f: Z' \to Z$ is an isomorphism off of a finite set of closed points of Z. The fiber $f^{-1}(p)$ of f over a point p in this finite set is a connected scheme of dimension 1.

Proposition 3.11 (Artin [1, Prop. 1.7]). *Let Z be a surface, and let $\hat{Z} = Spec(\hat{\mathcal{O}}_{Z,z})$ be its completion at a closed point z.*

(i) Suppose $f: Z' \to Z$ is a modification of Z. Then the induced morphism $\hat{f}: Z' \times_Z \hat{Z} \to \hat{Z}$ is a modification of $\hat{Z}$. There is a unique point $\hat{z}$ of $\hat{Z}$ lying over z. Z' is nonsingular above z iff $Z' \times_Z \hat{Z}$ is non-singular above $\hat{z}$.

(ii) Every modification $\hat{g}: \hat{Z}' \to \hat{Z}$ arises by base change from a modification $g: Z' \to Z$.

The residue field of a closed point of Y_L equals the algebraically closed field k. The following Lemma is clear from the definition of ordinary double points and the fact that a normal curve over Y_L is singular at only finitely many closed points.

Lemma 3.12. *Suppose that z is a regular closed point of a normal curve Z over Y_L. Let $\hat{\mathcal{O}}_z$ be the completion of the local ring $\mathcal{O}_z = \mathcal{O}_{Z,z}$ and let $\hat{M}_z$ be the maximal ideal of $\hat{\mathcal{O}}_z$. Let π be a uniformizing parameter at the point s of Y_L under z. Then z is a nonsingular point of the fiber Z_s if and only if π is not in $\hat{M}_z^2$. Z_s has an ordinary double point at z iff $\pi = f_1 f_2$ in $\hat{\mathcal{O}}_z$ for some f_1 and f_2 in $\hat{M}_z$ which are linearly independent mod $\hat{M}_z^2$. Finally, the irreducible component F of Z_s which contains z is reduced iff F contains a dense open set of nonsingular points which are regular on Z.*

Lemma 3.13. *Let $f: Z' \to Z$ be a modification, and suppose Z is a curve over $\mathcal{O}_L$. Then Z' is a curve over $\mathcal{O}_L$. Let z be a closed point of Z, and let w be a closed point of Z' over z. Then there is a unique point $\hat{z}$ (resp. $\hat{w}$) over z (resp. w) in $\hat{Z} = Spec(\hat{\mathcal{O}}_{Z,z})$ (resp. $Z' \times_Z \hat{Z}$). Let $\mathcal{O}_w$ (resp. $\mathcal{O}_{\hat{w}}$) be the local ring of w on Z' (resp. of $\hat{w}$ on $Z' \times_Z \hat{Z}$). Then the local homomorphism $\mathcal{O}_w \to \mathcal{O}_{\hat{w}}$ induced by projection morphism $Z' \times_Z \hat{Z} \to Z'$ induces an isomorphism $\hat{\mathcal{O}}_w = \hat{\mathcal{O}}_{\hat{w}}$ on completions.*

Proof. Let $\pi_{Z'}: Z' \to Y_L$ be the composition of $Z' \to Z$ with the structure morphism $\pi_Z: Z \to Y_L$. Since π_Z is of finite type and Z' is projective over Z, $\pi_{Z'}$ is of finite type. Because Z' is integral, the local rings of Z' are contained in the function field $K(Z') = K(Z)$. Hence these rings have no $\mathcal{O}_L$ torsion, so $\pi_{Z'}$ is flat because $\mathcal{O}_L$ is a Dedekind ring. The generic fibers of Z and Z' are isomorphic. Suppose F is an irreducible component of a closed fiber of Z'. If F is sent to a point on Z, then F has dimension 1 by Zariski's Main Theorem. Otherwise F is birational to its image on Z, so F again has dimension 1. Thus Z' is a curve over Y_L.

The uniqueness of $\hat{z}$ is contained in Proposition 3.11. The inclusion $\mathcal{O}_{Z,z} \to \mathcal{O}_{\hat{Z},\hat{z}}$ induces an isomorphism $k(z) = k(\hat{z}) = k$ of residue fields.

Hence the projection morphism $Z' \times_Z \hat{Z} \to Z'$ establishes a bijection between the fibers $\hat{f}^{-1}(\hat{z})$ and $f^{-1}(z)$. Thus there is a unique point $\hat{w}$ of $\hat{f}^{-1}(\hat{z})$ which lies over w. Since $\hat{z}$ is the unique point of $\hat{Z}$ over z, the fiber of $Z' \times_Z \hat{Z} \to Z'$ over w lies in $\hat{f}^{-1}(\hat{z})$. Thus $\hat{w}$ is the unique point of $Z' \times_Z \hat{Z}$ over w. Because surfaces are by definition noetherian, it follows from [6, EGA IV, Lemma 7.9.3.i] that $\mathcal{O}_w \to \mathcal{O}_{\hat{w}}$ induces an isomorphism $\hat{\mathcal{O}}_w = \hat{\mathcal{O}}_{\hat{w}}$.

Lemma 3.14. *With the notations of Corollary 3.6, there is an open neighborhood $\mathcal{V}''$ of p' inside $\mathcal{V}'$ for which there is a proper birational morphism $H'': \mathcal{U}'' \to \mathcal{V}''$ having the following properties. The Y_L-curve $\mathcal{U}''$ is regular. For each closed point s of Y_L, $\mathcal{U}''_s$ is reduced with at most ordinary double points. Both $\mathcal{V}''$ and $\mathcal{V}''_s$ are regular except possibly at p'. The morphism $H'': \mathcal{U}'' \to \mathcal{V}''$ is an isomorphism off of p'.*

Proof. By Corollary 3.6, $H: \overline{\mathcal{U}}' \to \mathcal{U}'$ is proper and birational, and $\overline{\mathcal{U}}'$ is an open dense subset of the regular complete curve $\mathcal{X}$ over Y_L. By [8, Theorem 2.8], $\mathcal{X}$ is projective over Y_L. Hence there is an ample Weil divisor on $\overline{\mathcal{U}}'$, and a sufficiently large multiple of this divisor will be very ample over $\mathcal{U}'$. This and the fact that H is proper imply that H is projective by [7, Remark II. 5.16.i]. Since $\overline{\mathcal{U}}'$ and $\mathcal{U}'$ are normal, we have now shown that $H: \overline{\mathcal{U}}' \to \mathcal{U}'$ is a modification.

Let q' be a point of $\mathcal{U}'$ over p'. Define $\hat{\mathcal{U}}' = Spec(\hat{\mathcal{O}}_{\mathcal{U}',q'})$. By Proposition 3.11, the morphism $\hat{H}: \overline{\mathcal{U}}' \times_{\mathcal{U}'} \hat{\mathcal{U}}' \to \hat{\mathcal{U}}'$ induced by H is a modification. Now $\mathcal{O}_{\mathcal{U}',q'}$ is étale over $\mathcal{O}_{\mathcal{V}',p'}$, and the residue fields $k(q')$ and $k(p')$ are isomorphic to k. Hence the inclusion $\mathcal{O}_{\mathcal{V}',p'} \to \mathcal{O}_{\mathcal{U}',q'}$ induces isomorphism $\hat{\mathcal{O}}_{\mathcal{V}',p'} = \hat{\mathcal{O}}_{\mathcal{U}',q'}$. Now Proposition 3.11 implies that there is a modification $H_0: \mathcal{U}_0 \to \mathcal{V}'$ for which there is a Cartesian diagram

$$\begin{array}{ccc}
\overline{\mathcal{U}'} \times_{\mathcal{U}'} \hat{\mathcal{U}}' = \mathcal{U}_0 \times_{\mathcal{V}'} \hat{\mathcal{U}}' & \longrightarrow & \mathcal{U}_0 \\
\hat{H} \Big\downarrow & & \Big\downarrow H_0 \\
\hat{\mathcal{U}}' = \mathrm{Spec}(\hat{\mathcal{O}}_{\mathcal{V}',p'}) & \longrightarrow & \mathcal{V}'
\end{array}$$

By Lemma 3.13, $\mathcal{U}_0$ is a curve over Y_L, and there is a unique point $\hat{p}'$ over

p' in $\hat{\mathcal{U}}' = Spec(\hat{\mathcal{O}}_{\mathcal{V}',p'})$. This $\hat{p}'$ is the unique point $\hat{q}'$ over q' with respect to the morphism $\hat{\mathcal{U}}' \to \mathcal{U}'$. Now Lemma 3.13 gives bijections between (i) the closed points of $\mathcal{U}_0$ over p', (ii) the closed points of $\mathcal{U}_0 \times_{\mathcal{V}'} \hat{\mathcal{U}}' = \overline{\mathcal{U}'} \times_{\mathcal{U}'} \hat{\mathcal{U}}'$ over $\hat{p}' = \hat{q}'$, and (iii) the closed points of $\overline{\mathcal{U}'}$ over q'. Let ϕ be the resulting bijection between (i) and (iii). Lemma 3.13 shows that for each closed point z of $\mathcal{U}_0$ over p', there is an isomorphism $\phi_z: \hat{\mathcal{O}}_{\mathcal{U}_0,z} \cong \hat{\mathcal{O}}_{\overline{\mathcal{U}'},\phi(z)}$ which is compatible with the inclusion of $\mathcal{O}_L$ into these completions.

Since $\overline{\mathcal{U}'}$ is regular, and a local ring is regular iff its completion is regular [7, Theorem I.5.4A], we conclude that $\mathcal{U}_0$ is regular over p'. Let s be a closed point of Y_L. The fiber $\overline{\mathcal{U}'}_s$ is reduced with at most ordinary double points by Corollary 3.6. Thus Lemma 3.12 and the isomorphisms ϕ_z above show that the components of $(\mathcal{U}_0)_s$ which lie over p' are reduced, and that the singular points of $(\mathcal{U}_0)_s$ which lie over p' are at most ordinary double points. Now $H_0: \mathcal{U}_0 \to \mathcal{V}'$ is an isomorphism off of a finite number of points of $\mathcal{V}'$. Hence there will be an open neighborhood $\mathcal{V}''$ of p' in $\mathcal{V}'$ such that H_0^{-1} is a morphism on $\mathcal{V}'' - \{p'\}$ and for which s is the unique closed point of Y under $\mathcal{V}''_L$. By Corollary 3.8, $\mathcal{V}'_s$ is reduced, so $\mathcal{V}''_s$ is reduced. By Corollary 3.9, we can further shrink $\mathcal{V}''$ so that both $\mathcal{V}''$ and $\mathcal{V}''_s$ are regular except possibly at p'. Now let $\mathcal{U}'' = H_0^{-1}(\mathcal{V}'')$ and let $H'': \mathcal{U}'' \to \mathcal{V}''$ be the morphism induced by H_0. Since H_0 is proper and birational, H'' is as well. A point of $\mathcal{U}''$ either lies in the fiber of $\mathcal{U}_0$ over p or maps isomorphically to its image on $\mathcal{V}''$. Hence by our construction of $\mathcal{U}_0$, $\mathcal{V}''$ and $\mathcal{U}''$, the unique closed fiber $\mathcal{U}''_s$ of $\mathcal{U}''$ is reduced with at most ordinary double points.

End of the proof of Artin's Theorem.

Let L and S be as in Corollary 3.9. For each $p' \in S$, let $H'': \mathcal{U}'' \to \mathcal{V}''$ be as in Lemma 3.14 for some open neighborhood $\mathcal{V}''$ of p'. We now construct a curve C^* over Y_L by patching $C_L^\sharp$ with the $\mathcal{U}''$ associated to $p' \in S$, this patching resulting from identifying $\mathcal{U}'' - H''^{-1}(p')$ with $\mathcal{V}'' - \{p'\} \subseteq C_L^\sharp$ via H''. This gives a morphism $J: C^* \to C_L^\sharp$ which is proper and birational because each of the morphisms $H'': \mathcal{U}'' \to \mathcal{V}''$ are proper and birational. The curve C^* is regular because S contained all the singular points of $C_L^\sharp$, and each $\mathcal{U}''$ is regular. Since the closed fibers $\mathcal{U}''_s$ are reduced, each C_s^* is reduced. Suppose b is a singular point of C_s^*. Then b lies over some $p' \in S$ by the definition of S, and b is an ordinary double point of the corresponding $\mathcal{U}''_s$ by Lemma 3.14. Since $\mathcal{U}''$ is an open neighborhood of b in C^*, it follows that C_s^* has at most ordinary double points. In view of Corollary 2.11, this completes the proof of Theorem 2.3.

References

[1] Artin, M.: Lipman's proof of resolution of singularities for surfaces, in Arithmetic Geometry, G. Cornell and J. Silverman editors, Springer-Verlag, New York Berlin Heidelberg (1986), p. 267-287.

[2] Bourbaki, N.: Algèbre Commutative, Eléments de Math. 27, 28, 30, 31, Hermann, Paris (1961-1965).

[3] Chinburg, T.: Minimal Models for Curves over Dedekind rings, in Arithmetic Geometry, G. Cornell and J. Silverman editors, Springer-Verlag, New York Berlin Heidelberg (1986), p. 309-326.

[4] Chinburg, T. and Rumely, R.: The Capacity Pairing, to appear.

[5] Deligne, P. and Mumford, D.: The irreducibility of curves of a given genus, Publ. Math. I.H.E.S., 36 (1969), p. 75-109.

[6] Grothendieck, A.: Eléments de géométrie algébrique (EGA) I-IV, Publ. Math. I.H.E.S., 4,8,11,17,20,24,28,32 (1960-1967).

[7] Hartshorne, R.: Algebraic Geometry, Graduate Texts in Mathematics No.52, third ed., Springer-Verlag, New York Heidelberg Berlin (1983).

[8] Lichtenbaum, S.: Curves over discrete valuation rings, Amer. J. Math. 15, no. 2(1968), p. 380-405.

[9] Lipman, J.: Desingularization of two-dimensional schemes, Ann. Math. 107 (1978), p. 151-207.

[10] Matsumura, H.: Commutative Algebra, W.A. Benjamin, New York (1970).

[11] Milne, J.S.: Étale Cohomology, Princeton Univ. Press, Princeton (1980).

[12] Mumford, D.: The topology of normal singularities on an algebraic surface and a criterion for simplicity, Publ. Math. I.H.E.S., 9 (1961), p. 5-22.

[13] Shafarevitch, I.: Lectures on Minimal Models and Birational Transformations of Two-dimensional Schemes, Tata Institute, Bombay (1966).

On the Manin constants of modular elliptic curves

Bas Edixhoven

1 Definitions and results

For M a positive integer, let $X_0(M)_{\mathbf{Q}}$ be the modular curve over $\mathbf{Q}$ classifying elliptic curves with a given cyclic subgroup of order M and let $J_0(M)_{\mathbf{Q}}$ be the jacobian of $X_0(M)_{\mathbf{Q}}$. An elliptic curve E over $\mathbf{Q}$ is said to be *modular* if it is an isogeny factor (isogenies over $\mathbf{Q}$) of some $J_0(M)_{\mathbf{Q}}$; the smallest M for which this happens is then called the level of E. The Shimura-Taniyama conjecture states that *every* elliptic curve E over $\mathbf{Q}$ is modular, and that the level of E equals the conductor of E. A modular elliptic curve of level M is called *strong* if there exists a closed immersion $E \hookrightarrow J_0(M)_{\mathbf{Q}}$. It follows from the multiplicity one principle for modular forms that such an immersion is unique up to sign.

Let E be a strong modular elliptic curve of level M. The corresponding *strong modular parametrization* $\phi : X_0(M)_{\mathbf{Q}} \to E$ is obtained as follows:

$$\begin{array}{ccc} J_0(M)_{\mathbf{Q}} & \to & E \\ \uparrow & \nearrow_{\phi} & \\ X_0(M)_{\mathbf{Q}} & & \end{array}$$

where $J_0(M)_{\mathbf{Q}} \to E$ is the dual of one of the two closed immersions $E \hookrightarrow J_0(M)_{\mathbf{Q}}$ and where $X_0(M)_{\mathbf{Q}} \to J_0(M)_{\mathbf{Q}}$ is the standard immersion sending the cusp ∞ to 0. Let $\mathcal{E}$ be the Néron model over $\mathbf{Z}$ of E. Then the $\mathbf{Z}$-module of translation invariant differentials on $\mathcal{E}$ is free of rank one; let ω be one of the two generators. We get the differential form $\phi^*\omega$ on $X_0(M)_{\mathbf{Q}}$.

Now there is also a second differential form on $X_0(M)_{\mathbf{Q}}$ related to E: the normalized newform $\sum a_n q^n \frac{dq}{q}$ corresponding to E (normalized means that $a_1 = 1$). The multiplicity one principle gives:

$$\phi^*\omega = c \sum_{n \geq 1} a_n q^n \frac{dq}{q}$$

for some $c \in \mathbf{Q}^*$. This number c is called the Manin constant of E (see [7], [8], [9] and [1]).

Conjecture 1 (Manin) $c = \pm 1$.

In [7] Mazur showed that $c \in \mathbf{Z}[\frac{1}{2m}]^*$, where m is the largest square dividing M. His proof uses results of Raynaud on specialization of finite group schemes (this explains why 2 plays a special rôle) and the description of Deligne and Rapoport [3] of the reduction of $X_0(M)_{\mathbf{Q}}$ over $\mathbf{Z}[\frac{1}{m}]$. The fact that $X_0(M)_{\mathbf{Q}}$ has stable reduction over $\mathbf{Z}[\frac{1}{m}]$ is crucial. It seems [14] [10] that in a letter to Mestre (Febr. 1985), Raynaud has stated general results implying that, for 2 not dividing m, one has $v_2(c) = 0$ or 1, where $v_2(c)$ is the valuation of c at 2. In this article we will be concerned with the primes p dividing m. First we prove the following proposition.

Proposition 2 *The number c is an integer.*

Proof. Let $X_0(M)$ be the compactified coarse moduli scheme $\overline{\mathrm{M}}([\Gamma_0(M)])$ (cf. [6] (8.6)). Let $X_0(M)^0$ be the open part of $X_0(M)$ where the projection to Spec($\mathbf{Z}$) is smooth. Let $\mathcal{E}$ be the Néron model of E over $\mathbf{Z}$. By the Néron property ϕ extends to $\phi : X_0(M)^0 \longrightarrow \mathcal{E}$. It follows immediately from [6] Thm.8.11.10 that the formal completion of $X_0(M)$ along the (unramified) cusp ∞ is Spf($\mathbf{Z}[[q]]$). We see that $\infty \in X_0(M)^0(\mathbf{Z})$ and that $\phi^*\omega_{\mathbf{Z}}$ is a differential form (without poles!) in a neighborhood of ∞. If c would not be integral, then $\phi^*\omega_{\mathbf{Z}}$ would have poles along the fibers of $X_0(M)^0$ in characteristics dividing the denominator of c.

The analogous result for parametrizations involving the modular curve $X_1(M)$ is proved in the paper [14] by Stevens. In that paper Stevens studies parametrizations of Weil curves by the modular curves $X_1(M)$. He conjectures (see [14] Conj. I) that *every* Weil curve E (not just the strong ones) admits a parametrization by $X_1(M)$ (where M is the conductor of E) with Manin constant equal to 1. He shows this conjecture to be compatible with quadratic twists over fields that are unramified at the primes where E has additive reduction. He also shows, using results of Rubin on special values of L-functions, that a weak form of his conjecture holds for certain elliptic curves with complex multiplication.

Using the classification by Mazur [7] of the rational torsion points of elliptic curves over $\mathbf{Q}$, one can easily show that up to the primes 2, 3, 5, and 7, Stevens's version of Manin's conjecture is equivalent to Manin's conjecture. For this one remarks that the difference between parametrizations by $X_1(M)_{\mathbf{Q}}$ and $X_0(M)_{\mathbf{Q}}$ lies in the so-called Shimura subgroup of $J_0(M)_{\mathbf{Q}}$, which is the Cartier dual of a constant group. It follows that the Manin constant c of a strong Weil curve E is not divisible by primes $p > 7$ where E has reduction type I_0^* or I_ν^* (we use Kodaira symbols to denote the possible reduction types of elliptic curves, see [15]).

The aim of this article is to attack Manin's conjecture for the remaining reduction types, these are: II, III, IV, IV^*, III^* and II^*. Our method to do this is in principle the same as Mazur's [7]: to study the valuation of c at a prime p by reducing ϕ modulo p. As remarked above, it is crucial to have stable models for $X_0(M)$ and E at p. This makes it necessary not to work with the model of $X_0(M)$ over $\mathbf{Z}$ as constructed in [6], but over a suitable extension of $\mathbf{Q}$ over which $X_0(M)$ and E acquire stable reduction. For a number of reasons, we suppose that $p > 7$. It has been proved by Carayol [2], that for modular elliptic curves the level and the conductor coincide. The conductors of the curves that we want to study are of the form p^2N, where p does not divide N. A stable model of $X_0(p^2N)$ at p is constructed in [4]. The application of these stable models to the study of Manin constants is given in [5] 4.6. The result that we state here is slightly stronger than [5] Thm.4.6.3.

Theorem 3 *Let $\phi : X_0(M)_{\mathbf{Q}} \to E$ be a strong modular parametrization, let c be its Manin constant and let $p > 7$ be a prime. Then p does not divide c, except possibly when E has potentially ordinary reduction at p of type II, III or IV. In that case, p divides c at most once.*

Of course, the cases where E has reduction type I_0, I_ν, I_0^* or I_ν^* are already proved by Mazur and Stevens. In the proof of the theorem for curves with complex multiplication we rely heavily on the work of Stevens. Combining our results with those of Stevens on complex multiplication curves, one can even show that elliptic curves over $\mathbf{Q}$ with complex multiplication by $\mathbf{Q}(\sqrt{-p})$, $p > 3$, have Manin constants (in the original sense as well as in the sense of Stevens) equal to 1 or 2. A lot of details of the proof of the theorem above are not given here; for them we refer to [5]. Finally I want to thank Ling San for giving his comments on a preliminary version of this article.

2 Stable models

The standard method to try to obtain a stable model of a curve over the spectrum S of a discrete valuation ring is as follows. One starts with a regular model X over S. After blowing up repeatedly in closed points of the special fibre one may assume X is a regular model in which the special fibre is a divisor with normal crossings. Now let n be the least common multiple of the multiplicities in this divisor and let π_0 be a uniformizer on S. Let T be $S[\pi]$, with $\pi^n = \pi_0$, thus we have $T \to S$ totally ramified of degree n. Let $\widetilde{X_T}$ be the normalization of the pullback X_T of X to T. Let t be the closed point of T. If n is invertible on S, then by easy explicit computations with the complete local rings of X, one can show that $\widetilde{X_T}$ is a stable model in the sense that the only singularities of X_t are ordinary double points.

Now suppose that n is invertible on S and that $\mathcal{O}_S(S)$ contains μ_n, the n-th roots of unity. Then μ_n acts on T by:

$$\forall \zeta \in \mu_n : \qquad \zeta^\# : \pi \mapsto \zeta\pi,$$

where $\zeta^\#$ is the automorphism of $\mathcal{O}_T$ corresponding to the automorphism ζ of T. The quotient for this action is the morphism $T \to S$. We let μ_n act on $X_T = X \times_S T$ via its action on T. It follows that μ_n also acts on its normalization $\widetilde{X_T}$. Since the morphism $\widetilde{X_T} \to T$ is equivariant and t is fixed (i.e., μ_n acts trivially on $k(t)$), μ_n acts $k(t)$-linearly on the special fibre $\widetilde{X_{T}}_{,t}$. The μ_n-action on the complete local rings of $\widetilde{X_{T}}_{,t}$ is easy to compute.

We will now apply the method described above to the modular curve $X_0(p^2N)$. Let $p > 3$ be a prime, and let N be a positive integer that

Figure 1: The stable reduction of $X_0(p^2N)$ at p.

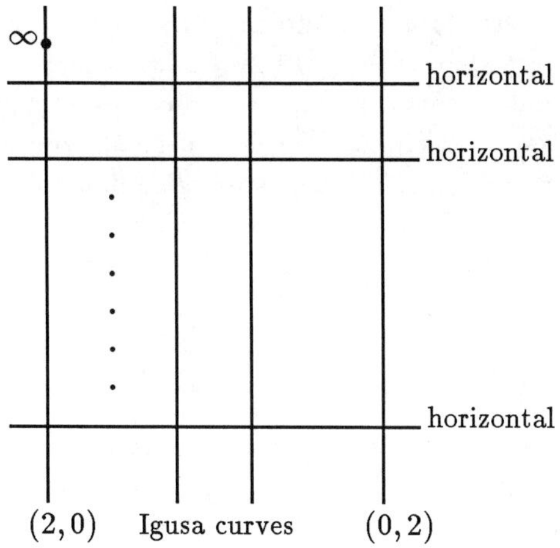

is not divisible by p. We start with the model over $\mathbf{Z}$ as contructed by Katz and Mazur in [6] 8.6. In their terminology, we let $X_0(p^2N)$ be the coarse compactified moduli scheme over $\mathbf{Z}$ associated to the moduli problem $\Gamma_0(p^2N)$. Let S be the spectrum of the ring of Witt vectors of $\overline{\mathbf{F}}_p$, and let $X_0(p^2N)_S$ be the pullback of $X_0(p^2N)$ to S. Then $X_0(p^2N)_S$ is a curve over S with a smooth geometrically irreducible generic fibre. The two-dimensional scheme $X_0(p^2N)$ is regular, except for some possible quotient singularities. There are two ways to get rid of these singularities. One way is to work with the minimal resolution of $X_0(p^2N)$ (see [4]). The other way is to introduce a suitable additional level structure (e.g., $[\Gamma(3)^{\mathrm{can}}]$); then to apply the method described above to the associated moduli scheme, and to obtain the stable model of $X_0(p^2N)$ as a quotient. The details of this computation can be found in [4]. It turns out that $X_0(p^2N)_S$ acquires stable reduction over $T = S[\pi]$, with $p = \pi^n$ and $n = \frac{p^2-1}{2}$. We denote the stable model by $\widetilde{X_T}$. The details of the computation of the action of μ_n on $\widetilde{X_{T,t}}$ can be found in [5]. The description that one finds of $\widetilde{X_{T,t}}$ together with its μ_n-action is as follows. Figure 1 contains a picture of $\widetilde{X_{T,t}}$.

The outer two vertical components are the (2,0) and (0,2) components (in the terminology of [6]) from $X_0(p^2N)_s$. These curves are both isomorphic to $X_0(N)_s$, where s denotes the closed point of S. The two central vertical components are isomorphic to the curve $\overline{\mathrm{M}}([\Gamma_0(N)],[Ig(p)/\{\pm1\}])$ (see [6] 12 for the details of the Igusa moduli problem). The supersingular points on the central vertical components correspond bijectively to the supersingular points on $X_0(N)_s$. The four vertical components are glued to the horizontal components at their supersingular points. The horizontal component corresponding to a supersingular point x of $X_0(N)_s$ is the smooth complete hyperelliptic curve given by the equation:

$$y^2 = x^{p+1} + 1,$$

if the automorphism group of x is $\{\pm1\}$ (by this we mean that the pair (E, G), corresponding to x via the moduli interpretation, has automorphism group $\{\pm1\}$). In general, the horizontal component at x is the quotient of this curve by some action of $\mathrm{Aut}_s(x)$. Each horizontal component is glued to the outer two vertical components at the points with $x = \infty$, $y = \infty$. At the points $x = 0$, $y = \pm1$ it is glued to the central vertical components.

The inertia group μ_n preserves the horizontal components; the action on each horizontal component is given by:

$$\forall\zeta \in \mu_n : \qquad \zeta^\# : \begin{cases} x & \mapsto & \zeta^{-\frac{p-1}{2}}x \\ y & \mapsto & \zeta^{\frac{n}{2}}. \end{cases}$$

The μ_n-action on the vertical components is as follows. The action is trivial on the outer two components. The two Igusa curves are interchanged. The stabilizer of either of them, $\mu_{\frac{n}{2}}$, acts on the Igusa curve via its action on the moduli problem $[Ig(p)/\{\pm1\}]$.

¿From this description, it is easy to make a list of all the points in $\widetilde{X_{T,t}}$ that are stabilized by $\mu_{\frac{n}{2}}$, together with the $\mu_{\frac{n}{2}}$-actions on the cotangent spaces at those points. This list is very useful in the study of the differential forms on $\widetilde{X_{T,t}}$.

Since we want to study modular parametrizations of elliptic curves by looking at their stable reductions, we have to study the stable reduction of elliptic curves too. This is done by the same method that we used for the modular curves. The results of these computations are summarized in the following proposition.

Proposition 4 *Let S be the spectrum of a complete discrete valuation ring, with algebraically closed residue field not of characteristic 2 or 3. Let E be an elliptic curve over the generic point of S. We attach a number n and a character $\chi_\omega : \mu_n \to \mu_n$ to E according to its reduction type as follows ($\chi : \mu_n \to \mu_n$ denotes the identity):*

type	I_0	I_ν	II	III	IV	I_0^*	I_ν^*	IV^*	III^*	II^*
n	1	1	6	4	3	2	2	3	4	6
χ_ω	χ	χ	χ	χ	χ	χ	χ	χ^{-1}	χ^{-1}	χ^{-1}

Let π_0 be a uniformizer on S. Then E acquires stable reduction over $T := S[\pi]$, where $\pi^n = \pi_0$. The Galois group of T over S is identified with μ_n by: $\zeta^\# \pi = \zeta\pi$ for all $\zeta \in \mu_n$. Let t be the closed point of T and $\widetilde{\mathcal{E}_T}$ the Néron model of E_T. The the action of μ_n on the tangent space at 0 of $\widetilde{\mathcal{E}_{T}}_{,t}$ is given by χ_ω.

3 Modular parametrizations

Let $\phi : X_0(p^2N) \to E$ be a strong modular parametrization where $p > 3$ is a prime that does not divide the positive integer N. We suppose that E has potentially good reduction at p (this means that we exclude the reduction type I_ν^*). Let $S := \operatorname{Spec} W(\overline{\mathbf{F}}_p)$, $n := \frac{p^2-1}{2}$, $U := S[\pi]$, where $\pi^n = p$. In the previous section we have seen that both $X_0(p^2N)$ and E acquire stable reduction over U. We denote these stable models by $\widetilde{X_U}$ and $\widetilde{\mathcal{E}_U}$, respectively. The parametrization ϕ gives, by pullback, a morphism (still denoted ϕ) between the generic fibres of $\widetilde{X_U}$ and $\widetilde{\mathcal{E}_U}$.

Proposition 5 *This ϕ extends to a morphism $\phi : \widetilde{X_U} \to \widetilde{\mathcal{E}_U}$.*

Proof. This follows from the facts that $\widetilde{\mathcal{E}_U}$ is an abelian scheme and $\widetilde{X_U}$ a stable curve.

The fact that the map $\phi : \widetilde{X_U}_{,u} \to \widetilde{\mathcal{E}_U}_{,u}$ (where u denotes the closed point of U) is μ_n-equivariant already gives a lot of information. For example, let us consider the orders of the μ_n-action on the various components of $\widetilde{X_U}_{,u}$: on a horizontal component, this order divides $p+1$; on the two outer vertical components it is 1 and on the two central vertical components together, it is $p-1$. Let d be the order of the μ_n-action on $\widetilde{\mathcal{E}_{U,u}}$ (see the table in proposition 4). It follows that all the components of $\widetilde{X_{U,u}}$ on which the order of the μ_n-action is not divisible by d will be contracted by ϕ. This leads us to the following proposition.

Proposition 6 *The morphism $\phi : \widetilde{X_{U,u}} \to \widetilde{\mathcal{E}_{U,u}}$ is constant on the two outer vertical components. On the other components we have:*

type	p	hor. components	central vert. comps.
I_0^*		contracted	
II, IV, IV^*, II^*	$p \equiv 1(3)$	contracted	
	$p \equiv -1(3)$		contracted
III, III^*	$p \equiv 1(4)$	contracted	
	$p \equiv -1(4)$		contracted

Remark. In order to prove that each horizontal component C is contracted in the case where E has reduction type I_0^*, one has to use that the quotient of C by $\mu_{\frac{n}{2}}$ has genus 0.

The remainder of this section will be concerned with the degree of inseparability of $\phi : \widetilde{X_{U,u}} \to \widetilde{\mathcal{E}_{U,u}}$. The next proposition shows that there will always be at least one component of $\widetilde{X_{U,u}}$ on which ϕ is either separable or the composition of a separable morphism with the Frobenius endomorphism of $\widetilde{\mathcal{E}_{U,u}}$. Let $\widetilde{J_U}^0$ denote the connected component of the Néron model of $J_0(p^2N)$ over U. Since $\widetilde{X_U}$ is a stable curve, $\widetilde{J_U}^0$ is the Picard scheme $\mathrm{Pic}^0_{\widetilde{X_U}/U}$ (see [12] 6.1.4, 6.1.6 and 8.2.1).

Proposition 7 *Let G be the kernel of $\phi^* : \widetilde{\mathcal{E}_{U,u}} \to \widetilde{J_{U,u}}^0$. If $p > 7$, then either $G = 0$ or $G = \ker(V : \widetilde{\mathcal{E}_{U,u}} \to \widetilde{\mathcal{E}_{U,u}})$, the kernel of the Verschiebung.*

Proof. Since $\phi : X_0(p^2N)_{\mathbf{Q}} \to E$ is strong, we have that $\phi^* : \widetilde{\mathcal{E}_U} \to \widetilde{J_U}^0$ is injective on the generic fibre. It follows from the semistability of $\widetilde{\mathcal{E}_U}$ and $\widetilde{J_U}^0$ that G is finite and consists of p-power torsion. Now we want to apply a result of Raynaud on the specialization of finite group schemes [13] Corollaire 3.3.6.1 (see also [7] Proposition 1.1) to study the kernel of $\phi^* : \widetilde{\mathcal{E}_U} \to \widetilde{J_U}^0$. We can not use this result of Raynaud directly, because the absolute ramification index of U is $n = \frac{p^2-1}{2} \geq p-1$.

Therefore let $T \to S$ be the minimal base change such that E acquires good reduction (see proposition 4, hence $T \to S$ is of degree 2, 3, 4 or 6). Taking the quotient of $\phi : \widetilde{X_U} \to \widetilde{\mathcal{E}_U}$ by the action of the subgroup $\mathrm{Gal}(U/T)$ of μ_n gives us a morphism $\phi : \widetilde{X_T} \to \widetilde{\mathcal{E}_T}$. By Picard functoriality we get: $\phi^* : \widetilde{\mathcal{E}_T} \to \widetilde{J_T}^0$, where $\widetilde{J_T}^0$ is defined to be $\mathrm{Pic}^0_{\widetilde{X_T}/T}$ ($\mathrm{Pic}^0_{\widetilde{X_T}/T}$ is representable by a smooth group scheme by [12] 8.2.1, we do not know whether $\mathrm{Pic}^0_{\widetilde{X_T}/T}$ is the Néron model). Since the absolute ramification index of T is less than $p-1$ (this explains the hypothesis "$p > 7$" in this proposition), we can now apply the result of Raynaud mentioned above to $\phi^* : \widetilde{\mathcal{E}_T}[p] \to H$, where H is the image of $\widetilde{\mathcal{E}_T}[p]$ in $\widetilde{J_T}^0$. The conclusion is that $\phi^* : \widetilde{\mathcal{E}_T} \to \widetilde{J_T}^0$ is a closed immersion.

By pullback to U we get a commutative diagram:

$$\begin{array}{ccc} \widetilde{\mathcal{E}_{T,U}} & \hookrightarrow & \widetilde{J_{T,U}}^0 \\ \downarrow & & \downarrow \\ \widetilde{\mathcal{E}_U} & \to & \widetilde{J_U}^0 \end{array}$$

where $\widetilde{\mathcal{E}_{T,U}} \to \widetilde{\mathcal{E}_U}$ is an isomorphism (since E has already good reduction over T) and where $\widetilde{J_{T,U}}^0 \to \widetilde{J_U}^0$ exists because of the Néron property of $\widetilde{J_U}^0$. It follows that:

$$G = \widetilde{\mathcal{E}_{T,t}} \cap \ker(\widetilde{J_{T,t}}^0 \to \widetilde{J_{U,u}}^0).$$

Using our knowledge of $\widetilde{X_T}$ and $\widetilde{X_U}$ it is not hard to show that $\ker(\widetilde{J_{T,t}}^0 \to \widetilde{J_{U,u}}^0)$ is purely additive: it is isomorphic to a direct sum of $\mathbf{G}_{a,t}$'s. The reason that this is true is that the multiplicities of the irreducible components in $\widetilde{X_{T,t}}$ are less than p (from this it follows that there is an exponential sequence).

It remains to find out what the intersection of $\widetilde{\mathcal{E}_T}_{,t}$ with a direct sum of additive groups can be. If $\widetilde{\mathcal{E}_T}_{,t}$ is an ordinary elliptic curve over t, then its p^m-torsion is isomorphic to $\mu_{p^m,t} \times \mathbf{Z}/p^m$ and it follows that the intersection is at most $\mathbf{Z}/p$, the kernel of V. If $\widetilde{\mathcal{E}_T}_{,t}$ is a supersingular elliptic curve, then its p-torsion subgroup is a group scheme called M_2 (see [11](15.5)). This M_2 cannot be embedded in a direct sum of additive groups, hence the intersection is at most the Frobenius kernel (which is isomorphic to $\alpha_{p,t}$). This completes the proof of the proposition.

4 Manin constants

We are now ready to apply the results of the preceding sections to the study of Manin constants. Let $\phi : X_0(p^2N) \to E$ be a strong modular parametrization, where $p > 7$ is a prime that does not divide the positive integer N. We suppose that E has potentially good reduction at p, not of type I_0^*. Let S, n, U and T be as in the preceding section: $S = \mathrm{Spec}W(\overline{\mathbf{F}}_p)$, $n = \frac{p^2-1}{2}$, $U = S[\pi]$ with $\pi^n = p$ and $T = S[\pi^{\frac{n}{d}}]$, where d is as in the table in proposition 4 ($T \to S$ is the minimal base change after which E acquires good reduction). Let $\phi : \widetilde{X_U} \to \widetilde{\mathcal{E}_U}$ be the strong modular parametrization extended to the stable models over U.

Let ω_S be a Néron differential of E over S. Then $\phi^*\omega_S$ is a meromorphic differential form on $X_0(p^2N)_S$. The valuation at p of the Manin constant of E is the multiplicity of the $(2,0)$-component (this is the component on which the cusp ∞ lies) in the divisor of $\phi^*\omega_S$. By explicit computations using minimal Weierstrass equations, it is easy to compute a Néron differential ω_U of E over U. It turns out that $\omega_U = \pi^a\omega_S$, where a is given in the following table:

type	II	III	IV	IV^*	III^*	II^*
a	$\frac{1}{6}\cdot n$	$\frac{1}{4}\cdot n$	$\frac{1}{3}\cdot n$	$\frac{2}{3}\cdot n$	$\frac{3}{4}\cdot n$	$\frac{5}{6}\cdot n$

Since the parametrization ϕ can be extended over the stable models (see proposition 5), $\phi^*\omega_U$ is a differential form on $\widetilde{X_U}$. In particular, $\phi^*\omega_U$ is a section of the invertible sheaf of regular differentials on $\widetilde{X_U}_{,u}$. We now have the following formula for the valuation $v_p(c)$ at p of the Manin constant c

of E:

$$v_p(c) = \tfrac{1}{n}(v_C(\phi^*\omega_U) - a),$$

where $v_C(\phi^*\omega_U)$ denotes the valuation of $\phi^*\omega_U$ (as a regular differential) along the $(2,0)$-component C of $\widetilde{X_{U,u}}$.

Proposition 8 *If $\phi : \widetilde{X_{U,u}} \to \widetilde{\mathcal{E}_{U,u}}$ is separable on some component of $\widetilde{X_{U,u}}$, then p does not divide c.*

Sketch of proof. According to the formula for $v_p(c)$ above, we have to show that $v_C(\phi^*\omega_U) < n + a$. By hypothesis, $\phi^*\omega_U$ gives a non-zero differential form on some component D of $\widetilde{X_{U,u}}$. Let D' be another component of $\widetilde{X_{U,u}}$ meeting D in a point x. Then $v_{D'}(\phi^*\omega_U)$ can be estimated in terms of $v_x(\phi^*\omega_U|_D)$, the order of vanishing at x of the restriction of $\phi^*\omega_U$ to D. One then proceeds in the same way with $(\pi^{-r}\phi^*\omega_U)|_{D'}$, where $r = v_{D'}(\phi^*\omega_U)$. It is clear from figure 1 that, after at most two steps, one arrives at the component C. It turns out that the estimates, if done properly, are sharp enough.

Proposition 9 *Under the hypotheses of this section, one has that $v_p(c) \leq 1$.*

Sketch of proof. The principle is the same as in the proof of the preceding proposition, but now one has to start at a component D on which ϕ has degree of inseparability at most 1. An easy computation shows that, in that case, $v_D(\phi^*\omega_U) \leq n$.

In order to get a stronger result, we will now consider E together with E', the quadratic twist of E over $\mathbf{Q}(\sqrt{(\frac{-1}{p})p})$. This twist acts on the reduction types at p as follows:

original type	II	III	IV	IV^*	III^*	II^*
type of twist	IV^*	III^*	III^*	II	III	IV

It is known that E' is again a modular elliptic curve, and it is easy to see that the conductor of E' is also p^2N. Let $\tilde{E}$ be the strong curve in the isogeny class of E', let $\tilde{\phi}$ be its strong modular parametrization and let $\tilde{c}$ be its Manin constant. It follows that we have a commutative diagram:

$$\begin{array}{ccc} X_0(p^2N)_{\mathbf{Q}} & \xrightarrow{\tilde{\phi}} & \tilde{E} \\ & {\scriptstyle \phi'} \searrow \quad \swarrow {\scriptstyle \alpha} & \\ & E' & \end{array}$$

with α a cyclic isogeny. Let $f\frac{dq}{q}$ and $\tilde{f}\frac{dq}{q}$ be the normalized newforms of E and $\tilde{E}$. Over $\mathbf{C}$, the modular parametrization ϕ is given as follows:

$$\phi : X_0(M)(\mathbf{C}) \to \mathbf{C}/\Lambda = E(\mathbf{C}) : \quad P \mapsto \int_\infty^P f\frac{dq}{q},$$

where Λ is the period lattice of $f\frac{dq}{q}$:

$$\Lambda = \left\{ \int_\gamma f\frac{dq}{q} \ ; \ \gamma \in \mathrm{H}_1(X_0(M)(\mathbf{C}), \mathbf{Z}) \right\}.$$

This gives the following formula for $\deg(\phi)$, the degree of ϕ:

$$\deg(\phi) = \|f\|^2/\mathrm{vol}(\mathbf{C}/\Lambda), \quad \text{where } \|f\|^2 = \tfrac{i}{2}\int_{X_0(M)(\mathbf{C})} |f|^2 \frac{dq \wedge d\bar{q}}{q\bar{q}}.$$

By a standard computation, see [17], one sees that

$$\delta\Lambda \subset \tilde{\Lambda} \subset \delta^{-1}\Lambda,$$

where $\tilde{\Lambda}$ is the period lattice of $\tilde{f}\frac{dq}{q}$ and δ is the Gauss sum $\sqrt{(\frac{-1}{p})p}$. Since $\tilde{\Lambda}$ is proportional to Λ, there are exactly three possibilities: $\tilde{\Lambda} = \delta\Lambda$, $\tilde{\Lambda} = \Lambda$ or $\tilde{\Lambda} = \delta^{-1}\Lambda$.

In the second case, E has complex multiplication. As stated in the introduction, a combination of our results and those of Stevens [14] implies that in this case p does not divide c. Now we assume that E does not have complex multiplication. The formula:

$$\|f^2\| = C \cdot [\mathrm{PSL}_2(\mathbf{Z}) : \Gamma_0(p^2N)] \cdot \mathrm{Res}_{s=2} \sum_{n=1}^{\infty} \frac{|a_n|^2}{n^s}$$

(where C is some fixed constant) in [17] implies that $||f||^2 = ||\tilde{f}||^2$. It now follows that $\deg(\tilde{\phi})$ equals $p^{-1}\deg(\phi)$ or $p\deg(\phi)$ in the first or third case, respectively. By comparing the stable reductions of ϕ and $\tilde{\phi}$, one can see that either ϕ or $\tilde{\phi}$ is separable on some component of $\widetilde{X_{U,u}}$. More precisely, we are in one of the two cases:

1. $\tilde{\Lambda} = \delta\Lambda$, $\deg(\tilde{\phi}) = p^{-1}\deg(\phi)$, ϕ is inseparable and $\tilde{\phi}$ is separable.

2. $\tilde{\Lambda} = \delta^{-1}\Lambda$, $\deg(\tilde{\phi}) = p\deg(\phi)$, ϕ is separable and $\tilde{\phi}$ is inseparable.

After replacing E, if necessary, by $\tilde{E}$, we may assume that E has reduction type II, III or IV at p. Let ω be a Néron differential of E over $\mathbf{Z}$, then $\omega' = \delta^{-1}\omega$ is a Néron differential on E'. Let c' be the Manin constant associated to ϕ': $\phi'^*\omega' = c'\tilde{f}\frac{dq}{q}$. Then we have the formulas:

$$\frac{\deg(\phi)}{c^2} = \frac{||f||^2}{\mathrm{vol}(E,\omega)} \qquad \frac{\deg(\phi')}{(c')^2} = \frac{||\tilde{f}||^2}{\mathrm{vol}(E',\omega')}$$

where

$$\mathrm{vol}(E,\omega) = \int_{E(\mathbf{C})} \tfrac{i}{2}\omega\wedge\overline{\omega} \quad \text{and} \quad \mathrm{vol}(E',\omega') = \int_{E'(\mathbf{C})} \tfrac{i}{2}\omega'\wedge\overline{\omega'}.$$

¿From these it follows directly that:

$$\deg(\tilde{\phi}) = p\deg(\phi)\frac{(c')^2}{c^2}\frac{1}{\deg(\alpha)}.$$

Since we are either in case 1 or case 2, we see that $\deg(\alpha)$ is a square. It follows from the classification of cyclic isogenies over $\mathbf{Q}$ [7] [16] that p does not divide $\deg(\alpha)$. This implies that $v_p(c) = v_p(c')$. We arrive at:

$$v_p(\deg\tilde{\phi}) = v_p(\deg\phi) + 1 + 2v_p(\frac{\tilde{c}}{c}).$$

In case 2 above, it follows that $v_p(\tilde{c}) = v_p(c) = 0$. In case 1 above, $v_p(c) = 1$ and $v_p(\tilde{c}) = 0$. Of course, if case 1 occurs, then Manin's conjecture is false.

If E has potentially supersingular reduction at p, then one can prove that only case 2 occurs. The reason for this is that then the μ_n-action on $\widetilde{\mathcal{E}_{U,u}}$ does not commute with the Frobenius endomorphism, giving a parity condition on the degree of inseparability of ϕ. If E has potentially ordinary reduction at p, then these two do commute, and we cannot prove that case 1 does not occur.

References

[1] B.J. Birch and H.P.F. Swinnerton-Dyer. *Elliptic curves and modular functions.* Modular functions of one variable IV, 2–32. Springer Lecture Notes in Mathematics 476 (1975).

[2] H. Carayol. *Sur les représentations ℓ-adiques associées aux formes modulaires de Hilbert.* Ann. scient. Éc. Norm. Sup., série 4, t.19, 409-468 (1986).

[3] P. Deligne and M. Rapoport. *Les schémas de modules des courbes elliptiques.* Modular functions of one variable II, 143–316. Springer Lecture Notes in Mathematics 349.

[4] S.J. Edixhoven. *Minimal resolution and stable reduction of $X_0(N)$.* Preprint Utrecht November 1986, accepted for publication in the Annales de l'Institut Fourier.

[5] S.J. Edixhoven. *Stable models of modular curves and applications.* Thesis Utrecht June 1989, to be published.

[6] N.M. Katz and B. Mazur. *Arithmetic moduli of elliptic curves.* Annals of Mathematics Studies 108, Princeton University Press (1985).

[7] B. Mazur. *Rational isogenies of prime degree.* Invent. Math. 44, 129–162 (1978).

[8] B. Mazur. *Courbes elliptiques et symboles modulaires.* Séminaire Bourbaki juin 1972. Springer Lecture Notes in Mathematics 317.

[9] B. Mazur and H.P.F. Swinnerton-Dyer. *Arithmetic of Weil curves.* Invent. Math. 25, 1–16 (1974).

[10] J.-F. Mestre and J. Oesterlé. Letter to B. Gross, May 1985.

[11] F. Oort. *Commutative group schemes.* Springer Lecture Notes in Mathematics 15 (1966).

[12] M. Raynaud. *Spécialisation du foncteur de Picard.* Publications Mathématiques de l'I.H.E.S. 38 (1970).

[13] M. Raynaud. *Schémas en groupes de type (p,...,p).* Bull. Soc. Math. France 102, 241-280 (1974).

[14] G. Stevens. *Stickelberger elements and modular parametrizations of elliptic curves.* To be published.

[15] J. Tate. *Algorithm for determining the type of a singular fibre in an elliptic pencil.* In Modular Functions of One Variable IV, 33-52. Springer Lecture Notes in Mathematics 476 (1975).

[16] M. Kenku. *On the number of* **Q**-*isomorphism classes of elliptic curves in each* **Q**-*isogeny class.* J. of Number Theory 15, 199-202 (1982).

[17] D. Zagier. *Modular parametrizations of elliptic curves.* Canad. Math. Bull. 28(3), 372-384 (1985).

The action of monodromy on torsion points of Jacobians

TORSTEN EKEDAHL

Let C_g be the generic curve of genus g over an algebraically closed field $\mathbf{k}$ with field of definition K_g. If $\overline{K}_g$ is a separable closure of K_g then we can consider the action of $Gal(\overline{K}_g/K_g)$ on the torsion of $Pic(C_{g,\overline{K}_g})$. One may ask for the image of this Galois group in the group of automorphisms of ${}_{tor}Pic(C_{g,\overline{K}_g})$. On the part which is of order prime to the characteristic the result is known, by transcendental methods, to be the full symplectic group. In this paper we will give an algebraic proof of this fact and also show that the action on the points of order a power of the characteristic p is the largest possible i.e. the full general linear group. Rather than dealing directly with the generic curve we will work with the corresponding moduli stacks. Using this reformulation Theorem 2.1 is the result giving the surjectivity for p-torsion points and Corollary 3.1 the result giving the surjectivity for all torsion points (this surjectivity is of course equivalent to the irreducibility statement of that theorem).

1. We need some preliminary results on Lang torsors (cf. [De: 1.4]). Let therefore G be an algebraic group over a finite field $\mathbf{F}$ and let $F : G \to G$ be the Frobenius map wrt $\mathbf{F}$. Then the map

$$\begin{aligned} G &\to G \\ g &\mapsto Fg \cdot g^{-1} \end{aligned}$$

is an étale Galois cover with Galois group $G(\mathbf{F})$.

Proposition 1.1. *i) Let X be an $\mathbf{F}$-variety with an action of G also defined over $\mathbf{F}$, S an $\mathbf{F}$-scheme, $g : S \to G$ an $\mathbf{F}$-morphism and $Y(\mathbf{F})$ the $X(\mathbf{F})$-bundle associated to the action of $G(\mathbf{F})$ on $X(\mathbf{F})$and the $G(\mathbf{F})$-torsor over S which is the pullback of the Lang torsor by g. Then $Y(\mathbf{F})$ is the kernel* *of the Frobenius map on $X \times_{\mathbf{F}} S$ wrt to $\mathbf{F}$ and the map given by the action of g on $X \times_{\mathbf{F}} S$.*

ii) Let S be an $\mathbf{F}_p$-scheme and $\phi : F^*\mathcal{O}_S^n \to \mathcal{O}_S^n$ an isomorphism so that $\mathcal{E} := \ker(\phi - 1)$ is a local system of n-dimensional $\mathbf{F}_p$-vector spaces. The matrix of ϕ gives a map $S \to GL_n$. The pullback of the Lang torsor is isomorphic to the $GL_n(\mathbf{F}_p)$-torsor of trivialisations of $\mathcal{E}$.

iii) Let S be an $\mathbf{F}_p$-scheme and $\phi : S \to GL_2$ a map given by the matrix

$$\begin{pmatrix} a & b \\ c & d \end{pmatrix}.$$

Then if $c \neq 0$, the $\mathbf{P}^1(\mathbf{F}_p)$-bundle $\mathcal{P}$ associated to the pullback by ϕ of the Lang torsor and the action of GL_2 on $\mathbf{P}^1$ is given by the equation $ct^{p+1} + dt^p - at - b = 0$.

Proof. To begin with i) is purely formal and true in awesome topos theoretic generality. Indeed, the T-points, where T is an S-scheme, of the pullback torsor consists of the $h \in G(T)$ for which $\mathrm{F}h \cdot h^{-1} = g$. We map a pair (h, x), where h is a point of the torsor and x a point of X with $Fx = x$, to a point hx of X with $F(hx) = FhFx = ghx$. In this way we obtain a map from the associated $X(\mathbf{F})$-torsor to the kernel of F and g on X. To verify that this is an isomorphism we may assume that there is an h with $Fh \cdot h^{-1} = g$ in which case translation by h makes it into the identity map. Furthermore, ii) follows directly from i) with $G = GL_n$ and $X = \mathbf{A}^n$ with the standard action. As for iii) we get from i) that $\mathcal{P}$ is defined by $(s^p : t^p) = (as + bt : cs + dt)$ and if $c \neq 0$ then we see that this subscheme of $\mathbf{P}_S^1$ does not meet $t = 0$. Dehomogenising gives the desired equation. Q.E.D.

2. Let us recall that if C is any smooth and proper curve over a field k of characteristic $p > 0$ then C is said to be ordinary if the group of divisor classes of order dividing p over the algebraic closure of k has the largest possible order namely p^g, where g is the genus. An equivalent condition is that the Frobenius map $F^*H^1(C, \mathcal{O}_C) \to H^1(C, \mathcal{O}_C)$ be an isomorphism. The last condition also makes sense for a singular (but still proper) curve and if it is fulfilled in such a case we will continue to call the curve ordinary. It is clear from the definition that ordinarity is an open condition, a nodal genus 1 curve is ordinary and a semistable curve is ordinary iff all its irreducible components are ordinary. For any n not divisible by p let $\mathcal{M}_{g,n}^{ord}$ be the moduli stack of ordinary curves with a proper Jacobian and a level n-structure. Pick any ordinary elliptic curve E over $\mathbf{k}$ and two points on it.

The stable curve D_g of genus g whose components are copies of E, whose dual graph is a line and whose double points lie under the copies of the two given points is then ordinary and gives rise to a k-point $\bar{s}$ of $\mathcal{M}_{g,n}^{ord}$. We may now consider the action of the fundamental group of $\mathcal{M}_{g,n}^{ord}$ wrt $\bar{s}$ on the Tate module $T_p^{g,n}$ of $Pic(D_g)$ which is a free $\mathbf{Z}_p$-module of rank g.

Theorem 2.1. *The map $\pi_1(\mathcal{M}_{g,n}^{ord}, \bar{s}) \to GL(T_p^{g,n})$ is surjective.*

Proof. Let $\mathcal{N}_{2,n}^{ord}$ be the moduli space of ordinary stable genus 2 curves with a level n-structure and 2 distinct points. Suppose we can prove that the fundamental group of $\mathcal{N}_{2,n}^{ord}$ maps onto $GL(T_p^{2,n})$, where $\bar{s}$ lifts using the chosen points to a point $\bar{s}'$ of $\mathcal{N}_{2,n}^{ord}$. Now $T_p(D_g)$ is naturally the direct sum of g copies of $T_p(E)$. For any $1 \leq i \leq g-1$ we can construct a stable curve of genus g over $\mathcal{N}_{2,n}^{ord}$; append D_{i-1} to the first point and D_{g-i-2} to the other one. This gives a map ϕ_i from $\mathcal{N}_{2,n}^{ord}$ to $\mathcal{M}_{g,n}^{ord}$. The action of $\pi_1(\mathcal{N}_{2,n}^{ord}, \bar{s}')$ on $T_p(D_g)$ then only moves the ith and $(i+1)$th factor and there it acts by the action of $\pi_1(\mathcal{N}_{2,n}^{ord}, \bar{s}')$ on $T_p(D_2)$. By assumption the image of $\pi_1(\mathcal{M}_{g,n}^{ord}, \bar{s}')$ on $T_p(D_g)$ contains the "diagonal" GL_2 moving the ith and $(i+1)$th factors. This is true for all i, and $GL(T_p(D_g))$ is clearly generated by these "diagonal" subgroups.

Before treating the case of genus 2 we will, however, do the genus 1 case (which is in fact well known cf. [K-M: Cor. 12.6.3]). Let E' be a super singular (i.e. non-ordinary) elliptic curve. We will then show that the local monodromy around E' gives the whole GL_1. Let us therefore consider a miniversal deformation $\mathcal{E} \to \mathbf{Spec}\ R$. The generic fibre of it is then ordinary. Choose a local parameter t and consider the formal group of $\mathcal{E}$. Its Verschiebung map V is then a power series ϕ without constant coefficient, with linear coefficient a generator of the maximal ideal of R (as the deformation is versal, we will do this calculation later) and whose first invertible coefficient is in degree p (as this is true for E'). The equation $\phi = 0$ then has the following Newton polygon (cf. [Ko:Ch. IV]):

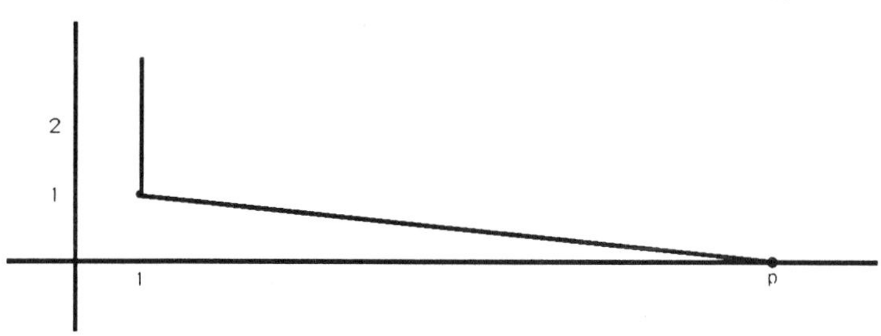

Hence we see that an element of order exactly p in the Picard group of a geometric generic fibre of $\mathcal{E} \longrightarrow \mathbf{Spec}\ R$ generates an extension of degree $p-1$, as the kernel of V is exactly the elements killed by p, and so the monodromy action on the points killed by p is surjective. If α is a point of order p its p-adic valuation is $1/p$ and so the Newton polygon of $\phi = \alpha$ is a straight line of slope $-1/p$ and therefore an element of order exactly p^2 generates an extension of order $p(p-1)$. Therefore, the monodromy action on the points of order p^2 is surjective. Continuing we get what we want for $g = 1$. We then also get the whole GL_1 for the family of elliptic curves with level n-structure and 2 distinct points as there is no problem in constructing points or level n-structures over a strictly Henselian base.

If we therefore turn to the case of genus 2, by deforming D_2 and using what we just proved for genus 1, we see that all diagonal matrices will be in the image of monodromy. Suppose we know that $\pi_1(\mathcal{N}^{ord}_{2,n}, \bar{s}')$ maps onto the general linear group of $T^{2,n}_p/p$. Then we can prove by induction that it maps onto $GL(T^{2,n}_p/p^m)$. Indeed, the kernel of

$$GL(T^{2,n}_p/p^{m+1}) \rightarrow GL(T^{2,n}_p/p^m)$$

is isomorphic to $End(T^{2,n}_p/p)$ with the action of $GL(T^{2,n}_p/p)$ given by conjugation being the adjoint action, as we have just seen $\pi_1(\mathcal{N}^{ord}_{2,n}, \bar{s}')$ hits the diagonal matrices and they generate $End(T^{2,n}_p/p)$ as $GL(T^{2,n}_p/p)$-module.

We are hence reduced to showing that the action of monodromy on the points killed by p is the largest possible. Recall that that if C is a smooth and proper curve of genus 2, then we may choose a basis e_1, e_2, f_1 and f_2 of the first de Rham cohomology such that they form a symplectic base i.e. $e_1 \wedge e_2 = f_1 \wedge f_2 = 0$ and $e_i \wedge f_j = \delta_{ij}$ and such that f_1 and f_2 form a basis for the regular 1-forms. Recall furthermore (cf. [B-O: 7.1]) that if $\mathcal{C}$ is a deformation of C of order less than p then e_1, e_2, f_1 and f_2 lift canonically to a basis for the first de Rham cohomology of $\mathcal{C}$ and there is a basis for the regular 1-forms of $\mathcal{C}$ of the form $\underline{f} - G\underline{e}$ where G is a symmetric matrix. The deformation is completely determined by G and any symmetric G can occur. Finally, the matrix for the action of the Frobenius on de Rham cohomology is the same for C and $\mathcal{C}$. Let now C be a curve of genus 2 for which the Frobenius on H^1 has square zero but is non-zero (which exists by [Mo-B]). There then are $e, f \in H^1_{DR}(C/k)$ s.t. e, Fe, F^2e and f form a symplectic basis as above. If the deformation $\mathcal{C}$ is determined by the matrix

$$G = \begin{pmatrix} a & b \\ b & c \end{pmatrix},$$

then the images of e and Fe in $H^1(\mathcal{C}, \mathcal{O}_{\mathcal{C}})$ form a basis h_1 and h_2 and the action of the Frobenius is given by $Fh_1 = h_2$ and $Fh_2 = ah_1 + bh_2$. If we now look at deformations over the dual numbers then, as deformations of curves are unobstructed it can be lifted to a deformation over $k[[t]]$ and then our calculation gives a matrix for the Frobenius modulo t_2. If we have a deformation $\mathcal{C}$ of C over $k[[x]]$ with Frobenius matrix

$$F = \begin{pmatrix} a & b \\ c & d \end{pmatrix}$$

such that the generic fibre is ordinary then we can consider the $\mathbf{P}^1(\mathbf{F}_p)$-cover of $k((t))$ whose fiber is the projective space associated to the $\mathbf{F}_p$-vector space of points of the Picard group of $\mathcal{C}$ killed by p. From (1.1) it follows that this cover has equation

$$cs^{p+1} + ds^p - as - b = 0. \tag{2.2}$$

(Strictly speaking the kernel of F-1 on $H^1(C, \mathcal{O}_C)$ is dual to the group of points of order p in the Picard group. However, the projective space of a 2-dimensional vector space is canonically isomorphic to that of its dual.) We now want to show that the action of local monodromy around C acts transitively on $\mathbf{P}^1(\mathbf{F}_p)$. Consider therefore a deformation of C over $k[[t]]$ which up to first order is described by the matrix

$$\begin{pmatrix} 0 & \delta \\ \delta & 0 \end{pmatrix},$$

where δ is the residue of t. Hence we get the following congruence modulo t^2 for the Frobenius matrix

$$F \equiv \begin{pmatrix} 0 & \delta \\ 1 & 0 \end{pmatrix}.$$

This shows that in this case the Newton polygon of (2.2) is a straight line of slope $-1/(p+1)$ and so the local monodromy for this particular deformation is already transitive. The next step is to show that the action of the global monodromy acts 2-transitively on $\mathbf{P}^1(\mathbf{F}_p)$. For this we will use the local monodromy around a stable curve C' whose components are two super singular elliptic curves. In this case there is a symplectic base for which $Fe_i = f_i$. Assume now that $p > 2$ and consider the deformation over $k[[t]]/(t^3)$ determined by the matrix

$$\begin{pmatrix} \epsilon & \epsilon^2 \\ \epsilon^2 & 0 \end{pmatrix},$$

where ϵ is the residue of t. Then the Frobenius is given by the same matrix modulo t^3 and the Newton polygon will be

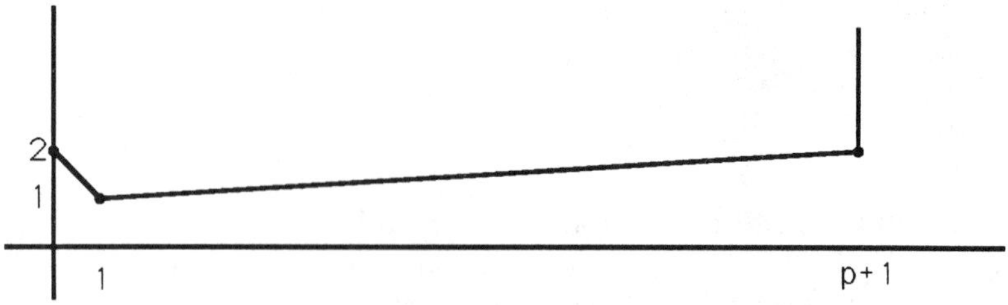

(2.3)

This then gives two orbits for the local monodromy; one of length 1 and one of length p and therefore the action is 2-transitive. Finally, the fact that the group of diagonal automorphisms lies in the image of monodromy shows that the action is 3-transitive. Hence the monodromy group maps onto the projective linear group and as it contains all the scalar matrices it maps onto the general linear group. We see that in all cases can we get two points in the families we have used by lifting two points on the starting curve to the whole deformation. Also the start curve has a proper Jacobian and so the whole deformation has a level n-structure. Hence the theorem is proved in the case of odd characteristic.

We are hence left with the case of $p = 2$. Consider then the family of double covers of $\mathbf{P}^1$ given by the equations

$$s^2 + (ax^3 + bx^2y + cxy^2 + dy^3)s = y,$$

where x and y are homogenous coordinates on $\mathbf{P}^1$. This defines a smooth curve C over an open subset of the (a, b, c, d)-space. We have a natural isomorphism $H^1(C, \mathcal{O}_C) = H^1(\mathbf{P}^1, \mathcal{O}_{\mathbf{P}^1}(-4)s)$ and hence a natural basis for $H^1(C, \mathcal{O}_C)$ consisting of the Čech cocycles, for the standard covering of $\mathbf{P}^1$, s/x^2y and s/xy^2. Hence we see that a matrix for the Frobenius map on $H^1(C, \mathcal{O}_C)$ is given by

$$\begin{pmatrix} c & a \\ d & b \end{pmatrix}.$$

If we put $a := t^2$, $b := 0$, $c := t$ and $d := 1$ then we get a family over **Spec** $k[[t]]$ whose fibre over $t = 0$ is smooth and so is smooth. Furthermore, the Newton polygon for (2.2), in this case, has a line of slope $1/p$ and we continue as before. Q.E.D.

Remark. i) If $p > 11$ then $PSL_2(\mathbf{F}_p)$ is generated by any pair of elements, one of order $p+1$ and one of order $p-1$. Hence, in that case, the last part of the argument is not necessary.

ii) The proof shows in fact that already the local monodromy around a curve whose components are super singular elliptic curves and whose dual graph is a line gives the whole GL, as all the types needed will be found in any neighbourhood of such a curve.

3. We now want to use the results and techniques of section 1 to show the following theorem.

Theorem 3.1. *Let $N \geq 2$ be an integer and let N' be the largest divisor of N prime to the characteristic of an algebraically closed field* $\mathbf{k}$. *Let $M_{g,N}$ be the moduli stack over* $\mathbf{k}$ *of stable curves of a fixed genus with a trivialisation of the kernel of multiplication by N on the Jacobian and compatible with the symplectic structure on this kernel given by the canonical polarisation and the choice of a primitive N' th root of unity. Then $M_{g,N}$ is irreducible.*

Proof. From Theorem 2.1 it follows that the theorem follows from the corresponding statement for N', so we may assume that N is prime to the characteristic and then that the characteristic is zero, by the argument of [D-M:§5]. The case $N = 1$ is known and admits a purely algebraic proof (cf. ?Ha]). The result for larger N is then equivalent to the monodromy action for some choice of base point being the largest possible. Assume that we know this result for genus 1. Then as before, using a line of elliptic curves as basepoint with symplectic base $e_1, \dots, e_g, f_1, \dots, f_g$ for the Tate module, we see that, for any i, all symplectic matrices which fix e_j and f_j, for $j \neq i$, and stabilise the space spanned by e_i and f_i belong to the image of monodromy. On the other hand, from theorem 2.1 it follows that for every $A \in GL_g$ a symplectic matrix of the form

$$\begin{pmatrix} A & C \\ 0 & (A^{-1})^t \end{pmatrix}$$

is in the image of the local monodromy around the prime factors of N. However, the diagonal matrices generate the symmetric matrices as a GL_g-representation (with action $S \mapsto ASA^t$) and the matrices of the form

$$\begin{pmatrix} I & C \\ 0 & I \end{pmatrix},$$

where C is diagonal, is in the image of monodromy, so we see that the image of monodromy contains all matrices of the form

$$\begin{pmatrix} A & C \\ 0 & (A^{-1})^t \end{pmatrix},$$

but they and the matrices we have already found generate the symplectic group.

We are hence left with proving the theorem in the case of genus 1. We will need the following "folklore" lemma (which I learned from J.-P. Serre).

Lemma 3.2. *Let G and H be groups. Then there is a bijection between the set of subgroups M of $G \times H$ which map surjectively by the two projections onto G and H and isomorphisms between quotient groups of G and H.*

Proof. Given a subgroup M of $G \times H$ with the desired properties consider $G' := M \bigcap G \times \{e\}$ and $H' := M \bigcap \{e\} \times H$. These subgroups are normal in G and H respectively. Indeed, for any $g \in G$ there is, by assumption, an $h \in H$ such that $(g, h) \in M$ and hence G' is stable under conjugation by g. By assumption the composite $M \hookrightarrow G \times H \to G$ is surjective and so G/G' is the quotient of M by $M \bigcap G' \times H = G' \times H'$. As the same thing is true of H/H' we get an isomorphism between G/G' and H/H'. On the other hand, starting with an isomorphism $\phi: \overline{G} \to \overline{H}$ between quotient groups $\overline{G}$ and $\overline{H}$ of G resp. H we associate to it the subgroup $\{(g, h) : \phi(\overline{g}) = \overline{h}\}$ of $G \times H$. It is clear that these two constructions are inverses of each other. Q.E.D.

Let $\rho : X_3 \to M_3$ be the modular surface of level 3 mapping to the modular curve of level 3. Assume we can prove that the monodromy of ρ maps onto $SL_2(\mathbf{Z}_p)$ for $p \neq 3$ and onto $1 + 3M_2^0$. Then we prove by induction that the monodromy maps onto $G := SL_2(\mathbf{Z}_2) \times (1 + 3M_3^0) \times SL_2(\mathbf{Z}_5) \times \ldots \times SL_2(\mathbf{Z}_p)$. In fact if q is the next prime and H is the image of monodromy in $G \times SL_2(\mathbf{Z}_q)$ then H maps onto both the factors under projection and by the lemma corresponds to an isomorphism between a quotient of H_1 of G and a quotient H_2 of $SL_2(\mathbf{Z}_q)$. If $q = 3$ then H_2 is a 3-group but $SL_2(\mathbf{Z}_2)$ has no nontrivial 3-group quotient. If $q > 3$ then H_2 is, by the same argument, not a q-group and as $SL_2(\mathbf{F}_q)$ is not a split central extension of $PSL_2(\mathbf{F}_q)$ and $PSL_2(\mathbf{F}_q)$ is a simple group, H_2 has $PSL_2(\mathbf{F}_q)$ as a quotient. However, G does not have $PSL_2(\mathbf{F}_q)$ as composition factor. Hence H corresponds to the the isomorphism of the trivial quotients i.e. H equals $G \times SL_2(\mathbf{Z}_q)$.

It therefore suffices to show that the monodromy of ρ maps onto $SL_2(\mathbf{Z}_p)$ (resp. $1 + 3M_2^0$). In the first case, as the image contains the diagonal matrices, it suffices as before to prove that it maps onto $SL_2(\mathbf{F}_p)$. Now M_3 is isomorphic to $\mathbf{P}^1$ so the monodromy is generated by local monodromy and as ρ is semistable the local monodromy is unipotent, so that the image of monodromy is generated by unipotent elements and contains the diagonal matrices. When $p \neq 2$ this implies that it equals $SL_2(\mathbf{F}_p)$. In case $p = 2$ assume that the monodromy map is not surjective. This and what we have shown implies that there is an element of order 2 in the Jacobian that is stable under monodromy and so an element of order 2 in the Jacobian of the generic fibre of ρ. Let H be the zero section of ρ, F a general fibre and $E_1, \ldots, E_8$ the horizontal curves not meeting H. As X_3 is rational these curves generate rationally the Picard of X_3. Hence, if S is the set of these curves, the intersection product gives an embedding $Pic(X_3)$ in $\mathbf{Z}^S$. Thus the group of rational points of the Jacobian of the generic fibre of ρ embeds in $\mathbf{Z}^S/M$, where M is the subgroup of $Pic(X_3)$ generated by the curves in S. However, this group is killed by 3 so there can be no element of order 2. Finally, to prove the surjectivity of the image of monodromy in $1 + 3M_3^0$ it is sufficient to show that the image of monodromy in $SL_2(\mathbf{F}_3)$ of the level 1 family $\rho : X_1 \to M_1$ is surjective which is proved as before if one notes that M_1 is simply connected as the covering $M_3 \to M_1$ is totally ramified (the fact that M_1 is a stack does not destroy the argument). Q.E.D.

BIBLIOGRAPHY

[B-O]: P. Berthelot, A. Ogus, "Notes on crystalline cohomology," Princeton University Press, Princeton, 1978.

[De]: P. Deligne, *Applications de la formule des traces aux sommes trigonométriques*, SLN **569** (1977), 168–232.

[D-M]: P. Deligne, D. Mumford, *The irreducibility of the space of curves of given genus*, Publ IHES **36** (1969), 75–110.

[K-M]: N. Katz, B. Mazur, "Arithmetic moduli of elliptic curves," Princeton University Press, Princeton, 1985.

[Ko]: N. Koblitz, "p-adic numbers, p-adic analysis and Zeta-functions," Springer-Verlag, Berlin, 1977.

[Mo-B]: L. Moret-Bailly, *Familles de courbes et de variétés abéliennes.*, Astérisque **86** (1981), 109–140.

An exceptional isomorphism between modular varieties

TORSTEN EKEDAHL, BERT VAN GEEMEN

This article aims to give a modular construction of a correspondence between modular varieties whose existence was suspected in [Ge-Ny]. The modular varieties in question are on the one hand a Siegel modular threefold X^o, that is a moduli space of 2 dimensional abelian varieties, which is a quotient of the moduli space with level 8 structure and on the other hand the self-product (over the base curve) V^o of $\mathcal{E} \to Y_0(8)$, the universal (smooth) elliptic curve with a cyclic subgroup of order 8. These varieties have smooth compactifications X and V respectively and it is shown in (loc. cit.) that the 2 dimensional Galois representation $G_{\mathbf{Q}} = Gal(\bar{\mathbf{Q}}/\mathbf{Q}) \to Aut(H^3(X, \mathbf{Q}_\ell))$ is a subrepresentation of $H^3(V, \mathbf{Q}_\ell)$. This fact can be explained by the theory of lifting of elliptic modular cusp forms to Siegel modular forms, but the Tate conjecture relating algebraic cycles and Galois invariant subspaces in the étale cohomology groups predicts the existence of an algebraic cycle Z on $X \times V$ inducing the injection of Galois representations $H^3(X) \to H^3(V)$. Using explicit equations, J. Stienstra found a dominant rational map $X \to V$ whose graph gives the desired cycle. In this paper we give in fact an isomorphism of stacks (cf. Cor 2.10) which also induces the desired inclusion. It seems however that our construction does not directly work for higher levels, whereas the lifting theory is quite general.

Roughly speaking the isomorphism is constructed as follows. One starts with the space of elliptic curves (i.e. a smooth genus 1 curves with a section) with 2 distinguished points. Given a point of that space one gets 6 points on a projective line; the line is the quotient of the curve by ± 1 and the 6 points are the 4 ramification points of the quotient map and the images of the 2 distinguished points. To these 6 points one associates the double cover ramified at them and finally one obtains a principally polarised abelian surface; the Jacobian of the double cover. This gives an element of $\mathcal{A}_2$ the moduli space of principally polarised 2-dimensional abelian varieties. In order to get an equivalence one first needs to keep track of various choices. We get for instance not just 6 points on a projective line but rather one point, corresponding to the zero section of the elliptic curve, one set of 3 and one of 2 points. Furthermore, only in the case when the 6 points are distinct do we get a smooth double cover, which corresponds to the condition that the Jacobian of the double cover being an abelian surface *and* to the theta divisor of this principally polarised abelian surface being irreducible. (It should also be noted that the genus 1 curve associated to a genus 2 curve C is simply the Prym variety associated to a double cover of C and so this article can also be seen as an attempt to compare certain level structures on C (or its Jacobian) with level structures on the associated Prym variety.)

The article is therefore divided into several parts. First we get an equivalence of data with the minimal number of choices and with a situation where the 6 points are distinct. The proof of the equivalence follows almost exactly the sketch given. The only difference is that we make the constructions in families of elliptic curves etc. This forces us to deal with one choice which was implicit in the above description. Namely that a double cover is not determined by the choice of a degree 6 divisor in a $\mathbf{P}^1$-bundle, one also needs the choice of a square root line bundle of the associated line bundle. The next step is to compare some further choices (i.e. level structures) on either side of the equivalence such as a level 2 structure on the elliptic curve. This is done by first comparing extra data for the elliptic curve and the double cover genus 2 curve and then to translate the data for the genus 2 curve with data involving only its Jacobian. Finally, we make a brief study of compactifications of moduli spaces. It turns that when we do this extension we obtain not an equivalence but only a map from the modular variety of genus 2 curves to the one for elliptic curves.

We generally formulate our results in terms of algebraic stacks rather than (coarse) moduli spaces. There are two reasons for this choice. Firstly, we have started with an elementary construction of an elliptic curve from the 6 Weierstrass points of a genus 2 curve, which gives us the searched for map between modular varieties on geometric points. With some care it is seen that this construction can be done in any family. Working with algebraic stacks immediately shows that the map on geometric points comes from an algebraic map. Secondly, the equivalence of stacks gives us the isomorphism of coarse moduli spaces by passing to the associated algebraic spaces. The equivalence of stacks, however, contains more information. To see an example of this let us recall that the moduli stack of elliptic curves is the quotient of the upper half plane by $SL_2(\mathbf{Z})$ as analytic *stacks* whereas the (coarse) moduli stack is the quotient as analytic *spaces*. The difference between those two quotients is that the former behaves as if the action of $SL_2(\mathbf{Z})$ is free whereas the latter doesn't as some elements do have fixed points. Hence the topological fundamental group of the stack is equal $SL_2(\mathbf{Z})$ with the upper half plane as universal cover while the fundamental group of the space is trivial, the space being simply the affine line. Similar phenomena will hold for many of the other stacks we will encounter (an example where the fundamental groups are equal is the stack of irreducible genus 1 curves with a section of the smooth locus; it as well as its associated analytic stack is simply-connected).

0. Preliminaries and conventions.

We will use the equality sign to denote a canonically defined isomorphism. Unless otherwise mentioned 2 will be invertible in all our schemes and S will denote such a scheme. A multisection of a map $T \to S$ of schemes is a closed subscheme V of T such that the composed map $V \to T \to S$ is finite and flat; a multisection everywhere of degree m will also be refered to as an m-section; a 1-section is then simply a section. If $\pi: T \to S$ is a map and V a subscheme of S then a section of π over V is a section of the restriction of π to V. Unless otherwise mentioned $\mathbf{P}$, $\mathcal{C}$, $\mathcal{D}$, $\mathcal{E}$, T, U and V will have the meaning given to them in Theorem 1.2 and the paragraph after it. By an *elliptic curve* over S we will mean a smooth proper map $\pi: \mathcal{E} \to S$ with fibers curves of genus 1 together with a chosen section of π. The *scheme of Weierstrass points* of an elliptic curve is then the fixed points of multiplication by -1 wrt the S-group structure on $\mathcal{E}$ with the chosen section as origin.

Definition 0.1. Let $\pi: X \to S$ be a smooth map everywhere of relative dimension 1, $\rho: T \to S$ a finite flat map and $\tau: T \to X$ an S-map. The *divisorial image* of τ is the subscheme defined by the invertible ideal which is the image under the norm map (i.e. the ideal generated by all norms of elements in the ideal) for the map $X \times_S T \to X$ induced by ρ of the ideal of the T-section of $X \times_S T$ induced by τ and π.

Recall that a finite flat map of degree 2 $\pi: T \to S$ is determined by a line bundle $\mathcal{L}$ on S and a map $\rho: \mathcal{L}^{\otimes -2} \to \mathcal{O}_S$; $\pi_*\mathcal{O}_T$ decomposes as $\mathcal{O}_S \bigoplus \mathcal{L}^{-1}$, where $\mathcal{L}^{-1}$ is the kernel of the trace map and the multiplication in $\pi_*\mathcal{O}_T$ determines and is determined by ρ. The cokernel of ρ is the structure sheaf of the branch locus R of the cover. If R is everywhere of codimension 1 then ρ gives also an isomorphism $\mathcal{L}^{\otimes 2} \cong \mathcal{O}_S(R)$ and conversely if R is any Cartier divisor on S giving a double cover with

R as its ramification locus is equivalent to giving a line bundle $\mathcal{L}$ and an isomorphism $\mathcal{L}^{\otimes 2} \cong \mathcal{O}_S(R)$. Furthermore, we have a global section s of $\mathcal{L}^{\otimes 2}$ given by the transpose of ρ and the double cover represents the problem of finding a section t of $\mathcal{L}$ such that $t^{\otimes 2} = s$. On the double cover itself there is therefore a canonical section of (the pullback of $\mathcal{L}$) which we will denote $t_{\mathcal{L}}$.

Lemma 0.2. *Suppose that U and V are two disjoint Cartier divisors, $\mathcal{L}$ and $\mathcal{M}$ line bundles on S and $\rho: \mathcal{L}^{\otimes 2} \to \mathcal{O}_S(U)$ and $\tau: \mathcal{M}^{\otimes 2} \to \mathcal{O}_S(V)$ isomorphisms. Let $\alpha: S_1 \to S$, $\beta: S_2 \to S$ and $\gamma: S_3 \to S$ be the double covers associated to ρ, τ and $\rho \otimes \tau$ respectively and $\delta: S_4 := S_1 \times_S S_2 \to S_3$ the map for which $\delta^* t_{\mathcal{L} \bigotimes \mathcal{M}} = t_{\mathcal{L}} \otimes t_{\mathcal{M}}$. Then δ is an étale double cover with associated line bundle equal to $\gamma^*\mathcal{L}(-U')$ (and $\gamma^*\mathcal{M}(-V')$) where U' is the intersection between the zero set of $t_{\mathcal{L} \bigotimes \mathcal{M}}$ and the inverse image of U. Furthermore, the étale cover $\beta^{-1}U \to U$ is isomorphic to the étale cover $\delta^{-1}U' \to U'$.*

Proof. The map δ induces a map $S_1 \times_S S_2 \to \gamma^*(\alpha)$ over S_3. Let us consider the situtation locally with U defined by $f = 0$ and V defined $g = 0$ so that the affine algebras of S_1, S_2, S_3 and S_4 are equal to $\mathcal{O}_S[s]/(s^2 - f)$, $\mathcal{O}_S[t]/(t^2 - g)$, $\mathcal{O}_S[y]/(y^2 - fg)$ and $\mathcal{O}_S[s,t]/(s^2 - f, t^2 - g)$ respectively and δ is given by $y \mapsto st$. Then the affine algebra of δ is given by $\mathcal{O}_{S_3}[s,t]/(s^2 - f, t^2 - g, y - st)$. If f is invertible then $\mathcal{O}_{S_4}$ is free as $\mathcal{O}_{S_3}$-module with a basis $\{1, s\}$ and $s^2 = f$ is invertible, if g is invertible then $\mathcal{O}_{S_4}$ is free as $\mathcal{O}_{S_3}$-module with a basis $\{1, t\}$ and $t^2 = g$ is invertible. This shows that δ is étale and it is clear that the annihilator of the cokernel of $\mathcal{O}_{S_3}[s]/(s^2 - f) \to \mathcal{O}_{S_4}$ is exactly the ideal defining U'. This last fact remains true globally and so if we write $\delta_*\mathcal{O}_{S_4} = \mathcal{O}_{S_3} \bigoplus \mathcal{N}^{-1}$ we see that the annihilator of the cokernel of the map $\gamma^*\mathcal{L}^{-1} \to \mathcal{N}^{-1}$ induced from $S_1 \times_S S_2 \to \gamma^*(\alpha)$ is the ideal defining U' and hence $\mathcal{N} \bigotimes \gamma^*\mathcal{L}^{-1} = \mathcal{O}_{S_3}(-U')$. Finally the last statement is clear since γ induces an isomorphism between U' and U and together with the projection $S_4 \to S_2$ it induces a map between étale covers $\delta^{-1}U' \to U'$ to $\beta^{-1}U \to U$. Q.E.D.

Definition 0.3. Let $\mathcal{A}$ be a principally polarised abelian S-scheme and α a section of order 2. On the scheme $2^{-1}\alpha := \{\beta \in \mathcal{A} : 2\beta = \alpha\}$ we have the equivalence relation $\beta_1 \sim \beta_2 \Longleftrightarrow e_2(\beta_1 - \beta_2, \alpha) = 1$, where $e_2(-,-)$ is the pairing on the kernel of multiplication by 2. A $\frac{1}{2}\alpha$-*structure* on $\mathcal{A}$ is the choice of one of the equivalence classes of this equivalence relation.

If $\mathcal{A}$ is a principally polarised abelian S-scheme, then a section α of order 2 of $\mathcal{A}$ gives rise to an étale cover of $\mathcal{A}$ in the following way. By the polarisation α corresponds to a line bundle $\mathcal{L}$ on $\mathcal{A}$ rigidified along the 0-section. As α is of order 2 so is the rigidified line bundle and hence there is a unique isomorphism $\mathcal{L}^{\otimes 2} \cong \mathcal{O}_S$ compatible with the rigidifications. This isomorphism is then used to construct a double cover (with affine algebra $\mathcal{O}_S \bigoplus \mathcal{L}$ and multiplication given by the isomorphism). An alternative way of constructing this cover is as follows. One considers the orthogonal complement G_α of α in ${}_2\mathcal{A}$ under the Weil pairing $e_2()$ induced by the polarisation. Multiplication by 2 on $\mathcal{A}$ factors as $\mathcal{A} \to \mathcal{A}/G_\alpha =: \mathcal{A}_\alpha \to \mathcal{A}$ for a unique map $A_\alpha \to \mathcal{A}$. This map is the required double cover. (This is seen for instance by considering the action G_α on the $\mathcal{O}_{\mathcal{A}}$-algebra $2_*\mathcal{O}_{\mathcal{A}}$ and dividing it into eigenspaces.) As the equivalence classes of $2^{-1}\alpha$ are exactly the orbits under G_α one gets the following lemma.

Lemma 0.4. *Using the notations of (0.3) and the preceding paragraph a $\frac{1}{2}\alpha$-structure is equivalent to choosing a section of the map $\mathcal{A}_\alpha \to \mathcal{A}$ over α.*

Proof. Q.E.D.

Lemma 0.5. *(No conditions on S.) Let $\tau: T \to S$ be a finite, flat S-scheme of degree 3 which is relatively Gorenstein (i.e. the dualising sheaf $\omega_{T/S}$ is a line bundle).*

i) There is up to unique isomorphism a unique S-embedding of T into a $\mathbf{P}^1$-bundle over S.

ii) If T has a section R then it and T itself give relative Cartier divisors in the $\mathbf{P}^1$-bundle $\mathbf{P}$ of i). Their difference is a 2-section of T which we will call the (schematic) complement of R in T. Its ideal is a line bundle $\mathcal{I}$ and there is a canonical choice of a line bundle $\mathcal{M}$ and an isomorphism $\mathcal{I} \cong \mathcal{M}^{\otimes 2}$.

Proof. Assume, to begin with, that T is embedded in a $\mathbf{P}^1$-bundle $\pi: \mathbf{P} \to S$. Then T is a relative Cartier divisor of degree 3 in $\mathbf{P}$ and so $\mathcal{L} := \omega_{\mathbf{P}/S}(T)$ is a line bundle which fiber by fiber is of degree 1 and so $\mathbf{P}$ can be identified with $\mathbf{P}(\mathcal{E})$, where $\mathcal{E} := \pi_*\mathcal{L}$. We have the adjunction exact sequence

$$0 \to \omega_{\mathbf{P}/S} \longrightarrow \mathcal{L} \longrightarrow \omega_T \to 0,$$

which gives the short exact sequence

$$0 \to \mathcal{E} \to \tau_*\omega_T \to R^1\pi_*\omega_{\mathbf{P}/S} \to 0.$$

The map $\tau_*\omega_T \to R^1\pi_*\omega_{\mathbf{P}/S}$ composed with the trace map $R^1\pi_*\omega_{\mathbf{P}/S} \to \mathcal{O}_S$ is the trace map for τ and so $\mathcal{E}$ is identified with the kernel of the trace map $\tau_*\omega_T \to \mathcal{O}_S$. The embedding of T into $\mathbf{P}$ is determined by a line bundle quotient of $\tau^*\mathcal{E}$; the correspondance is given by associating to an embedding the restriction of the map $\tau^*\tau_*\mathcal{O}_{\mathbf{P}(\mathcal{E})}(1) \to \mathcal{O}_{\mathbf{P}(\mathcal{E})}(1)$ to T. This shows that the line bundle quotient map is simply the composite of the pullback to τ of the inclusion $\mathcal{E} \to \tau_*\omega_T$ and the adjunction map $\tau^*\tau_*\omega_T \to \omega_T$. Hence $\mathbf{P}$ as well as the embedding of T into it is canonically determined by T itself (as an S-scheme). This proves the uniqueness part of i). To prove existence we need to show that, without assuming an embedding, the composite $\tau^*\mathcal{E} \to \tau^*\tau_*\omega_T \to \omega_T$ is surjective, where $\mathcal{E}$ is defined as the kernel of the trace map and then that the obtained S-map from T to $\mathbf{P}(\mathcal{E})$ is an embedding. This is something which can be checked geometric fiber by geometric fiber so we may assume that S is an algebraically closed field. As we have seen, what we want to show is true as soon as T embeds in any $\mathbf{P}^1$-bundle so what remains to be shown is that any T so embeds. If $\mathbf{k}$ is an algebraically closed field there are up to isomorphism exactly four 3-dimensional commutative $\mathbf{k}$-algebras; $\mathbf{k}^3$, $\mathbf{k}\bigoplus\mathbf{k}[x]/(x^2)$, $\mathbf{k}[x]/(x^3)$ and $\mathbf{k}[x,y]/(x,y)^2$. The last is not Gorenstein (having a 2-dimensional socle) and the others evidently embed into $\mathbf{P}^1$.

To prove ii) we note first that it follows from the construction that $\mathcal{O}_{\mathbf{P}}(T) = \omega_{\mathbf{P}/S}^{-1}(1) = \pi^*\det\mathcal{E}^{-1}(3)$. The next step is to compute $\mathcal{O}_{\mathbf{P}}(R)$. To do that we are required to compute the line bundle quotient of $\mathcal{E}$ which corresponds to R. This is the restriction to R of the map $\tau^*\mathcal{E} \to \tau^*\tau_*\omega_T \to \omega_T$ so that the searched for line bundle quotient is the restriction to R of ω_T. Now, by duality, ω_T can be identified with $\underline{Hom}_{\mathcal{O}_S}(\mathcal{O}_T,\mathcal{O}_S)$ whose restriction to R is $\underline{Hom}_{\mathcal{O}_S}(\mathcal{O}_R,\mathcal{O}_S) = \mathcal{O}_R$. Hence the restriction of ω_T to R is canonically trivial and $\mathcal{O}_{\mathbf{P}}(R) = \pi^*\det\mathcal{E}(1)$. Therefore if R' is the schematic complement of R in T then $\mathcal{O}_{\mathbf{P}}(R') = \mathcal{O}_{\mathbf{P}}(T-R) = \pi^*\det\mathcal{E}^{-2}(2) = (\pi^*\det\mathcal{E}^{-1}(1))^{\otimes 2}$. This proves ii). Q.E.D.

Remark. i) When T consists of 3 disjoint copies of S part i) simply says that up to projective equivalence there is a unique way of choosing 3 points on the projective line. Intuitively our result confirms that this remains valid even when the points may be infinitely close. In a case when all 3 points coincide up to first order this was proven by P. Deligne [De:I,5.2].

ii) As $\mathcal{O}_{\mathbf{P}}(T) = \pi^*\det\mathcal{E}^{-1}(3)$, T gives rise to (and is determined by) an everywhere non-zero section of $S^3\mathcal{E}\bigotimes\det\mathcal{E}^{-1}$. Conversely such a section gives a T. When 6 is invertible this is encompassed by a result of R. Miranda [Mi] who shows, more generally, that a section of $S^3\mathcal{E}\bigotimes\det\mathcal{E}^{-1}$ gives rise to a degree 3-cover, the zero locus of the section being the locus of "non-Gorenstein-ness" of the cover.

iii) Note that in general the choice of "square root" of $\mathcal{O}_{\mathbf{P}}(R')$ depends on R and not just on the embedding of R' in $\mathbf{P}$. For instance, when S is the spectrum of a field $\mathbf{k}$ and T consists of 3 distinct points then the choice of this square root amounts to a choice, up to the square of a scalar, of a quadratic form having $T \setminus R$ as its zero set. This form can then seen to be determined by the condition that its value on T (this value is well-defined up to a square) is 1 (again up to a square). In other words, the double cover associated to the square root is split over T.

In order to be able to compare levels properly we now want to formulate the well-known description of the kernel of multiplication by 2 on the Jacobian of a hyper-elliptic curve in terms of its Weierstrass in a way so that no choices choices are made. This leads to the following definition.

If P is a finite set of even cardinality we define an $\mathbf{F}_2$-vector space $A(P)$ associated to P as follows. We look at the set of subsets of P of even cardinality with the symmetric difference as group operation and then divide out by the subgroup consisting of the empty set and all of P. The group of permutations of P clearly acts as a group of group automorphisms $\mathcal{S}(P)$ of $A(P)$. Note also that the two representative subsets of an element of $A(P)$ are a set and its complement. Recall that a bilinear form on an abelian group is alternating if the scalar product of every element with itself is zero.

Remark. As we will see, if P is the set of Weierstrass points of a hyper-elliptic curve then $A(P)$ is indeed the kernel of multiplication by 2 on the Jacobian of the curve. The pairing to be defined will also be seen to be the Weil pairing.

Lemma 0.6. *The form* $\langle U, V\rangle_P := |U \bigcap V| \pmod 2$ *is the unique non zero alternating* $\mathbf{F}_2$*-valued form on* $A(P)$ *invariant under the action of* $\mathcal{S}(P)$.

Proof. One first verifies that the given form is well-defined and bilinear which is left to the reader. That it is alternating is clear and so it suffices to show that it is unique. As $A(P)$ is generated by 2-element subsets it is enough to show that it is unique on such. The only invariant under the action of $\mathcal{S}(P)$ on pairs of 2-element subsets is the cardinality of the intersection so there are only 3 cases. The case of cardinality of intersection equal to 2 is determined by the condition that the form be alternating. For a given 2-element subset we can write any 2-element subset as the sum of 2 other 2-element subsets meeting the given one in 1 point. Hence if the intersection is empty the scalar product is zero. In order for it to be non-trivial the scalar product of two 2-element subsets meeting in one point must be 1. Q.E.D.

Remark. For uniqueness we never used the fact that the form was alternating or even symmetric.

For each $s \in P$ we can define a quadratic form γ_s on $A(P)$ whose associated bilinear form is $\langle -, - \rangle_P$. Indeed, for an element α of $A(P)$ we let U be its unique representative which does not contain s. We then put $\gamma_s(\alpha) := |U|/2 \pmod 2$ and leave as an exercise to show that this is a quadratic form with $\langle -, - \rangle_P$ as its associated bilinear form. This gives a map from P into the set of quadratic forms on $A(P)$ whose associated bilinear form is $\langle -, - \rangle_P$ and it is easily verified that if $|P|$ is not divisible by 4 the this map is injective.

Remark. If P is the set of Weierstrass points of a hyperelliptic curve then γ_s is the quadratic form associated by Mumford (cf. [Mu:IIIa,Prop. 6.3 b]) to the theta characteristic $(g-1)s$, where g is the genus of the curve. Recall now that up to equivalence there are only 2 non-degenerate quadratic forms on an even-dimensional $\mathbf{Z}/2$-vector space (of dimension $2n$) and that the 2 possibilities are distinguished by the existence (the *split* case) or non-existence (the *non-split* case) of an isotropic subspace of dimension n. The split form can also be distinguished by there being $2^{2n-2} - 1 + (-1)^n 2^{n-1}$ isotropic elements in the vector space.

Lemma 0.7. *i) The quadratic form γ_s is split if $|P|/2$ is congruent to 1 or 2 modulo 4 and non-split if not.*

ii) If $|P| = 6$ then the association $s \mapsto \gamma_s$ gives a bijection between P and the set of non-split quadratic forms on $A(P)$ whose associated bilinear form is $\langle -, - \rangle_P$.

Proof. For i) one needs to compute the number of subsets of $P \setminus \{s\}$ of cardinality divisible by 4. Using the binomial theorem this is equal to

$$\frac{1}{4}\left((1+1)^{n-1} + (1-1)^{n-1} + (1+i)^{n-1} + (1-i)^{n-1}\right)$$

where $n := |P|$. Expanding this, using $(1+i)^2 = 2i$, gives the needed computation. As for ii), as the association is an injection it suffices to show that the set of such quadratic forms contains at most 6 elements. However the linear group of $A(P)$ acts transitively on all quadratic forms of given type and so the symplectic group of $\langle -,- \rangle_P$ acts transitively on quadratic forms of given type and associated bilinear form $\langle -,- \rangle_P$. However, $Sp(4, \mathbf{F}_2)$ has 720 elements and the stabiliser, in the symplectic group, of γ_s contains the symmetric group on $P \setminus \{s\}$ which contains 120 elements. Hence there are at most $720/120 = 6$ such forms. Q.E.D.

Remark. There are several ways to prove ii) (or equivalently that the map from the symmetric group on P to the symplectic group of $A(P)$ is surjective).

i) A transposition maps to a transvection and one gets all transvections this way. Transvections generate the symplectic group.

ii) Modeled on even theta characteristics one associates quadratic forms of Arf invariant 1 to 3-element subsets. By counting ($16 = \binom{6}{1} + \binom{6}{3}/2$) one sees that one gets all quadratic forms this way and hence all of Arf invariant -1 are of the form γ_s.

iii) The full monodromy action on the kernel of multiplication by 2 on the Jacobian of a curve is always the full symplectic group. In the case of genus 2 all curves are hyperelliptic so the monodromy group is the symmetric group.

As $A(P)$ is natural in P, for a finite étale cover of even degree $T \to S$ we can define another étale cover $A(T) \to S$ which is an étale group scheme with a canonical alternating form. If $T \to S$ is the branch locus of a double cover of a $\mathbf{P}^1$-bundle then we will want to identify $A(T) \to S$ with the kernel by multiplication by 2 on the Jacobian of the cover. This is straightforward if $T \to S$ is a split cover (and even if it has a section) but some care has to be taken in the general case.

Proposition 0.8. *Suppose $C \to S$ is a hyperelliptic curve and $R \to S$ its scheme of Weierstrass points. Then there is a canonical isomorphism of étale S-group schemes between $A(R)$ and ${}_2\underline{Pic}^0(C/S)$ taking the canonical alternating form to the Weil pairing.*

Proof. The line bundle of degree 2 giving the hyperelliptic linear system may not exist as such, it does, however, give a well-defined element ℓ in $\underline{Pic}(C/S)$. Let now $\mathcal{B}(R)$ be the étale cover of subsets of R of even cardinality. If M is a section of $\mathcal{B}(R)$ (over some S-scheme W) then we get an element $\mathcal{O}(M) - (\deg M/2)\ell$ of $\underline{Pic}^0(\mathcal{C}/S)(W)$. In this way we get a morphism of group schemes $\sigma: \mathcal{B}(R) \to {}_2Pic^0(\mathcal{C}/S)$. The line bundle associated to R is a multiple of ℓ in $\underline{Pic}(\mathcal{C}/S)$ and so σ factors to give a map $A(R) \to {}_2Pic^0(\mathcal{C}/S)$. To verify that it is an isomorphism it suffices to check on geometric fibers where it is well known. To see the correspondence on pairings one immediately reduces to the universal situation where S is the scheme of unordered $\deg R$-tuples of points on the projective line. As the space of ordered such tuples is irreducible, the monodromy action on the universal $\deg R$-divisor on the projective line over S is the full symmetric group and as the Weil pairing is non-trivial (non-degenerate even) we conclude by (0.6). Q.E.D.

1. An equivalence of stacks.

Definition 1.1. i) $\mathcal{F}$ is the stack for which the objects of $\mathcal{F}(S)$ consist of an elliptic curve $\mathcal{E}$ (with zero section T) over S together with a 2-section U of $\mathcal{P} := \mathcal{E}/\{\pm 1\}$ disjoint from the branch locus of $\mathcal{E} \to \mathcal{P}$. The morphisms are isomorphisms between curves preserving U.

ii) $\mathcal{M}_2^*$ is the stack for which the objects of $\mathcal{M}_2^*(S)$ consist of a smooth genus 2 curve $\mathcal{C}$ and a division of the branch locus of $\mathcal{C} \to \mathcal{C}/\iota =: \mathcal{P}$, where ι is the hyperelliptic involution, into three disjoint subschemes T, V, U which are multisections of degree 1, 3 and 2 respectively. The morphisms are isomorphisms between curves preserving the decompositions of the branch loci.

Remark. It is immediate from Artin's criterion [Ar:5.3] that $\mathcal{F}$ and $\mathcal{M}_2^*$ are algebraic stacks.

Theorem 1.2. *The two stacks $\mathcal{F}$ and $\mathcal{M}_2^*$ are equivalent. Under this equivalence the T's and U's correspond and V corresponds to the the complement of T in the Weierstrass points of $\mathcal{E}$.*

Proof. Using the discussion above on double covers we see that an object of $\mathcal{M}_2^*(S)$ is specified by a $\mathbf{P}^1$-bundle $\mathbf{P}$ over S, three disjoint multisections T, V and U of degree 1, 3 and 2 respectively, a line bundle $\mathcal{N}$

on $\mathbf{P}$ and an isomorphism $\mathcal{N}^{\otimes 2} \cong \mathcal{O}_{\mathbf{P}}(T+V+U)$. Using lemma 0.5 one sees that this is equivalent to giving a double étale cover $\tau: V \to S$ of S, a multisection U of $\mathbf{P}' \setminus T \bigcup V$, where $\mathbf{P}'$ is the $\mathbf{P}^1$-bundle associated to the disjoint union of V and a copy T of S, a line bundle $\mathcal{L}$ on S and an isomorphism $\mathcal{L}(3) \cong \mathcal{O}_{\mathbf{P}'}(V)$. On the other hand, an object of $\mathcal{F}(S)$ is specified by $\mathbf{P}^1$-bundle $\mathbf{P}$ over S, three disjoint multisections T, $V :=$ branch locus $\setminus T$ and U of degree 1, 3 and 2 respectively, a line bundle $\mathcal{N}$ on $\mathbf{P}$ and an isomorphism $\mathcal{N}^{\otimes 2} \cong \mathcal{O}_{\mathbf{P}}(T+V)$ which by a similar reasoning is equivalent to the same data. This correspondence clearly is natural for isomorphisms giving an equivalence of stacks. Q.E.D.

From the proof of the theorem it follows that if we have an object of $\mathcal{F}(S)$ (or equivalently of $\mathcal{M}_2^*(S)$) there is a canonical choice of square root of U on $\mathbf{P}$ (using the notations of the proof) and hence a canonical double cover of $\mathbf{P}$ ramified at U. Using the construction of section 0 and (0.2) we get a double étale cover $\mathcal{D} \to \mathcal{C}$. We will call $\mathcal{D}$ the *associated genus 3 curve* and the map $\mathcal{D} \to \mathcal{C}$ the *associated double cover*. By construction there are canonical sections of $\mathcal{C} \to \mathbf{P}$ over T, U and V and we will use the same letters to denote the images under these sections. Similarly for V and $\mathcal{E} \to \mathbf{P}$.

Proposition 1.3. *Let $(\mathcal{C}, T, U, V)$ be an element of $\mathcal{M}_2^*(S)$. The associated double cover is nontrivial and in particular the associated genus 3 curve is a curve of genus 3 (over S).*

Proof. To see this it suffices to check that it is non-trivial over one geometric point of S which is obvious. Q.E.D.

2. Comparing levels.

We will now see what happens with the equivalence of the previous section when we introduce various level structures on the two sides.

Definition 2.1. The superscript *sp* on $\mathcal{F}$ or $\mathcal{M}_2^*$ or any of the modifications to be defined presently means the stack obtained by adding the choice of an S-isomorphism between U and $S \times \mathbf{2}$ ($\mathbf{2} = \{\mathbf{0}, \mathbf{1}\}$). Similarly, the superscript *se* on $\mathcal{F}$ (and its modifications) will denote the choice of a section of $\mathcal{E} \to \mathbf{P}$ over U and on $\mathcal{M}_2^*$ the choice of a section of the associated double cover $\mathcal{D} \to \mathcal{C}$ over U.

i) $\mathcal{F}(n)$ is the stack, on schemes on which n is invertible, whose objects are those of $\mathcal{F}$ plus a level n-structure on the elliptic curve.

$\mathcal{F}_0(n)$ is the stack, on schemes on which n is invertible, whose objects are those of $\mathcal{F}$ plus a choice of a cyclic subgroup of order n in the elliptic curve.

$\mathcal{F}(2,4)$ is the stack whose objects are those of $\mathcal{F}$ plus a level 2-structure and a choice of one of the cyclic groups of order 4 of the elliptic curve containing the first of the elements of order 2.

ii) $\mathcal{M}_2^*(2)$ is the stack whose objects are those of $\mathcal{M}_2^*$ plus an identification of V with $S \times \mathbf{3}$.

$\mathcal{M}_2^*(0,4)$ is the stack whose objects are those of $\mathcal{M}_2^*$ plus the following data:

b) A section over the image R in $\mathcal{C}$ of $S \times \mathbf{0}$, of the double cover of $\mathcal{C}$ associated to the decomposition of the branch locus of $\mathcal{C} \to \mathbf{P}$ into T, $S \times \mathbf{0} \bigcup U$ and $S \times \{\mathbf{1},\mathbf{2}\}$ (which gives an element of $\mathcal{M}_2^*(S)$ different from the one associated to the decomposition T, V and U).

$\mathcal{M}_2^*(2,4)$ is the stack whose objects are those of $\mathcal{M}_2^*$ plus the following data:

a) An identification of V with $S \times \mathbf{3}$.
b) A section over the image R in $\mathcal{C}$ of $S \times \mathbf{0}$, of the double cover of $\mathcal{C}$ associated to the decomposition of the branch locus of $\mathcal{C} \to \mathbf{P}$ into T, $S \times \mathbf{0} \bigcup U$ and $S \times \{\mathbf{1},\mathbf{2}\}$ (which gives an element of $\mathcal{M}_2^*(S)$ different from the one associated to the decomposition T, V and U).

Remark. The stack $\mathcal{F}^{se,sp}$ is equivalent to an open substack of the fibre product of the moduli stack $\mathcal{X}$ classifying elliptic curves with a distinguished point with itself over the moduli stack $\mathcal{M}$ classifying elliptic curves. The equivalence is obtained by mapping $(\mathcal{E}, U, V, U \leftarrow S \times \mathbf{2}, U \to \mathcal{E})$ to $((\mathcal{E}, \mathrm{im}(S \times \mathbf{0} \to U \to \mathcal{E})), (\mathcal{E}, \mathrm{im}(S \times \mathbf{1} \to U \to \mathcal{E})))$.

Theorem 2.2. *The equivalence of Theorem 1.2 extends to an equivalence between*

$\mathcal{F}(2)$	and	$\mathcal{M}_2^*(2)$,
$\mathcal{F}_0(4)$	and	$\mathcal{M}_2^*(0,4)$,
$\mathcal{F}(2,4)$	and	$\mathcal{M}_2^*(2,4)$,
$\mathcal{F}^{se}$	and	$\mathcal{M}_2^{*se}$,
$\mathcal{F}^{sp}$	and	$\mathcal{M}_2^{*sp}$.

Proof. For the level 2-structure it is immediate by the construction of the equivalence. As for a cyclic subgroup of order 4 of $\mathcal{E}$, as every element of order 2 is orthogonal to itself under the Weil pairing it is clear that a subgroup of order 4 is equivalent to giving an element α of order 2 (the unique element of order 2 in the subgroup) and a $\frac{1}{2}\alpha$-structure. Giving α is the same thing as giving a section the scheme of elements of order 2 in $\mathcal{E}$ hence giving a section of V. By (0.4) a $\frac{1}{2}\alpha$-structure is equivalent to finding a section over α of the double cover associated to α. By (0.2) this covering $\mathcal{E}'$ is the fiber product, over $\mathbf{P}$, of the double cover $\mathbf{P}'$ ramified over the complement V' of α in V and the double cover ramified over the union of T and α (the first has a canonical meaning by (0.5) and the second as E is given and the first cover has canonical meaning). Equivalently, it is obtained as the double cover ramified over the inverse image of $T \bigcup \alpha$ in $\mathbf{P}'$. Hence choosing a section of $\mathcal{E}'$ over α is equivalent to choosing a section over α of the covering $\mathbf{P}' \to \mathbf{P}$. On the other hand, we get a decomposition of the branch locus of $\mathcal{C} \to \mathbf{P}$ as T, $S \times \mathbf{0} \bigcup U$ and $S \times \{1,2\}$ and by (0.2) the covering $\mathbf{P}' \to \mathbf{P}$ restricted to α is isomorphic to the covering $\mathcal{D} \to \mathcal{C}$ restricted to α. The case of a $(2,4)$-structure is the combination of the two previous cases, the case of an *sp*-structure is immediate, and the case of an *se*-structure follows again from (0.2). Q.E.D.

On the side of the genus 2 curve $\mathcal{C}$ it turns out that the different supplementary structures we have put on the curve may be interpreted in terms of more familiar structures on the Jacobian. Let us then first note that if $\mathcal{A} \to S$ is a principally polarised abelian scheme then as a defining line bundle is determined up to translation there is a well-defined $\mathcal{A}$-torsor $\mathcal{P}$ such that the polarisation gives a line bundle on $\mathcal{P}$. The theta divisor – the common support of the sections of this line bundle – is then a divisor in $\mathcal{P}$. Let us also say that a quadratic form on ${}_2\mathcal{A}$ is *compatible* if its associated bilinear form is the Weil pairing and its Arf invariant is that of (0.7) for $|P| = 2\dim\mathcal{A} + 2$. If the dimension of $\mathcal{A}$ is 2 and γ is a compatible form then for each non-isotropic (i.e. $\gamma(\alpha) = 1$) element α of ${}_2\mathcal{A}$ there is a unique (unordered) pair of isotropic elements $\{\alpha_1, \alpha_2\}$ such that $\alpha = \alpha_1 + \alpha_2$. We will call this pair the *isotropic pair associated to* α.

Lemma 2.3. If P is a set of 6 elements then for every choice of a compatible quadratic fom γ on $A(P)$ and every non-isotropic (i.e. $\gamma(\alpha) = 1$) element α of $A(P)$ there is a unique (unordered) pair of isotropic elements $\{\alpha_1, \alpha_2\}$ such that $\alpha = \alpha_1 + \alpha_2$.

By (0.7 ii) γ corresponds to an element s of P and from the definition it follows that an element of $A(P)$ considered as a 2-element subset of P is non-isotropic for γ iff it is disjoint from s. Hence what the lemma says is that any 2-element subset of P disjoint from s is the symmetric difference of a unique pair of 2-element subsets containing s which is obvious.Q.E.D.

Definition 2.4. Using the notations of (2.3) the pair $\{\alpha_1, \alpha_2\}$ will be called the *isotropic pair associated to* α.

Definition 2.5. With the notations of the preceding paragraph, a $(\gamma, \frac{1}{2}\alpha)$-structure is the choice of a section of $\mathcal{A}_\alpha \to \mathcal{A}$ over $\{\alpha_1, \alpha_2\}$.

Remark. The significance of a $\frac{1}{2}\alpha$-structure in this context is that if given it makes the problem of finding a section above α_1 equivalent to finding one above α_2; one simply requires that the sum in $\mathcal{A}_\alpha$ be equal to the given lifting of α.

We can now formulate the abelian surface analogs of the level structures we have defined for genus 2 curves.

Definition 2.6. $\mathcal{A}_2^*$ is the stack whose objects are principally polarised abelian surfaces $\mathcal{A}$ with smooth theta divisor and a choice of a section α of order 2 of $\mathcal{A}$ and a compatible quadratic form γ on ${}_2\mathcal{A}$ for which α is not isotropic.

The superscript *sp* on $\mathcal{A}_2^*$ or any of the modifications to be defined presently means the stack obtained by adding an ordering of the two isotropic elements whose sum is α. Similarly, the superscript *se* on $\mathcal{A}_2^*$ (and its modifications) will denote the choice of a $(\gamma, \frac{1}{2}\alpha)$-structure on $\mathcal{A}$.

$\mathcal{A}_2^*(2)$ is the stack whose objects are those of $\mathcal{A}_2^*$ plus three γ-isotropic elements $(\beta_1, \beta_2, \beta_3)$ all of which are orthogonal to α.

$\mathcal{A}_2^*(0,4)$ is the stack whose objects are those of $\mathcal{A}_2^*$ plus a choice of a section β of $\mathcal{A}$ different from but orthogonal, under the Weil pairing, to α and not isotropic for γ and a $(\gamma, \frac{1}{2}\beta)$-structure on $\mathcal{A}$.

$\mathcal{A}_2^*(2,4)$ is the stack whose objects are those of $\mathcal{A}_2^*$ plus the choices of both (2) and (0,4).

The following result then comes as no surprise.

Theorem 2.7. *There is an equivalence of stacks between* $\mathcal{M}_2^*$ *and* $\mathcal{A}_2^*$ *which extends to equivalences between*

$\mathcal{A}_2^*(2)$	and	$\mathcal{M}_2^*(2)$
$\mathcal{A}_2^*(0,4)$	and	$\mathcal{M}_2^*(0,4)$
$\mathcal{A}_2^*(2,4)$	and	$\mathcal{M}_2^*(2,4)$
$\mathcal{A}_2^{*se}$	and	$\mathcal{M}_2^{*se}$
$\mathcal{A}_2^{*sp}$	and	$\mathcal{M}_2^{*sp}$.

Proof. To begin with one passes back and forth between principally polarised abelian surfaces and curves of genus 2 using the Jacobian and the theta divisor respectively. From (0.8) it follows that 2-sections of the scheme of Weierstrass points R of the curve correspond to elements of order 2 of the abelian surface. The section T of R allows us to embed the curve in its Jacobian (and not just a torsor over it) so that T maps to the origin. Sections of R disjoint from T then map to elements of order 2 in the Jacobian which correspond to the 2-section which is the union of T and the section. On the other hand, the element α of order 2 corresponding to a 2-section of R disjoint from T equals the sum of the images α_1 and α_2 of the two sections of the 2-section (this sum is well-defined as it is independent of the order of the 2 sections). From the proof of (2.3) it follows that the pair $\{\alpha_1, \alpha_2\}$ is the isotropic pair associated to the isotropic element α. The equivalence between $\mathcal{M}_2^*$ and $\mathcal{A}_2^*$ follows because the choice of T and U corresponds, by (0.7 ii) and (0.8), to an element of order 2 of the Jacobian and a compatible quadratic form on the kernel of multiplication by 2.

To prove the (2)-equivalence one simply notices that sections of V map to elements of order 2, isotropic for γ and orthogonal to α and that they give all such elements (which is immediately seen from the $A(P)$-description).

Furthermore, from (0.2) it follows that the covering $\mathcal{D} \to \mathcal{C}$ is the pullback to $\mathcal{C}$ of the double cover of the Jacobian of $\mathcal{C}$ associated to the 2-section U and under the embedding. Hence a section of $\mathcal{D} \to \mathcal{C}$ over U corresponds precisely to a $(\gamma_T, \frac{1}{2}\alpha)$-structure. This gives the se part. The sp-part is obvious, and the (0,4)-part is similar to the se-part. Q.E.D.

We now summarise some of our results in the following corollary. Let us recall that a level n-structure on a principally polarised abelian variety A is a *symplectic* isomorphism between the kernel of n on A and $(\mathbf{Z}/n\mathbf{Z})^{2\dim A}$ (which has the standard symplectic structure).

Corollary 2.8. *i) Let $\mathcal{A}_2^\dagger$ be the algebraic stack whose objects are principally polarised abelian surfaces $\mathcal{A}$ together with; a quadratic form γ on ${}_2\mathcal{A}$ whose associated bilinear form is the Weil pairing and whose Arf invariant is -1, two sections α_1 and α_2 of ${}_2\mathcal{A}$ for which $\gamma(\alpha_1) = \gamma(\alpha_2) = 0$ and a $(\gamma, \frac{1}{2}\alpha)$-structure, where $\alpha := \alpha_1 + \alpha_2$, and whose morphisms are iso-*

morphisms preserving all structures. Let $\mathcal{M}$ be the algebraic stack whose objects are elliptic curves and whose morphisms are isomorphisms. Let $\mathcal{X}$ be the stack whose objects are elliptic curves with a section and whose morphisms are isomorphisms preserving the section and let $\mathcal{X} \to \mathcal{E}$ be the forgetful functor. Then $\mathcal{A}_2^{\dagger}$ is equivalent to the open substack $\mathcal{U}$ of $\mathcal{X} \times_{\mathcal{M}} \mathcal{X}$ obtained by removing $\{(x,y) : 2x = 0 \vee 2y = 0\} \bigcup \{(x,x)\} \bigcup \{(x,-x)\}$.

ii) Let $\mathcal{A}_2^{\sharp}$ be the algebraic stack whose objects are principally polarised abelian surfaces $\mathcal{A}$ together with:

a) A level 2-structure.

b) Sections over the first and second basis element of the double cover associated to their sum (which is an element of order 2).

c) Sections over the third and fourth basis element of the double cover associated to their sum.

Then $\mathcal{A}_2^{\sharp}$ is equivalent to the pullback over $\mathcal{M}(2,4) \to \mathcal{M}$ of $\mathcal{U}$, where $\mathcal{M}(2,4)$ is the algebraic stack whose objects are elliptic curves with a level 2-structure and a choice of one of the subgroups of order 4 containing the first element of order 2.

Proof. i) is simply a reformulation of the theorem using the remark after (2.1). As for ii) it will be proved as soon as we have identified $\mathcal{A}_2^{\sharp}$ with $\mathcal{A}_2^{(se,sp)}(2,4)$. To do this let us for a moment go back to $\mathcal{M}_2^{(se,sp)}(2,4)$. As part of the data we have a total ordering of the Weierstrass points of $\mathcal{C}$. Let us first show that such an ordering is equivalent to a level 2-structure. Indeed, one direction is clear and by (0.7 ii) one goes the other way by associating to a symplectic non-degenerate 4-dimensional space the set of compatible quadratic forms. Now the identification is obtained by interpreting the supplementary data on both sides. Q.E.D.

Remark. The topological fundamental group of the base extension of $\mathcal{A}_2^{\dagger}$ to $\mathbf{C}$, the complex numbers, is a subgroup of finite index of the mapping class group, in fact the inverse image of the appropriate congruence subgroup of $Sp(4,\mathbf{Z})$. As $\mathcal{X} \times_{\mathcal{M}} \mathcal{X}$ is normal and $\mathcal{A}_2^{\dagger}$ is open in it, this group maps surjectively onto the fundamental group of $\mathcal{X} \times_{\mathcal{M}} \mathcal{X}$ and, more precisely, of $\mathcal{X} \times_{\mathcal{M}} \mathcal{X} \setminus \{(x,y) : 2x = 0 \vee 2y = 0\}$ which is an extension of $SL_2(\mathbf{Z})$ by $\mathbf{Z}^2$ (resp. by a free group on 5 generators). In particular, as there are subgroups of finite index of $SL_2(\mathbf{Z})$ which are free of rank greater than 1 we see that there are subgroups of finite index of the mapping class group which map onto free groups of any finite rank and so onto free abelian groups of any finite rank. As all subgroups of $Sp_4(\mathbf{Z})$ of finite index have finite abelianisations we see that this map to $SL_2(\mathbf{Z})$ is very different from the natural map of the mapping class group onto $Sp_4(\mathbf{Z})$. In section 5 we will see a further vindication of this fact.

Lemma 2.9. *Let $\mathcal{M}_0(8)$ be the algebraic stack whose objects are elliptic curves with a chosen cyclic subgroup of order 8. Then $\mathcal{M}_0(8)$ is equivalent to $\mathcal{M}_0(2,4)$.*

Proof. Indeed, if (E, C_4, β) is an object of $\mathcal{M}_0(2,4)$ one associates to it $(E_\beta := E/\langle\beta\rangle, \ker : E_\beta \xrightarrow{{}^t\pi_\beta} E \to E/C_4)$, where ${}^t\pi_\beta$ is the transpose of the projection map $\pi_\beta \colon E \to E_\beta$. The inverse is given by associating to (E, C_8) the triple $(E/4C_8, C_8/4C_8, \beta)$, where β is the generator of ${}_2E/4C_8$. Q.E.D.

Combining we get

Proposition 2.10. *Let $\mathcal{X}_0'(8)$ be the algebraic stack of elliptic curves with a section, a cyclic subgroup C_8 of order 8 and a lifting of the section to the covering of the elliptic curve associated to $4C_8$. Then $\mathcal{A}_2^\natural$ is equivalent to an open substack of $\mathcal{X}_0'(8) \times_{\mathcal{M}_0(8)} \mathcal{X}_0'(8)$.*

Proof. Q.E.D.

3. Compactifications.

We will now extend the arguments presented to stable curves. It is easy to see that the picture must differ somewhat when the stable curve consists of 2 genus 1 curves. In fact when a smooth situation degenerates (in a 1-parameter family, say) to the union of 2 smooth genus 1 curves then 3 of the 6 Weierstrass points come together. The resulting 4 points (one of which is triple) have only 1 moduli whereas 2 smooth genus 1 curves have 2. Hence we can at most hope for a map from the space of stable genus 2 curves to the space of stable genus 1 curves with 2 distinguished points. This is in fact what we will obtain. To see this we will first need the following result on the bicanonical system of a stable curve.

Lemma 3.1. *Let C be a stable curve of genus 2 (over S). Then the (relative) bicanonical system gives a map of degree 2 onto a (fiber by fiber) reduced conic in a $\mathbf{P}^2$-bundle over S.*

Proof. As the bicanonical system commutes with base change [De-Mu:Thm. 1.2], this is something which can be checked fiber by fiber and then [Ca:Theorem A] shows that the bicanonical system is basepoint free. A simple case by case study then shows that the map is of degree 2 and maps onto a conic. Q.E.D.

This lemma already gives a clue to what is happening in the case mentioned above; the union of two elliptic curves is not the double cover of a projective line but rather of 2 projective lines meeting in a point and the triple point is blown up into 3 distinct points. Except for this things will work in a way close to the smooth situation. Our first vindication of this claim is the following definition-lemma.

Lemma-Definition 3.2. *Let $\pi: \mathcal{C} \to S$ be a family of stable genus 2 curves.*

i) $\mathbf{P}(\mathcal{C}) \to S$ is the conic fibration which is the image of $\mathcal{C}$ under the relative bicanonical map.

ii) If $\pi: \mathcal{C} \to S$ is a family of stable genus 2 curves then the disconnecting locus of $\mathbf{P}(\mathcal{C})$ is the closed subset of $\mathbf{P}(\mathcal{C})$ of points lying in the intersection of two components of any fiber of $\mathbf{P}(\mathcal{C}) \to S$. The disconnecting locus of $\mathcal{C}$ is the inverse image under the map $\mathcal{C} \to \mathbf{P}(\mathcal{C})$ of the disconnecting locus of $\mathbf{P}(\mathcal{C})$.

iii) The Weierstrass scheme of $\mathcal{C}$ is the the ramification locus of the map from the complement in $\mathcal{C}$ of its disconnecting locus to its image under the bicanonical map of $\mathcal{C}$. It is a 6-section of $\mathcal{C}$ as well as of $\mathbf{P}(\mathcal{C}) \to S$. When a fiber of π is reducible it meets each component three times.

Proof. Except for the last statement what needs to be shown is that the Weierstrass scheme is finite and flat of degree 6. It is flat and quasi-finite being the ramification locus of a double cover of a smooth S-scheme by a fiberwise reduced relatively Cohen-Macaulay scheme. We may therefore check that it is finite of degree 6 fiber by fiber. The only non-trivial case is when the stable curve canonical system of the curve has a base point (as in the other case the curve is a double cover of $\mathbf{P}^1$) and then it is clear as the bicanonical system reduces to the complete linear system given by twice the intersection point with the other fiber. This fact also gives the last statement. Q.E.D.

Definition 3.3. i) $\overline{\mathcal{F}}$ is the stack for which the objects of $\overline{\mathcal{F}}(S)$ consist of a stable genus 1 curve $\pi: \mathcal{E} \to S$ with a section T, not meeting the singular locus of π, together with a 2-section U of $\mathbf{P}(\pi_* \mathcal{O}_{\mathcal{E}}(2T)) \to S$, which is a $\mathbf{P}^1$-

bundle. The morphisms of $\overline{\mathcal{F}}(S)$ are the isomorphisms of curves preserving T and U.

ii) $\overline{\mathcal{M}_2^*}$ is the stack for which the objects of $\overline{\mathcal{F}}(S)$ consist of a stable genus 2 curve $\mathcal{C}$ and an expression of the Weierstrass scheme of $\mathcal{C}$, thought of as a Cartier divisor of $\mathcal{C}$, as the sum of three multisections T, V and U of degree 1, 3, and 2 respectively.

Remark. i) $\mathcal{F}$ and $\mathcal{M}_2^*$ corresponds to the cases when $\mathcal{E}$ resp. $\mathcal{C}$ are smooth and T and U resp. T, V and U are disjoint and étale.

ii) By the semi-stable reduction theorem $\overline{\mathcal{F}}$ and $\overline{\mathcal{M}_2^*}$ are proper algebraic stacks.

With the notations of i) we will use the V to denote the ramification locus of the map defined by the relative linear system of $2T$ minus T itself.

Theorem 3.4. *The map $\mathcal{M}_2^* \to \mathcal{F}$ of theorem 1.2 extends to a map $\overline{\mathcal{M}_2^*} \to \overline{\mathcal{F}}$.*

Proof. Ideally we would like to continue in the previous manner by associating to each element of $\overline{\mathcal{M}_2^*}(S)$ an element of $\overline{\mathcal{F}}(S)$ extending the earlier construction. For technical reasons we have not been able to do that. However, as $\overline{\mathcal{F}}$ and $\overline{\mathcal{M}_2^*}$ are algebraic stacks it is enough to give such an association when we restrict ourselves to versal (or in fact miniversal) families in $\overline{\mathcal{M}_2^*}$. Let therefore $\pi: \mathcal{C} \to S$ be a stable genus 2 curve which is versal at all its points. Our first step is to contract one of the $\mathbf{P}^1$'s of each reducible fiber of $\mathbf{P}(\mathcal{C}) \to S$ so as to obtain a $\mathbf{P}^1$-bundle. Note that any line bundle $\mathcal{L}$ on a conic bundle which has degree 1 on each fiber gives a contraction map onto a $\mathbf{P}^1$-bundle where the contracted components are those to which the restriction of $\mathcal{L}$ has degree 0. Furthermore, the $\mathbf{P}^1$-bundle and the contraction map only depend on the contracted component and not on $\mathcal{L}$ (as two line bundles contracting the same component in each fiber differ by a line bundle from the base). The rule for which component to contract will depend on how T, V and U meet the 2 components. In fact we want to contract the component not meeting T except when that component meets V 3 times. (The reason for the last exception is that we want the divisorial images of V and T under the contraction to be the Weierstrass scheme of a stable genus 1 fibration and hence the divisorial image of V may not have triple points.) In the first case we may contract using the relative linear system of $\mathcal{O}_{\mathbf{P}(\mathcal{C})}(2T)$ and in the second the relative linear system of $\omega_{\mathbf{P}(\mathcal{C})/S}(T)^{-1}$. Hence whenever only one of the two cases occur we can do the contraction. On the other hand, as the contraction – if it exists – is unique it is sufficient to show the existence locally. As the conditions distinguishing the two cases only involve intersection numbers, the locus of points of S over which the fibers of π are geometrically reducible can be divided up into two open and disjoint subsets such that on each such component one of the cases occur. Therefore the contraction is always possible locally and hence globally.

We therefore get a $\mathbf{P}^1$-bundle $\mathbf{P} \to S$ and an S-map $\rho: \mathbf{P}(\mathcal{C}) \to \mathbf{P}$. We let T, V and U denote the divisorial images of the subschemes T, V resp. U in $\mathbf{P}(\mathcal{C})$. It seems reasonable that contraction the component of $\mathcal{C}$ lying over the contracted component of $\mathbf{P}(\mathcal{C})$ should be a double cover of $\mathbf{P}$. Indeed, away from the image K of the contracted components this is clear. Note now that as π is versal K has codimension at least 2 at all its points. On the other hand this contraction is the normalisation over $\mathbf{P}$ of the double cover over the complement of K. Hence its affine algebra over $\mathbf{P}$ is a rank 2 reflexive sheaf. The trace map splits this sheaf as a sum of $\mathcal{O}$ and a rank 1 sheaf which is also reflexive. However, as $\mathcal{C} \to S$ is versal, S and hence $\mathbf{P}$ is regular and so a reflexive rank 1 sheaf is a line bundle and we have a double cover $\lambda: \mathcal{C}' \to \mathbf{P}$.

Lemma 3.5. *The ramification locus of λ equals the sum of T, V and U.*

Proof. Indeed, outside of K this is true by definition. However, K has codimension 2 and $\mathbf{P}$ is regular. Q.E.D.

We may now proceed exactly as we did before. The cover λ gives a line bundle $\mathcal{L}$ and an isomorphism $\mathcal{L}^{\otimes 2} \cong \mathcal{O}_{\mathbf{P}}(T+V+U)$ and by (0.5 ii) there is a canonical "square root" of $\mathcal{O}_{\mathbf{P}}(U)$ so by subtracting it off we get a line bundle $\mathcal{L}'$ and an isomorphism $\mathcal{L}'^{\otimes 2} \cong \mathcal{O}_{\mathbf{P}}(T+V)$. This gives a stable curve with a section T of the smooth locus except over the points where T and V meet. We will see that by a sequence of elementary transformations of $\mathbf{P}$ we may separate T and V keeping the data for a double cover. Let us first note that the schematic intersection of $T+V$ meets a fiber at most in double points. Indeed, as $\mathcal{C}$ is stable it is clear that this is true for the Weierstrass scheme in $\mathbf{P}(\mathcal{C})$. We have now chosen which component to contract so that we will not create a triple of $T+V$ during contraction (note that in general $T+U+V$ in $\mathbf{P}$ will, however, have triple points). This means that locally, in the étale topology, around the intersection of T and V there will be a branch V' of V meeting T and the rest of V will not meet T. Because of versality V and T meet transversally and their intersection maps isomorphically onto its image S' in S which is smooth. We can now blow up this intersection and then blow down the strict transform of the inverse image in $\mathbf{P}$ of S' to obtain the $\mathbf{P}^1$-bundle $\mathbf{P}'$. (Locally the picture is exactly that of an ordinary elementary transform with a 1-dimensional base crossed with a regular scheme.) Looking again at things locally we see that T and V' have been separated and neither of them meets the image of the strict transform of the inverse image in $\mathbf{P}$ of S'. Hence now S is disjoint from V and V itself meets fibers in at most double points. Furthermore, the isomorphism $\mathcal{L}'^{\otimes 2} \cong \mathcal{O}_{\mathbf{P}}(T+V)$ transforms into an isomorphism $\mathcal{L}'^{\otimes 2} \cong \mathcal{O}_{\mathbf{P}'}(T+V+2\mathbf{P}'_{S'})$ so that we may choose $\mathcal{L}'(-\mathbf{P}'_{S'})$ as our new square root. Q.E.D.

4. Levels and compactifications.

There seem to be some technical problems in defining our various level structures for the non-smooth curves (more precisely those which are the union of two genus 1 curves). We will therefore confine ourselves to the problem which interests us; namely of extending our results on principally polarised abelian surfaces with irreducible theta divisor to all principally polarised abelian surfaces. We will do this by bypassing the moduli stack of genus 2 curves and work directly with the abelian surfaces.

Definition 4.1. $\mathcal{A}_2$ is the stack whose objects are principally polarised abelian surfaces $\mathcal{A}$ a choice of a section α of order 2 of $\mathcal{A}$ and a compatible quadratic form γ on ${}_2\mathcal{A}$ for which α is not isotropic.

The superscript *sp* on $\mathcal{A}_2$ or any of the modifications to be defined presently means the stack obtained by adding an ordering of the two isotropic elements whose sum is α. Similarly, the superscript *se* on $\mathcal{A}_2$ (and its modifications) will denote the choice of a $(\gamma, \frac{1}{2}\alpha)$-structure on $\mathcal{A}$.

$\mathcal{A}_2(2)$ is the stack whose objects are those of $\mathcal{A}_2$ plus three γ-isotropic elements $(\beta_1, \beta_2, \beta_3)$ all of which are orthogonal to α.

$\mathcal{A}_2(0,4)$ is the stack whose objects are those of $\mathcal{A}_2$ plus a choice of a section β of $\mathcal{A}$ different from but orthogonal, under the Weil pairing, to α and not isotropic for γ and a $(\gamma, \frac{1}{2}\beta)$-structure on $\mathcal{A}$.

$\mathcal{A}_2(2,4)$ is the stack whose objects are those of $\mathcal{A}_2$ plus the choices of both (2) and (0, 4).

The following result, of course comes as no surprise.

Theorem 4.2. *The equivalences of (2.7) and (2.2) extend to a map from*

$$\mathcal{A}_2(2) \quad \text{to} \quad \overline{\mathcal{F}}(2),$$

$\mathcal{A}_{20}(4)$	*to*	$\overline{\mathcal{F}}(0,4)$,
$\mathcal{A}_2(2,4)$	*to*	$\overline{\mathcal{F}}(2,4)$,
$\mathcal{A}_2{}^{se}$	*to*	$\overline{\mathcal{F}}^{se}$,
$\mathcal{A}_2{}^{sp}$	*to*	$\overline{\mathcal{F}}^{sp}$.

Proof. Indeed, it is enough to notice that all the variants of $\overline{\mathcal{F}}$ map by a finite map to $\overline{\mathcal{F}}$ itself and that all the variants of $\mathcal{A}_2$ are normal (in fact smooth over $\mathbf{Z}$). Q.E.D.

Remark. i) What stops us from defining extensions of the level structures in the the case of stable genus 2 curves is that one needs to copy the previous arguments and in particular one needs the definition of the associated double cover and needs to express it as a fibered product. In the case of a versal deformation this causes no problem which would be enough to prove things about the extension but not to define it. In the general case one needs only to define the associated double cover away from the disconnecting locus (as T, U and V never meet it). We assume that this would not cause any essential problems but have not felt motivated to carry such an argument through.

ii) The image of the map from $\mathcal{A}_2$ to $\overline{\mathcal{F}}$ does not lie in $\mathcal{F}$. Indeed, following the constructions through one sees that the genus 1 curve associated to a principally polarised abelian surface A is smooth exactly when both points of U meet the same component of $\mathbf{P}(C)$, where C is the theta divisor of A. (This is true as it is equivalent to V on $\mathbf{P}$ having no double points.)

5. Genus 1 levels vs. genus 2 levels.

We will here take the opportunity to note that in general putting a level structure on an elliptic curve is a condition completely independent from a level structure on the associated genus 2 curve. To simplify, let us show this for an odd prime order level p for the elliptic curve and an odd prime order level q (not necessarily distinct from p) for the genus 2 curve. We will use the following lemma (which the first named author learned from J.-P. Serre).

Lemma 5.1. *Let G and H be groups. Then there is a bijection between the set of subgroups M of $G\times H$ which map surjectively by the two projections onto G and H and isomorphisms between quotient groups of G and H.*

Proof. Given a subgroup M of $G \times H$ with the desired properties consider $G' := M \bigcap G \times \{e\}$ and $H' := M \bigcap \{e\} \times H$. These subgroups are normal in G and H respectively. Indeed, for any $g \in G$ there is, by assumption, an $h \in H$ such that $(g, h) \in M$ and hence G' is stable under conjugation by g. By assumption the composite $M \hookrightarrow G \times H \to G$ is surjective and so G/G' is the quotient of M by $M \bigcap G' \times H = G' \times H'$. As the same thing is true of H/H' we get an isomorphism between G/G' and H/H'. On the other hand, starting with an isomorphism $\phi: \overline{G} \to \overline{H}$ between quotient groups $\overline{G}$ and $\overline{H}$ of G resp. H we associate to it the subgroup $\{(g, h) : \phi(\overline{g}) = \overline{h}\}$ of $G \times H$. It is clear that these two constructions are inverses of each other. Q.E.D.

We assume that we are placed over an algebraically closed field of characteristic different from 2, p and q. Let Γ be the algebraic fundamental group of $\mathcal{F} \cong \mathcal{M}_2^*$ with respect to a fixed base point. We have 2 surjective maps $\Gamma \to SL_2(\mathbf{Z}/p)$ and $\Gamma \to Sp_4(\mathbf{Z}/q)$ given by the action on the kernel of multiplication by p on the Jacobian of the base point as an elliptic curve resp. by multiplication by p on the Jacobian of the base point as a genus 2 curve. That the 2 level conditions are independent is equivalent to saying that the induced map $\Gamma \to SL_2(\mathbf{Z}/p) \times Sp_4(\mathbf{Z}/q)$ is surjective. As the projection on both factors is surjective the image Λ is,by the lemma, described by giving a common quotient group of the 2 factors. However, except for $SL_2(\mathbf{Z}/3)$ which is solvable, $SL_2(\mathbf{Z}/p)$ and $Sp_4(\mathbf{Z}/q)$ modulo their center are simple and $PSL_2(\mathbf{Z}/p)$ and $PSp_4(\mathbf{Z}/q)$ are non-isomorphic groups (their Sylow 2-groups are different). Hence the only possible common quotient is the trivial group and $\Gamma \to SL_2(\mathbf{Z}/p) \times Sp_4(\mathbf{Z}/q)$ is surjective.

In particular, the pullback of the equivalence $\mathcal{F} \cong \mathcal{M}_2^*$, considered as a correspondence, to the spaces with odd level structures added is irreducible and hence induces the zero map on "new" modular forms. This shows that the fact that our equivalence induces a non-trivial correspondence in the case mentioned in the introduction is indeed coincidental.

BIBLIOGRAPHY

[Ar]: M. Artin, *Versal Deformations and Algebraic Stacks*, Inventiones. math. **27** (1974), 165–189.

[Ca]: F. Catanese, *Pluricanonical Gorenstein curves*, in "Enumerative Geometry and Classical Algebraic Geometry. Progr. in Math., vol 42," Birkhäuser, Boston, 1982, pp. 51–95.

[De]: P. Deligne, *Equations différentielles à points singuliers réguliers*, SLN 163, Springer-Verlag.

[De-Mu]: P. Deligne, D. Mumford, *The irreducibility of the space of curves of given genus*, Publ IHES **36** (1969), 75–110.

[Ge-Ny]: B. v. Geemen, N. Nygaard, *L-Functions of some Siegel Modular 3-Folds, Preprint 546*, (preprint) Rijksuniversiteit Utrecht, 1988.

[Mi]: R. Miranda, *Triple covers in Algebraic Geometry*, Am. J. of Math. **107, No. 5** (1985), 1123–1158.

[Mu]: D. Mumford, "Tata lectures on theta II. Progr. in Math. vol. 43.," Birkhäuser, Boston, 1984.

Chern Functors

J. Franke*

This is the second of four papers in which we try to come to terms with Deligne's problem of constructing a functorial Riemann-Roch isomorphism for the determinant line bundle of the cohomology of a proper smooth morphism $p: X \to S$

$$\det \mathbb{R}p_*\mathcal{E} \to (I_{X/S}\mathfrak{ch}(\mathcal{E})\mathfrak{Td}(T_{X/S}))^{(1)}. \tag{1}$$

The first step in such a construction is to give life to the right hand side of (1). This was done by Deligne and Elkik ([D],[E]), who treated (1) as a global expression. It is our approach to give life to each ingredient of the right hand side of (1), i.e., we can not only integrate the Chern functors along the fibres, we can also say what the Chern functors themselves are. This allows us to approach (1) by copying Grothendieck's proof of Riemann-Roch via embeddings into projective spaces, as we shall see in a forthcoming paper.

As the first step in this program, Chow categories as target categories for the Chern functors have been introduced in [F1]. Here we study the Chern functors themselves. Because of difficulties with the intersection product for non-smooth schemes over Spec($\mathbb{Z}$), we introduce $\mathfrak{c}_k(\mathcal{E})$ not as a mere object of the Chow category $\mathfrak{CH}^k(X)$, but as a whole intersection product functor

$$\mathfrak{c}_k(\mathcal{E}) \cap -: \tilde{\mathfrak{CH}}^p(X) \to \tilde{\mathfrak{CH}}^{p+k}(X). \tag{2}$$

In the first five paragraphs of §1, we introduce $\mathfrak{c}_1(\mathcal{L}) \cap A$ for a line bundle $\mathcal{L}$, using a functorial version of the product

$$H^1(X, K_1) \otimes E_2^{p,q}(X) \to E_2^{p+1,q-1}(X),$$

where E_2 is the E_2-term of Quillen's spectral sequence. Starting from this point, in the remaining paragraphs of §1 we construct (2), copying Grothendieck's definition of the Chern classes. We also prove a Whitney isomorphism for the Chern functors. The second paragraph considers further properties of the Chern functors (like relation to the Gysin functor $\underline{f}^!$

*Karl-Weierstraß-Institut für Mathematik, DDR-1080 Berlin, Mohrenstraße 39. Current address: The Institute for Advanced Study, Princeton, NJ 08540,USA. Partially supported by NSF Grant DMS 8610730 constructed in [F1]), which are useful both for §3 and for the proof of functorial Riemann-Roch. In §3 we give an axiomatic characterisation of the Chern bundles by means of six natural isomorphisms (3.2.1.-4 and **AX 0**, **AX 1**) and four compatibilities **AX 2–5** (of which the last one, **AX 5**, is very likely to be redundant) between these six isomorphisms. This is similar to the axioms for IC_2 in [D]. Finally we compare our functor $\underline{p}_*(\mathfrak{c}_2(\mathcal{E}))$ with Deligne's functor IC_2 and indicate how a similar comparison can be carried out for Elkik's line bundles.

The first paragraph almost coincides with §6 of [F2] (save for the correction of some sign errors) and has been announced in [F3]. I owe thanks to A.A. Beilinson, Ju.I. Manin, and A.N. Parchin for a number of helpful discussions. This paper has been finished during the author's stay at the Max-Planck-Institute in Spring 1989, with some final technical corrections made at the Institute of Advanced Study. I want thank both institutes for their hospitality. Also, I am indebted to E. Gustafsson from the IAS and to M.-P. van der Hulst from the University of Amsterdam for their assistance in retyping the manuscript in TEX.

Notations We use all the notations of [F1] for the Chow categories $\mathfrak{C}\mathfrak{H}^k$ and $\tilde{\mathfrak{C}\mathfrak{H}}^k$, the functors $\underline{f}^*, \underline{f}^!, \underline{g}_*$, and $\mathfrak{sp}_\lambda$ between them, and for the E_2-term of Quillen's spectral sequence. In particular, $CH^k(X) = E_2^{k,-k}(X)$ and $G_k(X) = E_2^{k-1,-k}(X)$. The product in the higher algebraic K-theory is Waldhausen's (cf. [W]). As we did in [F1], we suppose schemes to be noetherian, separated, and universally catenary. The notion of a bifibred Picard category over a bicategory and related concepts have been defined in [F1]. Recall that for a category K and multiplicative systems S^*, S_* of arrows in K we denote by (K, S_*, S^*) the bicategory which has the same objects as K, elements of S_* as horizontal arrows, elements of S^* as vertical arrows, and Cartesian squares as bimorphisms. In our definition of families S_*, we will often use the abbreviation c.r.d. for "constant relative dimension". We also use NT to mark squares of a diagram which are commutative just because they consist of Natural Transformations.

For a global section a of the multiplicative group of the structure sheaf of a scheme X, we denote by $[a] \in K_1(X)$ the class defined by the one by one matrix a. The most important example for this paper is $[-1]$, defined by the invertible section -1 of the structure sheaf.

For the reader's convenience, I have been asked to summarise some of the results of [F1]. The Chow categories $\mathfrak{C}\mathfrak{H}^k$ and $\tilde{\mathfrak{C}\mathfrak{H}}^k$ are intended to be in a close relation to the set of algebraic cycles of codimension k, similar to the relation between line bundles and Weil divisors on a sufficiently regular scheme. The construction to such categories is based on Quillen's localisation spectral sequence (cf. [Q] for details)

$$E_1^{p,q} = K_{-p-q}(\mathfrak{M}^p/\mathfrak{M}^{p+1}) \Rightarrow K_{-p-q}(\mathfrak{M}(X)),$$

where $\mathfrak{M}(X)$ is the category of coherent sheaves on X and $\mathfrak{M}^p$ is its descending filtration by the subcategories of coherent sheaves supported in codimension $\geq p$. The construction of this spectral sequence is similar to the construction of the spectral sequence of a filtered complex, using Quillen's exact sequence for localisation modulo a Serre class instead of the long exact homology sequence.

The tail of the E_1-term is

$$\coprod_{x \in X_{p-2}} K_2(k(x)) \to \coprod_{x \in X_{p-1}} k(x)^* \to \coprod_{x \in X_p} \mathbb{Z}, \tag{3}$$

where X_k is the set of codimension k points and $k(x)$ is the residue field of x. The positions of these groups in the bigrading of the E_1-term are $(p-2, -p)$, $(p-1, -p)$, and $(p, -p)$. The right arrow maps a rational function on a subvariety of codimension $p-1$ to its divisor, and the left arrow maps an element of $K_2(k(x))$ for a codimension $p-2$ point to minus the collection of its tame symbols at the codimension $p-1$ points. It is clear that the cokernel of the right arrow is $CH^p(X)$, the group of cycles of codimension p modulo rational equivalence. Let $G_p(X) = E_2^{p-1,-p}(X)$ be the cohomology at the second of these three terms. Then it is clear that for a normal scheme X the group $G_1(X)$ coincides with the group of invertible global sections of the structure sheaf.

Since the category of line bundles and isomorphisms of line bundles can easily be identified with the category of $\mathcal{O}_X^*$-principal homogeneous sheaves on X_{zar}, one might hope that $\mathfrak{CH}^p$ is somehow related to a category of principal homogeneous sheaves for G_p. In fact, G_p is easily seen to be a presheaf on the Zariski topology. However, it is usually not a Zariski sheaf if $p > 1$. For instance, if X is a regular scheme satisfying Gersten's conjecture, then

$$G_p(X) \cong H^{p-1}(X_{zar}, \mathcal{K}_p)$$

where $\mathcal{K}_p$ is the Zariski sheaf associated to the Zariski presheaf K_p. It is clear that if $p > 1$ every cohomology class can be killed on a Zariski covering.

The solution to this problem is to use a coarser topology $X_{(p)}$ instead of the Zariski topology. If X is irreducible, an open subset for $X_{(p)}$ is a Zariski open subset U which is either empty or contains all points of codimension less than p. For reducible schemes X the definition may be found in [F1]. Then $\mathfrak{CH}^p(X)$ can be defined as the category of G_p-principal homogeneous sheaves on $X_{(p)}$. This is a Picard category with the well-known sum operation of principal homogeneous sheaves. The zero object is G_p itself. $\mathfrak{CH}^p(X)$ works fine for regular schemes satisfying Gersten's conjecture, but in some singular cases or in cases where Gersten's conjecture is not known it has some bad properties. The reason is that the set of isomorphism classes of objects of $\mathfrak{CH}^p(X)$ is not $CH^p(X)$, but only the set of those rational equivalence classes which are locally trivial with respect to the Zariski topology on X. To remove this obstacle, it is necessary to replace $X_{(p)}$ by a certain Grothendieck topology $\tilde{X}_{(p)}$. It has the same open subsets as $X_{(p)}$, but the condition for a collection U_i to be a covering of the $X_{(p)}$-open subset is relaxed to the assumption that U is the union of the U_i and a closed subset of codimension larger than p. Let $\tilde{\mathfrak{CH}}^p(X)$ be the category of G_p-principal homogeneous sheaves on $\tilde{X}_{(p)}$. This is also a Picard category, and $\mathfrak{CH}^p(X)$ is a full subcategory of $\tilde{\mathfrak{CH}}^p(X)$. The group of isomorphism classes of objects of $\tilde{\mathfrak{CH}}^p(X)$ is $CH^p(X)$. We also introduce the graded Picard categories

$$\begin{aligned}\tilde{\mathfrak{CH}}^{\cdot}(X) &= \coprod_{i=0}^{\infty} \tilde{\mathfrak{CH}}^i(X)\\ \mathfrak{CH}^{\cdot}(X) &= \coprod_{i=0}^{\infty} \mathfrak{CH}^i(X),\end{aligned}$$

where $\mathfrak{CH}^0(X) = \tilde{\mathfrak{CH}}^0(X)$ is defined to be the Picard category which has the free group generated by the codimension zero points of X as group of objects and in which all morphisms are identities.

If X is normal and if $\mathcal{L}$ is a line bundle on X, then the sheaf $\mathcal{L}^*$ of invertible sections of $\mathcal{L}$ is a principal homogeneous sheaf on $X_{zar} = X_{(1)}$ for $\mathcal{O}_X^* = G_1$. This identifies $\mathfrak{CH}^1(X)$ with the category of line bundles on X, and $\mathfrak{CH}^1(X) = \tilde{\mathfrak{CH}}^1(X)$ if and only if X is locally factorial. For the sake of simplicity, we will sometimes denote $\mathcal{L}^*$ by $\mathcal{L}$ if it is considered as an object of the Chow category.

It is possible to define the notion of a rational section of an object of the Chow category. For $A \in \tilde{\mathfrak{CH}}^p(X)$, let

$$A_r(X) = \bigcup_{U \text{ open and dense in } X_{(p)}} A(U).$$

This is a set on which $(G_p)_r(X)$, the group of rational sections of the zero object G_p, acts transitively. It is easy to see that $(G_p)_r(X)$ can be identified with the cokernel of the left arrow in (3). Since (3) is a complex, the right arrow induces a map

$$\mathfrak{c}: (G_p)_r(X) \to Z^p(X) = \coprod_{x \in X_p} \mathbb{Z}.$$

For general objects $A \in \tilde{\mathfrak{CH}}^p(X)$, there is a unique map

$$\mathfrak{c}: A_r(X) \to Z^p(X)$$

with the following two properties:

- We have $\mathfrak{c}(g+a) = \mathfrak{c}(g) + \mathfrak{c}(a)$ for $g \in (G_p)_r(X)$ and $a \in A_r(X)$.
- If $U \subseteq X_{(p)}$ is open, then $a \in A_r(X)$ comes from an element of $A(U)$ if and only if the restriction of $\mathfrak{c}(a)$ to U vanishes.

This notion of rational sections of objects of the Chow categories and the cycles defined by them is a generalisation of the notion of non-vanishing rational sections of line bundles on a normal locally factorial scheme and the Weil divisors defined by them.

The considerations in [F1] proceed by defining three elementary functors between the Chow categories. For a flat morphism $f: X \to Y$, we have a pull-back functor

$$\underline{f}^*: \tilde{\mathfrak{CH}}^k(Y) \to \tilde{\mathfrak{CH}}^k(X).$$

This is easily constructed as the composition of the sheaf-theoretic inverse image functor f^{-1} with pushout for the pull-back morphism $f^{-1}G_{p,Y} \to G_{p,X}$. The second elementary functor is

$$\underline{g}_*: \tilde{\mathfrak{CH}}^k(X) \to \tilde{\mathfrak{CH}}^{k-d}$$

for a proper map $g: X \to Y$ of constant relative dimension d. Its definition is not so short but still straightforward. This functor generalises the direct image of algebraic cycles by proper maps.

The third elementary functor is used for giving an axiomatic characterisation for the restriction of objects of the Chow categories to closed subsets defined locally by a regular sequence. It is based on the observation that if the closed subset is a divisor with trivial normal bundle, such restrictions can be defined quite easily (cf. [FU, Remark 2.3.]). Thus, let us assume that $D \subset X$ is of codimension one and suppose that λ is a section of $\mathcal{O}_X$ in some neighbourhood of D such that the closed subscheme defined by the vanishing of λ is D. The specialisation functor is a functor

$$\mathfrak{sp}_\lambda: \tilde{\mathfrak{CH}}^k(X - D) \to \tilde{\mathfrak{CH}}^k(D).$$

A similar homomorphism for the terms of the Quillen spectral sequence is defined by composing multiplication by $[\lambda]$ with the differential of the localisation sequence. Using this specialisation homomorphism, the definition of a specialisation functor is quite straightforward.

These three elementary functors satisfy an abundance of compatibilities. One of them is that we have canonical isomorphisms $\underline{f}^*\underline{g}^* \to \underline{(gf)}^*$ and

$\underline{f}_*\underline{g}_* \to \underline{(fg)}_*$ and that if $f : X \to Y$ is flat and if $g : Z \to Y$ is proper of constant relative dimension and if f' and g' denote the base-change of f by g and the base-change of g by f, then we have a canonical functor-isomorphism $\underline{f}^*\underline{g}_* \cong \underline{g'}_*\underline{f'}^*$. These three functor-isomorphism satisfy a number of conditions, which are reflected by the notion of a bifibred Picard category over a bicategory. There are also notions of a biadmissible functor between bifibred categories and of a biadmissible functor-isomorphism between biadmissible functors.

The only non-trivial result of [F1] is the construction of a Gysin functor

$$\underline{f}^!: \tilde{\mathfrak{CH}}^k(X) \to \tilde{\mathfrak{CH}}^k(Y)$$

for a local complete intersection $f: Y \to X$ which is a closed submorphism of a smooth morphism. It is no longer possible to define this functor directly. Instead, this functor is required to come with five natural transformations:

1. If $h : Z \to X$ is flat and if f' is the base-change of f by h and h' the base-change of h by g, we have an isomorphism $\underline{f'}^!\underline{h}^* \cong \underline{h'}^*\underline{f}^!$.
2. If $g : Z \to X$ is proper of constant relative dimension and if f' is the base-change of f by g and g' the base-change of g by f and if f' is again a local complete intersection of the same relative dimension as f, we have an isomorphism $\underline{f}^!\underline{g}_* \cong \underline{g'}_*\underline{f'}^!$.
3. If $D \subset X$ and λ are the same as above and if $D' \subset Y$ is the preimage of D and if λ' is the inverse image of λ and if D' is defined by the vanishing of λ', then we have an isomorphism $\underline{f}^!_D \mathfrak{sp}_\lambda \cong \mathfrak{sp}_{\lambda'}\underline{f}^!$.
4. If $h: Z \to Y$ is a morphism such that fh (and hence h itself too) is a local complete intersection which allows a closed embedding into a smooth morphism, then we have an isomorphism $\underline{h}^!\underline{f}^! \cong \underline{(fh)}^!$.
5. If $h: X \to Z$ is flat such that hf is also flat then we have an isomorphism $\underline{f}^!\underline{h}^* \cong \underline{(hf)}^*$.

These natural transformations have to satisfy a number of conditions. Besides requiring certain stuff to be biadmissible, they enforce the commutativity of five diagrams. In what follows, we just call them 'the conditions'. They just mean that all diagrams involving these transformations are commutative if they ought to be commutative.

Then the main theorem of [F1] consists of two parts:

- There exists a family of Gysin-functors $\underline{f}^!$ for each local complete intersection f which has a closed embedding into a smooth morphism, together with natural transformations 1.–5. satisfying the conditions.

- If $\underline{f}^?$ is another family of Gysin-functors together with natural transformations 1.–5. satisfying the conditions, then there exists a unique family of functor-isomorphisms $\underline{f}^! \cong \underline{f}^?$ which respect the natural transformations 1.–5.

For a precise formulation of this result, we refer to [F1,§4].

In the remaining section of [F1], we use this result to construct an intersection product functor, which is then applied to Bloch's generalisation of the autoduality of the Jacobian of a curve. However, these things are not needed in this paper.

In the following cases, the subcategory $\mathfrak{CH}^{\cdot}$ is preserved by the functors described above:

- If $f : X \to Y$ is flat (resp. lci), then $\underline{f}^*$ (resp. $\underline{f}^!$) maps $\mathfrak{CH}^p(Y)$ to $\mathfrak{CH}^p(X)$.
- If $g : X \to Y$ is a closed immersion of codimension d, then $\underline{g}_*$ maps $\mathfrak{CH}^p(X)$ to $\mathfrak{CH}^{p+d}(Y)$. However, if $p : X \to Y$ is the resolution of singularities for a variety which is smooth but not locally factorial, then it is easy to see that $\underline{p}_*$ does not map $\mathfrak{CH}^1(X)$ to $\mathfrak{CH}^1(Y)$. If X can be embedded into $g : \mathbb{P}^k_Y \to Y$, then $\underline{g}_*$ does not map $\mathfrak{CH}^{k+1}(\mathbb{P}^k_Y)$ to $\mathfrak{CH}^1(Y)$.
- If X, D, and λ are such that $\mathfrak{sp}_\lambda$ is defined, then the composition of $\mathfrak{sp}_\lambda$ and the restriction from X to $X - D$ preserves $\mathfrak{CH}^p$. However, this does not seem to be true for $\mathfrak{sp}_\lambda$ itself.
- For $k > 0$ and any vector bundle $\mathcal{E}$ on X, the functor $\mathfrak{c}_k(\mathcal{E})\cap$- (which is to be defined in this paper) maps $\tilde{\mathfrak{CH}}^p(X)$ to $\mathfrak{CH}^{p+d}(X)$.

1 Construction of the Chern functors

1.1 Some preparations For a topological space X, a sheaf F on X, and a covering $\mathcal{U} = \{U_i\}_{i\in I}$ of X, let

$$\check{C}^*(\mathcal{U}, F) = \coprod_{t=0} \prod_{i_0,\ldots,i_t\in I} F(U_{i_0} \cap \ldots \cap U_{i_t})$$

be the complex of Čech cochains. We denote by $\check{Z}^i(\mathcal{U}, D)$ and $\check{B}^i(\mathcal{U}, F)$ the groups of closed and exact cochains and by

$$(\mathrm{d}_0\, \boldsymbol{\alpha})_{i_0,\ldots,i_q} = \sum_{k=0}^{q} (-1)^k \alpha_{i_0,\ldots,\widehat{i_k},\ldots,i_q}$$

the Čech differential d_0. We will often denote q-cycles by bold and their evaluation on open subsets by usual letters.

For a complex of sheaves F^* with differential d_1 we put

$$\begin{aligned} \check{C}^i(\mathcal{U}, F^*) &= \bigoplus_{k+l=i} \check{C}^k(\mathcal{U}, F^l) \\ \mathrm{d} &= (-1)^l\, \mathrm{d}_0 + \mathrm{d}_1 \end{aligned}$$

and define $\check{B}^i$ and $\check{Z}^i$ by means of d.

Let $\mathcal{U}$ be a covering of X_{zar}. To an element $\boldsymbol{\alpha}$ of $\check{Z}^q(\mathcal{U}, E_1^{*,-q})$ we associate an object $\mathbb{O}(\boldsymbol{\alpha})$ of $\tilde{\mathfrak{Ch}}^q(X)$ as follows. For an open subset W in $X_{(q)}$, we denote by $\mathcal{U} \cap W$ the covering of W by the $U_i \cap W$ and define

$$\mathbb{O}(\boldsymbol{\alpha})(W) = \{\boldsymbol{x} \in \check{C}^{q-1}(\mathcal{U} \cap W, E_1^{*,-q}) \,|\, \mathrm{d}(\boldsymbol{x}) = -\boldsymbol{\alpha}\,|_W\}/\check{B}^{q-1}(\mathcal{U} \cap W, E_1^{*,-q}). \tag{1}$$

Since the sheaves $E_1^{p,q}$ are flabby (cf. the proof of [Q,5.8.]) , every $g \in G_p(W)$ defines a Čech cohomology class $\tilde{g} \in \check{H}^{q-1}(\mathcal{U} \cap W, E_1^{*,-q})$ and acts on the set (1) by the rule $x \to x + \tilde{g}$. It is easy to see that $\mathbb{O}(\boldsymbol{\alpha})$ is an object of $\tilde{\mathfrak{Ch}}^q(X)$.

Let $\boldsymbol{\alpha}, \boldsymbol{\alpha}' \in \check{Z}^q(\mathcal{U}, E_1^{*,-q})$ and $\boldsymbol{\gamma} \in (\check{C}^{q-1}/\check{B}^{q-1})(\mathcal{U}, E_1^{*,-q})$ such that $\mathrm{d}\,\boldsymbol{\gamma} = \boldsymbol{\alpha}' - \boldsymbol{\alpha}$. Then there is an isomorphism $\mathbb{O}(\boldsymbol{\gamma})\colon \mathbb{O}(\boldsymbol{\alpha}) \to \mathbb{O}(\boldsymbol{\alpha}')$ which sends $\boldsymbol{x}$ in (1) to $\boldsymbol{x} - \boldsymbol{\gamma}\,|_W$.

Let $\mathcal{V} = \{V_j\}_{j\in J}$ be a refinement of $\mathcal{U}$, and let $\Phi\colon J \to I$ be a function with $V_i \subseteq U_{\Phi(i)}$. It defines a homomorphism

$$\xi_\Phi\colon \check{C}^*(\mathcal{U}, E_1^{*,q}) \to \check{C}^*(\mathcal{V}, E_1^{*,q})$$

and a canonical isomorphism

$$\underline{\xi}_\Phi\colon \mathbb{O}(\boldsymbol{\alpha}) \to \mathbb{O}(\xi_\Phi(\boldsymbol{\alpha})) \tag{2}$$

by the rule $x \to \xi_\Phi(x)$ in (1). If $\mathcal{W}$ is a refinement of $\mathcal{V}$ indexed by K and $\Psi: K \to J$ an admissible function, then $\xi_{\Phi\Psi} = \xi_\Psi \xi_\Phi$ and $\underline{\xi}_{\Phi\Psi} = \underline{\xi}_\Psi \underline{\xi}_\Phi$.

Let E^* and F^* be complexes of sheaves on X, G a presheaf on X, and $\{.,.\}: E^* \otimes G \to F^*$ a homomorphism of complexes. If $x \in \check{C}^p(\mathcal{U}, E^*)$ and $y \in \check{C}^q(\mathcal{U}, G)$, we define $\{x, y\}$ by the usual formula

$$\{x,y\}_{i_1,\dots,i_r} = \{x_{i_0,\dots,i_{r-q}} \mid U_{i_0} \cap \dots \cap U_{i_r}, y_{i_{r-q},\dots,i_r} \mid U_{i_0} \cap \dots \cap U_{i_r}\}. \quad (3)$$

We have

$$d(\{x,y\}) = \{d(x), y\} + (-1)^p \{x, d(y)\}. \quad (4)$$

1.2 The functor $\mathfrak{c}_1(\mathcal{L}) \cap A$ Let $A \in \mathrm{Ob}(\tilde{\mathfrak{C}\mathfrak{H}}^k(X))$, $k \geq 0$, and $\mathcal{L}$ be a line bundle on X. We choose a covering $\mathcal{U}$ of X_{zar} and non-vanishing sections ℓ_i of $\mathcal{L}$ on U_i. Let $\varphi = (\varphi_{ij}) = (\ell_j/\ell_i) \in \mathcal{O}_X^*(U_i \cap U_j) \subseteq K_1(U_i \cap U_j)$ be the 1-cycle defined by the ℓ_i.

Let the product $\{.,.\}: E_1^{p,-q}(X) \otimes K_i(X) \to E_1^{p,-q-i}$ be defined by

$$\{(a_x)_{x \in X_p}, \varphi\} = (a_x \varphi)_{x \in X_p}.$$

Note that up to a sign the corresponding pairing of Čech cohomology groups coincides with the product described in [G,p.181]. Now (3) defines

$$\{.,.\}: \check{C}^p(\mathcal{U}, E_1^{*,-k}) \otimes \check{C}^q(\mathcal{U}, K_1) \to \check{C}^{p+q}(\mathcal{U}, E_1^{*,-k-1}).$$

Let $a \in A_r(X)$ be a rational section. We put

$$(\mathfrak{c}_1(\mathcal{L}) \cap A)_{\mathcal{U},\ell_i,a} = \mathfrak{O}((-1)^k \{\mathfrak{c}(a), \varphi\}) \subset \mathrm{Ob}(\tilde{\mathfrak{C}\mathfrak{H}}^{k+1}(X)) \quad (5)$$

where $\mathfrak{c}(a)$ is the cycle defined by a (cf. [F1,§3.3.]).

If a' is another rational section of A, then $a - a' \in E_1^{k-1,-k}(X)/\mathrm{im}(d_1)$ and $d_1(a - a') = \mathfrak{c}(a) - \mathfrak{c}(a')$. Since $d(\varphi) = 0$, we have by (4)

$$\mathfrak{O}((-1)^k \{a' - a, \varphi\}): (\mathfrak{c}_1(\mathcal{L}) \cap A)_{\mathcal{U},\ell_i,a} \to (\mathfrak{c}_1(\mathcal{L}) \cap A)_{\mathcal{U},\ell_i,a'}. \quad (6)$$

If $\tilde{\ell}_i$ are other trivialisations of $\mathcal{L}$ on U_i, we put $\tilde{\varphi}_{ij} = \tilde{\ell}_j \tilde{\ell}_i^{-1}$ and $\psi_i = \tilde{\ell}_i \ell_i^{-1}$. Then $d(\psi) = \tilde{\varphi}\varphi^{-1}$, hence

$$\mathfrak{O}(\{\mathfrak{c}(a), \psi\}): (\mathfrak{c}_1(\mathcal{L}) \cap A)_{\mathcal{U},\ell_i,a} \to (\mathfrak{c}_1(\mathcal{L}) \cap A)_{\mathcal{U},\tilde{\ell}_i,a}. \quad (7)$$

If $\mathcal{V}$ is a refinement of $\mathcal{U}$ and Φ as in (2), we have

$$\xi_\Phi: (\mathfrak{c}_1(\mathcal{L}) \cap A)_{\mathcal{U},\ell_i,a} \to (\mathfrak{c}_1(\mathcal{L}) \cap A)_{\mathcal{V},\ell_{\Psi(i)},a}. \quad (8)$$

Of course, there are several compatibilities which must be checked. For instance, if we replace a by a' and ℓ_i by $\tilde{\ell}_i$, we have

$$\begin{aligned}
&(-1)^k\{a'-a,\varphi\}+\{\mathfrak{c}(a'),\psi\} \\
&= (-1)^k\{a'-a,\tilde{\varphi}\}+\{\mathfrak{c}(a),\psi\}-(-1)^k\{a'-a,\mathrm{d}(\psi)\} \\
&\quad +\{\mathrm{d}(a'-a),\psi\} \\
&= (-1)^k\{a'-a,\tilde{\varphi}\}+\{\mathfrak{c}(a),\psi\}+\mathrm{d}(\{a'-a,\psi\})
\end{aligned}$$

hence

$$\begin{aligned}
&\mathbb{O}(\{\mathfrak{c}(a'),\psi\})\circ\mathbb{O}((-1)^k\{a'-a,\varphi\}) = \\
&= \mathbb{O}((-1)^k\{a'-a,\tilde{\varphi}\})\circ\mathbb{O}(\{\mathfrak{c}(a),\psi\}),
\end{aligned}$$

which proves the compatibility of (6) and (7). The two other cases (i.e., compatibility between (6) and (8) and between (7) and (8)) are verified in a similar manner. By means of (6), (7), and (8), the objects $(\mathfrak{c}_1(\mathcal{L})\cap A)_{\mathcal{U},\ell_i,a}$ can be glued to one object $\mathfrak{c}_1(\mathcal{L})\cap A$. It defines a biadditive functor $\mathfrak{Pic}(X)\times\tilde{\mathfrak{CH}}^k(X)\to\mathfrak{CH}^{k+1}(X)$, where $\mathfrak{Pic}(X)$ is the groupoid of line bundles on X. Biadditivity means that there are canonical isomorphisms

$$\begin{aligned}
\mathfrak{c}_1(\mathcal{L}\otimes\mathcal{M})\cap A &\to \mathfrak{c}_1(\mathcal{L})\cap A\oplus\mathfrak{c}_1(\mathcal{M})\cap A \\
\mathfrak{c}_1(\mathcal{L})\cap(A\oplus B) &\to \mathfrak{c}_1(\mathcal{L})\cap A\oplus\mathfrak{c}_1(\mathcal{L})\cap B
\end{aligned}$$

which satisfy the additivity conditions of [DM,§1.8] in each of the two variables and make the diagram

$$\begin{array}{ccc}
\mathfrak{c}_1(\mathcal{L}\otimes\mathcal{M})\cap(A\oplus B) & \longrightarrow & \begin{array}{c}\mathfrak{c}_1(\mathcal{L})\cap(A\oplus B)\oplus \\ \mathfrak{c}_1(\mathcal{M})\cap(A\oplus B)\end{array} \\
\downarrow & & \downarrow \\
\begin{array}{c}\mathfrak{c}_1(\mathcal{L}\otimes\mathcal{M})\cap A\oplus \\ \mathfrak{c}_1(\mathcal{L}\otimes\mathcal{M})\cap B\end{array} & \longrightarrow & \begin{array}{c}\mathfrak{c}_1(\mathcal{L})\cap A\oplus\mathfrak{c}_1(\mathcal{L})\cap B \\ \oplus\mathfrak{c}_1(\mathcal{M})\cap A\oplus\mathfrak{c}_1(\mathcal{M})\cap B\end{array}
\end{array} \tag{9}$$

commutative.

Let V and W be Zariski-open subsets of X, $a\in A(V)$, ℓ a non-vanishing section of $\mathcal{L}$ on V. We want to define

$$\ell\cap a\in(\mathfrak{c}_1(\mathcal{L})\cap A)(V\cup W). \tag{10}$$

Without loosing generality we may assume $X=V\cup W$. For a moment we also assume that V is open and dense in $X_{(k)}$, later we can get rid of this assumption. In the notations of (5),

$$(\{\mathfrak{c}(f),\ell/\ell_i\})_{i\in I}\ \mathrm{mod}\ (\check{B}^{k-1}(\mathcal{U},E_1^{*,-k})) \tag{11}$$

defines an element of $\mathfrak{O}((-1)^k\{\mathfrak{c}(a),\varphi_{ij}\})(X) = (\mathfrak{c}_1(\mathcal{L})\cap A)_{\mathcal{U},\ell_i,a}(X)$. The product in (11) is well-defined because the supports of $\mathfrak{c}(a)$ and $\operatorname{div}(\ell)$ are disjoint. It is easy to see that (11) is compatible with (7) and (8), hence it defines (10).

If g is a rational function on X which has no zeros or poles intersecting the support of $\mathfrak{c}(a)$, then there is a well-defined cap-product $\mathfrak{c}(a)\cap g \in G_{k+1}(X)$ which is defined as the equivalence class of $\left(n_{\mathfrak{c}(a)}(x) g \mid_{\operatorname{Spec}(x)}\right)_{x\in X_k}$ in $G_{k+1}(X)$, where $n_{\mathfrak{c}(a)}(x)$ is the multiplicity of x in $\mathfrak{c}(a)$, and we have

$$(g\ell)\cap a - \ell\cap a = \mathfrak{c}(a)\cap g. \tag{12}$$

Let $g \in G_k(U)$, where U is a Zariski-open subset of X containing $Z = \operatorname{supp}(\operatorname{div}(\ell))$. Since the sheaves $E_1^{p,q}$ are flabby (cf. the proof of [Q,5.8.]), g defines a hypercohomology class in $\check{H}^{k-1}(U, E_1^{*,-k})$ which we denote by the same letter g. The section ℓ defines a cohomology class

$$(\mathcal{L},\ell) \in H^1_Z(X,\mathcal{O}^*_X) \to H^1_Z(X,\mathcal{K}_1)$$

with support in Z. The product

$$\cap\colon H^p_Z(X,\mathcal{K}_1)\otimes \check{H}^q(U,E_1^{*,-k}) \to \check{H}^{p+q}(X,E_1^{*,-k-1}) \tag{13}$$

defines

$$(\mathcal{L},\ell)\cap g \in G_{k+1}(X). \tag{14}$$

If U is open and dense in $X_{(k)}$, then

$$\ell\cap(g+a) - \ell\cap a = (\mathcal{L},\ell)\cap g. \tag{15}$$

Since in the definition of (14) we do not assume that U is open and dense in $X_{(k)}$, formula (15) may be used to define $\ell\cap a$ for $a\in A(U)$ without the assumption that the Zariski-open subset U is also open in $X_{(k)}$.

If $A \in \operatorname{Ob}(\mathfrak{C}\mathfrak{H}^0(X))$, then we define $\mathfrak{c}_1(\mathcal{L})\cap A$ by formula (5) with $a = \mathfrak{B}$ (cf. [F1,3.5]). We also put $\mathfrak{c}_1(\mathcal{L}) = \mathfrak{c}_1(\mathcal{L})\cap \mathbf{1}$, where $\mathbf{1}$ is the object of $\mathfrak{C}\mathfrak{H}^0$ defined by the cycle which has multiplicity one on every irreducible component. Since there is only one rational section, we do not need (6). The transformations (7) and (8) are defined by the same formulas as above.

If $k < 0$, the $\mathfrak{c}_1(\mathcal{L})\cap\cdot\colon \tilde{\mathfrak{C}\mathfrak{H}}^k(X) \to \tilde{\mathfrak{C}\mathfrak{H}}^{k+1}(X)$ is defined to be the only additive functor between these categories (the only object of the source category is mapped to zero of the target category, and there are no non-trivial automorphisms in both categories).

1.3 Example Let X be a smooth curve over a normal base scheme S, and let $\mathcal{L}$ and $\mathcal{M}$ be line bundles on X. We assume that ℓ and m are rational sections of $\mathcal{L}$ and $\mathcal{M}$ on X whose divisors do not intersect. We put

$$<\ell,m> = p_*(\ell\cap m) \in \underline{p}_*(\mathfrak{c}_1(\mathcal{L})\cap\mathfrak{c}_1(\mathcal{M}))(S),$$

where $p: X \to S$ is the projection. In this case, (12) and (15) imply that $< \ell, m >$ satisfies the transformation rules [D,(6.1.2)], and consequently $\underline{p}_*(\mathfrak{c}_1(\mathcal{L}) \cap \mathfrak{c}_1(\mathcal{M}))$ can be identified with the line bundle $<\mathcal{L}, \mathcal{M}>$ defined in [D,§6].

1.4 Compatibility with direct and inverse images Let $p: Y \to X$ be a flat morphism, $\mathcal{U}$ a covering of X_{zar}, and $\alpha \in \check{Z}^k(\mathcal{U}, E_1^{*,-k})$. Let $p^{-1}(\mathcal{U})$ be the covering of Y by the sets $p^{-1}(U_i)$. There is a natural morphism $p^*: \check{C}^*(\mathcal{U}, E_{1,X}^{*,-k}) \to \check{C}^*(p^{-1}(\mathcal{U}), E_{1,Y}^{*,-k})$ which on the cohomology groups defines the homomorphism p^* of [F1,§1]. There is an isomorphism in $\tilde{\mathfrak{CH}}^k(Y)$

$$\underline{p}^*(\mathbb{O}(\alpha)) \to \mathbb{O}(p^*(\alpha)) \tag{16}$$

sending $\boldsymbol{x}$ in (1) to $p^*(x)$.

Let $q: Z \to X$ be proper of relative dimension d. Formula [F1,1.(7)] defines a homomorphism of complexes

$$q_*: \check{C}^*(q^{-1}(\mathcal{U}), E_{1,Z}^{*,-k}) \to \check{C}^{*-d}(\mathcal{U}, E_{1,X}^{*,d-k})$$

of degree $-d$. If $\beta \in \check{Z}^k(p^{-1}(\mathcal{U}), E_{1,Z}^{*,-k})$, then there is an isomorphism

$$\underline{q}_*(\mathbb{O}(\beta)) \to \mathbb{O}(q_*(\beta)) \tag{17}$$

sending $\boldsymbol{x}$ in (1) to $q_*(\boldsymbol{x})$.

Let $\mathcal{L}$ be a line bundle on X, $\mathcal{U}$ a covering of X_{zar} on which $\mathcal{L}$ is trivialised by sections ℓ_i, $\varphi_{ij} = \ell_j/\ell_i$, $p: Y \to X$ a flat morphism, $q: Z \to X$ a proper morphism of relative dimension d, $A \in \text{Ob}(\tilde{\mathfrak{CH}}^k(X))$, $B \in \text{Ob}(\tilde{\mathfrak{CH}}^k(Z))$, a and b rational sections of A and B.

Then $p^*(\{\mathfrak{c}(a), \varphi\}) = \{\mathfrak{c}(a), p^*\varphi\}$, hence (16) defines

$$\underline{p}^*((\mathfrak{c}_1(\mathcal{L}) \cap A)_{\mathcal{U},\ell_i,a}) \to (\mathfrak{c}_1(p^*\mathcal{L}) \cap \underline{p}^* A)_{p^{-1}(\mathcal{U}),p^*(\ell_i),p^*(a)} \,. \tag{18}$$

It is easy to see that (18) is compatible with (6), (7), and (8), hence it defines

$$\underline{p}^*(\mathfrak{c}_1(\mathcal{L}) \cap A) \to \mathfrak{c}_1(p^*\mathcal{L}) \cap \underline{p}^* A. \tag{19}$$

In a similar manner one constructs

$$\underline{q}_*((\mathfrak{c}_1(q^*\mathcal{L}) \cap A)_{q^{-1}(\mathcal{U}),q^*(\ell_i),a}) \to (\mathfrak{c}_1(\mathcal{L}) \cap \underline{q}_* A)_{\mathcal{U},\ell_i,q_*(a)} \tag{20}$$

using (17) and the adjunction formula. We get

$$\underline{q}_*(\mathfrak{c}_1(q^*\mathcal{L}) \cap A) \to \mathfrak{c}_1(\mathcal{L}) \cap \underline{q}_* A. \tag{21}$$

The isomorphisms (19) and (21) are compatible with composition of flat and proper morphisms and with the base change isomorphism of [F1,§3.12].

More precisely, this means the following. If X-schemes are denoted $p: Y \to X$, then $\tilde{\mathfrak{CH}}^{\cdot}(Y)$ is a bifibred Picard category over (X-schemes, proper morphisms of constant relative dimension, flat morphisms). Then it is easy to see that $\mathfrak{c}_1(p^*\mathcal{L}) \cap - : \tilde{\mathfrak{CH}}^{\cdot}(Y) \to \tilde{\mathfrak{CH}}^{\cdot}(Y)$, equipped with the transformations (19) and (21), is a biadmissible functor (in the sense of [F1,3.11.]) between bifibred Picard categories.

Let $p: Y \to X$ be flat, $\mathcal{L}$ a line bundle on X, $A \in \mathrm{Ob}(\tilde{\mathfrak{CH}}^k(X))$, ℓ and a rational sections of $\mathcal{L}$ and A. It is easy to see that the image of $p^*(\ell \cap a)$ by (19) is $p^*(\ell) \cap p^*(a)$. If p is proper, $B \in \mathrm{Ob}(\tilde{\mathfrak{CH}}^k(Y))$, $b \in B_r(Y)$, then $p_*(p^*(\ell) \cap b)$ is mapped to $\ell \cap p_*(b)$ by (21).

1.5 Commutativity We want to define an isomorphism

$$\sigma_{\mathcal{L},\mathcal{M}}: \mathfrak{c}_1(\mathcal{L}) \cap \mathfrak{c}_1(\mathcal{M}) \cap A \to \mathfrak{c}_1(\mathcal{M}) \cap \mathfrak{c}_1(\mathcal{L}) \cap A. \tag{22}$$

Let $\mathcal{U}$ be a covering of X_{zar} on which $\mathcal{L}$ is trivialised by ℓ_i. Our first step is to define an isomorphism

$$\mathfrak{c}_1(\mathcal{L}) \cap \mathfrak{O}(\alpha) \to \mathfrak{O}((-1)^k\{\alpha, \varphi\}) \tag{23}$$

for $\alpha \in \check{Z}^k(\mathcal{U}, E_1^{*,-k})$, where $\varphi_{ij} = \ell_j/\ell_i$. It will identify $\mathfrak{c}_1(\mathcal{L}) \cap \mathfrak{O}(\gamma)$ with $\mathfrak{O}((-1)^k\{\gamma, \varphi\})$ if $\gamma \in \check{C}^{k-1}(\mathcal{U}, E_1^{*,-k})$ and $\mathrm{d}(\gamma) = \alpha' - \alpha$.

Let $g \in \mathfrak{O}(\alpha)_r(X)$ be a rational section. By definition (1), one checks easily that g has a representative $\mathfrak{g} \in \check{C}^{k-1}(\mathcal{U}, E_1^{*,-k})$ with the property $\mathrm{d}(\mathfrak{g}) + \alpha = \mathfrak{c}(g) \in E_1^{k,-k}(X) \subseteq \check{C}^{k-1}(\mathcal{U}, E_1^{*,-k})$. If x is a section on $W \subseteq X$ of $(\mathfrak{c}_1(\mathcal{L}) \cap \mathfrak{O}(\alpha))_{\mathcal{U},\ell_i,g}$, i.e., $x \in \check{C}^k(\mathcal{U}, E_1^{*,-k-1})$ and $\mathrm{d}(x) = -(-1)^k\{\mathfrak{c}(g), \varphi\}$ on W, then $y = x + (-1)^k\{\mathfrak{g}, \varphi\}$ satisfies

$$\begin{aligned} \mathrm{d}(y) &= \mathrm{d}(x) + (-1)^k\{\mathrm{d}(\mathfrak{g}), \varphi\} = \\ &= -(-1)^k\{\mathfrak{c}(g), \varphi\} + (-1)^k\{\mathfrak{c}(g), \varphi\} - (-1)^k\{\alpha, \varphi\}, \end{aligned}$$

hence $y \in \mathfrak{O}((-1)^k\{\alpha, \varphi\})(W)$. It is easy to see that the function $x \to y$ commutes with (6), hence it defines the isomorphism (23). The transformation (23) is compatible with the isomorphisms (7), (8), (19), and (21).

Now we assume that $\mathcal{M}$ too is trivialised on $\mathcal{U}$, by sections m_i, with transition functions $\psi_{ij} = m_j/m_i$. Our next step is to construct an isomorphism

$$\chi_{\mathcal{U},\mathcal{L},\ell_i,\mathcal{M},m_i,\alpha}: \mathfrak{O}(-\{\alpha, \psi, \varphi\}) \to \mathfrak{O}(-\{\alpha, \varphi, \psi\}) \tag{24}$$

for $\alpha \in \check{Z}^k(\mathcal{U}, E_1^{*,-k})$. In (24), $\{.,.,.\}$ denotes the iteration of $\{.,.\}$. An easy calculation, using repeatedly the fact that $\mathrm{d}(\varphi) = \mathrm{d}(\psi) = 0$, shows

$$\begin{aligned} (\varphi\psi - \psi\varphi)_{ijk} &= \varphi_{ij}\psi_{jk} - \psi_{ij}\varphi_{jk} = \varphi_{ij}\psi_{jk} + \varphi_{jk}\psi_{ij} \\ &= \varphi_{ik}\psi_{jk} - \varphi_{jk}\psi_{jk} + \varphi_{jk}\psi_{ij} \\ &= \varphi_{ik}\psi_{ik} - \varphi_{jk}\psi_{jk} - \varphi_{ij}\psi_{ij} \\ &= -d(\gamma)_{ijk}, \end{aligned} \tag{25}$$

where $\gamma_{\alpha,\beta} = \psi_{\alpha,\beta}\psi_{\alpha,\beta}$. By (25), $\chi_{\mathcal{U},\mathcal{L},\ell_i,\mathcal{M},m_i,\boldsymbol{\alpha}}$ may be defined by $\mathfrak{O}((-1)^k\{\boldsymbol{\alpha},\gamma\})$. In the special case $\mathcal{L} = \mathcal{M}$ and $\ell_i = m_i$, the well-known identity between Steinberg symbols

$$\varphi_{\alpha,\beta}\varphi_{\alpha,\beta} = \varphi_{\alpha,\beta}[-1] \quad \text{in} \quad K_2(k(x))$$

(cf. [M,Lemma 9.8]) can be used to compute $\chi_{\mathcal{U},\mathcal{L},\ell_i,\mathcal{L},\ell_i,\boldsymbol{\alpha}}$ on $\mathfrak{O}(\{\boldsymbol{\alpha},\varphi,\varphi\})$:

$$\chi_{\mathcal{U},\mathcal{L},\ell_i,\mathcal{L},\ell_i,\boldsymbol{\alpha}} = \boldsymbol{\alpha} \cap \varphi \cap [-1], \tag{26}$$

where the first product is (13) with $X = S = U$ and the second product is

$$\check{H}^1(X,\mathcal{K}_1) \otimes K_1(X) \to \check{H}^1(X,\mathcal{K}_2).$$

Now we are ready to define (22). For a rational section $a \in A_r(X)$, consider the isomorphism

$$\begin{aligned} \mathfrak{c}_1(\mathcal{L}) \cap \mathfrak{O}((-1)^k\{\mathfrak{c}(a),\psi\}) &\to \mathfrak{O}(-\{\mathfrak{c}(a),\psi,\varphi\}) \to \\ \mathfrak{O}(-\{\mathfrak{c}(a),\varphi,\psi\}) &\to \quad \mathfrak{c}_1(\mathcal{M}) \cap \mathfrak{O}((-1)^k\{\mathfrak{c}(a),\varphi\}), \end{aligned} \tag{27}$$

where the first and the third arrows are of type (23) and the middle one is (24). We want to check that it commutes with the isomorphisms (6), (7), and (8). For (6), we have to prove the commutativity of

$$\begin{array}{ccc} \mathfrak{O}(-\{\mathfrak{c}(a),\psi,\varphi\}) & \xrightarrow{\mathfrak{O}((-1)^k\{\mathfrak{c}(a),\gamma\})} & \mathfrak{O}(-\{\mathfrak{c}(a),\varphi,\psi\}) \\ \Big\downarrow{\scriptstyle \mathfrak{O}(-\{b-a,\psi,\varphi\})} & & \Big\downarrow{\scriptstyle \mathfrak{O}(-\{b-a,\varphi,\psi\})} \\ \mathfrak{O}(-\{\mathfrak{c}(b),\psi,\varphi\}) & \xrightarrow[\mathfrak{O}((-1)^k\{\mathfrak{c}(b),\gamma\})]{} & \mathfrak{O}(-\{\mathfrak{c}(b),\varphi,\psi\}). \end{array}$$

But

$$\begin{aligned} &(-1)^k\{\mathfrak{c}(a),\gamma\} - \{b-a,\varphi,\psi\} - (-1)^k\{\mathfrak{c}(b),\gamma\} + \{b-a,\psi,\varphi\} \\ &= \quad \{b-a,\mathrm{d}(\gamma)\} - (-1)^k\{\,\mathrm{d}(b-a),\gamma\} = -(-1)^k\,\mathrm{d}(\{b-a,\gamma\}), \end{aligned}$$

and the diagram commutes. For (7) the compatibility is verified in a similar manner, and for (8) it is trivial.

We have seen that the isomorphisms (27) fit together, defining (23). Since $\varphi\psi = -\psi\varphi$, we have $\sigma_{\mathcal{L},\mathcal{M}}\sigma_{\mathcal{M},\mathcal{L}} = \mathrm{Id}$. By (26), the action of $\sigma_{\mathcal{L},\mathcal{L}}$ on $\mathfrak{c}_1(\mathcal{L}) \cap \mathfrak{c}_1(\mathcal{L}) \cap A$ is given by

$$\sigma_{\mathcal{L},\mathcal{L}} = [\mathcal{L}] \cap [-1] \cap [A], \tag{28}$$

where the products are the same as in (26). It is easy to see that $\sigma_{\mathcal{L},\mathcal{M}}$ is compatible with (19) and (21).

Let ℓ, m, and a rational sections of $\mathcal{L}, \mathcal{M}, A$ on X whose divisors and cycles meet properly, i.e., such that $\ell \cap a$, $m \cap a$, $\ell \cap m \cap a$, $m \cap \ell \cap a$ are rational sections of $\mathfrak{c}_1(\mathcal{L}) \cap A, \ldots, \mathfrak{c}_1(\mathcal{M}) \cap \mathfrak{c}_1(\mathcal{L}) \cap A$. We want to prove

$$\sigma_{\mathcal{L},\mathcal{M}}(\ell \cap m \cap a) = m \cap \ell \cap a. \tag{29}$$

The first thing we have to do is to compute the image $\ell \cap m \cap a$ under the isomorphism (23):

$$\mathfrak{c}_1(\mathcal{L}) \cap (\mathfrak{c}_1(\mathcal{M}) \cap A)_{\mathcal{U},m_i,a} \to \mathcal{O}(-\{\mathfrak{c}(a), \psi, \varphi\}).$$

$\mathcal{O}((-1)^k\{\mathfrak{c}(a), \psi\})$ has a rational section $\mathfrak{g} = m \cap a$ given by $\mathfrak{g} = \{\mathfrak{c}(a), m/m_i\}$. Applying the definition of (23), we find that the image of $\ell \cap m \cap a$ in $\mathcal{O}(-\{\mathfrak{c}(a), \psi, \varphi\})$ is given by the class of

$$\{\mathfrak{c}(m \cap a), \ell/\ell_i\} - (-1)^k\{\mathfrak{c}(a), m/m_i, \varphi_{ij}\} \tag{30}$$

modulo $\check{B}^{k+1}(\mathcal{U}, E_1^{*,-k-2})$. In a similar manner we find that the image of $m \cap \ell \cap a$ in $\mathcal{O}(-\{\mathfrak{c}(a), \varphi, \psi\})$ is given by

$$\{\mathfrak{c}(\ell \cap a), m/m_i\} - (-1)^k\{\mathfrak{c}(a), \ell/\ell_i, \psi_{ij}\}. \tag{31}$$

By (30), (31), and the definition of $\sigma_{\mathcal{L},\mathcal{M}}$ the proof of (29) is reduced to the investigation of the difference

$$\begin{aligned}&(-1)^k(\{\mathfrak{c}(a), \ell/\ell_i, \psi_{ij}\} - \{\mathfrak{c}(a), m/m_i, \varphi_{ij}\} \\ &\quad -\{\mathfrak{c}(a), \varphi_{ij}, \psi_{ij}\}) + \{\mathfrak{c}(m \cap a), \ell/\ell_i\} - \{\mathfrak{c}(\ell \cap a), m/m_i\},\end{aligned} \tag{32}$$

where the third term comes from our definition of (24). Since the supports of div(ℓ), div(m), and $\mathfrak{c}(a)$ intersect properly, m/m_i and ℓ/ℓ_i have residue classes $(m/m_i)(x) \in k(x)^*$, $(\ell/\ell_i)(x) \in k(x)^*$ for $x \in (U_i)_k \cap \operatorname{supp}(\mathfrak{c}(a))$. Consequently, there is a well-defined element of $E_1^{k,-k-2}(U_i)$:

$$\lambda_i = \{\mathfrak{c}(c), \ell/\ell_i, m/m_i\} = (n_x\{(\ell/\ell_i)(x), (m/m_i)(x)\})_{x \in (U_i)_k} \tag{33}$$

where $\mathfrak{c}(a) = (n_x)_x$. Since the divisors of ℓ and m have no common component intersecting the support of $\mathfrak{c}(a)$, the tame symbol of (33) is given by

$$\mathrm{d}(\lambda_i) = \{\mathfrak{c}(\ell \cap a), m/m_i\} - \{\mathfrak{c}(m \cap a), \ell/\ell_i\}.$$

For the Čech differential of λ we find

$$\mathrm{d}_0(\lambda) = \{\mathfrak{c}(a), m/m_i, \varphi_{ij}\} - \{\mathfrak{c}(a), \ell/\ell_i, \psi_{ij}\} + \{\mathfrak{c}(a), \varphi_{ij}, \psi_{ij}\}.$$

Consequently, the Čech hyperdifferential of λ is

$$\begin{aligned}&(-1)^k(\{\mathfrak{c}(a), m/m_i, \varphi_{ij}\} - \{\mathfrak{c}(a), \ell/\ell_i, \psi_{ij}\} + \{\mathfrak{c}(a), \varphi_{ij}, \psi_{ij}\}) \\ &+ \quad \{\mathfrak{c}(\ell \cap a), m/m_i\} - \{\mathfrak{c}(m \cap a), \ell/\ell_i\},\end{aligned}$$

and (32) is a complete differential. The proof of (29) is complete.

Let $\mathcal{L},\mathcal{M},\mathcal{N}$ be line bundles on X, and $A \in \tilde{\mathfrak{Ch}}^k(X)$. We want to prove the commutativity of

$$\begin{array}{ccc}
\mathfrak{c}_1(\mathcal{L})\cap\mathfrak{c}_1(\mathcal{M})\cap\mathfrak{c}_1(\mathcal{N})\cap A & \longrightarrow & \mathfrak{c}_1(\mathcal{M})\cap\mathfrak{c}_1(\mathcal{L})\cap\mathfrak{c}_1(\mathcal{N})\cap A \\
\downarrow & & \downarrow \\
\mathfrak{c}_1(\mathcal{L})\cap\mathfrak{c}_1(\mathcal{N})\cap\mathfrak{c}_1(\mathcal{M})\cap A & & \mathfrak{c}_1(\mathcal{M})\cap\mathfrak{c}_1(\mathcal{N})\cap\mathfrak{c}_1(\mathcal{L})\cap A \\
\downarrow & & \downarrow \\
\mathfrak{c}_1(\mathcal{N})\cap\mathfrak{c}_1(\mathcal{L})\cap\mathfrak{c}_1(\mathcal{M})\cap A & \longrightarrow & \mathfrak{c}_1(\mathcal{N})\cap\mathfrak{c}_1(\mathcal{M})\cap\mathfrak{c}_1(\mathcal{L})\cap A
\end{array} \qquad (34)$$

If $A,\mathcal{L},\mathcal{M},\mathcal{N}$ have rational sections a,ℓ,m,n whose cycles and divisors meet properly (i.e., $\ell\cap m\cap n\cap a$ etc. are rational sections), then (34) follows from (29). In the general case, let $p: E \to X$ be the fibre space of the bundle $\mathcal{L}\oplus\mathcal{M}\oplus\mathcal{N}$. Because $\underline{p}^*$ is an equivalence of categories, it suffices to prove (34) after base-change to E. On E, there are tautological sections ℓ,m,n of $p^*\mathcal{L}$, $p^*\mathcal{M}$, $p^*\mathcal{N}$. If a is any rational section of A on X, then the previous remark can be applied to ℓ,m,n, and $p^*(a)$. The proof of (34) is complete.

Example Let us return to the situation of 1.3. By (29), $\underline{p}_*(\sigma_{\mathcal{L},\mathcal{M}})$ is the isomorphism $<\mathcal{L},\mathcal{M}>\to<\mathcal{M},\mathcal{L}>$ which sends $<\ell,m>$ to $<m,\ell>$. Integrating (28) along the fibres, we get the well-known identity

$$<\ell,\tilde{\ell}>=(-1)^{\deg(\mathcal{L})}<\tilde{\ell},\ell>$$

in $<\mathcal{L},\mathcal{L}>$ (cf. [D,6.2.]).

1.6 Lemma *Let $\mathcal{E}$ be a vector bundle of dimension e on X, $p:\mathbb{P}(\mathcal{E})\to X$ its projective fibration, and $\mathcal{O}(-1)^*_p\mathcal{E}$ the tautological line bundle. It has a first Chern class $\mathfrak{c}_1(\mathcal{O}(1))\in\check{H}^1(\mathbb{P}(\mathcal{E}),\mathcal{K}_1)$. Then the homomorphism*

$$\prod_{j=0}^{e-1} E_e^{p-j,q+j}(X) \quad\to\quad E_2^{p,q}(\mathbb{P}(\mathcal{E})) \qquad (35)$$

$$(\alpha_j) \quad \to \quad \sum_{j=0}^{e-1} c_1(\mathcal{O}(1))^j p^*(\alpha_j)$$

is an isomorphism, where the product in (35) is the product

$$\check{H}^p(Y, \mathcal{K}_q) \times E_2^{k,1}(Y) \to E_2^{p+k,q-1}(Y)$$

defined in [G,p.281].

Proof This is [G, Note (i) on p.287]. The proof is similar to the proof of [G,Theorem 8.10.].

1.7 Corollary *The functor*

$$\begin{array}{rcl} \times_{j=0}^{e-1} \tilde{\mathfrak{CH}}^{p-j}(X) & \to & \tilde{\mathfrak{CH}}^p(\mathbb{P}(\mathcal{E})) \\ (A_j) & \to & \oplus_{j=0}^{e-1} \mathfrak{c}_1(\mathcal{O}(1)^j \cap \underline{p}^*(A_j)) \end{array} \tag{36}$$

is an equivalence of categories. Here the symbol $\mathfrak{c}_1(\mathcal{L})^j \cap B$ denotes the iteration $\mathfrak{c}_1(\mathcal{L}) \cap \mathfrak{c}_1(\mathcal{L}) \cap \ldots \cap B$.

Proof It follows from 1.6 that (36) induces isomorphisms between π_1 and π_0 of the Picard categories on both sides of (36).

1.8 Definition of the Chern functors We may compose the functor

$$\begin{array}{rcl} \tilde{\mathfrak{CH}}^p(X) & \to & \tilde{\mathfrak{CH}}^{p+e}(\mathbb{P}(\mathcal{E})) \\ A & \to & \mathfrak{c}_1(\mathcal{O}(1))^e \cap \underline{p}^*(A) \end{array}$$

with the inverse of (36) to obtain additive functors

$$\mathfrak{c}_i(\mathcal{E}) \cap A \colon \tilde{\mathfrak{CH}}^p(X) \to \tilde{\mathfrak{CH}}^{p+i}(X), \quad 1 \leq i \leq e$$

$$\mathfrak{c}_0(\mathcal{E}) \cap A = A$$

and an additive functor-isomorphism

$$\oplus_{j=0}^{e} \mathfrak{c}_1(\mathcal{O}(1))^{e-j} \cap \underline{p}^*(\mathfrak{c}_j(\mathcal{E}) \cap A) \to \mathbf{0}. \tag{37}$$

The Chern functors are unique up to unique functor-isomorphism: If $\tilde{\mathfrak{c}}_j(\mathcal{E}) \cap \cdot$ are other additive functors with $\tilde{\mathfrak{c}}_0(\mathcal{E}) \cap A$ and an isomorphism (37), then there exists a unique functor-isomorphism $\mathfrak{c}_j(\mathcal{E}) \cap A \to \tilde{\mathfrak{c}}_j(\mathcal{E}) \cap A$ compatible with (37) which is the identity if $j = 0$.

If $\mathcal{E}$ is a line bundle, we have $\mathbf{P}(\mathcal{E}) = X$ and $\mathcal{O}(1) = \mathcal{E}^{-1}$, and it follows easily that (37) is solved by the functor $\mathfrak{c}_1(\mathcal{E}) \cap A$ defined in 1.2. Let $\phi\colon \mathcal{E} \to \mathcal{E}'$ be an isomorphism of vector bundles. It induces an isomorphism $\mathbf{P}(\mathcal{E}) \to \mathbf{P}(\mathcal{E}')$, hence there is a unique isomorphism $\mathfrak{c}_j(\mathcal{E}) \cap A \to \mathfrak{c}_j(\mathcal{E}') \cap A$ which is compatible with (37) and is the identity if $j = 0$.

Let $\mathcal{E}$ be a vector bundle on X, $f\colon Y \to X$ a flat morphism and $g\colon Z \to X$ a proper morphism of relative dimension d. Using the results of 1.4. it is not hard to construct natural isomorphisms

$$\begin{aligned} \underline{f}^*(\mathfrak{c}_j(\mathcal{E}) \cap A) &\to \mathfrak{c}_j(f^*\mathcal{E}) \cap \underline{f}^* A, \quad A \in \widetilde{\mathfrak{CH}}^{\cdot}(X) \\ \underline{g}_*(\mathfrak{c}_j(g^*\mathcal{E}) \cap B) &\to \mathfrak{c}_j(\mathcal{E}) \cap \mathfrak{g}_* B, \quad B \in \widetilde{\mathfrak{CH}}^{\cdot}(Z). \end{aligned} \tag{38}$$

They satisfy the following compatibility with composition of flat and proper morphisms and the base-change isomorphism [F1,3.12.]: If for an X-scheme $f\colon Y \to X$ we consider the functor

$$\mathfrak{c}_j(f^*\mathcal{E}) \cap \cdot\colon \widetilde{\mathfrak{CH}}^{\cdot}(X) \to \widetilde{\mathfrak{CH}}^{\cdot}(Y),$$

then (38) defines on it the structure of a biadmissible functor between bifibred Picard categories over (X-schemes, proper morphisms, flat morphisms).

Our next steps aim at proving the functorial version of the Whitney sum formula. First we need the following isomorphisms:

1.9 Let

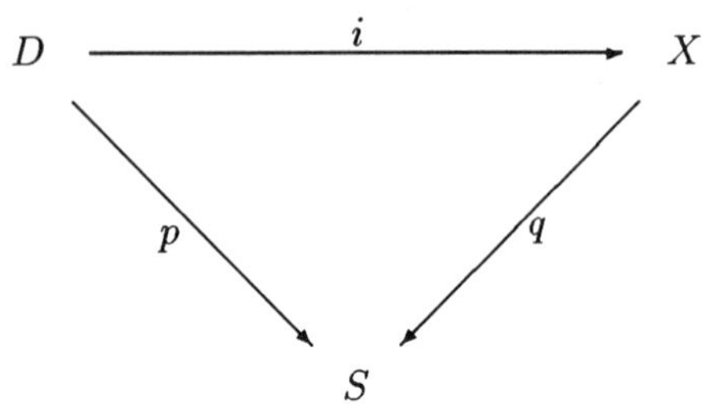

be a commutative diagram in which i is a regular immersion of codimension one and p and q are flat. We put $\mathcal{O}(D) = \mathcal{I}^{-1}$, where $\mathcal{I}$ is the sheaf of ideals defining D. There is a natural isomorphism

$$\mathfrak{c}_1(\mathcal{O}(D)) \cap \underline{q}^* A \to \underline{i}_* \underline{p}^* A \tag{39}$$

which sends "**1**" $\cap\, q^*(a)$ to $i_* p^*(a)$, where a is a rational section of A on S and "**1**" is the canonical section of $\mathcal{O}(D)$. This isomorphisms is (in an obvious sense) compatible with flat base-changes $X' \to X$, flat maps $S \to S'$, and proper base-changes $S' \to S$. If $\mathcal{L}$ is a line bundle on X, the diagram

$$\begin{array}{ccc}
\mathfrak{c}_1(\mathcal{O}(D)) \cap q^*(\mathfrak{c}_1(\mathcal{L}) \cap A) & \longrightarrow & \underline{i}_*\underline{p}^*(\mathfrak{c}_1(\mathcal{L}) \cap A) \\
\Big\downarrow & & \Big\downarrow \\
\mathfrak{c}_1(q^*\mathcal{L}) \cap \mathfrak{c}_1(\mathcal{O}(D)) \cap \underline{q}^*A & \longrightarrow & \mathfrak{c}_1(q^*\mathcal{L}) \cap \underline{i}_*\underline{p}^*A
\end{array} \qquad (40)$$

commutes.

Proof We have only to check that the above definition of (39) is independent of the choice of a, i.e., that

$$\mathfrak{c}_1(\mathcal{O}(D)) \cap \underline{q}^*(\gamma) = \underline{i}_*\underline{p}^*(\gamma)$$

for $\gamma \in \mathrm{Aut}_{\widetilde{\mathfrak{CH}}^k(S)}(A) = G_k(S)$. If $\ell_i \in \mathcal{O}_X(U_i - D)$ are trivialisations of $\mathcal{O}(D)$ on an open covering $\mathcal{U}$ of X, then the l.h.s. of the last equality is given by the cohomology class of $(-1)^k\{q^*\gamma, \varphi\} \in \check{Z}^k(\mathcal{U}, E_1^{*,-k-1})$, where $\gamma \in E_1^{k-1,-k}(S)$ is a representative for γ, and $\varphi_{ij} = \ell_j/\ell_i$. If $q^*(\gamma) = (g_x)_{x \in X_k}$, then because D is flat over S and of codimension one in X, $g_x \neq 1$ implies that x does not belong to D, hence the image of ℓ_i in $k(x)$ is well-defined. Consequently, the product

$$\alpha = (q^*(\gamma)\ell_i)_{i\in I} = (g_x\ell_i)_{i\in I, x\in X_k}$$

is well-defined. It satisfies

$$\begin{aligned}
\mathrm{d}(\alpha) &= \{(-1)^k q^*(\gamma), \varphi\} + \mathrm{d}_1((q^*(\gamma)\ell_i)_{i\in I}) \\
&= \{(-1)^k q^*(\gamma), \varphi\} - \mathrm{d}_1((\ell_i q^*(\gamma))_{i\in I}).
\end{aligned}$$

By [F1,Lemma 1.6.], we have

$$\mathrm{d}_1(\ell_i q^*(\gamma)) = i_*p^*(\gamma)\,|_{U_i} \quad \text{in} \quad E_1^{k,-k-1}(U_i),$$

proving the desired identity.

1.10 Now we are ready to construct the Whitney isomorphism for exact sequences of the form

$$0 \to \mathcal{F} \to \mathcal{E} \to \mathcal{L} \to 0$$

where $\mathcal{L}$ is a line bundle. We apply 1.9. to the diagram

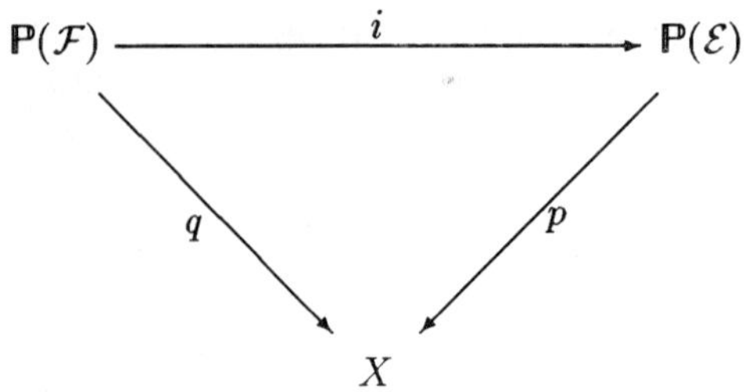

There is a natural isomorphism $i^*\mathcal{O}(1)_{\mathcal{E}} \cong \mathcal{O}(1)_{\mathcal{F}}$, so we may denote both line bundles by the same symbol $\mathcal{O}(1)$. Restricting $\mathcal{E} \to \mathcal{F}$ to the subbundle $\mathbb{O}(-1) \subset p^*\mathcal{E}$ we obtain a section ξ of $Hom(\mathbb{O}(-1), p^*\mathcal{L}) = p^*\mathcal{L} \otimes \mathcal{O}(1)$ on $\mathbb{P}(\mathcal{E})$. The subscheme defined by the vanishing of ξ is $\mathbb{P}(\mathcal{F})$. By 1.9., there is a canonical isomorphism

$$\underline{i}_*\underline{q}^*B \to \mathfrak{c}_1(p^*\mathcal{L} \otimes \mathcal{O}(1)) \cap p^*B, \quad B \in \widetilde{\mathfrak{C}\mathfrak{H}}^{\cdot}(X)$$

which sends $i_*q^*(b)$ to $\xi \cap p^*(b)$. Applying this to (37) (for the bundle $\mathcal{F}$), we find an isomorphism

$$\begin{aligned}
0 \quad &\to \quad \underline{i}_*\Big(\bigoplus_{j=0}^{f} \mathfrak{c}_1(\mathcal{O}(1))^{f-j} \cap \underline{q}^*(\mathfrak{c}_j(\mathcal{F}) \cap A)\Big) &(41)\\
&\to \quad \bigoplus_{j=0}^{f} \mathfrak{c}_1(\mathcal{O}(1))^{f-j} \cap \underline{i}_*\underline{q}^*(\mathfrak{c}_j(\mathcal{F}) \cap A)) \\
&\to \quad \bigoplus_{j=0}^{f} \mathfrak{c}_1(\mathcal{O}(1))^{f-j} \cap \\
&\qquad \big(\mathfrak{c}_1(p^*\mathcal{L}) \cap \underline{p}^*(\mathfrak{c}_j(\mathcal{F}) \cap A) \oplus \mathfrak{c}_1(\mathcal{O}(1)) \cap \underline{p}^*(\mathfrak{c}_j(\mathcal{F}) \cap A)\big) \\
&\to \quad \underline{p}^*(\mathfrak{c}_1(\mathcal{L}) \cap \mathfrak{c}_f(\mathcal{F}) \cap A) \oplus \mathfrak{c}_1(\mathcal{O}(1))^e \cap \underline{p}^*(\mathfrak{c}_0(\mathcal{F}) \cap A) \oplus \\
&\qquad \oplus \bigoplus_{j=0}^{f} \mathfrak{c}_1(\mathcal{O}(1))^{e-j} \cap \underline{p}^*(\mathfrak{c}_1(\mathcal{L}) \cap \mathfrak{c}_{j-1}(\mathcal{F})) \cap A \oplus \mathfrak{c}_j(\mathcal{F}) \cap A).
\end{aligned}$$

Note that the first arrow is (37), the second is derived from (21), the third from (39), and the fourth arrow in (41) comes from (19). The composition of the arrows in (41) is of the form (37) for the vector bundle $\mathcal{E}$. Since (37) defines the Chern functors up to unique isomorphism, (41) defines the Whitney isomorphism we are looking for:

$$\mathfrak{c}_j(\mathcal{E}) \cap A \to \begin{cases} A = \mathfrak{c}_0(\mathcal{F}) \cap A & \text{if } j = 0 \\ \mathfrak{c}_1(\mathcal{L}) \cap \mathfrak{c}_{j-1}(\mathcal{F}) \cap A \oplus \mathfrak{c}_j(\mathcal{F}) \cap A & \text{if } 1 \leq j \leq f \\ \mathfrak{c}_1(\mathcal{L}) \cap \mathfrak{c}_f(\mathcal{F}) \cap A & \text{if } j = e \end{cases}. \tag{42}$$

We can write this in the shorter form

$$\mathfrak{c}.(\mathcal{E}) \cap A \to \mathfrak{c}.(\mathcal{L}) \cap \mathfrak{c}.(\mathcal{F}) \cap A,$$

where $\mathfrak{c}.(\mathcal{E}) \cap A = \bigoplus_{j \geq 0} \mathfrak{c}_j(\mathcal{E}) \cap A$ in $\widetilde{\mathfrak{C}\mathfrak{H}}^{\cdot}(X)$.

1.11 Symmetry Before we can construct the analogue of (42) in the general case we have to define a symmetry isomorphism between the Chern functors and to explain its relation to (42).

Let $\mathcal{E}$ and $\mathcal{F}$ be vector bundles of dimension e and f on X. Let $r: \mathbf{P}(\mathcal{E}) \times \mathbf{P}(\mathcal{F}) \to X$ be the projection. If $B(i,j)$ $(0 \leq i \leq e, 0 \leq j \leq f)$ are objects of a commutative Picard category, we put

$$\tilde{\bigoplus_{i,j}} B(i,j) = \bigoplus_{i=1}^{e} \bigoplus_{j=1}^{f} B(i,j) \ominus B(0,0)$$

By applying 1.7. twice, we find that the following isomorphism in $\tilde{\mathfrak{CH}}^{e+f}(\mathbf{P}(\mathcal{E}) \times \mathbf{P}(\mathcal{F}))$, together with the identity $\mathfrak{c}_0(\mathcal{E}) \cap \mathfrak{c}_0(\mathcal{F}) \cap A = A$, characterises $\mathfrak{c}_j(\mathcal{E}) \cap \mathfrak{c}_k(\mathcal{F}) \cap A$ up to unique isomorphism:

$$\tilde{\bigoplus_{j,k}} \mathfrak{c}_1(q^*\mathcal{O}(1)_{\mathcal{F}})^{f-k} \cap \mathfrak{c}_1(p^*\mathcal{O}(1)_{\mathcal{E}})^{e-j} \cap \underline{r}^*(\mathfrak{c}_j(\mathcal{E}) \cap \mathfrak{c}_k(\mathcal{F}) \cap A) \to 0$$

(p and q are the projections of $\mathbf{P}(\mathcal{E}) \times \mathbf{P}(\mathcal{F})$ to $\mathbf{P}(\mathcal{E})$ and $\mathbf{P}(\mathcal{F})$). In a similar manner $\mathfrak{c}_k(\mathcal{F}) \cap \mathfrak{c}_j(\mathcal{E}) \cap A$ is characterised by the isomorphism

$$\tilde{\bigoplus_{j,k}} \mathfrak{c}_1(p^*\mathcal{O}(1)_{\mathcal{E}})^{e-j} \cap \mathfrak{c}_1(q^*\mathcal{O}(1))_{\mathcal{F}})^{f-k} \cap \underline{r}^*(\mathfrak{c}_k(\mathcal{F}) \cap \mathfrak{c}_j(\mathcal{E}) \cap A) \to 0.$$

Let $\mathcal{L}$ and $\mathcal{M}$ be line bundles on a scheme and a and b integers. We define an isomorphism $\mathfrak{c}_1(\mathcal{L})^a \cap \mathfrak{c}_1(\mathcal{M})^b \to \mathfrak{c}_1(\mathcal{M})^b \cap \mathfrak{c}_1(\mathcal{L})^a$ by the following permutation of the factors:

$$\begin{array}{ccccccccc} 1 & 2 & 3 & \dots & a & a+1 & a+2 & \dots & a+b \\ \downarrow & \downarrow & \downarrow & & \downarrow & \downarrow & \downarrow & & \downarrow \\ b+1 & b+2 & b+2 & \dots & b+1 & 1 & 2 & \dots & b \end{array} \qquad (43)$$

In (43), $1; \dots; a$ are the factors of the product $\mathfrak{c}_1(\mathcal{L})^a$ and $a+1; \dots; a+b$ are the factors of $\mathfrak{c}_1(\mathcal{M})^b$. In other words, we apply $\sigma_{\mathcal{L},\mathcal{M}}$ ab times but we never apply $\sigma_{\mathcal{L},\mathcal{L}}$ or $\sigma_{\mathcal{M},\mathcal{M}}$. If an isomorphism $\mathfrak{c}_1(\mathcal{L})^a \cap \mathfrak{c}_1(\mathcal{M})^b \to \mathfrak{c}_1(\mathcal{M})^b \cap \mathfrak{c}_1(\mathcal{L})^a$ is used without comment, it is supposed to be of the form (43).

There is a unique isomorphism $\sigma_{\mathcal{E},\mathcal{F}}: \mathfrak{c}_j(\mathcal{E}) \cap \mathfrak{c}_k(\mathcal{F}) \cap A \to \mathfrak{c}_k(\mathcal{F}) \cap \mathfrak{c}_j(\mathcal{E}) \cap A$ which is the identity if $jk = 0$ and makes the diagram

$$\begin{array}{ccc}
\tilde{\bigoplus\limits_{j,k}} \mathfrak{c}_1(q^*\mathcal{O}(1)_{\mathcal{F}})^{f-k} \cap \mathfrak{c}_1(p^*\mathcal{O}(1)_{\mathcal{E}})^{e-j} \cap \underline{r}^*(\mathfrak{c}_j(\mathcal{E}) \cap \mathfrak{c}_k(\mathcal{F}) \cap A) & & \\
\downarrow & \searrow & \\
\tilde{\bigoplus\limits_{j,k}} \mathfrak{c}_1(p^*\mathcal{O}(1)_{\mathcal{E}})^{e-j} \cap \mathfrak{c}_1(q^*\mathcal{O}(1)_{\mathcal{F}})^{f-k} \cap \underline{r}^*(\mathfrak{c}_j(\mathcal{E}) \cap \mathfrak{c}_k(\mathcal{F}) \cap A) & & 0 \\
\downarrow \sigma_{\mathcal{E},\mathcal{F}} & \nearrow & \\
\tilde{\bigoplus\limits_{j,k}} \mathfrak{c}_1(p^*\mathcal{O}(1)_{\mathcal{E}})^{e-j} \cap \mathfrak{c}_1(p^*\mathcal{O}(1)_{\mathcal{F}})^{f-k} \cap \underline{r}^*(\mathfrak{c}_k(\mathcal{F}) \cap \mathfrak{c}_j(\mathcal{E}) \cap A) & &
\end{array} \tag{44}$$

commutative.

We have an analogue of (34) and the identity $\sigma_{\mathcal{E},\mathcal{F}}\sigma_{\mathcal{F},\mathcal{E}} = 1$ because these properties are satisfied for line bundles. It is also clear that $\sigma_{\mathcal{E},\mathcal{F}}$ is compatible with flat pull-back and proper push-forward. Let $0 \to \mathcal{F} \to \mathcal{E} \to \mathcal{L} \to 0$ be an exact sequence with $\dim(\mathcal{L}) = 1$, and let $\mathcal{G}$ be a vector bundle on X. It follows from (40) that the diagram

$$\begin{array}{ccc}
\mathfrak{c}.(\mathcal{L}) \cap \mathfrak{c}.(\mathcal{F}) \cap \mathfrak{c}_j(\mathcal{G}) \cap A) & \longrightarrow & \mathfrak{c}.(\mathcal{E}) \cap \mathfrak{c}_j(\mathcal{G}) \cap A \\
\downarrow \sigma_{\mathcal{F},\mathcal{G}} & & \\
\mathfrak{c}.(\mathcal{L}) \cap \mathfrak{c}_j(\mathcal{G}) \cap \mathfrak{c}.(\mathcal{F}) \cap A & & \downarrow \sigma_{\mathcal{E},\mathcal{F}} \\
\downarrow \sigma_{\mathcal{L},\mathcal{G}} & & \\
\mathfrak{c}_j(\mathcal{G}) \cap \mathfrak{c}.(\mathcal{L}) \cap \mathfrak{c}.(\mathcal{F}) \cap A & \longrightarrow & \mathfrak{c}_j(\mathcal{G}) \cap \mathfrak{c}.(\mathcal{E}) \cap A
\end{array} \tag{45}$$

commutes.

1.12 Let $\mathcal{E}$ and $\mathcal{F}$ be vector bundles on X of dimensions e and f, and let $A = (A_l)_{0 \leq l < \infty} \in \mathrm{Ob}(\widetilde{\mathfrak{C}\mathfrak{H}}^{\cdot}(X))$. We define an element

$$\tau.(\mathcal{E}, \mathcal{F}, A) = (\tau_k(\mathcal{E}, \mathcal{F}, A))_{1 \leq k < \infty} \in \coprod_{k \geq 1} G_k(X) \tag{46}$$

by

$$\tau_k(\mathcal{E}, \mathcal{F}, A) = \sum_{l=0}^{\infty} \sum_{i+j=k-l-1} (e-i)(f-j) c_i(\mathcal{E}) c_j(\mathcal{F})[A_l][-1], \tag{47}$$

where $c_i(\mathcal{E})[A]$ is defined as the isomorphism class of $\mathfrak{c}_i(\mathcal{E}) \cap A$, and $[-1]$ is $-1 \in K_1$.

The aim of 1.12. is to prove the following formula: Let

$$0 \to \mathcal{F} \to \mathcal{E} \xrightarrow{\pi} \mathcal{L} \oplus \mathcal{M} \to 0$$

be an exact sequence of vector bundles, with $\dim(\mathcal{L}) = \dim(\mathcal{M}) = 1$. We put $\mathcal{G} = \pi^{-1}(\mathcal{M})$, $\mathcal{H} = \pi^{-1}(\mathcal{L})$. Let $A \in \mathrm{Ob}(\widetilde{\mathfrak{C}\mathfrak{H}}^{k}(X))$. Then the following diagram commutes:

$$\begin{array}{ccccc}
\mathfrak{c}.(\mathcal{E}) \cap A & \xrightarrow{\alpha} & \mathfrak{c}.(\mathcal{L}) \cap .(\mathcal{G}) \cap A & \xrightarrow{\beta} & \mathfrak{c}.(\mathcal{L}) \cap \mathfrak{c}.(\mathcal{M}) \cap \mathfrak{c}.(\mathcal{F}) \cap A \\
\downarrow{\scriptstyle \tau.(\mathcal{L}, \mathcal{M}, \mathfrak{c}.(\mathcal{E}) \cap A)} & & & & \downarrow{\scriptstyle \sigma_{\mathcal{L},\mathcal{M}}} \\
\mathfrak{c}.(\mathcal{E}) \cap A & \xrightarrow[\gamma]{} & \mathfrak{c}.(\mathcal{M}) \cap .(\mathcal{H}) \cap A & \xrightarrow[\delta]{} & \mathfrak{c}.(\mathcal{M}) \cap \mathfrak{c}.(\mathcal{L}) \cap \mathfrak{c}.(\mathcal{F}) \cap A
\end{array} \tag{48}$$

where the horizontal arrows are in an obvious manner constructed from the Whitney sum isomorhism.

Let $t: Y \to X$ be the fibre space of the bundle $\mathcal{L} \oplus \mathcal{M} \oplus c\check{E}^{\oplus(e-1)}$, where $\check{\mathcal{E}}$ is the dual of $\mathcal{E}$. The $e-2$ $\check{\mathcal{E}}$-coordinates define sections $\lambda_3, \ldots, \lambda_e$ of $t^*\check{\mathcal{E}}$. The L- and M-coordinates define sections ℓ and m of $t^*\mathcal{L}$ and $t^*\mathcal{M}$ on Y. We define rational sections λ_1, λ_2 of $t^*\check{\mathcal{E}}$ by $\lambda_1(\mathcal{G}) = 0$, $\lambda_1(\ell) = 1$ and $\lambda_2(\mathcal{H}) = 0$, $\lambda_2(m) = 1$.

We have a Cartesian diagram of projective fibrations over Y

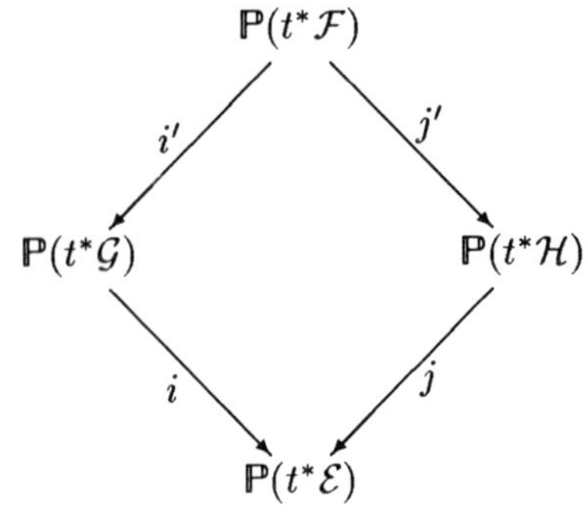

and denote the projections from $\mathbb{P}(\mathcal{E})$, $\mathbb{P}(\mathcal{H})$, $\mathbb{P}(\mathcal{G})$, and $\mathbb{P}(\mathcal{F})$ to Y by p, q, r, and s. By the isomorphism between $\check{\mathcal{E}}$ and $(p_*\mathcal{O}(1)_\mathcal{E})$, the λ_i define sections Λ_i of $\mathbb{P}(1)$ on $\mathbb{P}(\mathcal{E})$ (rational sections if $i = 1$ or $i = 2$). To avoid awkward expressions, we denote the restrictions of Λ_i and of $\mathcal{O}(1)_\mathcal{E}$ to one of the projective subspaces of $\mathbb{P}(\mathcal{E})$ by the same letters Λ_i and $\mathcal{O}(1)$.

Let a_i be a rational section of $\mathfrak{c}_i(\mathcal{F}) \cap A$ on X. Then

$$\varepsilon = \bigoplus_{n=0}^{e-2} \Lambda_e \cap \ldots \cap \Lambda_{n+3} \cap s^*t^*(a_n)$$

is a rational section of

$$\bigoplus_{n=0}^{e-2} \mathfrak{c}_1(\mathcal{O}(1))^{e-2-n} \cap \underline{s}^*\underline{t}^*(\mathfrak{c}_n(\mathcal{F}) \cap A)$$

on $\mathbb{P}(\mathcal{F})$. Let $b \in (G_{k+e-2})_r(\mathbb{P}(t^*\mathcal{F}))$ be the image of ε by (37). We denote the isomorphism $\mathfrak{c}.(\mathcal{M}) \cap \mathfrak{c}.(\mathcal{F}) \to \mathfrak{c}.(\mathcal{G})$ by ζ. It is easy to see that $\Lambda_2 \otimes m$ is the canonical section ξ of $p^*t^*\mathcal{M} \otimes \mathcal{O}(1)$ which has a simple zero along $\mathbb{P}(\mathcal{H})$. Using this and (41), we see that the image of

$$\begin{aligned} &\bigoplus_{n=0}^{e-2} (\Lambda_e \cap \ldots \Lambda_{n+3} \cap \Lambda_2 \cap \zeta(a_n) \oplus \ldots \\ &\quad \Lambda_e \ldots \cap \Lambda_{n+3} \cap \zeta(m \cap a_n)) \in \\ &\quad \in \left(\bigoplus_{n=0}^{e-1} \mathfrak{c}_1(\mathcal{O}(1))^{e-1-n} \cap \underline{r}^*\underline{t}^*(\mathfrak{c}_n(\mathcal{G}) \cap A) \right)_r (\mathbb{P}(\mathcal{G})) \end{aligned} \tag{49}$$

in $(G_{l+e-1})_r(\mathbb{P}(t^*\mathcal{G}))$ is $i'_*(b)$. The lax notation (49) means more precisely

$$\begin{aligned} \zeta' \Bigg(&\bigoplus_{n=0}^{e-2} (\Lambda_e \cap \ldots \cap \Lambda_{n+3} \cap \Lambda_2 \cap r^*t^*(a_n) \oplus \\ &\oplus \Lambda_e \cap \ldots \cap \Lambda_{n+3} \cap r^*t^*(m \cap a_n))\Bigg), \end{aligned}$$

where

$$\zeta' \colon \bigoplus_{n=0}^{e-2} \Big(\mathfrak{c}_1(\mathcal{O}(1))^{e-1-n} \cap \underline{r}^* \underline{t}^* (\mathfrak{c}_1(\mathcal{M}) \cap \mathfrak{c}_n(\mathcal{F}) \cap A) \oplus$$
$$\oplus \mathfrak{c}_1(\mathcal{O}(1))^{e-2-n} \cap \underline{r}^* \underline{t}^* (\mathfrak{c}_1(\mathcal{M}) \cap \mathfrak{c}_n(\mathcal{F}) \cap A) \Big)$$
$$\rightarrow \bigoplus_{m=0}^{e-1} \mathfrak{c}_1(\mathcal{O}(1))^{e-1-n} \cap \underline{r}^* \underline{t}^* (\mathfrak{c}_m(\mathcal{G}) \cap A)$$

is the isomorphism derived from ζ. A similar computation can be applied to the image of (49) by $\underline{i}_*$. The result is that the image of

$$\begin{aligned} \bigoplus_{n=0}^{e-2} \Big(& \Lambda_e \cap \ldots \cap \Lambda_{n+3} \cap \Lambda_2 \cap \Lambda_1 \cap \alpha^{-1}\beta^{-1}(a_n) \oplus \\ & \Lambda_e \cap \ldots \cap \Lambda_{n+3} \cap \Lambda_2 \cap \alpha^{-1}\beta^{-1}(\ell \cap a_n) \oplus \\ & \oplus \Lambda_e \cap \ldots \cap \Lambda_{n+3} \cap \Lambda_1 \cap \alpha^{-1}\beta^{-1}(m \cap a_n) \oplus \\ & \oplus \Lambda_e \cap \ldots \Lambda_{n+3} \cap \alpha^{-1}\beta^{-1}(\ell \cap m \cap a_n) \Big) \in \\ \in \Big(& \oplus_{m=0}^{e} \mathfrak{c}_1(\mathcal{O}(1))^{e-n} \cap \underline{r}^* \underline{t}^* (\mathfrak{c}_m(\mathcal{E}) \cap A) \Big)_r (\mathbb{P}(t^*\mathcal{E})). \end{aligned} \tag{50}$$

in $(G_{k+e})_r(\mathbb{P}(t^*\mathcal{E}))$ is $i_* i'_*(b)$. The meaning of (50) is similar to that of (49), α and β are isomorphisms in (48). Applying the same method to the embeddings j and j', we find that the image of

$$\begin{aligned} \bigoplus_{n=0}^{e-2} \Big(& \Lambda_e \cap \ldots \Lambda_{n+3} \cap \Lambda_1 \cap \Lambda_2 \cap \gamma^{-1}\delta^{-1}(a_n) \oplus \\ & \oplus \Lambda_e \cap \ldots \cap \Lambda_{n+3} \cap \Lambda_1 \cap \gamma^{-1}\delta^{-1}(m \cap a_n) \oplus \\ & \oplus \Lambda_e \cap \ldots \cap \Lambda_{n+3} \cap \Lambda_2 \cap \gamma^{-1}\delta^{-1}(\ell \cap a_n) \oplus \\ & \oplus \Lambda_e \cap \ldots \cap \Lambda_{n+3} \cap \gamma^{-1}\delta^{-1}(m \cap \ell \cap a_n) \Big) \end{aligned} \tag{51}$$

in $(G_{k+e})_r(\mathbb{P}(t^*\mathcal{E}))$ is $j_* j'_*(b) = i_* i'_*(b)$. It follows that (50) and (51) are equal. By (29) and 1.7., this reduces the proof of (48) to the proof of

$$\begin{aligned} & \Lambda_e \cap \ldots \cap \Lambda_{n+3} \cap \Lambda_1 \cap \Lambda_2 \cap p^*t^*(a_n) - \Lambda_e \cap \ldots \\ & \quad \ldots \cap \Lambda_{n+3} \cap \Lambda_2 \cap \Lambda_1 \cap p^*t^*(a_n) = \\ & = \mathfrak{c}_1(\mathcal{O}(1))^{e-n-1} \cap \underline{p}^* \underline{t}^* (\tau_{k+n+1}(\mathcal{L}, \mathcal{M}, \mathfrak{c}.(\mathcal{F}) \cap A)). \end{aligned} \tag{52}$$

By (28) and (29), the difference in (52) is

$$\mathfrak{c}_1(\mathcal{O}(1))^{e-2-j} \cap \Big([-1] \cap [\mathcal{O}(1)] \cap p^*t^*([\mathfrak{c}_j(\mathcal{F}) \cap A]) \Big)$$
$$= \mathfrak{c}_1(\mathcal{O}(1))^{e-1-j} \cap \underline{p}^* \underline{t}^* (\tau_{k+n+1}(\mathcal{L}, \mathcal{M}, \mathfrak{c}.(\mathcal{F}) \cap A)),$$

and the proof of (48) is complete.

1.13 Let

$$\Sigma \colon 0 \rightarrow \mathcal{F} \rightarrow \mathcal{E} \xrightarrow{\pi} \mathcal{G} \rightarrow 0$$

be an exact sequence of vector bundles, and let $0 = \mathcal{G}_0 \subset \mathcal{G}_1 \subset \ldots \subset \mathcal{G}_g = \mathcal{G}$ be a filtration of $\mathcal{G}$ with $\dim(\mathcal{G}_i/\mathcal{G}_{i-1}) = 1$. Let $\mathcal{E}_i = \pi^{-1}\mathcal{G}_i$ and $\mathcal{L}_i = \mathcal{G}_i/\mathcal{G}_{i-1}$. A successive application of 1.10. gives us isomorphisms

$$\begin{aligned} \mathfrak{c}.(\mathcal{G}) \cap A &\to \mathfrak{c}.(\mathcal{L}_g) \cap c.(\mathcal{G}_{g-1}) \cap A \to \ldots \\ &\to \mathfrak{c}.(\mathcal{L}_g) \cap \ldots \cap \mathfrak{c}.(\mathcal{L}_1) \cap A \\ \mathfrak{c}.(\mathcal{F}) \cap A &\to \mathfrak{c}.(\mathcal{L}_g) \cap \mathfrak{c}.(\mathcal{E}_{g-1}) \cap A \to \ldots \\ &\to \mathfrak{c}.(\mathcal{L}_g) \cap \ldots \cap \mathfrak{c}.(\mathcal{L}_1) \cap \mathfrak{c}.(\mathcal{F}) \cap A. \end{aligned}$$

We want to prove that the resulting isomorphism

$$\Phi_{\mathcal{G}.} : \mathfrak{c}.(\mathcal{G}) \cap \mathfrak{c}.(\mathcal{F}) \cap A \to \mathfrak{c}.(\mathcal{E}) \cap A \tag{53}$$

is independent of the filtration $\mathcal{G}.$ of $\mathcal{G}$. We proceed by induction on the dimension of $\mathcal{F}$.

1.13.1 Let $\mathcal{F}$ be a line bundle. The sheaf $\mathcal{M}$ of splittings of the exact sequence Σ is a principal homogeneous sheaf for $\mathit{Hom}(\mathcal{G}, \mathcal{F})$, hence it is representable by a smooth X-scheme M and [G, Theorem 8.3] asserts that pull-back to M is an isomorphism on $E_2^{p,q}$ of the Quillen spectral sequence, such that it is sufficient to prove our assertion after pull-back to M.

We may thus achieve that Σ has a splitting $s: \mathcal{G} \to \mathcal{E}$. Let

$$\Sigma' : 0 \to \mathcal{G} \to \mathcal{E} \to \mathcal{F} \to 0$$

be the exact sequence defined by this splitting. We want to prove that

$$\begin{array}{ccc} \mathfrak{c}.(\mathcal{G}) \cap \mathfrak{c}.(\mathcal{F}) \cap A & \xrightarrow{\sigma_{\mathcal{F},\mathcal{G}}} \mathfrak{c}.(\mathcal{F}) \cap \mathfrak{c}.(\mathcal{G}) \cap A \longrightarrow & \mathfrak{c}.(\mathcal{E}) \cap A \\ \downarrow & & \downarrow {\scriptstyle \tau.(\mathcal{F},\mathcal{G},A)} \\ \mathfrak{c}.(\mathcal{L}_g) \cap \ldots \cap \mathfrak{c}.(\mathcal{L}_1) \cap \mathfrak{c}.(\mathcal{F}) \cap A & \longrightarrow & \mathfrak{c}.(\mathcal{E}) \cap A \end{array} \tag{54}$$

commutes. Since the arrows in the upper row are independent of $\mathcal{G}.$, we conclude that (53) is independent of $\mathcal{G}.$.

We prove (54) by induction on g. If $\mathcal{G}$ is a line bundle, (54) is (48) in the special case where the bundle $\mathcal{F}$ occurring in (48) is zero. If $g > 1$, we put $\mathcal{H} = \mathcal{G}_{g-1}$ and consider the following diagram:

(55)

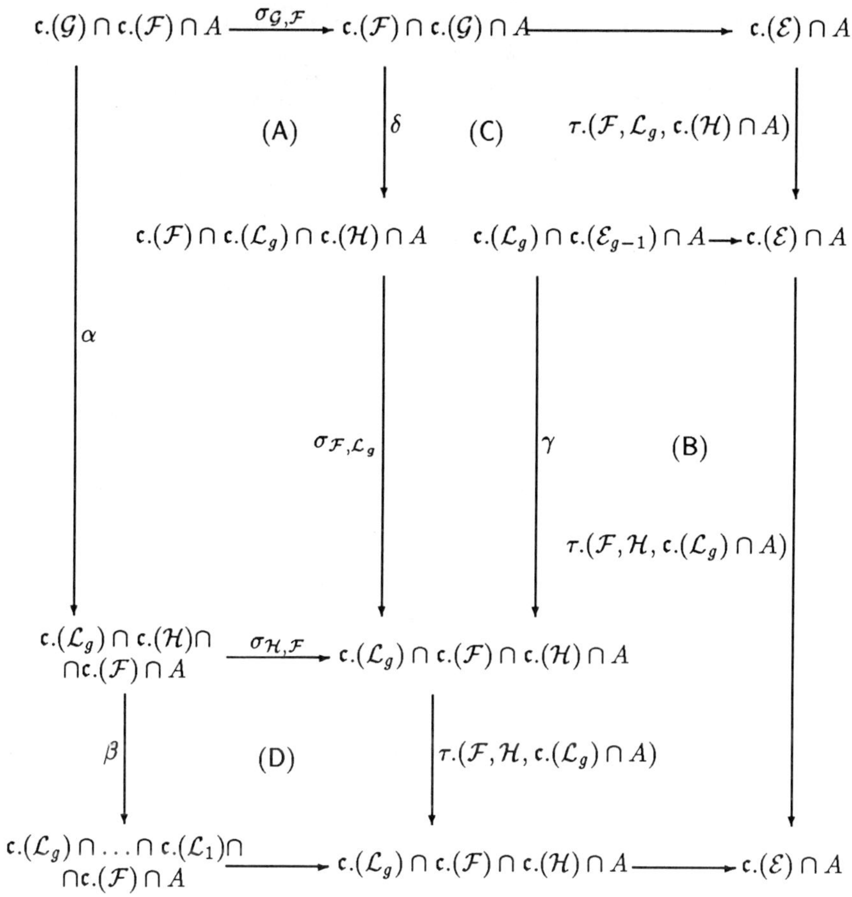

The arrows α and δ in (55) are defined by $0 \to \mathcal{H} \to \mathcal{G} \to \mathcal{L}_g \to 0$, β is defined by the ascending filtration $(\mathcal{G})_{k, 0 \le k \le g-1}$ of $\mathcal{H}$, γ is defined by the sequence $0 \to \mathcal{H} \xrightarrow{s} \mathcal{E}_{g-1} \to \mathcal{F} \to 0$. The commutativity of (A) is consequence of (45), (D) is the induction assumption, (C) is (48), and (B) is trivial. An easy calculation shows

$$\tau.(\mathcal{F}, \mathcal{L}_g, \mathfrak{c}.(\mathcal{H}) \cap A) + \tau.(\mathcal{E}, \mathcal{H}, \mathfrak{c}.(\mathcal{L}_g) \cap A) = \tau.(\mathcal{F}, \mathcal{G}, A).$$

It follows that the outer contour of (55) is (54), and the proof of (54) is complete.

1.13.2 The splitting principle *Let $\mathcal{E}$ be an e-dimensional vector bundle on X, and let $p: Y \to X$ be its flag fibration parametrising maximal flags.*

(a) $\underline{p}^*: \tilde{\mathfrak{CH}}^p(X) \to \tilde{\mathfrak{CH}}^p(Y)$ *is a faithful functor.*
(b) Let $p_{1,2}$ *be the projections of* $Y \times_X Y$ *to its factors and* $r = pp_1 = pp_2$*. If* A *and* B *are objects of* $\tilde{\mathfrak{CH}}^p(X)$ *and if* $\varphi: \underline{p}^* A \to \underline{p}^* B$ *is an isomorphism, then* f *is of the form* $\underline{p}^*(\psi)$ *for a (unique)* $\psi: A \to B$ *if and only if*

$$\underline{p}_1^*(\varphi) = \underline{p}_2^*(\varphi) \tag{56}$$

in $\mathrm{Hom}(\underline{r}^* A, \underline{r}^* B)$.

Proof The projection p admits a factorisation

$$Y = Y_0 \stackrel{p^{(1)}}{\longrightarrow} Y_1 \longrightarrow \ldots \stackrel{p^{(e-1)}}{\longrightarrow} Y_{e-1} = X$$

into j-dimensional projective fibrations $p^{(j)}$. By 1.7., $\underline{p}^{(j)*}$ is faithful, and *(a)* follows. Condition (54) in (b) is certainly necessary. If f exists we have $p^*([A]) = p^*([B])$ in $\mathrm{CH}^k(Y)$. By 1.6., p^* is injective on the Chow groups, and we conclude that there is a homomorphism $h: A \to B$. Then $f' = f - \underline{p}^*(h) \in G_k(Y)$ satisfies

$$p_1^*(f') = p_2^*(f') \quad \text{in } G_k(Y \times Y). \tag{57}$$

Let $0 = \mathcal{E}_0 \subset \mathcal{E}_1 \subset \ldots \subset \mathcal{E}_e = p^*\mathcal{E}$ be the universal flag of $p^*\mathcal{E}$, and let $\mathcal{L}_i = \mathcal{E}_i / \mathcal{E}_{i-1}$. Iterating 1.6., we have

$$G_k(Y) = \bigoplus_{j_1=0}^{e-1} \cdots \bigoplus_{j_{e-1}=0}^{1} c_1(\mathcal{L}_1)^{j_1} \cap \ldots \cap c_1(\mathcal{L}_{e-1})^{j_{e-1}} \cap p^* \left(G_{k-\sum j_i}(X)\right) \tag{58}$$

$$G_k(Y \times Y) = \bigoplus_{j_1; j_1'=0}^{e-1} \cdots \bigoplus_{j_{e-1}; j_{e-1}'=0}^{1} c_1(p_1^*\mathcal{L}_1)^{j_1} \cap c_1(p_2^*\mathcal{L}_1)^{j_1'} \cap \ldots$$
$$\ldots \cap c_1(p_1^*\mathcal{L}_{e-1})^{j_{e-1}} \cap c_1(p_2^*\mathcal{L}_{e-1})^{j_{e-1}'} \cap r^* \left(G_{k-\sum j_i - \sum j_i'}(X)\right). \tag{59}$$

If we represent f' in the form (58), then (59) implies that (57) is valid if and only is all components of f' are zero save for the component belonging to $(j_1, \ldots, j_{e-1}) = (0, \ldots, 0)$ in (58), i.e., if and only if $f' = p^*(g')$, and *(b)* follows.

1.13.3 Now we are ready to perform the induction argument announced at the beginning of 1.13.. Let $\dim(\mathcal{F}) = f > 1$, and assume that our claim (i.e., that (53) is independent of the filtration) has already been verified for bundles of dimension less than f. By part *(a)* of the splitting principle we may assume that $\mathcal{F}$ has a subbundle $\mathcal{H}$ of dimension $f - 1$. We consider the following commutative diagram, in which each arrow is in an obvious manner constructed from (42):

$$
\begin{array}{ccccc}
\mathfrak{c}.(\mathcal{G})\cap\mathfrak{c}.(\mathcal{F})\cap A & \xrightarrow{\alpha} & \mathfrak{c}.(\mathcal{L}_g)\cap\ldots\cap\mathfrak{c}.(\mathcal{L}_1)\cap\mathfrak{c}.(\mathcal{F})\cap A & \xrightarrow{\beta} & \mathfrak{c}.(\mathcal{E}\cap A) \\
\Big\downarrow{\scriptstyle\gamma} & & \Big\downarrow & & \Big\uparrow{\scriptstyle\zeta} \\
\begin{array}{c}\mathfrak{c}.(\mathcal{G})\cap\mathfrak{c}.(\mathcal{F}/\mathcal{H})\cap\\ \cap\mathfrak{c}.(\mathcal{H})\cap A\end{array} & \xrightarrow[\delta]{} & \mathfrak{c}.(\mathcal{L}_g)\cap\ldots\cap\mathfrak{c}.(\mathcal{L}_1)\cap\mathfrak{c}.(\mathcal{F}/\mathcal{H})\cap\mathfrak{c}.(\mathcal{H})\cap A & & \\
 & & \Big\uparrow{\scriptstyle\varepsilon} & & \\
 & & \mathfrak{c}.(\mathcal{E}/\mathcal{H})\cap\mathfrak{c}.(\mathcal{H})\cap A & &
\end{array}
$$

By the induction assumption, $\zeta\varepsilon$ is independent of the filtration $\mathcal{G}_i$. By the result of 1.13.1., the same is true about $\varepsilon^{-1}\delta$. It follows that $\beta\alpha$ is independent of the $\mathcal{G}_i$. Since $\beta\alpha$ is (53), we are through.

1.14 The Whitney isomorphism Let

$$\Sigma\colon 0 \to \mathcal{F} \to \mathcal{E} \xrightarrow{\pi} \mathcal{G} \to 0$$

be an exact sequence of vector bundles on X. We denote by $p\colon Y \to X$ the flag manifold of $\mathcal{G}$, by $p_{1,2}\colon Y \times_X Y \to Y$ the projections to the factors and put $r = pp_1 = pp_2$. Let $\mathcal{G}. = (0 = \mathcal{G}_0 \subset \mathcal{G}_1 \subset \ldots \subset \mathcal{G}_g = p^*\mathcal{G})$ be the universal flag of $\mathcal{G}$. It defines an isomorphism (53)

$$\Phi_{\mathcal{G}.}\colon \mathfrak{c}.(p^*\mathcal{G}) \cap \mathfrak{c}.(p^*\mathcal{F}) \cap \underline{p}^* A \to \mathfrak{c}.(p^*\mathcal{E}) \cap \underline{p}^* A.$$

Since (as one proves easily) (53) is compatible with flat base change, we have

$$\underline{p}_i^*(\Phi_{\mathcal{G}.}) = \Phi_{p_i^*\mathcal{G}.}\colon \mathfrak{c}.(r^*\mathcal{G}) \cap \mathfrak{c}.(r^*\mathcal{F}) \cap \underline{r}^* A \to \mathfrak{c}.(r^*\mathcal{E}) \cap \underline{r}^* A.$$

By the main result of 1.13., this is independent of $i \in \{1,2\}$. By part *(b)* of the splitting principle 1.13.2., we conclude that there exists a unique

$$\Phi_\Sigma\colon \mathfrak{c}.(\mathcal{G}) \cap \mathfrak{c}.(\mathcal{F}) \cap A \to \mathfrak{c}.(\mathcal{E}) \cap A \tag{60}$$

with $\Phi_{\mathcal{G}} = \underline{p}^*(\Phi_\Sigma)$. If $\mathcal{G}$ is a line bundle, we have $Y = X$ and (60) coincides with (42).

If $\mathcal{G}$ has a flag $0 = \Gamma_0 \subset \Gamma_1 \subset \ldots \subset \Gamma_g = \mathcal{G}$ with one-dimensional quotients $\Lambda_i = \Gamma/\Gamma_{i-1}$, then the diagram

$$
\begin{array}{ccc}
\mathfrak{c}.(\mathcal{G}) \cap \mathfrak{c}.(\mathcal{F}) \cap A & \xrightarrow{\alpha} & \mathfrak{c}.(\Lambda_g) \cap \ldots \cap \mathfrak{c}.(\Lambda_1) \cap \mathfrak{c}.(\mathcal{E}) \cap A \\
& {\scriptstyle \Phi_\Sigma} \searrow \quad \swarrow {\scriptstyle \beta} & \\
& \mathfrak{c}.(\mathcal{E}) \cap A &
\end{array}
\qquad (61)
$$

commutes. Indeed, $\beta\alpha = \Phi_{\Gamma.}$, hence $\underline{p}^*(\beta\alpha) = \Phi_{p^*\Gamma.} = \Phi_{\mathcal{G}.} = \underline{p}^*(\Phi_\Sigma)$ by the main result of 1.13., and (61) follows from the splitting principle.

Let $0 \subset \mathcal{G} \subset \mathcal{F} \subset \mathcal{E}$ be a filtration of $\mathcal{E}$. Then the diagram

$$
\begin{array}{ccc}
\mathfrak{c}.(\mathcal{E}) \cap A & \longrightarrow & \mathfrak{c}.(\mathcal{E}/\mathcal{F}) \cap \mathfrak{c}.(\mathcal{F}) \cap A \\
\downarrow & & \downarrow \\
\mathfrak{c}.(\mathcal{E}/\mathcal{G}) \cap \mathfrak{c}.(\mathcal{G}) \cap A & \longrightarrow & \mathfrak{c}.(\mathcal{E}/\mathcal{F}) \cap \mathfrak{c}.(\mathcal{F}/\mathcal{G}) \cap \mathfrak{c}.(\mathcal{G}) \cap A
\end{array}
\qquad (62)
$$

commutes. By the splitting principle, it suffices to prove (62) in the case that $\mathcal{E}/\mathcal{F}$ and $\mathcal{F}/\mathcal{G}$ have flags with one-dimensional quotients, in which (62) follows from (61).

It is easy to see that (64) is compatible with isomorphisms $\Sigma \to \Sigma'$ of short exact sequences and with flat and proper base-changes $Y \to X$. By (61) and the splitting principle it is possible to extend (45) to the case $\dim(\mathcal{L}) \geq 1$.

Let $\mathcal{E}$ and $\mathcal{F}$ be vector bundles on X. We have sequences

$$
\begin{aligned}
\Sigma_1 &: 0 \to \mathcal{E} \to \mathcal{E} \oplus \mathcal{F} \to \mathcal{F} \to 0 \\
\Sigma_2 &: 0 \to \mathcal{F} \to \mathcal{E} \oplus \mathcal{F} \to \mathcal{E} \to 0.
\end{aligned}
$$

The diagram

$$
\begin{array}{ccc}
\mathfrak{c}.(\mathcal{F}) \cap \mathfrak{c}.(\mathcal{E}) \cap A & \xrightarrow{\varphi_{\Sigma_1}} & \mathfrak{c}.(\mathcal{E} \oplus \mathcal{F}) \cap A \\
{\scriptstyle \sigma_{\mathcal{E},\mathcal{F}}} \downarrow & & \downarrow {\scriptstyle \tau.(\mathcal{E},\mathcal{F},A)} \\
\mathfrak{c}.(\mathcal{E}) \cap \mathfrak{c}.(\mathcal{F}) \cap A & \xrightarrow[\varphi_{\Sigma_2}]{} & \mathfrak{c}.(\mathcal{E} \oplus \mathcal{F}) \cap A
\end{array}
\qquad (63)
$$

commutes. This is a consequence of 1.12., (63) and the splitting principle.

1.15 If $i > 0$ and $A \in \mathrm{Ob}(\tilde{\mathfrak{CH}}^k(X))$, then $\mathfrak{c}_i(\mathcal{E}) \cap A \in \mathrm{Ob}(\mathfrak{CH}^{k+i}(X))$. For X has a Zariski covering on which $\mathcal{E}$ and hence $\mathfrak{c}_i(\mathcal{E}) \cap A$ are trivial, and we apply [F1, 3.8.].

1.16 *Let*

$$\begin{array}{ccc} Y & \xrightarrow{\quad i \quad} & X \\ & {\scriptstyle q} \searrow \quad \swarrow {\scriptstyle p} & \\ & S & \end{array}$$

be a commutative diagram with p and q flat and i a regular closed immersion of codimension one. If $\mathcal{E}$ is a vector bundle on S, then the diagram

$$\begin{array}{ccccc} \underline{i}_*\underline{q}^*(\mathfrak{c}_k(\mathcal{E}) \cap A) & \longrightarrow & \underline{i}_*(\mathfrak{c}_k(q^*\mathcal{E}) \cap \underline{q}^* A) & \longrightarrow & \mathfrak{c}_k(p^*\mathcal{E}) \cap \underline{i}_*\underline{q}^* A \\ \downarrow & & & & \downarrow \\ \mathfrak{c}_1(\mathcal{O}(D)) \cap \underline{p}^*(\mathfrak{c}_k(\mathcal{E}) \cap A) & & & & \mathfrak{c}_k(p^*\mathcal{E}) \cap \mathfrak{c}_1(\mathcal{O}_X(D)) \cap \underline{p}^* A \\ & {\scriptstyle \alpha} \searrow & & \swarrow {\scriptstyle \beta} & \\ & & \mathfrak{c}_1(\mathcal{O}_X(D)) \cap \mathfrak{c}_k(p^*\mathcal{E}) \cap \underline{p}^* A & & \end{array} \qquad (64)$$

commutes, where the two vertical arrows are given by (39).

Proof If $\mathcal{E}$ is a line bundle, (64) coincides with (40). Let

$$0 \to \mathcal{F} \to \mathcal{E} \to \mathcal{G} \to 0$$

be an exact sequence such that (64) is true for $\mathcal{F}$ and $\mathcal{G}$. Then we have the diagram

$$
\begin{array}{ccc}
\underline{i}_*\underline{q}^*(\mathfrak{c}.(\mathcal{E}) \cap A) & \longrightarrow & \mathfrak{c}_1(\mathcal{O}_X(D)) \cap \underline{p}^*(\mathfrak{c}.(\mathcal{E}) \cap A) \\
\Big\downarrow & \text{NT} & \Big\downarrow \\
\underline{i}_*\underline{q}^*(\mathfrak{c}.(\mathcal{G}) \cap \mathfrak{c}.(\mathcal{F}) \cap A) & \longrightarrow & \mathfrak{c}_1(\mathcal{O}_X(D)) \cap \underline{p}^*(\mathfrak{c}.(\mathcal{G}) \cap \mathfrak{c}.(\mathcal{F}) \cap A) \\
\Big\downarrow & & \Big\downarrow \\
\underline{i}_*\left(\mathfrak{c}.(q^*\mathcal{G}) \cap \underline{q}^*(\mathfrak{c}.(\mathcal{F}) \cap A)\right) & \text{(A)} & \mathfrak{c}_1(\mathcal{O}_X(D)) \cap \mathfrak{c}.(p^*\mathcal{G}) \cap \underline{p}^*(\mathfrak{c}.(\mathcal{F}) \cap A) \\
\Big\downarrow & & \Big\downarrow \\
\mathfrak{c}.(p^*\mathcal{G}) \cap \underline{i}_*\underline{q}^*(\mathfrak{c}.(\mathcal{F}) \cap A) & \longrightarrow & \mathfrak{c}.(p^*\mathcal{G}) \cap \mathfrak{c}_1(\mathcal{O}(D)) \cap \underline{p}^*(\mathfrak{c}.(\mathcal{F}) \cap A) \\
\Big\downarrow & & \Big\downarrow \\
\mathfrak{c}.(p^*\mathcal{G}) \cap \underline{i}_*\left(\mathfrak{c}(q^*\mathcal{F}) \cap \underline{q}^*A\right) & \text{(B)} & \mathfrak{c}.(p^*\mathcal{G}) \cap \mathfrak{c}_1(\mathcal{O}(D)) \cap \mathfrak{c}.(p^*\mathcal{F}) \cap \underline{p}^*A \\
\Big\downarrow & & \Big\downarrow \\
\mathfrak{c}.(p^*\mathcal{G}) \cap \mathfrak{c}.(p^*\mathcal{F}) \cap \underline{i}_*\underline{q}^*A & \longrightarrow & \mathfrak{c}.(p^*\mathcal{G}) \cap \mathfrak{c}.(p^*\mathcal{F}) \cap \mathfrak{c}_1(\mathcal{O}(D)) \cap \underline{p}^*A \\
\Big\downarrow & \text{NT} & \Big\downarrow \\
\mathfrak{c}.(\mathcal{E}) \cap \underline{i}_*\underline{q}^*A & \longrightarrow & \mathfrak{c}.(p^*\mathcal{E}) \cap \mathfrak{c}_1(\mathcal{O}_X(D)) \cap \underline{p}^*A\,.
\end{array}
$$

(A) and (B) are (64) for $\mathcal{G}$ and $\mathcal{F}$. As we did in [F1], we used the label NT to denote squares which commute just because the arrows involved in them are natural transformations. By the biadmissibility of the Whitney isomorphisms $\mathfrak{c}.(\mathcal{E}) \cap A \to \mathfrak{c}.(\mathcal{G}) \cap \mathfrak{c}.(\mathcal{F}) \cap A$, the composition of the left column is the top row of (64). By the generalisation of (45), the composition of the right column is $\beta^{-1}\alpha$. It follows that (64) is true for $\mathcal{E}$. Thus it is possible to prove (64) by induction on $\dim(\mathcal{E})$, using the splitting principle.

2 Further Properties of the Chern functors.

2.1 Relation between $\mathfrak{c}_1$ and specialisation. Let $D \subset X$ be a closed subscheme on X whose sheaf of ideals is in some neighbourhood of D generated by f. For a line bundle $\mathcal{L}$ on X and $A \in \mathrm{Ob}(\tilde{\mathfrak{CH}}^k(X-D))$, we want to construct an isomorphism

$$\alpha_{\mathcal{L},f}\colon \mathfrak{c}_1(\mathcal{L}\big|_D) \cap \mathfrak{sp}_f(A) \to \mathfrak{sp}_f(\mathfrak{c}_1(\mathcal{L}\big|_{X-D}) \cap A). \tag{1}$$

To this end we fix a covering $\mathcal{U}\colon X = \bigcup_i U_i$ of X on which $\mathcal{L}$ is trivialised by non-vanishing sections ℓ_i. We denote by $\mathcal{U}\big|_{X-D}$, $\mathcal{U}\big|_D$ the coverings $D = \bigcup(D \cup U_i)$ and $X - D = \bigcup((X-) \cap U_u)$. Let $C^*(\mathcal{U}, E_1^{\cdot,-p})$ be the total Čech complex with differential $\mathfrak{d}$. The closed and exact Čech chains are denoted $Z^*(\mathcal{U}, E_1^{\cdot,-p})$ and $B^*(\mathcal{U}, E_1^{\cdot,-p})$. For $c \in Z^p(\mathcal{U}, E_1^{\cdot,-p})$ $\mathfrak{O}(c) \in \tilde{\mathfrak{CH}}^p(X)$ has been defined in 1.(1). In our situation, we have a homomorphism

$$sp_f\colon E_1^{p,q}(X-D) \to E_1^{p,q}$$

(cf. [F2,§1.5]). The induced homomorphism

$$sp_f\colon C^*(\mathcal{U}\big|_{X-D}) \to C^*(\mathcal{U}\big|_D)$$

turns easily out to be a homomorphism of complexes. Consequently we have an homomorphism

$$\mathfrak{sp}_f(\mathfrak{O}(c)) \to \mathfrak{O}(sp_f(c)) \tag{2}$$

for $c \in C^p(\mathcal{U}, E_1^{*,-p})$. If $V \subseteq (X-D)_p$ is an open subset and $a \in \mathfrak{O}(C)(V)$ is given by x as in 1.(1)., then (2) maps $sp_f(a) \in \mathfrak{sp}_f(\mathfrak{O}(c)(D - \overline{(X-D-V)})$ to the section of $\mathfrak{O}(sp_f(c))$ defined by the Čech cycle $sp_f(x) \in Z^p(\mathcal{U}\Big|_{D-\overline{(X-D-V)}})$.

Now we are ready to define the isomorphisms (1). If $a \in A_r(X-D)$ and ϕ_{ij} denotes the Čech cycle ℓ_i/ℓ_j, then

$$(\mathfrak{c}_1(\mathcal{L}) \cap A)_{\mathcal{U},\ell,a} = \mathfrak{O}(\{\phi_{ij}, \mathfrak{c}(a)\}).$$

Since $sp_f(\{\phi_{ij}, \mathfrak{c}(a)\}) = \{\phi_{ij}\big|_D, sp_f(\mathfrak{c}(a))\} = \{\phi_{ij}\big|_D, \mathfrak{c}(sp_f(a))\}$, (2) defines an isomorphism

$$\mathfrak{sp}_f(\mathfrak{c}_1(\mathcal{L} \cap A)_{\mathcal{U},l,a}) = \mathfrak{sp}_f(\mathfrak{O}(\{\phi_{ij}, \mathfrak{c}(a)\})) \to$$

$$\to \mathfrak{O}(\{\phi_{ij}\big|_D, \mathfrak{c}(sp_f(a))\}) = (\mathfrak{c}_1(\mathcal{L}\big|_D) \cap \mathfrak{sp}_f(A))_{\mathcal{U}|_D,\ell|_D,sp_f(a)}.$$

It is easy to see that these isomorphism are compatible with the isomorphisms for changing ℓ or a and refining $\mathcal{U}$ (cf. §1.2.). Consequently they define (1).

2.2 Relation between $\mathfrak{c}_k$ and specialisation. Let X, D and f be the same as before, and let $\mathcal{E}$ be a vector bundle of dimension e on X. We denote by $\mathbb{P}(\mathcal{E})$ the corresponding projective fibration and by $p: \mathbb{P}(\mathcal{E}) \to X$ the projection. If $A \in \tilde{\mathfrak{CH}}^p(X - D)$, then from the isomorphism

$$\bigoplus_{k=0}^{e} \mathfrak{c}_1(\mathcal{O}(1))^{e-k} \cap \underline{p}^*(\mathfrak{c}_k(\mathcal{E}) \cap A) \to 0$$

in $\tilde{\mathfrak{CH}}^{k+e}(\mathbb{P}(\mathcal{E}))$ we derive by (1) an isomorphism

$$\begin{aligned} &\bigoplus_{k=0}^{e} \mathfrak{c}_1(\mathcal{O}(1)\big|_D)^{e-k} \cap \underline{p}^*(\mathfrak{sp}_f(\mathfrak{c}_k(\mathcal{E}) \cap A)) \to \\ &\quad \to \bigoplus_{k=0}^{e} \mathfrak{c}_1(\mathcal{O}(1))\big|_D)^{e-k} \cap \mathfrak{sp}_{p^*(f)}(\underline{p}^*(\mathfrak{c}_k(\mathcal{E}) \cap A)) \to \\ &\quad \to \mathfrak{sp}_{p^*(f)}\Big(\bigoplus_{k=0}^{e} \mathfrak{c}_1(\mathcal{O}(1)\big|_{X-D})^{e-k} \cap \underline{p}^*(\mathfrak{c}_k(\mathcal{E}) \cap A))\Big) \to 0 \end{aligned}$$

Since $\mathfrak{sp}_f(\mathfrak{c}_0(\mathcal{E}) \cap A) = \mathfrak{sp}_f(A) = \mathfrak{c}_0(\mathcal{E}\big|_D) \cap \mathfrak{sp}_f(A)$, this isomorphism and the definition of the Chern functors in §1.8. give an isomorphism

$$\alpha_{\mathcal{E},f}: \mathfrak{sp}_f(\mathfrak{c}_k(\mathcal{E}) \cap A) \to \mathfrak{c}_k(\mathcal{E}\big|_D) \cap \mathfrak{sp}_f(A). \tag{3}$$

2.3 Properties of the isomorphism (3). The following properties are easily verified:

2.3.1 Compatibility with pull-back and push-forward. Let K_{sp} be the category defined in [F2,§3.13]. Let objects of $K_{sp} \setminus (X, D, f)$ (cf. [Q,§1] for a definition of that category) be denoted by $(q: Y \to X, q^{-1}(D), q^*(f))$, and let $\mathbb{K}_{sp} \setminus (X, D, f)$ be $(K_{sp} \setminus (X, D, f)$, flat morphisms, proper morphisms of constant relative dimension). Then $\tilde{\mathfrak{CH}}^{\cdot}(Y)$ and $\tilde{\mathfrak{CH}}^{\cdot}(q^{-1}(D))$ are bifibred over $\mathbb{K}_{sp} \setminus (X, D, f)$, and the functors

$$\begin{aligned} \mathfrak{sp}_{q^*(f)} &: \tilde{\mathfrak{CH}}^{\cdot}(Y) \to \tilde{\mathfrak{CH}}^{\cdot}(q^{-1}(D)) \\ \mathfrak{c}_k(q^*\mathcal{E}) \cap . &: \tilde{\mathfrak{CH}}^{\cdot}(Y) \to \tilde{\mathfrak{CH}}^{\cdot}(Y) \\ \mathfrak{c}_k(q^*\mathcal{E}\big|_D) \cap . &: \tilde{\mathfrak{CH}}^{\cdot}(q^{-1}(Y)) \to \tilde{\mathfrak{CH}}^{\cdot}(q^{-1}(Y)) \end{aligned}$$

are biadmissible. The property is that the isomorphism

$$\mathfrak{sp}_{q^*(f)}(\mathfrak{c}_k(q^*\mathcal{E}) \cap .) \to \mathfrak{c}_k(q^*\mathcal{E}\big|_D) \cap \mathfrak{sp}_{q^*(f)}$$

is biadmissible.

2.3.2 Compatibility with the Whitney sum isomorphism. If

$$0 \to \mathcal{F} \to \mathcal{E} \to \mathcal{G} \to 0$$

is an exact sequence of vector bundles on X, then the diagram

$$\begin{array}{ccc}
\mathfrak{c}.(\mathcal{F}\,|_D)\cap\mathfrak{c}.(\mathcal{G}\,|_D)\cap\mathfrak{sp}_f(A) & \longrightarrow & \mathfrak{c}.(\mathcal{E}\,|_D)\cap\mathfrak{sp}_f(A) \\
\downarrow & & \\
\mathfrak{c}.(\mathcal{F}\,|_D)\cap\mathfrak{sp}_f(\mathfrak{c}.(\mathcal{G})\cap A) & & \downarrow \\
\downarrow & & \\
\mathfrak{sp}_f(\mathfrak{c}.(\mathcal{F})\cap\mathfrak{c}.(\mathcal{G})\cap A) & \longrightarrow & \mathfrak{sp}_f(\mathfrak{c}(\mathcal{E})\cap A)
\end{array} \qquad (4)$$

commutes.

2.3.3 Let $D_i \subset X$ ($i \in \{1;2\}$) be regular closed immersions of codimension one, with sheaf of ideals trivialised by f_i. We assume that the sequence $\{f_1; f_2\}$ is regular in a neighbourhood of $D_1 \cap D_2$. If $\mathcal{E}$ is a vector bundle on X and $A \in \mathrm{Ob}(\tilde{\mathfrak{CH}}^{l}(X - D_1 - D_2))$, then the diagram

$$\begin{array}{ccc}
\mathfrak{sp}_{f_1}(\mathfrak{sp}_{f_2}(\mathfrak{c}_k(\mathcal{E})\cap A)) & \longrightarrow & \mathfrak{sp}_{f_2}(\mathfrak{sp}_{f_1}(\mathfrak{c}_k(\mathcal{E})\cap A)) \\
\downarrow & & \downarrow \\
\mathfrak{sp}_{f_1}(\mathfrak{c}_k(\mathcal{E}\,|_{D_2})\cap\mathfrak{sp}_{f_2}(A)) & & \mathfrak{sp}_{f_2}(\mathfrak{c}_k(\mathcal{E}\,|_{D_1})\cap\mathfrak{sp}_{f_1}(A)) \\
\downarrow & & \downarrow \\
\mathfrak{c}_k(\mathcal{E}\,|_{D_1\cap D_2})\cap\mathfrak{sp}_{f_1}(\mathfrak{sp}_{f_2}(A)) & \longrightarrow & \mathfrak{c}_k(\mathcal{E}\,|_{D_1\cap D_2})\cap\mathfrak{sp}_{f_2}(\mathfrak{sp}_{f_1}(A))
\end{array} \qquad (5)$$

commutes. The horizontal arrows have been defined in [F,§3.15].

2.3.4 If in the commutative triangle

$$\begin{array}{ccc} D & \longrightarrow & X \\ & p \searrow \quad \swarrow q & \\ & Z & \end{array}$$

p and q are flat and $\mathcal{E}$ is a vector bundle on Z, then the diagram

$$\begin{array}{ccc} \mathfrak{sp}_f\left(\underline{q}^*(\mathfrak{c}_k(\mathcal{E}) \cap A)\right) & \longrightarrow & \underline{p}^*\mathfrak{c}_k(\mathcal{E}) \cap A \\ \downarrow & & \\ \mathfrak{sp}_f(\mathfrak{c}_k(q^*\mathcal{E}) \cap \underline{q}^*A) & & \downarrow \\ \downarrow {\scriptstyle \alpha_{f,q^*\mathcal{E}}} & & \\ \mathfrak{c}_k(p^*\mathcal{E}) \cap \mathfrak{sp}_f \underline{q}^*A & \longrightarrow & \mathfrak{c}_k(p^*\mathcal{E}) \cap \underline{p}^*A \end{array} \qquad (6)$$

commutes.

2.3.5 Let $\mathcal{L}$ be a line bundle on X, $A \subset (\tilde{\mathfrak{CH}}^k(X-D))$, $a \in A_r(X-D)$. We assume that ℓ is a rational section of $\mathcal{L}$ on X whose divisor meets $\mathfrak{c}(a)$, D, and $D \cap \mathrm{supp}(\mathfrak{c}(a))$ properly. Then

$$\alpha_{\mathcal{L},\lambda}(sp_\lambda(\ell \cap a)) = \ell\big|_D \cap sp_\lambda(a) \in (\mathfrak{c}_1(\mathcal{L}\big|_D) \cap \mathfrak{sp}_\lambda A)_r(D).$$

2.4 Relation between $\mathfrak{c}_k$ and $f^!$.

Proposition: *Let $\underline{f}^!$ be the Gysin functor constructed in [F1,§4]. There exists a unique collection of isomorphisms*

$$\beta_{f,\mathcal{E}}: \underline{f}^!(\mathfrak{c}_k(\mathcal{E}) \cap A) \to \mathfrak{c}_k(f^*\mathcal{E}) \cap \underline{f}^!A \qquad (7)$$

for a local complete intersection morphism $f: X \to Y$ which admits an immersion into a smooth Y-scheme (abbreviated: an slci-morphism $f: X \to Y$), a vector bundle $\mathcal{E}$ on Y, and $A \in \tilde{\mathfrak{CH}}^k(Y)$ such that the following properties are satisfied:

2.4.1 Compatibility with pull-back and push-forward. *Let S be a scheme and $\mathcal{E}$ be a vector bundle on S. Let $\mathbf{K}_{\mathrm{lci},S}$ be defined by replacing "scheme" by "S-scheme" in the definition of $\mathbf{K}_{\mathrm{lci}}$ (cf. [F1,§4.7.]). If objects of this bicategory are denoted $f: X \to Y$, then $\mathcal{E}_X$ and $\mathcal{E}_Y$ refer to the pull-backs of $\mathcal{E}$ to X and Y. Then*

$$\mathfrak{c}_k(\mathcal{E}_X)\cap . \;:\; \tilde{\mathfrak{CH}}^{\cdot}(X) \to \tilde{\mathfrak{CH}}^{\cdot}(X),$$
$$\mathfrak{c}_k(\mathcal{E}_Y)\cap . \;:\; \tilde{\mathfrak{CH}}^{\cdot}(Y) \to \tilde{\mathfrak{CH}}^{\cdot}(Y),$$

and

$$\underline{f}^! \;:\; \tilde{\mathfrak{CH}}^{\cdot}(Y) \to \tilde{\mathfrak{CH}}^{\cdot}(X)$$

are biadmissible functors between bifibred Picard categories over $\mathbf{K}_{\mathrm{lci},S}$. The condition is that

$$\beta_{f,\mathcal{E}_Y}: \underline{f}^!(\mathfrak{c}_k(\mathcal{E}_Y)\cap -) \to \mathfrak{c}_k(\mathcal{E}_X)\cap \underline{f}^!$$

is a biadmissible functor-isomorphism.

2.4.2 Compatibility with composition. *If $X \xrightarrow{f} Y \xrightarrow{g} Z$ are lci-morphisms such that g and gf (and hence f too) are slci, then the diagram*

$$\begin{array}{ccc}
\underline{f}^!\underline{g}^!(\mathfrak{c}_k(\mathcal{E})\cap A) & \longrightarrow & \underline{(gf)}^!(\mathfrak{c}_k(\mathcal{E})\cap A) \\
\downarrow & & \\
\underline{f}^!(\mathfrak{c}_k(g^*\mathcal{E})\cap \underline{g}^! A) & & \downarrow \\
\downarrow & & \\
\mathfrak{c}_k(f^*g^*\mathcal{E})\cap \underline{f}^!\underline{g}^! A & \longrightarrow & \mathfrak{c}_k\Big((gf)^*\mathcal{E}\Big)\cap \underline{(gf)}^! A
\end{array} \qquad (8)$$

commutes for every vector bundle $\mathcal{E}$ on Z.

2.4.3 Compatibility with specialisation. *Let (f, X, Y, D, λ) be an object of $K_{\mathrm{lci},sp}$ (cf. [F2,§4.7.]). It is given by a Cartesian diagram*

$$\begin{array}{ccc}
f^{-1}(D) & \subset & X \\
f_D \downarrow & & \downarrow f \\
D & \subset & Y
\end{array}$$

and a function λ in a neighbourhood of D defining D. If $\mathcal{E}$ is a vector bundle on Y and $A \in \mathrm{Ob}(\tilde{\mathfrak{CH}}^l(Y))$, then the diagram

$$
\begin{array}{ccc}
\mathfrak{sp}_{f^*(\lambda)}\underline{f}^!(\mathfrak{c}_k(\mathcal{E})\cap A) & \longrightarrow & \underline{f}_D^!(\mathfrak{sp}_\lambda(\mathfrak{c}_k(\mathcal{E})\cap A)) \\
{\scriptstyle \mathfrak{sp}_{f^*(\lambda)}(\beta_{f,\mathcal{E}})}\big\downarrow & & \big\downarrow{\scriptstyle \underline{f}^!(\alpha_{\mathcal{E},\lambda})} \\
\mathfrak{sp}_{f^*(\lambda)}\left(\mathfrak{c}_k(f^*\mathcal{E})\cap \underline{f}^! A\right) & & \underline{f}_D^!\left(\mathfrak{c}_k(\mathcal{E}\,|_D)\cap \mathfrak{sp}_\lambda A\right) \\
{\scriptstyle \alpha_{f^*\mathcal{E},f^*(\lambda)}}\big\downarrow & & \big\downarrow{\scriptstyle \beta_{f_D,\mathcal{E}|_D}} \\
\mathfrak{c}_k(f_D^*\mathcal{E}\,|_D)\cap \mathfrak{sp}_{f^*(\lambda)}\underline{f}^! A & \longrightarrow & \mathfrak{c}_k(f_D^*\mathcal{E}\,|_D)\cap \underline{f}_D^!\mathfrak{sp}_\lambda(A)
\end{array}
\qquad (9)
$$

commutes.

2.4.4 *If in*

$$
\begin{array}{ccc}
X & \xrightarrow{\ f\ } & Y \\
 & {\scriptstyle q}\searrow \quad \swarrow{\scriptstyle p} & \\
 & Z &
\end{array}
$$

p and q are flat and f is slci, *then the diagram*

$$
\begin{array}{ccc}
\underline{f}^!\underline{p}^*\left(\mathfrak{c}_k(\mathcal{E})\cap A\right) & \longrightarrow & \underline{q}^*\left(\mathfrak{c}_k(\mathcal{E})\cap A\right) \\
\downarrow & & \\
\underline{f}^!\left(\mathfrak{c}_k(p^*\mathcal{E})\cap \underline{p}^*A\right) & & \downarrow \\
\downarrow{\scriptstyle \beta_{f,p^*\mathcal{E}}} & & \\
\mathfrak{c}_k(q^*\mathcal{E})\cap \underline{f}^!\underline{p}^*A) & \longrightarrow & \mathfrak{c}_k(q^*\mathcal{E})\cap \underline{q}^*A
\end{array}
\qquad (10)
$$

commutes for every vector bundle $\mathcal{E}$ on Z. Note that the long vertical arrow in (10) is given by [F1,4.7.1].

2.5 Proof of 2.4. We proceed in four steps. In 2.5.1.–3. we prove that $\beta_{i,\mathcal{E}}$ exists and is unique for regular immersions i. In 2.5.4., we extend this to the general case.

2.5.1 Let $i: X_0 \to X_1$ be a regular closed immersion. We denote by $m: M_0 \to M_1$ its deformation to the normal bundle (cf. [Fu,§5]). This is a regular closed immersion

$$
\begin{array}{ccc}
M_0 & \xrightarrow{\ m\ } & M_1 \\
\downarrow{\scriptstyle \pi_0} & & \downarrow{\scriptstyle \pi_1} \\
X_0 \times \mathbb{P}^1 & \xrightarrow[i \times \mathbb{P}^1]{} & X_1 \times \mathbb{P}^1 .
\end{array}
$$

The following properties are satisfied (cf. [F1, §4.2.]):

(i) π_0 is an isomorphism. Let the superscript $^{(a)}$ denote the restriction of morphisms with source M_i to $M_i^{(a)} = \pi_i^{-1}(X \times \mathbb{A}^1)$. Then $\pi_1^{(a)}: M_1^{(a)} \to X_1 \times \mathbb{A}^1$ is an isomorphism.

(ii) Let p_i denote the composition $M_i \xrightarrow{\pi_i} X_i \times \mathbb{P}^1 \to X_i$.

(iii) Let the superscript $^{(\infty)}$ denote the restriction of morphisms with source

M_i to $M_i^{(\infty)} = \pi_i^{-1}(X \times \infty) \subset M_i$. Then $p_1^{(\infty)}$ factors over a map $p_\infty: M_1^{(\infty)} \to X_0$, and p_∞ is the projection of a vector bundle with zero section $m^{(\infty)}: M_0^{(\infty)} \to M_1^{(\infty)}$. Hence $\underline{p}_\infty^*: \tilde{\mathfrak{CH}}^{\cdot}(X_0) \to \tilde{\mathfrak{CH}}^{\cdot}(M_1^{(\infty)})$ is an equivalence of categories.
(iv) The formation of M is compatible with any base change $Y_1 \to X_1$ after which i remains regular of the same codimension.

Let $\lambda \in \Gamma(\mathbb{P}^1 - \{0\}, \mathcal{O}_{\mathbb{P}^1})$ be the inverse of the coordinate function. For the sake of simplicity it is denoted by the same letter λ for all projective lines over an arbitrary scheme. There is a canonical isomorphism

$$\begin{aligned} \underline{i}^! A \to \mathfrak{sp}_\lambda \underline{p}_0^{(a)*} \underline{i}^! A &\to \mathfrak{sp}_\lambda \underline{m}^{(a)!} \underline{p}_1^{(a)*} A \\ &\to \underline{m}^{(\infty)!} \mathfrak{sp}_\lambda \underline{p}_1^{(a)*} A \\ &\to (\underline{p}_\infty^*)^{-1} \mathfrak{sp}_\lambda \underline{p}_1^{(a)*} A. \end{aligned} \tag{11}$$

(cf. [F2,§4.4]) for $A \in \mathrm{Ob}(\tilde{\mathfrak{CH}}^{\cdot}(X_1))$. For a vector bundle $\mathcal{E}$ on X_1, we define $\beta_{i,\mathcal{E}}$ by the composition

$$\begin{aligned} \underline{i}^! [\mathfrak{c}_k(\mathcal{E}) \cap A] &\to (\underline{p}_\infty^*)^{-1} (\mathfrak{sp}_\lambda (\underline{p}_1^{(a)*} (\mathfrak{c}_k(\mathcal{E}) \cap A))) \to \\ &\to (\underline{p}_\infty^*)^{-1} (\mathfrak{sp}_\lambda (\mathfrak{c}_k(p_1^* \mathcal{E}) \cap \underline{p}_1^{(a)*} A) \overset{\alpha_{\lambda, p_1^* \mathcal{E}}}{\longrightarrow} \\ &\to (\underline{p}_\infty^*)^{-1} (\mathfrak{c}_k(p_\infty^* \mathcal{E} \mid_{X_0}) \cap \mathfrak{sp}_\lambda (\underline{p}_1^{(a)*} A)) \to \\ &\to \mathfrak{c}_k(\mathcal{E} \mid_{X_0}) \cap (\underline{p}_\infty^*)^{-1} (\mathfrak{sp}_\lambda \underline{p}_1^{(a)*} A) \to \\ &\to \mathfrak{c}_k(\mathcal{E} \mid_{X_0}) \cap \underline{i}^! A. \end{aligned} \tag{12}$$

By applying 2.4.1., (9), and (10) to the isomorphism in (11), we see that a system of isomorphisms $\beta_{i,\mathcal{E}}$ satisfying 2.4.1., 2.4.3., and 2.4.4. for regular closed immersions must be given by (12). Conversely, since (12) contains only transformations compatible with flat and proper base change and with specialisation, 2.4.1. and 2.4.3. are consequences of (iv). 2.4.4. follows from 2.3.4. by an easy computation.

2.5.2 It remains to prove that $\beta_{i,\mathcal{E}}$ satisfies 2.4.2. in the case of regular closed immersions. First we prove (8) in the following case: A and B are the bundle spaces of vector bundles $\mathcal{A}$ and $\mathcal{B}$ on X, $f: X \to A$ is the zero section, $g: A \to B$ is an injective homomorphism of vector bundles, and $\mathcal{E} = r^* \mathcal{F}$, where $r: B \to X$ is the bundle projection and $\mathcal{F}$ is a vector bundle on X.

Without loosing generality we may assume that there is a projection $p: B \to A$ of vector bundles. Otherwise we consider the X-scheme

$$\pi: Z = \{\text{ projections from } \mathcal{A} \text{ to } \mathcal{B}\} \to X,$$

which is a principal homogeneous space for the vector bundle $Hom_{\mathcal{O}_X}(\mathcal{B}/\mathcal{A}, \mathcal{A})$. Since $\pi^*: \tilde{\mathfrak{CH}}^{\cdot}(X) \to \tilde{\mathfrak{CH}}^{\cdot}(Z)$ is an equivalence of categories, it suffices to verify (8) after base-change to Z, where the desired projection p exists.

Now we consider the projections $B \xrightarrow{p} A \xrightarrow{q} X$. Then $\underline{f}^! \cong (\underline{q}^*)^{-1}$, $\underline{q}^! \cong (\underline{p}^*)^{-1}$, $\underline{(gf)}^! \cong ((\underline{qp})^*)^{-1}$, and the diagrams

$$
\begin{array}{ccc}
\underline{f}^!\underline{g}^! & \longrightarrow & (\underline{q}^*)^{-1}(\underline{p}^*)^{-1} \\
\downarrow & & \downarrow \\
\underline{(gf)}^! & \longrightarrow & ((\underline{pq})^*)^{-1}
\end{array}
\qquad
\begin{array}{ccc}
\underline{f}^!(\mathfrak{c}_k(\mathcal{E}) \cap .) & \longrightarrow & \mathfrak{c}_k(f^*\mathcal{E}) \cap \underline{f}^!(.) \\
\downarrow & & \downarrow \\
(\underline{q}^*)^{-1}(\mathfrak{c}_k(q^*.) \cap .) & & \mathfrak{c}_k(\mathcal{F}) \cap (\underline{q}^*)^{-1}
\end{array}
$$

commute (for the right one, this is 2.4.4.). Since the analogues of the right diagram for g and gf are also commutative, our claim follows from the properties of the isomorphisms 1.(38).

2.5.3 To prove (8) in the case of arbitrary regular immersions $X_0 \xrightarrow{f} X_1 \xrightarrow{g} X_2$, we consider the deformation to the normal bundle

$$
\begin{array}{ccccc}
M_0 & \xrightarrow{m_0} & M_1 & \xrightarrow{m_1} & M_2 \\
\downarrow{\scriptstyle \pi_0} & & \downarrow{\scriptstyle \pi_1} & & \downarrow{\scriptstyle \pi_2} \\
X_0 \times \mathbb{P}^1 & \longrightarrow & X_1 \times \mathbb{P}^1 & \longrightarrow & X_2 \times \mathbb{P}^1
\end{array}
$$

with the following properties (cf. [F2,§4.2.]):
(i) π_0 is an isomorphism, and if the superscript $^{(a)}$ denotes restriction of morphisms to $M_i^{(a)} = \pi_i^{-1}(X_i \times \mathbb{A}^1)$, then $\pi_i^{(a)}: M_i^{(a)} \to X_i \times \mathbb{A}^1$ is an isomorphism.
(ii) We denote by p_i the projections $M_i \xrightarrow{\pi_i} X_i \times \mathbb{P}^1 \to X_i$.
(iii) 2.5.2. is applicable to the composition

$$M_0^{(\infty)} \xrightarrow{m_0^{(\infty)}} M_1^{(\infty)} \xrightarrow{m_1^{(\infty)}} M_2^{(\infty)}.$$

Let λ be the same as in 2.5.1. By the construction of the isomorphism $\underline{f}^!\underline{g}^! \to \underline{(gf)}^!$ in [F2,§4.8.], the diagram

$$\begin{array}{ccc}
\underline{f}^!\underline{g}^!A & \longrightarrow \mathfrak{sp}_\lambda \underline{p}_0^{(a)*}\underline{f}^!\underline{g}^!A \longrightarrow & \underline{m}_0^{(\infty)!}\mathfrak{sp}_\lambda \underline{p}_1^{(a)*}\underline{g}^!A \\
\downarrow & & \downarrow \\
\underline{(gf)}^!A & & \underline{m}_0^{(\infty)!}\underline{m}_!^{(\infty)!}\mathfrak{sp}_\lambda \underline{p}_2^{(a)*}A \\
\downarrow & & \downarrow \\
\mathfrak{sp}_\lambda \underline{p}_0^{(1)*}\underline{(gf)}^!A & \longrightarrow & (\underline{m}_0^{(\infty)}\underline{m}_1^{(\infty)})^!\mathfrak{sp}_\lambda \underline{p}_2^{(a)*}A
\end{array}$$

commutes. Using this, we can deduce (8) from 2.5.2., 2.4.1., and 2.4.3..

2.5.4 We have proven that $\beta_{i,\mathcal{E}}$ exists and is unique for regular immersions i. Let $f: X \to Y$ be an slci-morphism, and let $\mathcal{E}$ be a vector bundle on Y. We choose a factorisation of f

$$\sigma: X \xrightarrow{\iota} S \xrightarrow{p} Y,$$

where p is smooth. Then $\underline{f}^! \cong \underline{f}_\sigma^! = \underline{i}^!\underline{p}^*$ (cf. [F2,§4.10.]). We define $\beta_{f,\mathcal{E}}$ by

$$\begin{aligned}
\underline{i}^!\underline{p}^*\mathfrak{c}_k(\mathcal{E}) \cap A \quad &\to \quad \underline{i}^!(\mathfrak{c}_k(p^*\mathcal{E}) \cap \underline{p}^*A) \\
&\xrightarrow{\beta_{i,p^*\mathcal{E}}} \quad \mathfrak{c}_k(f^*\mathcal{E}) \cap \underline{i}^!\underline{p}^*A. \qquad (13)
\end{aligned}$$

By 2.4.4., 2.4.2., and our result about the uniqueness of $\beta_{i,p^*\mathcal{E}}$, a system of isomorphisms $\beta_{f,\mathcal{E}}$ satisfying 2.4.1.-4. must be given by (13) if it exists.

Our first task is to prove that (13) is independent of σ. This follows from 2.4.4. (applied in the case of regular immersions) and the construction of the "change of factorisation" isomorphism $\underline{f}_{\sigma_1}^! \to \underline{f}_{\sigma_2}^!$ in [F2,§4.9–10]. Now 2.4.1., 2.4.3., and 2.4.4 can immediately be reduced to the case of regular immersions.

The proof of (8) can be split up into the following four cases:
(α) f and g are regular closed immersions. This case has already been dealt

with.
(β) f is a regular closed immersion, and g is smooth. This case follows from definition (13) and [F2,§4.12.,Sublemma 1]. We note that this is the only case of 2.4.2. which does not follow from the other points of 2.4..
(γ) f and g are smooth. This case follows immediately from (13).
(δ) f is smooth, and $g = i$ is a regular immersion. By our assumption, it factors over a smooth Z-scheme S. Consider the diagram

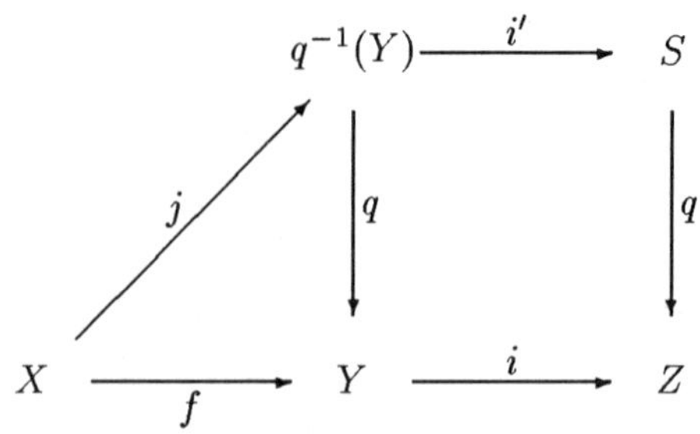

The square is Cartesian; i, i', and j are regular immersions, p, q, and f are smooth. By the definition of the isomorphism (?): $\underline{f}^! \underline{g}^! \to \underline{(gf)}^!$ in [F2,§4.11.], the following diagram commutes:

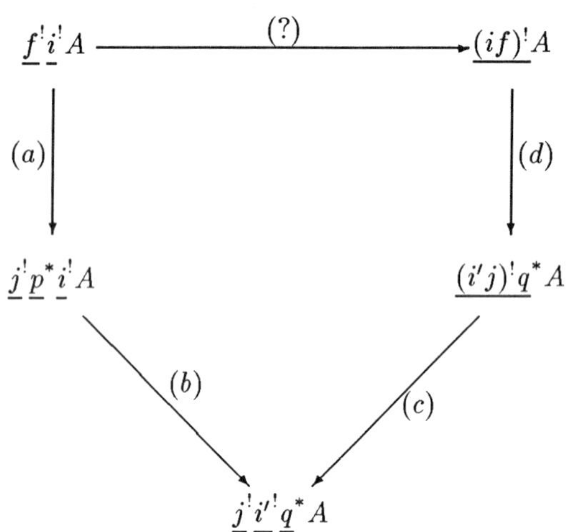

The compatibility of the isomorphisms $\beta_{\cdot,\cdot *\mathcal{E}}$ with the arrows (a), (b), (c), (d) follows from case (β), 2.4.1., case (α), and case (β). It follows that these isomorphisms are compatible with (?), which is (11). The proof of 2.4 is complete.

2.6 For our axiomatic characterisation of Chern functors we need some further properties of the isomorphisms $\beta_{f,\mathcal{E}}$.

2.6.1 Let $i: Y \to X$ be a regular closed immersion, $\mathcal{L}$ a line bundle on X, and $A \in \tilde{\mathfrak{CH}}^{k}(X)$. We assume that a and ℓ are rational sections of A and $\mathcal{L}$ in X such that $\mathfrak{c}(a)$, $\operatorname{div}(\ell)$ and Y meet properly. Then $\ell \cap a \in (\mathfrak{c}_1(\mathcal{L}) \cap A)_r(X)$, $i^!(\ell \cap a) \in (\mathfrak{c}_1(\mathcal{L}) \cap A)_r(Y)$, and $\underline{i}^!(\mathfrak{c}_1(\mathcal{L}) \cap A) \to \mathfrak{c}_1(i^*\mathcal{L}) \cap \underline{i}^! A$ maps $i^!(\ell \cap a)$ to $i^*(\ell) \cap i^!(a)$.

Proof: *Step 1:* First we assume that we are in the following situation
- X is a vector bundle over Y, with bundle projection p, and i corresponds to the zero section of that bundle.
- $\mathcal{L} = p^*\mathcal{L}_1$, $\ell = p^*(\ell_1)$, $A = \underline{p}^* A_1$, $a = p^*(a_1)$ for some $\mathcal{L}_1$ and A_1 on Y.
Then the claim follows from 2.4.4.

Step 2: In the general case we consider the deformation to the normal bundle

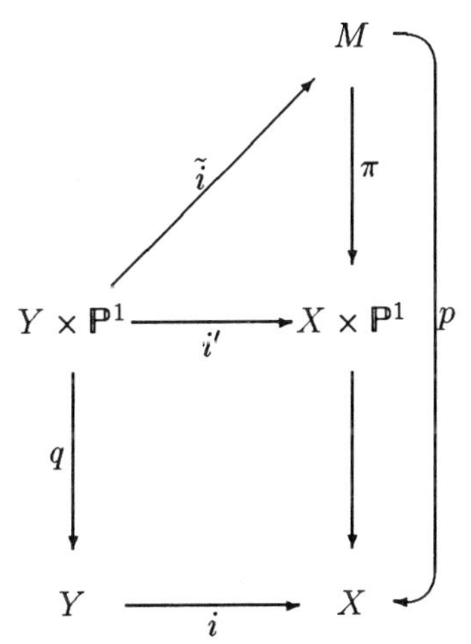

$M^{(\infty)} = \pi^{-1}(\infty)$, $M^{(a)} = \pi^{-1}(X \times \mathbb{A}^1)$, superscripts $^{(a)}$ or $^{(\infty)}$ = restriction to $M^{(a)}$ or $M^{(\infty)}$ and denote by λ a coordinate function on $\mathbb{P}^1$ as in 2.5.1.. We consider the rational section $sp_\lambda(p^{(a)*}(a))$ of $\mathfrak{sp}_\lambda \underline{p}^{(a)*} A$. In the following computation we will use the canonical isomorphism $\mathfrak{sp}_\lambda \underline{q}^{(a)*} A \simeq A$ without warning. By the axioms of 2.4., we have

$$
\begin{aligned}
\beta_{i,\mathcal{L}}(i^!(\ell \cap a)) &= sp_\lambda p^{(a)*}(\beta_{i,\mathcal{L}}(i^!(\ell \cap a))) \qquad (14)\\
&= sp_\lambda \beta_{\tilde{i},(p^{(a)*}\mathcal{L})}(\tilde{i}^!(p^{(a)*}\ell \cap p^{(a)*}(a)))\\
&= \beta_{\tilde{i}^{(\infty)},p^{(\infty)*}\mathcal{L}}(\tilde{i}^{(\infty)!}p^{(\infty)*}(\ell) \cap \tilde{i}^{\infty !} sp_\lambda p^{(a)*} a).
\end{aligned}
$$

We have used 2.4.1. in line 2 and 2.4.3. and 2.3.4. in line 3. Now we note that $sp_\lambda p^{(a)*}(a) = p^{(\infty)*}i^!(a)$, where $p^{(\infty)}: M \to Y$ is the restriction of p. This allows us to apply step 1, hence (14) is equal to

$$i^*\ell \cap \tilde{i}^{(\infty)!}(sp_\lambda p^{(a)*}(a)) = i^*(\ell) \cap i^!(a),$$

where the last equality holds for similar reasons as in (14). This proves 2.6.1..

2.6.2 Let

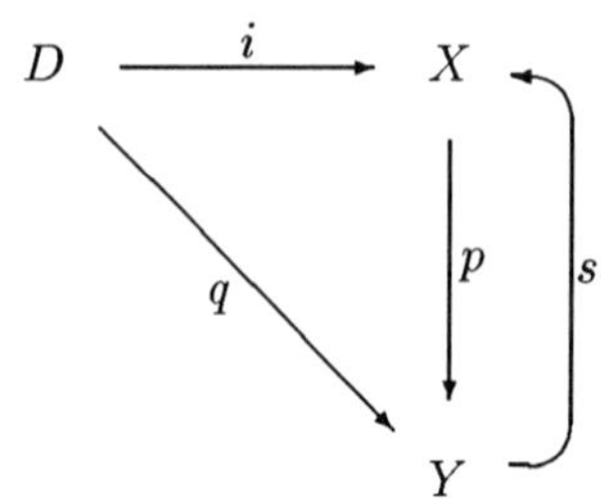

(15)

be a commutative diagram with p and q smooth, i a regular immersion of codimension one, and s a section of p. We assume that $s^{-1}(D)$ is regular of codimension one in Y and denote by $i_Y: s^{-1}(D) \to Y$, $s_D: s^{-1}(D) \to D$ the restrictions of i and s. Furthermore we assume that there is a flat map $r: Y \to Z$ whose restriction r_D to D remains flat:

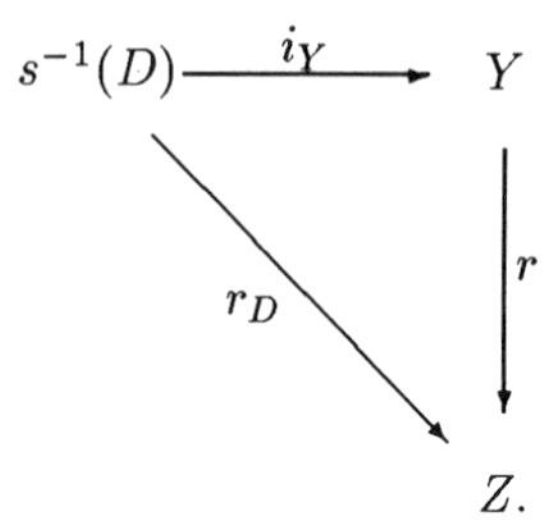

Note that there are isomorphisms

$$\begin{aligned} \mathfrak{c}_1(\mathcal{O}(D)) \cap \underline{p}^*\underline{r}^*A &\to \underline{i}_*\underline{q}^*\underline{r}^*A \\ \mathfrak{c}_1(\mathcal{O}(s^{-1}(D))) \cap \underline{r}^*A &\to \underline{i}_{Y*}\underline{r}_D^*A. \end{aligned}$$

(cf. equation 1.(39).). We assert that the diagram

$$
\begin{array}{ccc}
\underline{s}^!(\mathfrak{c}_1(\mathcal{O}(D)) \cap \underline{p}^*\underline{r}^* A) & \longrightarrow & \underline{s}^!\underline{i}_*\underline{q}^*\underline{r}^* A \\
\downarrow & & \downarrow \\
\mathfrak{c}_1(\mathcal{O}(s^{-1}(D))) \cap \underline{s}^!\underline{p}^*\underline{r}^* A & & \underline{i}_{Y*}\underline{s}^!_D\underline{q}^*\underline{r}^* A \\
\downarrow & & \downarrow \\
\mathfrak{c}_1(\mathcal{O}(s^{-1}D)) \cap \underline{r}^* A & \longrightarrow & \underline{i}_{Y*}\underline{r}^*_D A
\end{array}
\qquad (16)
$$

commutes. The lower vertical arrows are of type [F1,4.7.1.], and the right upper vertical isomorphism is the base-change isomorphism for $\underline{s}^!$ provided by [F1,4.7.], where this functor is required to be biadmissible over $\mathbb{K}_{lci}$.

Proof: Let a be a rational section of A, and let "**1**" be the canonical section of $\mathcal{O}(D)$ (resp. of $\mathcal{O}(s^{-1}(D))$) which has a zero along D (resp. $s^{-1}(D)$). Then by 2.6.1., the construction of 1.(39), and the construction of the remaining arrows in [F1], the diagram (16) acts on $s^!(\text{“}\mathbf{1}\text{''} \cap p^*r^*(a))$ as follows:

$$
\begin{array}{ccc}
s^!(\text{”}\mathbf{1}\text{”} \cap p^*r^*(a)) & \longrightarrow & s^!(i_*q^*r^*(a)) \\
\downarrow & & \downarrow \\
\text{”}\mathbf{1}\text{”} \cap s^!p^*r^*(a) & & i_{Y*}s^!_D q^*r^*(a) \\
\downarrow & & \downarrow \\
\text{”}\mathbf{1}\text{”} \cap r^*(a) & \longrightarrow & i_{Y*}r^*_D(a)
\end{array}
$$

which proves our claim.

2.6.3 We consider again (15) under the assumption that p and q are smooth, i is a regular immersion of codimension one, and s is a section of p. Now we assume that the images of i and s are disjoint. Then $\underline{s}^!\underline{i}_*$ has a canonical trivialisation. On the other side, $s^*\mathcal{O}(D)$ is trivialised by $s^*(\text{“}\mathbf{1}\text{””})$, and we obtain another trivialisation

$$\begin{aligned} \underline{s}^!\underline{i}_*\underline{q}^*A &\to \underline{s}^!(\mathfrak{c}_1(\mathcal{O}(D)\cap \underline{p}^*A) \to \\ &\to \mathfrak{c}_1(s^*\mathcal{O}(D)) \cap \underline{s}^!\underline{p}^*A \xrightarrow{s^*(\text{“}\mathbf{1}\text{''})} 0. \end{aligned}$$

We claim that these trivialisations coincide.

Proof: This is similar to 2.6.2.. The first trivialisation maps $s^!i_*q^*(a)$ to zero, while the second one maps it to

$$s^!i_*q^*(a) \to s^!(\text{“}\mathbf{1}\text{''} \cap p^*(a)) \to s^*(\text{“}\mathbf{1}\text{''}) \cap s^!p^*(a) \to 0.$$

2.6.4 Let $f\colon X \to Y$ be slci, $\mathcal{E}$ and $\mathcal{F}$ be vector bundles on Y, and $A \in \mathrm{Ob}(\tilde{\mathfrak{CH}}^k(Y))$. Then the diagram

$$\begin{array}{ccc} \underline{f}^!(\mathfrak{c}_k(\mathcal{E}) \cap \mathfrak{c}_l(\mathcal{F}) \cap A) & \xrightarrow{\sigma_{\mathcal{E},\mathcal{F}}} & \underline{f}^!(\mathfrak{c}_l(\mathcal{F}) \cap \mathfrak{c}_k(\mathcal{E}) \cap A) \\ \downarrow & & \downarrow \\ \mathfrak{c}_k(f^*\mathcal{E}) \cap \underline{f}^!(\mathfrak{c}_1(\mathcal{F}) \cap A) & & \mathfrak{c}_l(f^*\mathcal{F})\underline{f}^!(\mathfrak{c}_k(\mathcal{E}) \cap A) \\ \downarrow & & \downarrow \\ \mathfrak{c}_k(f^*\mathcal{E}) \cap \mathfrak{c}_l(f^*\mathcal{F}) \cap \underline{f}^!A & \xrightarrow{\sigma_{\mathcal{E},\mathcal{F}}} & \mathfrak{c}_l(f^*\mathcal{F}) \cap \mathfrak{c}_k(f^*\mathcal{E}) \cap \underline{f}^!A \end{array}$$

commutes.

Proof: By deformation to the normal bundle, we can reduce this to the case of the zero section of a vector bundle, which is clear from 2.4.4. because $\sigma_{\mathcal{E},\mathcal{F}}$ is compatible with flat pull-back.

2.6.5 Compatibility with the Whitney isomorphism. Let $f: X \to Y$ be slci, and let $0 \to \mathcal{F} \to \mathcal{E} \to \mathcal{G} \to 0$ be an exact sequence of vector bundles on Y. Then for $A \in \mathrm{Ob}(\tilde{\mathfrak{C}\mathfrak{H}}^k(Y))$ the diagram

$$\begin{array}{ccc}
\underline{f}^!(\mathfrak{c}.(\mathcal{E}) \cap A) & \longrightarrow & \underline{f}^!(\mathfrak{c}.(\mathcal{G}) \cap \mathfrak{c}.(\mathcal{F}) \cap A) \\
\Big\downarrow & & \Big\downarrow \\
 & & \mathfrak{c}.(f^*\mathcal{G}) \cap \underline{f}^!(\mathfrak{c}.(\mathcal{F}) \cap A) \\
 & & \Big\downarrow \\
\mathfrak{c}.(f^*\mathcal{E}) \cap \underline{f}^! A & \longrightarrow & \mathfrak{c}.(f^*\mathcal{G}) \cap \mathfrak{c}.(f^*\mathcal{F}) \cap \underline{f}^! A
\end{array}$$

commutes.

Proof: If f is smooth, this is clear. This reduces us to the case of a regular closed immersion f. In this case the diagram commutes because (12) contains only transformations which are compatible with the Whitney isomorphism (cf. for instance 2.3.3.).

2.7 Relation between $\mathfrak{c}_1$ and the determinant. For a vector bundle $\mathcal{E}$ of dimension e, we denote by $\det(\mathcal{E}) = \overset{e}{\Lambda}\, \mathcal{E}$ its determinant line bundle. If $\Sigma: 0 \to \mathcal{F} \to \mathcal{E} \to \mathcal{G} \to 0$ is an exact sequence of vector bundles, then there is a canonical isomorphism

$$i_\Sigma: \det(\mathcal{F}) \otimes \det(\mathcal{G}) \to \det(\mathcal{E}).$$

Proposition: *There is a unique system of isomorphisms*

$$\iota_{\mathcal{E}}: \mathfrak{c}_1(\det(\mathcal{E})) \cap A \to \mathfrak{c}_1(\mathcal{E}) \cap A \tag{17}$$

for $A \in \mathrm{Ob}(\tilde{\mathfrak{C}\mathfrak{H}}^k(X))$ and a vector bundle $\mathcal{E}$ on X such that the following properties are satisfied:

2.7.1 Compatibility with pull-back and push-forward. *If X-schemes are denoted $p: Y \to X$, then $\tilde{\mathfrak{CH}}^{\cdot}(Y)$ is a bifibred Picard category over the bicategory (X-schemes, proper morphisms of constant relative dimension, flat morphisms) and $\mathfrak{c}_1(\det(p^*\mathcal{E})) \cap \cdot$ and $\mathfrak{c}_1(p^*\mathcal{E}) \cap \cdot: \tilde{\mathfrak{CH}}^{\cdot}(Y) \to \tilde{\mathfrak{CH}}^{\cdot}(Y)$ are biadmissible functors. The condition is that*

$$\iota_{p^*\mathcal{E}}: \mathfrak{c}_1(\det(p^*\mathcal{E})) \cap \cdot \to \mathfrak{c}_1(p^*\mathcal{E}) \cap \cdot$$

is a biadmissible functor-isomorphism.

2.7.2 Compatibility with the Whitney isomorphism. *If*

$$\Sigma: 0 \to \mathcal{F} \to \mathcal{E} \to \mathcal{G} \to 0$$

is an exact sequence of vector bundles on X, then the diagram

$$\begin{array}{ccc}
\mathfrak{c}_1(\mathcal{G}) \cap A \oplus \mathfrak{c}_1(\mathcal{F}) \cap A & \xrightarrow{\iota_{\mathcal{G}} \cap \iota_{\mathcal{F}}} & \mathfrak{c}_1(\det(\mathcal{G})) \cap A \oplus \mathfrak{c}_1(\det(\mathcal{F})) \cap A \\
& & \downarrow \\
\varphi_\Sigma \Big\downarrow & & \mathfrak{c}_1(\det(\mathcal{G}) \otimes \det(\mathcal{F})) \cap A \\
& & \Big\downarrow i_\Sigma \\
\mathfrak{c}_1(\mathcal{E}) \cap A & \xrightarrow{\iota_{\mathcal{E}}} & \mathfrak{c}_1(\det(\mathcal{E})) \cap A
\end{array}$$

commutes.

2.7.3 Normalisation. *If $\mathcal{L}$ is a line bundle, $\iota_{\mathcal{L}}$ is the identity. These conditions characterise $\iota_{\mathcal{E}}$ uniquely. In addition, the following properties are satisfied:*

2.7.4 *If $\mathcal{E}$ and $\mathcal{F}$ are vector bundles on X, then the diagram*

$$\begin{array}{ccc}
\mathfrak{c}_1(\det(\mathcal{E})) \cap \mathfrak{c}_1(\det(\mathcal{F})) \cap A & \xrightarrow{\sigma_{\det(\mathcal{E}),\det(\mathcal{F})}} & \mathfrak{c}_1(\det(\mathcal{F})) \cap \mathfrak{c}_1(\det(\mathcal{E})) \cap A \\
\Big\downarrow{\scriptstyle \iota_{\mathcal{E}} \cap \iota_{\mathcal{F}}} & & \Big\downarrow{\scriptstyle \iota_{\mathcal{F}} \cap \iota_{\mathcal{E}}} \\
\mathfrak{c}_1(\mathcal{E}) \cap \mathfrak{c}_1(\mathcal{F}) \cap A & \xrightarrow[\sigma_{\mathcal{E},\mathcal{F}}]{} & \mathfrak{c}_1(\mathcal{F}) \cap \mathfrak{c}_1(\mathcal{E}) \cap A
\end{array}$$

commutes.

2.7.5 *The isomorphisms $\iota_{\mathcal{E}}$ and $\beta_{f,\mathcal{E}}$ (for a* lci-*morphism f) are compatible.*

Proof: By the splitting principle it is clear that 2.7.1–3. characterise $\iota_{\mathcal{E}}$ uniquely and that 2.7.4. and 2.7.5. can be reduced to the case of line bundles in which they are clear.

It remains to construct an isomorphism $\iota_{\mathcal{E}}$ with 2.7.1.–3.. Let $\mathcal{E}_.: 0 = \mathcal{E}_0 \subset \mathcal{E}_1 \subset \cdots \subset \mathcal{E}_e = \mathcal{E}$ be a full flag of $\mathcal{E}$ with quotients $\mathcal{L}_i = \mathcal{E}_i/\mathcal{E}_{i-1}$. We have an isomorphism

$$\mathfrak{c}_1(\det(\mathcal{E})) \cap A \to \bigoplus_{i=1}^{e} \mathfrak{c}_1(\mathcal{L}_i) \cap A \to \mathfrak{c}_1(\mathcal{E}) \cap A, \tag{18}$$

where the first isomorphism is derived from ι_Σ and the second isomorphism is derived from the isomorphisms φ_Σ. It suffices to prove that (18) is independent of the filtration $\mathcal{E}$, for then we can use 1.13.2.(b) to descent (18) from the flag manifold of $\mathcal{E}$ to X (cf. the construction of 1.(60)). Because A is isomorphic to an object $\underline{i}_* B$ for $B \in \mathrm{Ob}(\widetilde{\mathfrak{C}\mathfrak{H}}^0(Z))$ and $i: Z \to X$ a closed subscheme of codimension k and since (18) contains only biadmissible transformations, we may assume $A \in \mathrm{Ob}(\widetilde{\mathfrak{C}\mathfrak{H}}^0(X))$. Then the restriction functor

$$\widetilde{\mathfrak{C}\mathfrak{H}}^1(X) \to \underset{\eta \in X_0}{\times} \widetilde{\mathfrak{C}\mathfrak{H}}^1(\operatorname{Spec} k(\eta))$$

is faithful, so we may assume X is the spectrum of a field. Let $p: F \to X$ be the full flag manifold of $\mathcal{E}$. Because (18) contains only transformations which are compatible with the functor $\underline{s}^!$ for $s: X \to F$ a section of p, it suffices to prove that the isomorphism between line bundles

$$\mathfrak{c}_1(p^*(\det(\mathcal{E}))) \cap \mathbf{1} \to \mathfrak{c}_1(p^*\mathcal{E}) \cap \mathbf{1}$$

is constant on F. This is clear because F is a proper variety.

2.8 Transition to the virtual category. For an exact category $\mathfrak{P}$, we denote by $\mathfrak{K}(\mathfrak{P})$ its virtual category in the sense of [D,§4]. For a scheme X, $\mathfrak{K}(X) = \mathfrak{K}(\mathfrak{P}(X))$ is the category of virtual vector bundles on X. By the universal properties of the virtual category ([D,§4.3.]), there exist unique (up to unique functor-isomorphism) additive functors

$$\mathfrak{c}_i(\mathcal{E}) \cap A \colon \mathfrak{K}(X) \times \widetilde{\mathfrak{CH}}^k(X) \to \widetilde{\mathfrak{CH}}^{k+1}(X), \quad i > 0$$
$$\mathfrak{c}_0(\mathcal{E}) \cap A = A$$

together with additive (in A) functor isomorphisms

$$\begin{array}{rcl} \mathfrak{c}_1(0) \cap A & \to & 0 \ \text{ if } \ i > 0 \\ \mathfrak{c}.(\mathcal{E} \oplus \mathcal{F}) \cap A & \to & \mathfrak{c}.(\mathcal{E}) \cap \mathfrak{c}.(\mathcal{F}) \cap A \end{array}$$

such that

(i) $\mathfrak{c}_i(\mathcal{E}) \cap A = \mathfrak{c}_i([\mathcal{E}]) \cap A$ if $\mathcal{E}$ is a vector bundle and $[\mathcal{E}]$ the corresponding virtual bundle.

(ii) If $\Sigma\colon 0 \to \mathcal{F} \to \mathcal{E} \to \mathcal{G} \to 0$ is a short exact sequence of vector bundles, then the following diagram involving the Whitney sum isomorphism and the isomorphism $[\mathcal{E}] \to [\mathcal{G}] \oplus [\mathcal{F}]$ induced by Σ commutes:

$$\begin{array}{ccc} \mathfrak{c}.(\mathcal{E}) \cap A & \longrightarrow & \mathfrak{c}.(\mathcal{G}) \cap \mathfrak{c}.(\mathcal{F}) \cap A \\ \| & & \| \\ \mathfrak{c}.([\mathcal{E}]) \cap A & \longrightarrow & \mathfrak{c}.([\mathcal{G}]) \cap \mathfrak{c}.([\mathcal{F}]) \cap A\,. \end{array}$$

In the rest of this paper, we will for the sake of simplicity not distinguish between vector bundles themselves and the virtual vector bundles defined by them. Using the universal property of the virtual category, we get isomorphisms

$$\begin{array}{rcl} \mathfrak{c}_i(\mathcal{E}) \cap \mathfrak{c}_j(\mathcal{F}) \cap A & \to & \mathfrak{c}_j(\mathcal{F}) \cap \mathfrak{c}_i(\mathcal{E}) \cap A \\ \mathfrak{c}_i(f^*\mathcal{E}) \cap \underline{f}^* A & \to & \underline{f}^*(\mathfrak{c}_i(\mathcal{E}) \cap A) \\ \mathfrak{c}_i(f^*\mathcal{E}) \cap \underline{f}^! A & \to & \underline{f}^!(\mathfrak{c}_i(\mathcal{E}) \cap A) \\ \mathfrak{c}_i(\mathcal{E}) \cap \underline{g}_* A & \to & \underline{g}_*(\mathfrak{c}_i(g^*\mathcal{E}) \cap A) \\ \mathfrak{sp}_\lambda(\mathfrak{c}_i(\mathcal{E}) \cap A) & \to & \mathfrak{c}_i(\mathcal{E}\,|_D) \cap \mathfrak{sp}_\lambda A \\ \mathfrak{c}_1(\mathcal{E}) \cap A & \to & \mathfrak{c}_1(\det(\mathcal{E})) \cap A \end{array}$$

because the corresponding isomorphisms for "real" bundles are compatible with the Whitney sum isomorphism. These isomorphisms for virtual vector

2.9 Polynomials in the Chern functors. Let $P(c_i(\mathcal{E}_j))$ be a polynomial with integral coefficients in the Chern classes of vector bundles $\mathcal{E}_j$, $j \in J$, on X. For a total ordering $j_1 < j_2 < \ldots < j_N$ of J and virtual vector bundles $\mathcal{E}_j$ we put

$$[P(\mathfrak{c}_i(\mathcal{E}_j))]_< \cap A = \bigoplus_\alpha n_\alpha \mathfrak{c}_1(\mathcal{E}_{i_1})^{\alpha_{1,i_1}} \cap \mathfrak{c}_2(\mathcal{E}_{i_1})^{\alpha_{2,i_1}} \cap \ldots \tag{19}$$
$$\ldots \cap \mathfrak{c}_1(\mathcal{E}_{i_2})^{\alpha_{1,i_2}} \cap \mathfrak{c}_2(\mathcal{E}_{i_2})^{\alpha_{2,i_2}} \cap \ldots \cap A,$$

where

$$P = \sum_\alpha n_a \prod_{i,j} c_i(\mathcal{E}_j)^{\alpha_{ij}}.$$

This means, all monomials of the polynomial P are ordered lexicographically according to the indices j (coming first) and i. If α and β are multiindices, then there exists a unique isomorphism

$$\begin{array}{c}
\mathfrak{c}_1(\mathcal{E}_{i_1})^{\alpha_{1,i_1}} \cap \mathfrak{c}_2(\mathcal{E}_{i_1})^{\alpha_{2,i_1}} \cap \ldots \cap \mathfrak{c}_1(\mathcal{E}_{i_2})^{\alpha_{1,i_2}} \cap \mathfrak{c}_2(\mathcal{E}_{i_2})^{\alpha_{2,i_2}} \cap \ldots \cap \\
\mathfrak{c}_1(\mathcal{E}_{i_1})^{\beta_{1,i_1}} \cap \mathfrak{c}_2(\mathcal{E}_{i_1})^{\beta_{2,i_1}} \cap \ldots \cap \mathfrak{c}_1(\mathcal{E}_{i_2})^{\beta_{1,i_2}} \cap \mathfrak{c}_2(\mathcal{E}_{i_2})^{\beta_{2,i_2}} \cap \ldots \cap A \\
\downarrow \\
\mathfrak{c}_1(\mathcal{E}_{i_1})^{\alpha_{1,i_1}+\beta_{1,i_1}} \cap \mathfrak{c}_2(\mathcal{E}_{i_1})^{\alpha_{2,i_1}+\beta_{2,i_1}} \cap \ldots \\
\ldots \cap \mathfrak{c}_1(\mathcal{E}_{i_2})^{\alpha_{1,i_2}+\beta_{1,i_2}} \cap \mathfrak{c}_2(\mathcal{E}_{i_2})^{\alpha_{2,i_2}+\beta_{2,i_2}} \cap \ldots \cap A
\end{array}$$

defined by applying the transformations $\sigma_{\cdot,\cdot}$ to the permutation which brings all factors into the right order with the minimal number of transpositions, i.e., without interchanging identical factors $\mathfrak{c}_i(\mathcal{E}_j) \cap \mathfrak{c}_i(\mathcal{E}_j)$. From the isomorphism (20) we derive a canonical isomorphism

$$[P(\mathfrak{c}_j(\mathcal{E}_j))]_< \cap [Q(\mathfrak{c}_i(\mathcal{E}_j))]_< \cap A \to [(PQ)(\mathfrak{c}_i(\mathcal{E}_j))]_< \cap A. \tag{20}$$

The diagram

$$\begin{array}{ccc}
[(PQR)(\mathfrak{c}_i(\mathcal{E}_j))]_< \cap A & \longrightarrow & [P(\mathfrak{c}_i(\mathcal{E}_j))]_< \cap [(QR)(\mathfrak{c}_i(\mathcal{E}_j))]_< \cap A \\
\downarrow & & \\
[(PQ)(\mathfrak{c}_i(\mathcal{E}_j))]_< \cap [R(\mathfrak{c}_i(\mathcal{E}_j))]_< \cap A & & \downarrow \\
\downarrow & & \\
\multicolumn{3}{c}{[P(\mathfrak{c}_i(\mathcal{E}_j))]_< \cap [Q(\mathfrak{c}_i(\mathcal{E}_i))]_< \cap [R(\mathfrak{c}_i(\mathcal{E}_j))]_< \cap A}
\end{array} \tag{22}$$

commutes.

Let $\ll$ be another ordering of J and $\pi: J \to J$ be the permutation with $\pi(i) \ll \pi(j)$ iff $i < j$. For each monomial there exists a unique isomorphism

$$\mathfrak{c}_1(\mathcal{E}_{i_1})^{\alpha_{1,i_1}} \cap \mathfrak{c}_2(\mathcal{E}_{i_1})^{\alpha_{2,i_1}} \cap \ldots \cap \mathfrak{c}_1(\mathcal{E}_{i_2})^{\alpha_{1,i_2}} \cap \mathfrak{c}_2(\mathcal{E}_{i_2})^{\alpha_{2,i_2}} \cap \ldots \cap A$$

$$\downarrow$$

$$\mathfrak{c}_1(\mathcal{E}_{\pi(i_1)})^{\alpha_{1,\pi(i_1)}} \cap \mathfrak{c}_2(\mathcal{E}_{\pi(i_1)})^{\alpha_{2,\pi(i_1)}} \cap \ldots$$
$$\ldots \cap \mathfrak{c}_1(\mathcal{E}_{\pi(i_2)})^{\alpha_{1,\pi(i_2)}} \cap \mathfrak{c}_2(\mathcal{E}_{\pi(i_2)})^{\alpha_{2,\pi(i_2)}} \cap \ldots \cap A$$

defined by the permutation which brings all factors to the right order with the minimal number of transpositions. We get a canonical isomorphism

$$[P(\mathfrak{c}_i(\mathcal{E}_j))]_< \cap A \to [P(\mathfrak{c}_i(\mathcal{E}_j))]_\ll \cap A. \tag{22}$$

These isomorphisms satisfy the necessary compatibility to glue the objects $[P(\mathfrak{c}_i(\mathcal{E}_j))]_< \cap A$ to one object $P(\mathfrak{c}_i(\mathcal{E}_j)) \cap A$. If confusions are impossible, we will also write $\mathfrak{P}(\mathcal{E}_j) \cap A$ for $P(\mathfrak{c}_i(\mathcal{E}_j)) \cap A$. The isomorphisms (21) and (23) commute, giving a canonical isomorphism

$$\mathcal{N}: P(\mathfrak{c}_i(\mathcal{E}_j)) \cap Q(\mathfrak{c}_i(\mathcal{E}_j)) \cap A \to (PQ)(\mathfrak{c}_i(\mathcal{E}_j)) \cap A \tag{23}$$

satisfying the analogue of (22).

Let us stress that $\mathfrak{P}(\mathcal{E}_j) \cap A$ behaves badly if we identify some of the vector bundles $\mathcal{E}_i$. For instance, if $P(c_i(\mathcal{E}), c_j(\mathcal{F}))$ is a polynomial in the Chern classes of two vector bundles and if $Q(c_i(\mathcal{E})) := P(c_i(\mathcal{E}), c_j(\mathcal{E}))$, then there is no canonical isomorphism

$$\mathfrak{P}(\mathcal{E}, \mathcal{E}) \cap A \to \mathfrak{Q}(\mathcal{E}) \cap A$$

unless we fix an order of the two arguments of P.

There is, however, the following substitution principle: Let $\mathcal{F}(\mathcal{G}_l)$ be a functor in virtual bundles $\mathcal{G}_l$, and let a functor-isomorphism

$$\alpha: \mathfrak{c}_k(\mathcal{F}(\mathcal{G}_l)) \cap A \to Q_k(\mathfrak{c}_m(\mathcal{G}_l)) \cap A$$

be given. If $P(c_k(\mathcal{F}), c_i(\mathcal{E}_j))$ is a polynomial in Chern classes, then α induces a canonical isomorphism

$$P(\mathfrak{c}_k(\mathcal{F}(\mathcal{G}_l)), \mathfrak{c}_i(\mathcal{E}_j)) \cap A \to R(\mathfrak{c}_i(\mathcal{E}_j), \mathfrak{c}_m(\mathcal{G}_l)) \cap A \tag{24}$$

where

$$R(c_i(\mathcal{E}_j), \mathfrak{c}_m(\mathcal{G}_l)) = P(Q_k(c_m(\mathcal{G}_l)), c_i(\mathcal{E}_j)).$$

The isomorphism (25) is independent of the choice of order of the variables $\mathcal{G}_k, \mathcal{E}_l$.

If our polynomials have the more general size $P(\dim(\mathcal{E}_j), c_i(\mathcal{E}_j))$ (like Hirzebruch's polynomials *ch* and *Td*, which are polynomials not only in the Chern classes but also in the dimension of the bundles), then these methods also apply. We get a functor

$$P(\dim(\mathcal{E}_j), \mathfrak{c}_i(\mathcal{E}_j)) \cap A = \mathfrak{P}(\mathcal{E}_j) \cap A$$

in virtual vector bundles $\mathcal{E}_j$ and $A \in \mathrm{Ob}(\tilde{\mathfrak{CH}}^k(X))$ satisfying similar properties as above.

2.10 Twist by a line bundle. Let

$$P_j(\dim(\mathcal{E}), c_k(\mathcal{E}), c_1(\mathcal{L})) = \sum_{l=0}^{j} \binom{\dim(\mathcal{E}) + l - j}{l} c_1(\mathcal{L})^{j-l} \cap c_l(\mathcal{E}) \quad (25)$$

be the polynomial with the property

$$c_j(\mathcal{E} \otimes \mathcal{L}) = P_j(\dim(\mathcal{E}), c_k(\mathcal{E}), c_1(\mathcal{L})).$$

We have the obvious identities

$$\begin{aligned} & P_j(\dim(\mathcal{E}), c_k(\mathcal{E}), c_1(\mathcal{L}) + c_1(\mathcal{M})) = \\ = \ & P_j(\dim(\mathcal{E}), P_i(\dim(\mathcal{E}), c_k(\mathcal{E}), c_1(\mathcal{M})), c_1(\mathcal{L})). \end{aligned} \quad (26)$$

$$\begin{aligned} & P_i(\dim(\mathcal{E}) + \dim(\mathcal{E}'), c_1(\mathcal{L}), \sum_{k+l-i} c_k(\mathcal{E}'')c_l(\mathcal{E}')) = \\ = \ & \sum_{k+l=j} P_k(\dim(\mathcal{E}'), c_m(\mathcal{E}'), c_1(\mathcal{L})) P_l(\dim(\mathcal{E}''), c_n(\mathcal{E}''), c_1(\mathcal{L})). \end{aligned} \quad (27)$$

Theorem: *There exists a unique functor isomorphism*

$$\mathfrak{c}_j(\mathcal{E} \otimes \mathcal{L}) \cap A \to \mathfrak{P}_j(\mathcal{L}, \mathcal{E}) \cap A \quad (28)$$

with the following properties:

2.10.1 Compatibility with direct sum and inverse images. *If $\mathcal{L}$ and $\mathcal{E}$ are a line bundle and a virtual bundle on S and if S-schemes are denoted $p: X \to S$, then*

$$\mathfrak{c}_j(p^*\mathcal{E} \otimes p^*\mathcal{L}) \cap \cdot : \tilde{\mathfrak{CH}}^{\cdot}(X) \to \tilde{\mathfrak{cH}}^{\cdot}(X)$$

and

$$\mathfrak{P}_j(\mathcal{L}, \mathcal{E}) \cap \cdot : \tilde{\mathfrak{CH}}^{\cdot}(X) \to \tilde{\mathfrak{CH}}^{\cdot}(X)$$

are biadmissible functors between bifibred Picard categories over S-schemes. Then (29) is supposed to be biadmissible.

2.10.2 Normalisation. *If $\mathcal{E}$ is a line bundle, then (29) in dimension zero is the identity of A, (29) in dimension one is the canonical isomorphism $\mathfrak{c}_1(\mathcal{L}\otimes\mathcal{E})\cap A \to \mathfrak{c}_1(\mathcal{L})\cap A \oplus \mathfrak{c}_1(\mathcal{E})\cap A$, and (29) in dimension larger than one is the identity of the zero object.*

2.10.3 Compatibility with the Whitney sum isomorphism. *If $\mathcal{E}$ and $\mathcal{F}$ are virtual bundles, then the diagram*

$$\begin{array}{ccc} \mathfrak{c}_k((\mathcal{E}\oplus\mathcal{F})\otimes\mathcal{L})\cap A & \longrightarrow & \bigoplus_{i+j=k} \mathfrak{c}_i(\mathcal{E}\otimes\mathcal{L})\cap \mathfrak{c}_j(\mathcal{F}\otimes\mathcal{L})\cap A \\ \downarrow & & \downarrow \\ \mathfrak{P}_k(\mathcal{E}\oplus\mathcal{F},\mathcal{L})\cap A & \longrightarrow & \bigoplus_{i+j=k} \mathfrak{P}_i(\mathcal{E},\mathcal{L})\cap \mathfrak{P}_j(\mathcal{F},\mathcal{L})\cap A \end{array} \tag{30}$$

commutes up to a correcting sign

$$\begin{aligned} \delta_k(\mathcal{E},\mathcal{F},\mathcal{L},A) &= c_1(\mathcal{L})\cap\tau_{k-1}(\mathcal{E}\otimes\mathcal{L},\mathcal{F}\otimes\mathcal{L},A) \qquad (30)\\ &= c_1(\mathcal{L})\cap \sum_{\substack{n+m=\\=k-2}} (\dim(\mathcal{E})-n)(\dim(\mathcal{F})-m)c_n(\mathcal{E})\cap c_m(\mathcal{F})\cap[A]\cap[-1]. \end{aligned}$$

The lower horizontal arrow is defined by (28), (25), (24) and the Whitney sum isomorphism.

These properties suffice to characterise (29). The following properties are also satisfied:

2.10.4 *If $\mathcal{M}$ is another line bundle, then the diagram*

$$
\begin{array}{ccc}
P_i(\dim(\mathcal{E}), \mathfrak{c}_1(\mathcal{L}) \oplus \mathfrak{c}_1(\mathcal{M}), \mathfrak{c}_j(\mathcal{E})) \cap A & \longrightarrow & \mathfrak{c}_i(\mathcal{E} \otimes \mathcal{L} \otimes \mathcal{M}) \cap A \\
\downarrow & & \\
P_i(\dim(\mathcal{E} \otimes \mathcal{L}), \mathfrak{c}_1(\mathcal{M}), \mathfrak{c}_j(\mathcal{E} \otimes \mathcal{L})) \cap A & & \downarrow \\
\downarrow & & \\
\multicolumn{3}{c}{P_i(\dim(\mathcal{E}), \mathfrak{c}_1(\mathcal{M}), P_j(\dim(\mathcal{E}), \mathfrak{c}_1(\mathcal{L}), \mathfrak{c}_k(\mathcal{E}))) \cap A}
\end{array}
\tag{32}
$$

commutes.

2.10.5 Compatibility with $f^!$. *The diagram*

$$
\begin{array}{ccc}
\mathfrak{c}_k(f^*(\mathcal{E} \otimes \mathcal{L})) \cap \underline{f}^! A & \longrightarrow & \underline{f}^!(\mathfrak{c}_k(\mathcal{E} \otimes \mathcal{L}) \cap A) \\
\downarrow & & \downarrow \\
\mathfrak{P}_k(f^*\mathcal{E}, f^*\mathcal{L}) \cap \underline{f}^! A & \longrightarrow & \underline{f}^!(\mathfrak{P}_k(\mathcal{E}, \mathcal{L}) \cap A)
\end{array}
\tag{33}
$$

commutes.

Proof: *Step 1:* It follows from the splitting principle that 2.10.1.–2.10.3. characterise (29) uniquely. 2.10.5. for a line bundle $\mathcal{E}$ follows from 2.10.2., and the general case of 2.10.5. follows from this case and 2.10.3 by the splitting principle. It remains to construct an isomorphism with the properties 2.10.1.–4.. It suffices to consider "real" vector bundles $\mathcal{E}$ and to consider short exact sequences $0 \to \mathcal{F} \to \mathcal{H} \to \mathcal{E} \to 0$ in 2.10.3..

Step 2: To construct (29) for a vector bundle $\mathcal{E}$ we use the identifications $\mathbb{P}(\mathcal{E} \otimes \mathcal{L}) = \mathbb{P}(\mathcal{E})$ and $\mathcal{O}(1)_{\mathcal{E}\otimes\mathcal{L}} = \mathcal{O}(1)_{\mathcal{E}} \otimes \mathcal{L}^{-1}$. Let $p\colon \mathbb{P}(\mathcal{E}) \to X$ be the projection. We have canonical isomorphisms

$$
\begin{array}{c}
\bigoplus_{j=0}^{e} \mathfrak{c}_1(\mathcal{O}(1)_{\mathcal{E}\otimes\mathcal{L}})^{e-j} \cap \underline{p}^*\left(\bigoplus_{k=0}^{j} \binom{e+k-j}{k} \mathfrak{c}_1(\mathcal{L})^k \cap \mathfrak{c}_{j-k}(\mathcal{E}) \cap A\right) \\
\Big\downarrow \\
\bigoplus_{j=0}^{e} \bigoplus_{k=0}^{j} \binom{e+k-j}{k} \mathfrak{c}_1(\mathcal{O}(1)_{\mathcal{E}\otimes\mathcal{L}})^{e-j} \cap \mathfrak{c}_1(p^*\mathcal{L})^k \cap \underline{p}^*(\mathfrak{c}_{j-k}(\mathcal{E}) \cap A) \\
\Big\downarrow \\
\bigoplus_{l=0}^{e} \mathfrak{c}_1(\mathcal{O}(1)_{\mathcal{E}})^{e-l} \cap \underline{p}^*(\mathfrak{c}_l(\mathcal{E}) \cap A) \qquad \xrightarrow{\ 1.(37)\ } \quad 0
\end{array}
$$

defining (29). The proofs of 2.10.1., 2.10.2., and 2.10.4. are straightforward. It remains to prove 2.10.3..

Step 3: First we prove (30) in the case of an exact sequence

$$0 \to \mathcal{F} \to \mathcal{H} \to \mathcal{E} \to 0 \tag{33}$$

of vector bundles with $\dim(\mathcal{E}) = 1$. We have the diagram

$$
\begin{array}{ccc}
\mathbb{P}(\mathcal{F}) & \xrightarrow{\quad i \quad} & \mathbb{P}(\mathcal{H}) \\
 & {\scriptstyle q}\searrow \quad \swarrow{\scriptstyle p} & \\
 & X &
\end{array}
$$

of projective fibrations. We consider the diagram

$$\underline{i}_*\left(\bigoplus_{j=0}^{f} \mathfrak{c}_1(\mathcal{O}(1)_{\mathcal{F}\otimes\mathcal{L}})^{f-j}\cap \underline{q}^*(\mathfrak{P}_j(\mathcal{F},\mathcal{L})\cap A)\right)$$

(35)

$$\downarrow$$

$$\underline{i}_*\left(\bigoplus_{j=0}^{f} \mathfrak{c}_1(\mathcal{O}(1)_{\mathcal{F}})^{f-j}\cap \underline{q}^*(\mathfrak{c}_j(\mathcal{F})\cap A)\right)$$

$$\downarrow \alpha \qquad \text{(A)} \qquad \downarrow \beta$$

$$\bigoplus_{j=0}^{f} \mathfrak{c}_1(\mathcal{O}(1)_{\mathcal{H}\otimes\mathcal{L}})^{f-j}\cap \mathfrak{c}_1(\mathcal{E}\otimes\mathcal{L}(1))\cap \underline{p}^*(\mathfrak{P}_j(\mathcal{F},\mathcal{L})\cap A)$$

$$\downarrow \gamma$$

$$\bigoplus_{j=0}^{f} \mathfrak{c}_1(\mathcal{O}(1)_{\mathcal{H}})^{f-j}\cap \mathfrak{c}_1(\mathcal{E}\otimes\mathcal{L}(1))\cap \underline{p}^*(\mathfrak{c}_j(\mathcal{F})\cap A)$$

$$\downarrow \delta \qquad \text{(B)} \qquad \downarrow \varepsilon$$

$$\bigoplus_{j\,|\,0}^{e} \mathfrak{c}_1(\mathcal{O}(1)_{\mathcal{H}\otimes\mathcal{L}})^{e-j}\cap \underline{p}^*(\mathfrak{P}_j(\mathcal{H},\mathcal{L})\cap A)$$

$$\downarrow$$

$$\bigoplus_{j=0}^{e} \mathfrak{c}_1(\mathcal{O}(1)_{\mathcal{H}})^{e-j}\cap \underline{p}^*(\mathfrak{c}_j(\mathcal{E})\cap A)\,.$$

The isomorphisms α and β interchange $\underline{i}_*$ and $\mathfrak{c}_1(\mathcal{O}(1))$ and apply 1.(39). The two arrows δ and ϵ are built of (24), (25), and the Whitney sum isomorphism. The isomorphism γ interchanges $\mathfrak{c}_1(\mathcal{L}^{-1})^k$ and $\mathfrak{c}_1(\mathcal{E}\otimes\mathcal{L}(1))$ and applies (24). The other two arrows are of type (24).

The commutativity of (A) follows from 1.(40). If $\mathfrak{c}_1(\mathcal{L}^{-1})^l$ occurs (with multiplicity $\binom{f-j}{l}$) in the binomial resolution of the power $(\mathfrak{c}_1(\mathcal{E}(1)\otimes\mathcal{L}))^{f-j}$, then γ involves interchanging $\mathfrak{c}_1(\mathcal{L})$ l times with $\mathfrak{c}_1(\mathcal{L}^{-1})$.

Since the arrows in (B) use only minimal permutations, (B) commutes up to the sign

$$\sum_{j=0}^{f}\sum_{l=0}^{f-j} l \binom{f-j}{l} c_1(\mathcal{O}(1)_{\mathcal{H}})^{f-j-1} \cap c_1(\mathcal{L}^{-1})^l \cap$$
$$\cap p^*(P_j(\dim(\mathcal{F}), c_i(\mathcal{F}), c_1(\mathcal{L})) \cap [-1] \cap [A]) =$$

$$= \sum_{j=0}^{f-1}(f-j)(c_1(\mathcal{O}(1)_{\mathcal{H}\otimes\mathcal{L}})^{f-j-1} \cap c_1(\mathcal{L}) \cap p^*(c_j(\mathcal{F}\otimes\mathcal{L}) \cap [-1] \cap [A]).$$

By 1.6., the definition of (29) in step 2, and the definition of the Whitney isomorphism in 1.10. we conclude that (30) commutes up to the sign

$$(f+2-k)c_1(\mathcal{L}) \cap c_{k-2}(\mathcal{F}\otimes\mathcal{L}) \cap [-1] \cap [A] = \delta_k(\mathcal{E}, \mathcal{F}, \mathcal{L}, A),$$

proving (30) in the special case of an exact sequence (34) with a line bundle $\mathcal{E}$. For an arbitrary exact sequence (34), (30) follows by induction on the dimension of $\mathcal{E}$, using the splitting principle. For arbitrary virtual bundles $\mathcal{E}$, $\mathcal{F}$ (30) follows by the universal property of Deligne's virtual category. The proof of 2.10. is complete.

3 Axiomatic Characterisation of the Chern functors

3.1 The trivialisation $T_{\mathcal{E},s}$. Let $\mathcal{E}$ be a vector bundle of dimension e on X. A non-vanishing global section s of $\mathcal{E}$ defines a short exact sequence

$$0 \to \mathcal{O}_X \to \mathcal{E} \to \mathcal{G} \to 0 \tag{1}$$

and hence a trivialisation

$$T_{\mathcal{E},s}: \mathfrak{c}_e(\mathcal{E}) \cap A \to \mathfrak{c}_{e-1}(\mathcal{G}) \cap \mathfrak{c}_1(\mathcal{O}_X) \cap A \to 0.$$

Proposition: (i) *Let $\sigma: X \to \mathbb{P}(\mathcal{E})$ be the section of $\mathbb{P}(\mathcal{E})$ defined by s. Because $\underline{\sigma}^*\mathcal{O}(1) \simeq \mathcal{O}_X$ canonically, $\underline{\sigma}^!(\mathfrak{c}_1(\mathcal{O}(1)^k \cap \underline{p}^*) \simeq \mathfrak{c}_1(\mathcal{O}_X)^k \cap A$ has a canonical trivialisation if $k > 0$. Hence by applying $\underline{\sigma}^!$ to the morphism (1.(37))*

$$\bigoplus_{j=0}^{e} \mathfrak{c}_1(\mathcal{O}(1)^{e-j} \cap \underline{p}^*(\mathfrak{c}_j(\mathcal{E}) \cap A) \to 0$$

defining $\mathfrak{c}.(\mathcal{E}) \cap A$, we obtain a trivialisation

$$\begin{aligned} 0 \to \underline{\sigma}^!\Big(\bigoplus_{j=0}^{e} \mathfrak{c}_1(\mathcal{O}(1))^{e-j} \cap \underline{p}^*(\mathfrak{c}_j(\mathcal{E}) \cap A)\Big) \to \\ \to \quad \underline{\sigma}^!\underline{p}^*(\mathfrak{c}_e(\mathcal{E}) \cap A) \to \mathfrak{c}_e(\mathcal{E}) \cap A \end{aligned} \tag{2}$$

of $\mathfrak{c}_e(\mathcal{E}) \cap A$. We claim that this trivialisation coincides with $T_{\mathcal{E},s}$.
(ii) *Let $0 \to \mathcal{F} \to \mathcal{E} \xrightarrow{\pi} \mathcal{L} \to 0$ be an exact sequence of vector bundles on X, with* dim$(\mathcal{L}) = 1$. *We assume that $\mathcal{E}$ has a* non-vanishing *section s such that $i: D \to X$ is a regular immersion of codimension one, where D is the subscheme defined by the vanishing of $\pi(s)$. Then $\pi(s)$ defines an isomorphism $\mathcal{L} \simeq \mathcal{O}_X(D)$. We assume also that $r: X \to Z$ is a flat morphism whose restriction r_D to D is also flat.*
Then

$$\begin{aligned} \mathfrak{c}_e(\mathcal{E}) \cap \underline{r}^*A \quad &\longrightarrow \quad \mathfrak{c}_1(\mathcal{L}) \cap \mathfrak{c}_f(\mathcal{F}) \cap \underline{r}^*A \\ &\xrightarrow{\sigma_{\mathcal{L},\mathcal{F}}} \quad \mathfrak{c}_f(\mathcal{F}) \cap \mathfrak{c}_1(\mathcal{L}) \cap \underline{r}^* \\ &\xrightarrow{\pi(s)} \quad \mathfrak{c}_f(\mathcal{F}) \cap \mathfrak{c}_1(\mathcal{O}_X(D)) \cap \underline{r}^*A \quad (f = e-1) \\ &\xrightarrow{1.(39)} \quad \mathfrak{c}_f(\mathcal{F}) \cap \underline{i}_*\underline{r}_D^*A \\ &\longrightarrow \quad \underline{i}_*(\mathfrak{c}_f(\mathcal{F}\,|_D) \cap \underline{r}_D^*A) \\ &\xrightarrow{T_{\mathcal{F}|_D, s|_D}} \quad \underline{i}_*(0) = 0 \end{aligned} \tag{3}$$

defines a trivialisation of $\mathfrak{c}_e(\mathcal{E}) \cap \underline{r}^*A$. We claim that this trivialisation coincides with $T_{\mathcal{E},s}$.

Proof of (i). Without loosing generality we may assume that (1) splits, because this can be achieved by passing to a certain principal homogeneous sheaf for the dual of $\mathcal{G}$. By 1.(63)., the diagram

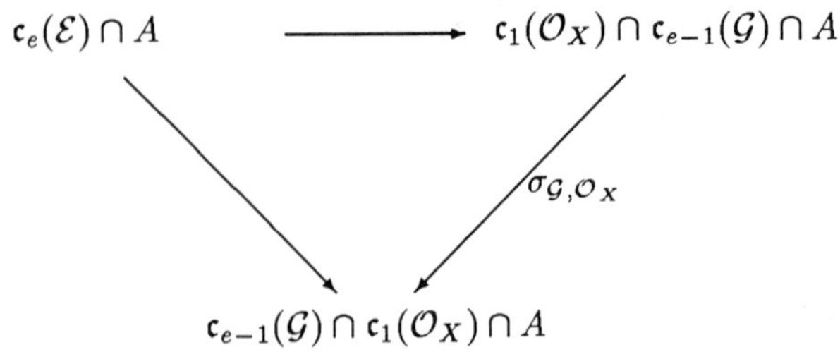

commutes up to the sign

$$(\dim(\mathcal{G}) - (e-1)) \cap \mathfrak{c}_{e-1}(\mathcal{G}) \cap [-1] \cap [A] = 0.$$

Hence, $T_{\mathcal{E},s}$ coincides with the trivialisation defined by the complementary sequence

$$0 \to \mathcal{G} \to \mathcal{E} \to \mathcal{L} = \mathcal{O}_X \to 0 \tag{4}$$

and $\mathfrak{c}_1(\mathcal{O}_X) \cap \mathfrak{c}_{e-1}(\mathcal{G}) \cap A \to 0$.

Now we consider the diagram

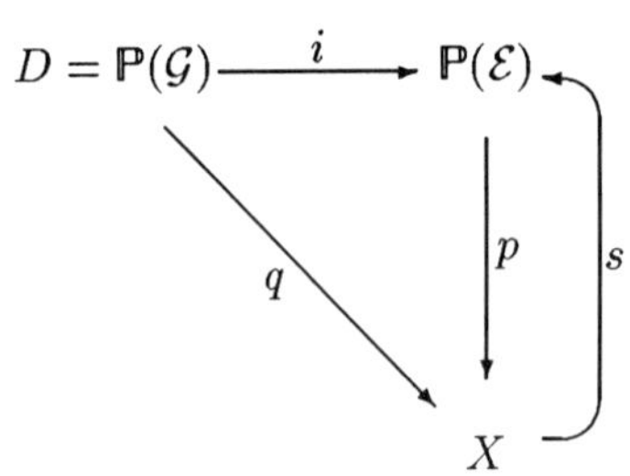

By our construction of the Whitney isomorphism associated to (4) (cf. 1.10.), the diagram

$$
\begin{array}{ccc}
\underline{\sigma}^!\big(\bigoplus_{j=0}^{e} \mathfrak{c}_1(\mathcal{O}(1))^{e-j} \cap \underline{p}^*(\mathfrak{c}_j(\mathcal{E}) \cap A)\big) & \xrightarrow[\underline{\sigma}^!(1.(37))]{} & 0 \\
\Big\downarrow {\scriptstyle 1.(42)} & & \Big\| \\
\underline{\sigma}^!\big(\bigoplus_{j=0}^{e-1} \mathfrak{c}_1(\mathcal{O}(1))^{e-1-j} \cap \mathfrak{c}_1(\mathcal{O}(1)) \cap \underline{p}^*(\mathfrak{c}_j(\mathcal{F}) \cap A) \oplus & & \\
\oplus \bigoplus_{j=0}^{e-1} \mathfrak{c}_1(\mathcal{O}(1))^{e-1-j} \cap \underline{p}^*(\mathfrak{c}_1(\mathcal{L}) \cap \mathfrak{c}_j(\mathcal{G}) \cap A)\big) & & \\
\Big\downarrow {\scriptstyle 1.(39)} & & \\
\underline{\sigma}^!\big(\bigoplus_{j=0}^{e-1} \mathfrak{c}_1(\mathcal{O}(1))^{e-1-j} \cap \underline{i}_*\underline{q}^*(\mathfrak{c}_j(\mathcal{F}) \cap A)\big) & & \\
\Big\downarrow {\scriptstyle 1.(38)} & & \\
\underline{\sigma}^!\underline{i}_*\big(\bigoplus_{j=0}^{e-1} \mathfrak{c}_1(\mathcal{O}(1))^{e-1-j} \cap \underline{q}^*(\mathfrak{c}_j(\mathcal{F}) \cap A)\big) & \xrightarrow[\underline{\sigma}^!\underline{i}_*(1.(37))]{} & 0
\end{array}
$$

commutes. By the additivity of the canonical functor-isomorphism $\underline{\sigma}^!\underline{i}_* \to 0$, the right vertical isomorphism $\underline{\sigma}^!\underline{i}_*(1.(37))$ coincides with the canonical trivialisation of $\underline{\sigma}^!\underline{i}_*$ defined by the disjointness of the images of i and σ. Using this and 2.6.3., we conclude that the composition

$$
\begin{array}{c}
\bigoplus_{j=0}^{e-1} \mathfrak{c}_1(\underline{\sigma}^*\mathcal{O}(1))^{e-1-j} \cap \mathfrak{c}_1(\upsilon^*\mathcal{L}(1)) \cap \underline{\upsilon}^!\underline{p}^*(\mathfrak{c}_j(\mathcal{G}) \cap A) \\
\Big\downarrow \\
\underline{\sigma}^!\big(\bigoplus_{j=0}^{e-1} \mathfrak{c}_1(\mathcal{O}(1))^{e-1-j} \cap \mathfrak{c}_1(\mathcal{O}(1)) \cap \underline{p}^*(\mathfrak{c}_j(\mathcal{G}) \cap A) \oplus \\
\oplus \bigoplus_{j=0}^{e-1} \mathfrak{c}_1(\mathcal{O}(1))^{e-1-j} \cap \underline{p}^*(\mathfrak{c}_1(\mathcal{L}) \cap \mathfrak{c}_j(\mathcal{G}) \cap A)\big) \\
\Big\downarrow {\scriptstyle \underline{\sigma}^!\underline{i}_*(1.(37)) \circ 1.(38) \circ 1.(39)} \\
0
\end{array}
$$

coincides with the trivialisation defined by the canonical isomorphism $\sigma^*\mathcal{L}(1) \simeq \mathcal{O}_X$. We get the commutative diagram

$$\begin{array}{ccc}
\underline{\sigma}^!\big(\bigoplus_{j=0}^{e} \mathfrak{c}_1(\mathcal{O}(1))^{e-j} \cap \underline{p}^*(\mathfrak{c}_j(\mathcal{E}) \cap A)\big) & \xrightarrow{\ \underline{\sigma}^!(1.(37))\ } & 0 \\
\Big\downarrow {\scriptstyle 1.(42)} & \nearrow {\scriptstyle (\underline{\sigma}^*(\mathcal{L}(1)) \xrightarrow{\text{canonical}} 0)} & \\
\bigoplus_{j=0}^{e-1} \mathfrak{c}_1(\underline{\sigma}^*\mathcal{O}(1))^{e-1-j} \cap \mathfrak{c}_1(\sigma^*\mathcal{L}(1)) \cap \underline{\sigma}^!\underline{p}^*(\mathfrak{c}_j(\mathcal{F}) \cap A) & &
\end{array} \qquad (5)$$

(we have not yet used the fact that $\mathcal{L} \simeq \mathcal{O}_X$ in (4)). Now, by the canonical isomorphism $\mathcal{L} \simeq \mathcal{O}_X$ and $\sigma^*\mathcal{O}(1) \simeq \mathcal{O}_X$, the vertical arrow in the last diagram is (2), and the composition of the two other ones is $T_{\mathcal{E},s}$.

Proof of (ii): We consider the fibre square

$$\begin{array}{ccc}
D & \xrightarrow{\ i\ } & X \\
{\scriptstyle \sigma_D}\Big\downarrow & & \Big\downarrow{\scriptstyle \sigma} \\
\mathbb{P}(\mathcal{F}) & \xrightarrow[\ j\ ]{} & \mathbb{P}(\mathcal{E}).
\end{array}$$

The projections from $\mathbb{P}(\mathcal{E})$ and $\mathbb{P}(\mathcal{F})$ to X are denoted by p and q. The assertion is proved by applying 2.6.2. to this situation. In fact, let us consider the following diagram:

Diagram (6):

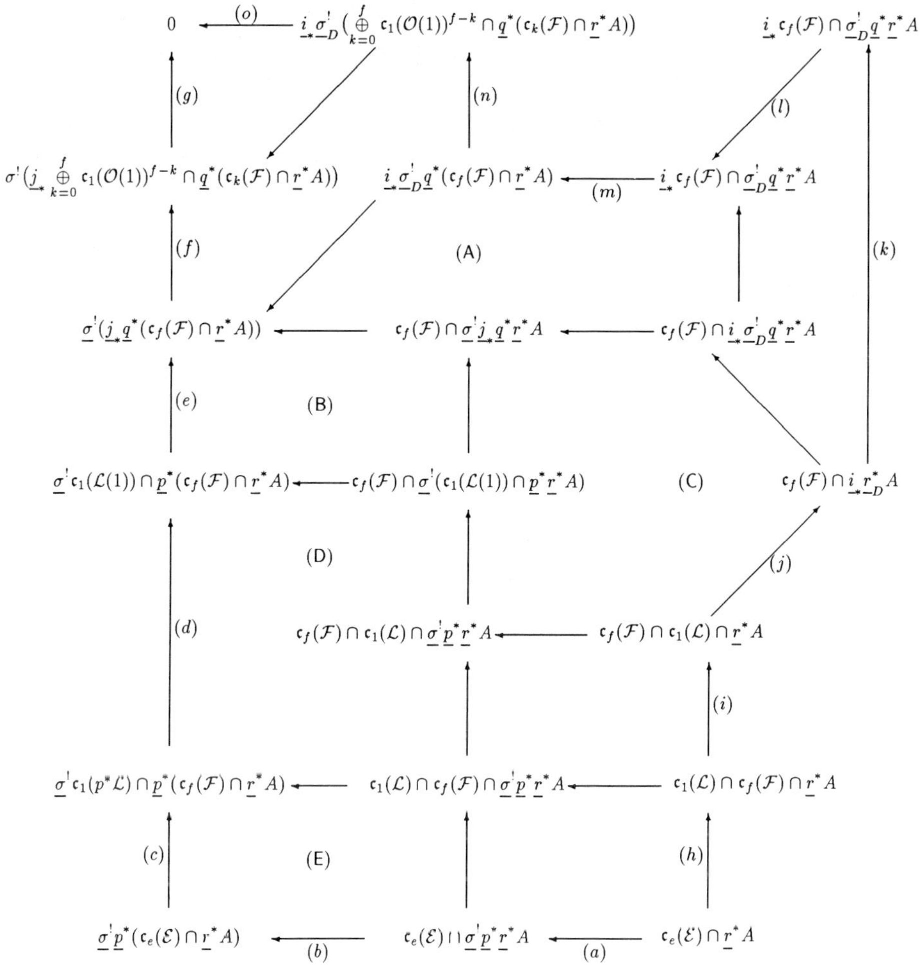

0
(o)
$\underline{i}_*\underline{\sigma}^!_D(\bigoplus_{k=0}^{f} \mathfrak{c}_1(\mathcal{O}(1))^{f-k} \cap \underline{q}^*(\mathfrak{c}_k(\mathcal{F}) \cap \underline{r}^*A))$
$\underline{i}_*\mathfrak{c}_f(\mathcal{F}) \cap \underline{\sigma}^!_D\underline{q}^*\underline{r}^*A$
(g)
(n)
(l)
$\sigma^!(\underline{j}_*\bigoplus_{k=0}^{f} \mathfrak{c}_1(\mathcal{O}(1))^{f-k} \cap \underline{q}^*(\mathfrak{c}_k(\mathcal{F}) \cap \underline{r}^*A))$
$\underline{i}_*\underline{\sigma}^!_D\underline{q}^*(\mathfrak{c}_f(\mathcal{F}) \cap \underline{r}^*A)$
(m)
$\underline{i}_*\mathfrak{c}_f(\mathcal{F}) \cap \underline{\sigma}^!_D\underline{q}^*\underline{r}^*A$
(f)
(A)
(k)
$\underline{\sigma}^!(\underline{j}_*\underline{q}^*(\mathfrak{c}_f(\mathcal{F}) \cap \underline{r}^*A))$
$\mathfrak{c}_f(\mathcal{F}) \cap \underline{\sigma}^!\underline{j}_*\underline{q}^*\underline{r}^*A$
$\mathfrak{c}_f(\mathcal{F}) \cap \underline{i}_*\underline{\sigma}^!_D\underline{q}^*\underline{r}^*A$
(e)
(B)
$\underline{\sigma}^!\mathfrak{c}_1(\mathcal{L}(1)) \cap \underline{p}^*(\mathfrak{c}_f(\mathcal{F}) \cap \underline{r}^*A)$
$\mathfrak{c}_f(\mathcal{F}) \cap \underline{\sigma}^!(\mathfrak{c}_1(\mathcal{L}(1)) \cap \underline{p}^*\underline{r}^*A)$
(C)
$\mathfrak{c}_f(\mathcal{F}) \cap \underline{i}_*\underline{r}^*_DA$
(D)
(j)
(d)
$\mathfrak{c}_f(\mathcal{F}) \cap \mathfrak{c}_1(\mathcal{L}) \cap \underline{\sigma}^!\underline{p}^*\underline{r}^*A$
$\mathfrak{c}_f(\mathcal{F}) \cap \mathfrak{c}_1(\mathcal{L}) \cap \underline{r}^*A$
(i)
$\underline{\sigma}^!\mathfrak{c}_1(p^*\mathcal{L}) \cap \underline{p}^*(\mathfrak{c}_f(\mathcal{F}) \cap \underline{r}^*A)$
$\mathfrak{c}_1(\mathcal{L}) \cap \mathfrak{c}_f(\mathcal{F}) \cap \underline{\sigma}^!\underline{p}^*\underline{r}^*A$
$\mathfrak{c}_1(\mathcal{L}) \cap \mathfrak{c}_f(\mathcal{F}) \cap \underline{r}^*A$
(c)
(E)
(h)
$\underline{\sigma}^!\underline{p}^*(\mathfrak{c}_e(\mathcal{E}) \cap \underline{r}^*A)$
(b)
$\mathfrak{c}_e(\mathcal{E}) \cap \underline{\sigma}^!\underline{p}^*\underline{r}^*A$
(a)
$\mathfrak{c}_e(\mathcal{E}) \cap \underline{r}^*A$

In the definition of the arrows (d), (f), and (n) we have used the canonical isomorphism $\sigma^*\mathcal{O}(1) \simeq \mathcal{O}_X$ defined by the section s. The commutativity of (A) is essentially 2.4.1., (B) follows from 1.16., (C) is 2.6.2., (D) is essentially 2.6.4. and (E) is a consequence of 2.6.5.. The commutativity of the other squares is more or less obvious.

By part (i) of the proposition, the composition $(o) \circ (n) \circ (m) \circ (l)$ coincides with $T_{\mathcal{F}|_D, s|_D}$. Consequently, the composition $(o) \circ \ldots \circ (h)$ is (3). On the other side, it follows from the explicit description of the Whitney isomorphism in 1.10. and part (i) of the proposition that the composition $(g) \circ \ldots \circ (a)$ is $T_{\mathcal{E},s}$. It follows that (3) and $T_{\mathcal{E},s}$ coincide. The proof of the proposition is complete.

3.2 The axioms. We assume that for a vector bundle $\mathcal{E}$ on X

$$\tilde{\mathfrak{c}}_j(\mathcal{E}) \cap \cdot : \tilde{\mathfrak{CH}}^k(X) \to \tilde{\mathfrak{CH}}^{k+j}(X) \tag{7}$$

is an additive functor, and that the following natural transformations are given:

3.2.1 A symmetry isomorphism

$$\sigma_{\mathcal{E},\mathcal{F}}: \tilde{\mathfrak{c}}_j(\mathcal{E}) \cap \tilde{\mathfrak{c}}_k(\mathcal{F}) \cap A \to \tilde{\mathfrak{c}}_k(\mathcal{F}) \cap \tilde{\mathfrak{c}}_j(\mathcal{E}) \cap A \tag{8}$$

3.2.2 For a flat morphism f, an isomorphism

$$\tilde{\mathfrak{c}}_j(f^*\mathcal{E}) \cap \underline{f}^* A \to \underline{f}^*(\tilde{\mathfrak{c}}_j(\mathcal{E}) \cap A). \tag{9}$$

3.2.3 For a proper morphism g of constant relative dimension, an isomorphism

$$\tilde{\mathfrak{c}}_j(\mathcal{E}) \cap \underline{g}_* A \to \underline{g}_*(\tilde{\mathfrak{c}}_j(g^*\mathcal{E}) \cap A). \tag{10}$$

3.2.4 A Whitney sum isomorphism

$$\tilde{\mathfrak{c}}.(\mathcal{E}) \cap A \to \tilde{\mathfrak{c}}.(\mathcal{G}) \cap \tilde{\mathfrak{c}}.(\mathcal{F}) \cap A \tag{11}$$

for every short exact sequence

$$0 \to \mathcal{F} \to \mathcal{E} \to \mathcal{G} \to 0 \tag{12}$$

of vector bundles (the case $\mathcal{F} = 0$ or $\mathcal{G} = 0$ is not excluded).

The following axioms must be satisfied:

AX 0 (Vanishing) $\tilde{\mathfrak{c}}_j(\mathcal{E}) \cap .$ is the zero functor if $j < 0$ or $j > \dim(\mathcal{E})$, and $\tilde{\mathfrak{c}}_0(\mathcal{E}) \cap A = A$. For $\mathcal{G} = 0$, this implies $\tilde{\mathfrak{c}}.(\mathcal{G}) \cap A = A$.

AX 1 (Normalisation) $\tilde{\mathfrak{c}}_1(\mathcal{L}) \cap A = \mathfrak{c}_1(\mathcal{L}) \cap A$ for a line bundle $\mathcal{L}$, where $\mathfrak{c}_1(\mathcal{L}) \cap .$ is the additive functor introduced in 1.2.. If $\mathcal{F}$ and $\mathcal{G}$ are line bundles, then (8) coincides with the symmetry introduced in 1.5.. Also, in the case of line bundles (9) and (10) coincide with the isomorphisms introduced in 1.4..

AX 2 (Compatibilities for (9) and (10)) If $\mathcal{E}$ is a vector bundle on X and if X-schemes are denoted by $p: Y \to X$, then (9) and (10) define for the additive functors

$$\tilde{\mathfrak{c}}_j(p^*\mathcal{E}) \cap .: \tilde{\mathfrak{CH}}^{\cdot}(Y) \to \tilde{\mathfrak{CH}}^{\cdot}(Y)$$

the structure of a biadmissible functor between bifibred Picard categories over (X-schemes). If $\mathcal{E}$ and $\mathcal{F}$ are vector bundles on X, then the isomorphism (8)

$$\tilde{\mathfrak{c}}_j(p^*\mathcal{E}) \cap \tilde{\mathfrak{c}}_k(p^*\mathcal{F}) \cap . \to \tilde{\mathfrak{c}}_k(p^*\mathcal{F}) \cap \tilde{\mathfrak{c}}_j(p^*\mathcal{E}) \cap$$

is a biadmissible functor morphism. Similar, if (12) is an exact sequence of vector bundles on X, then the isomorphism (11)

$$\tilde{\mathfrak{c}}.(p^*\mathcal{E}) \cap . \to \tilde{\mathfrak{c}}.(p^*\mathcal{G}) \cap \tilde{\mathfrak{c}}.(p^*\mathcal{F}) \cap .$$

is biadmissible.

AX 3 The analogues of 1.(45). (for vector bundles of arbitrary dimension) and of 1.(62). for the isomorphisms (8) and (11) commute. Note that this would allow us to apply 2.8. to the functors $\tilde{\mathfrak{c}}_j(\mathcal{E}) \cap .$, but it will not be necessary to do so.

Corollary: $\sigma_{\mathcal{E},\mathcal{F}}\sigma_{\mathcal{F},\mathcal{E}} = \mathrm{Id}$, and the analogue of 1.(34). for the isomorphisms (8) commutes. Note that this will enable us to apply 2.9. to the functors $\tilde{\mathfrak{c}}_j$. Hence for a polynomial P in Chern classes the polynomials $P(\dim(\mathcal{E}_j), \tilde{\mathfrak{c}}_i(\mathcal{E}_j)) \cap A = \tilde{\mathfrak{P}}(\mathcal{E}_j) \cap A$ in Chern functors is well-defined. This will be important in the formulation of **AX 4**.

Proof: This is clear from **AX 1** if all the vector bundles involved are line bundles. The general case follows by induction on the dimension of the vector bundles, using **AX 3** and the splitting principle.

AX 4 (Twist by a line bundle) The analogue of 2.10.1.–4. for the functors $\tilde{\mathfrak{c}}_i$ and the isomorphisms (8)–(11) is true, i.e., there exists an isomorphism

$$\tilde{\mathfrak{c}}_j(\mathcal{E} \otimes \mathcal{L}) \cap A \to \tilde{\mathfrak{P}}(\mathcal{E}, \mathcal{L}) \cap A \qquad (13)$$

satisfying 2.10.1.–2.10.4. (The uniqueness of such an isomorphism is clear).

AX 5 By the previous axioms, the definition of the trivialisation $T_{\mathcal{E},s}$ of $\mathfrak{c}_e(\mathcal{E})\cap .$ defined by a global non-vanishing section s of $\mathcal{E}$ works for the functors $\tilde{\mathfrak{c}}_j$. We assume that Proposition 3.1.(ii). remains true for the functors $\tilde{\mathfrak{c}}_j$.

Remark It seems very likely that **AX 5** is a consequence of the other axioms. I hope I will be able to return to that object in the forthcoming paper on functorial Riemann-Roch.

3.3 Theorem: *If $\tilde{\mathfrak{c}}_j(\mathcal{E})\cap .$ satisfies the properties listed in 3.2., then there is a unique additive functor-isomorphism*

$$\tilde{\mathfrak{c}}_j(\mathcal{E})\cap A \to \mathfrak{c}_j(\mathcal{E})\cap A \tag{14}$$

which commutes with the transformations 3.2.1.–4. and is the identity of $j \le 0$, $j > \dim(\mathcal{E})$, or if $\mathcal{E}$ is a line bundle.

Proof: The uniqueness of (14) is clear. To prove the existence, we first consider some consequences of the axioms:
Step 1: Let $\mathcal{L}$ and $\mathcal{M}$ be line bundles and $\mathcal{E}$ an arbitrary vector bundle. We want to check that the diagram

$$\begin{array}{ccc}
\tilde{\mathfrak{c}}_j(\mathcal{E}\otimes\mathcal{L})\cap\tilde{\mathfrak{c}}_1(\mathcal{M})\cap A & \xrightarrow{\sigma_{\mathcal{E}\mathcal{F}}} & \tilde{\mathfrak{c}}_1(\mathcal{M})\cap\tilde{\mathfrak{c}}_j(\mathcal{E}\otimes\mathcal{L})\cap A \\
\downarrow & & \downarrow \\
\tilde{\mathfrak{P}}(\mathcal{E},\mathcal{L})\cap\tilde{\mathfrak{c}}_1(\mathcal{M})\cap A & \longrightarrow & \tilde{\mathfrak{c}}_1(\mathcal{M})\cap\tilde{\mathfrak{P}}_j(\mathcal{E},\mathcal{L})\cap A
\end{array}$$

commutes (cf. **AX 4**. The lower horizontal arrow is 2.(24)). By 2.10.2., this is clear if $\mathcal{E}$ is a line bundle. The general case follows by induction on the dimension of $\mathcal{E}$, using 2.10.3. and the splitting principle. Here 1.(45). is used again, because the induction argument involves using the Whitney isomorphism.
Step 2: Let

$$\begin{array}{ccc}
D & \xrightarrow{\;i\;} & X \\
& {\scriptstyle q}\searrow \quad \swarrow{\scriptstyle p} & \\
& S &
\end{array}$$

be a commutative diagram with p and q flat and i a regular closed immersion of codimension one. For a vector bundle $\mathcal{E}$ on S, we want to check that the diagram

$$
\begin{array}{ccccc}
\underline{i}_*\underline{q}^*(\tilde{\mathfrak{c}}_j(\mathcal{E})\cap A) & \longrightarrow & \underline{i}_*(\tilde{\mathfrak{c}}_j(q^*\mathcal{E})\cap \underline{q}^*A) & \longrightarrow & \tilde{\mathfrak{c}}_j(p^*\mathcal{E})\cap \underline{p}^* \\
\Big\downarrow{\scriptstyle 1.(39)} & & & & \Big\downarrow{\scriptstyle 1.(39)} \\
\tilde{\mathfrak{c}}_1(\mathcal{O}_X(D))\cap \underline{p}^*(\tilde{\mathfrak{c}}_j(\mathcal{E})\cap A) & & & & \tilde{\mathfrak{c}}_j(p^*\mathcal{E})\cap \tilde{\mathfrak{c}}_1(\mathcal{O}_X(D))\cap \underline{p}^*A \\
& \searrow & & \swarrow & \\
& & \tilde{\mathfrak{c}}_1(\mathcal{O}_X(D))\cap \tilde{\mathfrak{c}}_j(p^*\mathcal{E})\cap \underline{p}^*A & &
\end{array}
$$

commutes.

If $\mathcal{E}$ is a line bundle, this is 1.(40). The general case follows by the splitting principle from 1.(45) (cf. **AX 3**) and **AX 2**. The details have been presented in 1.16..

Step 3: Now we are ready to construct (14). For a vector bundle $\mathcal{E}$ of dimension e on X, we denote by $p: \mathbb{P}(\mathcal{E}) \to X$ the projective fibration. Then

$$\mathcal{E}(1) := p^*\mathcal{E} \otimes \mathcal{O}(1)$$

has a canonical non-vanishing global section s, defining a trivialisation

$$
\begin{array}{ccc}
\bigoplus_{j=0}^{e} \mathfrak{c}_1(\mathcal{O}(1))^{e-j} \cap \underline{p}^*(\tilde{\mathfrak{c}}_j(\mathcal{E})\cap A) & \xrightarrow{\ \textbf{AX 1, AX 4}\ } & \tilde{\mathfrak{c}}_e(\mathcal{E}(1))\cap A \\
& & \Big\downarrow{\scriptstyle T_{\mathcal{E}(1),s}} \\
& & 0
\end{array}
\qquad (15)
$$

Because the isomorphism 1.(37). characterises $\mathfrak{c}_j(\mathcal{E})\cap A$ up to unique isomorphism, (15) defines an isomorphism (14). The compatibility of (14) with the isomorphisms 3.2.2. and 3.2.3. is clear because (15) contains only biadmissible transformations. It remains to prove that (14) commutes with 3.2.1. and 3.2.4..

Step 4: The hard part is the compatibility with (11). By the splitting principle and because 1.(62). was supposed to be commutative, it suffices to consider short exact sequences (12) with $\dim(\mathcal{G}) = 1$. We consider the following diagram.

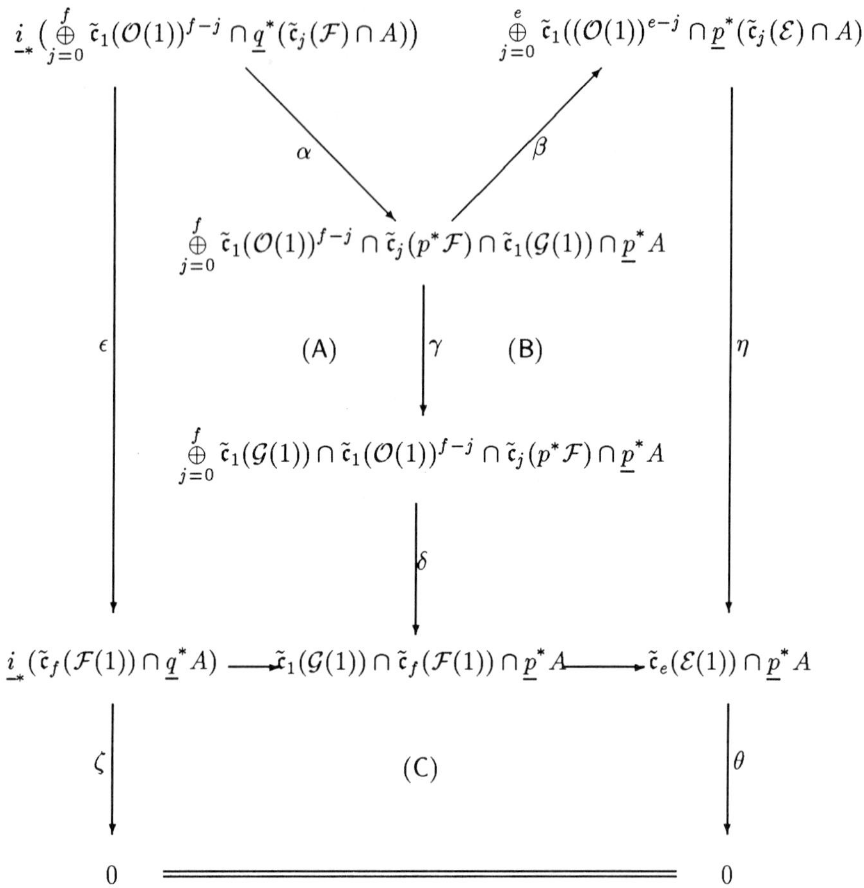

Here α is given by the first interchanging $\tilde{\mathfrak{c}}_j(\mathcal{F}) \cap .$ and $\underline{g}^*$, then interchanging both $\tilde{\mathfrak{c}}_j(\mathcal{F}) \cap .$ and $\tilde{\mathfrak{c}}_1(\mathcal{O}(1)) \cap .$ with $\underline{i}_*$, and finally applying 1.(37). By step 3, the result of first interchanging $\tilde{\mathfrak{c}}_1(\mathcal{O}(1))$ with $\underline{i}_*$, applying $\underline{i}_*\underline{q}^*(\tilde{\mathfrak{c}}_j(\mathcal{F}) \cap A) \to \tilde{\mathfrak{c}}_1(\mathcal{G}(1)) \cap \underline{p}^*(\tilde{\mathfrak{c}}_j(\mathcal{F}) \cap A)$, and then bringing $\tilde{\mathfrak{c}}_j(\mathcal{F})$ to the left side would be the same. Consequently, $\beta\alpha$ coincides with the composition of the arrows

$$\underset{-*}{i}\Big(\bigoplus_{j=0}^{f}\tilde{\mathfrak{c}}_1(\mathcal{O}(1))^{f-j}\cap \underline{q}^*(\tilde{\mathfrak{c}}_j(\mathcal{F})\cap A)\Big)$$

$$\Big\downarrow$$

$$\bigoplus_{j=0}^{e}\tilde{\mathfrak{c}}_j((\mathcal{O}(1))^{e-j}\cap \underline{p}^*(\tilde{\mathfrak{c}}_j(\mathcal{F})\cap A\oplus \tilde{\mathfrak{c}}_1(\mathcal{G})\cap\tilde{\mathfrak{c}}_{j-1}(\mathcal{F})\cap A)$$

$$\Big\downarrow$$

$$\bigoplus_{j=0}^{e}\tilde{\mathfrak{c}}_1((\mathcal{O}(1))^{e-j}\cap \underline{p}^*(\tilde{\mathfrak{c}}_j(\mathcal{E})\cap A).$$

where the first arrow is defined similar to 1.(41) and the second arrow is given by (11). The pentagon (A) commutes by a combination of 2.10.1. (cf. **AX 4**) and step 2. The pentagon (B) would by 2.10.3. (i.e., **AX 4**) commute up to the sign

$$\begin{aligned} c_1(\mathcal{O}(1))\cap c_{f-1}(\mathcal{F}(1))\cap[-1]\cap[\underline{p}^*A] &= \\ &= \sum_{l=0}^{f-1}(l+1)c_1(\mathcal{O}(1))^{l+1}\cap c_{f-l-1}(p^*\mathcal{F})\cap[-1]\cap[\underline{p}^*A] \end{aligned} \tag{16}$$

if $\delta\gamma$ was the bottom horizontal arrow in 2.(30). However, γ involves $(f-j)$ times interchanging $\tilde{\mathfrak{c}}_1(\mathcal{O}(1))$ with itself, whereas the arrow in 2.(30) uses only minimal permutations. Hence γ produces the additional sign

$$\sum_{j=0}^{f}(f-j)c_1(\mathcal{O}(1))^{f-j}\cap c_j(p^*\mathcal{F})\cap[-1]\cap[\underline{p}^*A]$$

cancelling (16). Hence (B) commutes.

By **AX 5**, (C) also commutes. Now $\zeta\epsilon$ is (15) for $\mathcal{F}$ and $\theta\eta$ is (15) for $\mathcal{E}$, whereas $\beta\alpha$ is 1.(41) with $\mathfrak{c}$ replaced by $\tilde{\mathfrak{c}}$. This proves compatibility between (14) and the Whitney isomorphism.
Step 5: The compatibility between (14) and $\sigma_{\mathcal{E},\mathcal{F}}$ follows now by induction on $\dim(\mathcal{E})$ and $\dim(\mathcal{F})$, using **AX 1** for the start and the result of step 4, 1.(45) (cf. **AX 3**), and the splitting principle for the induction argument. The proof of 3.3. is complete.

3.4 Comparison with Deligne's IC_2. Let $p: X \to S$ be a proper smooth morphism of relative dimension one, where S is normal and locally factorial. For line bundles $\mathcal{L}$, $\mathcal{M}$ on X, put

$$<\mathcal{L},\mathcal{M}>= \underline{p}_*(\mathfrak{c}_1(\mathcal{L})\cap \mathfrak{c}_1(\mathcal{M})). \tag{17}$$

Note that by 1.3. this is the line bundle on S constructed in [D]. If ℓ and m are rational sections of $\mathcal{L}$ and $\mathcal{M}$ whose divisors do not intersect, then

$$<\ell,m>= p_*(l\cap m) \tag{18}$$

is a section of $<\mathcal{L},\mathcal{M}>$ on S satisfying the transformation rules of [D]. For a virtual vector bundle $\mathcal{E}$ on X, put

$$IC_2(\mathcal{E}) = \underline{p}_*(\mathfrak{c}_2(\mathcal{E})). \tag{19}$$

This functor has the following two structures:

3.4.1 For a line bundle $\mathcal{L}$ on X, a canonical isomorphism $IC_2(\mathcal{L})\simeq \mathcal{O}_X$ defined by the canonical trivialisation $\mathfrak{c}_2(\mathcal{L})\to 0$ in $\tilde{\mathfrak{CH}}^2(X)$.

3.4.2 A Whitney isomorphism

$$IC_2(\mathcal{E}\oplus\mathcal{F})\to IC_2(\mathcal{E})\oplus IC_2(\mathcal{F})\otimes <\det(\mathcal{E}),\det(\mathcal{F})> \tag{20}$$

defined by the following arrows:

$$\begin{aligned} \underline{p}_*(\mathfrak{c}_2(\mathcal{E}\oplus\mathcal{F})) &\to \underline{p}_*(\mathfrak{c}_2(\mathcal{E}))\otimes \underline{p}_*(\mathfrak{c}_2(\mathcal{F}))\otimes \underline{p}_*(\mathfrak{c}_1(\mathcal{E})\cap \mathfrak{c}_1(\mathcal{F})) \\ &\to IC_2(\mathcal{E})\otimes IC_2(\mathcal{F})\otimes \underline{p}_*(\mathfrak{c}_1(\det(\mathcal{E}))\cap \mathfrak{c}_1(\det(\mathcal{F}))) \\ &\to IC_2(\mathcal{E})\otimes IC_2(\mathcal{F})\otimes <\det(\mathcal{E}),\det(\mathcal{F})>, \end{aligned} \tag{21}$$

where the first arrow is the Whitney isomorphism for the functors $\mathfrak{c}_k$, the second arrow is 2.7. plus definition (19), and the third one *differs by the sign*

$$(-1)^{\deg(\det(\mathcal{E}))\dim(\mathcal{F})}$$

from the tautological arrow given by (17).

Proposition. *IC_2, together with the isomorphisms 3.4.1. and 3.4.2., satisfies the axioms of [D, Proposition 9.4.]. Consequently, it is canonically isomorphic to the functor which Deligne named IC_2.*

Proof: *Step 1:* It is immediately verified that if $T(\mathcal{E})$ is defined as in [D,9.5.], then T becomes an additive functor between Picard categories. In particular the compatibility of $T(\mathcal{E})$ with the symmetries of its source and target categories follows from 1.(63) in view of the sign convention we made

in (21). It remains to verify assumption (iii) of [D,9.5.]. We start with a preparation for this.
Step 2: By 2.10., we have an isomorphism

$$IC_2(\mathcal{E}\otimes\mathcal{L}) \simeq IC_2(\mathcal{E})\otimes <\det(\mathcal{E}),\mathcal{L}>^{e-1} \otimes <\mathcal{L},\mathcal{L}>^{e(e-1)/2}, \qquad (22)$$

(with $e = \dim(\mathcal{E})$). Let us consider the square

$$\begin{array}{ccc}
IC_2((\mathcal{E}\oplus\mathcal{F})\otimes\mathcal{L}) & \longrightarrow & \begin{array}{c} IC_2(\mathcal{E}\otimes\mathcal{L})\otimes IC_2(\mathcal{F}\otimes\mathcal{L})\otimes \\ \otimes <\det(\mathcal{E},\otimes\mathcal{L}),\det(\mathcal{F}\otimes\mathcal{L})> \end{array} \\
\downarrow & & \downarrow \qquad (23) \\
\begin{array}{l} IC_2(\mathcal{E}\oplus\mathcal{F})\otimes \\ \otimes <\det(\mathcal{E}\oplus\mathcal{F}),\mathcal{L}>^{e+f-1}\otimes \\ \otimes <\mathcal{L},\mathcal{L}>^{(e+f)(e+f-1)/2} \end{array} & \longrightarrow & \begin{array}{r} IC_2(\mathcal{E})\otimes IC_2(\mathcal{F})\otimes <\det(\mathcal{E}),\det(\mathcal{F})>\otimes \\ \otimes <\det(\mathcal{E}),\mathcal{L}>^{e-1}\otimes <\det(\mathcal{F}),\mathcal{L}>^{f-1}\otimes \\ \otimes <\det(\mathcal{E}),\mathcal{L}>^{f}\otimes <\det(\mathcal{F}),\mathcal{L}>^{e}\otimes \\ \otimes <\mathcal{L},\mathcal{L}>^{ef+f(f-1)/2+e(e-1)/2} \end{array}
\end{array}$$

By 2.10.3., the correct sign for (23) would be

$$(-1)^{\deg(\mathcal{L})ef} \qquad (24)$$

if there was no sign convention in (20). However, the sign convention modifies the top arrow by

$$(-1)^{f(\deg(\det(\mathcal{L}))+\deg(\mathcal{L})e)}$$

and the bottom arrow by

$$(-1)^{f\,\deg(\det(\mathcal{E}))},$$

cancelling (24). Consequently, (23) commutes on the nose.

If $\mathcal{E}$ is a line bundle, (22) reduces to a canonical isomorphism $IC_2(\mathcal{E}\otimes\mathcal{L}) \simeq IC_2(\mathcal{E})$, and this isomorphism respects the canonical trivialisations of both sides.
Step 3: Now we are ready to verify assumption (iii) in [D,9.5.]. We recall that this signifies the following:

Let $\mathcal{Q}'$ be a line bundle on S, s a section of p, $\mathcal{Q} = s_*\mathcal{Q}'$ in $\mathfrak{K}(X)$, $\mathcal{L}$ a line bundle on X together with an isomorphism $\mathcal{Q}' \simeq s^*\mathcal{L}$. Then the isomorphisms

$$\begin{array}{rcl}
\mathcal{O}_S & \to & IC_2(\mathcal{L}) \\
 & \to & IC_2(\mathcal{L}(-s(S)))\otimes IC_2(\mathcal{Q})\otimes <\det(\mathcal{Q}),\mathcal{L}(-s(S)) \\
 & \to & IC_2(\mathcal{Q})\otimes s^*\mathcal{L}(-s(S)),
\end{array}$$

where the second arrow is given by (20) and the exact sequence

$$0 \to \mathcal{L}(-s(S)) \to \mathcal{L} \to \mathcal{Q} \to 0,$$

define

$$I_{\mathcal{L}}: IC_2(\mathcal{Q}) \to \mathcal{Q}'^{-1} \otimes s^*\mathcal{O}_X(-s(S)).$$

The condition is that $I_{\mathcal{L}}$ is independent of $\mathcal{L}$.

Because S is integral, we may assume $S = \mathrm{Spec}(k)$ for some field k. Then $\mathcal{Q} = \mathcal{O}_S$, and $\mathcal{L}$ has a trivialisation at s. If $\mathcal{M}$ is another line bundle on S, we have to check $I_{\mathcal{L}} = I_{\mathcal{M}}$. We claim that this condition depends only on $\mathcal{N} = \mathcal{M} \otimes \mathcal{L}^{-1}$. Indeed, by step 2 we have the commutative square

$$\begin{array}{ccc}
IC_2(\mathcal{L}\otimes\mathcal{N}) & \longrightarrow & \begin{array}{c} IC_2(\mathcal{L}\otimes\mathcal{N}(-s))\otimes \\ \otimes <\mathcal{L}\otimes\mathcal{N}(-s), \mathcal{O}_X(s)> \otimes \\ \otimes IC_2(\mathcal{Q}\otimes\mathcal{N}) \end{array} \\
\swarrow \quad \downarrow & & \downarrow \\
0 \qquad\quad & & \\
\searrow \quad \downarrow & & \\
IC_2(\mathcal{L}) & \longrightarrow & \begin{array}{c} IC_2(\mathcal{L}(-s))\otimes <\mathcal{L}(-s), \mathcal{O}_X(s)> \otimes \\ \otimes IC_2(\mathcal{Q})\otimes <\mathcal{N}, \mathcal{O}_X(S)> \otimes \\ \otimes <\mathcal{N}, \mathcal{O}_X(s)>^{-1} \end{array}
\end{array},$$

reducing the proof of $I_{\mathcal{L}} = I_{\mathcal{M}}$ to the commutativity of

$$\begin{array}{ccc}
IC_2(\mathcal{Q}) & \longrightarrow & IC_2(\mathcal{Q}\otimes\mathcal{N}) \\
& \searrow \quad \swarrow & \\
& IC_2(\mathcal{Q})\otimes <\mathcal{N}, \mathcal{O}_X(s)>^{-1}, &
\end{array}$$

where the horizontal arrow is given by the trivialisation of $\mathcal{N}$ at the point s, the right slanted arrow is (22), and the left slanted arrow is given by the trivialisation of $<\mathcal{N}, \mathcal{O}_X(s)>$ which sends $<n, \text{“}\mathbf{1}\text{''}>$ to $n(s)$, where “**1**'' is the canonical section of $\mathcal{O}_X(s)$ with a zero at s, and n is any rational section of $\mathcal{N}$ whose order at s is zero. It is clear that the last diagram depends only on $\mathcal{N}$, which proves our claim.

Because the condition we have to verify depends only on $\mathcal{L}\otimes\mathcal{M}^{-1}$, we may assume that $\mathcal{L}$ and $\mathcal{M}$ have global section ℓ and m with $\ell(s) = m(s) = 1$ and whose divisors do not intersect. Then $\sigma = (\ell, -m)$ is a global non-vanishing section of $\mathcal{E} = \mathcal{L}\oplus\mathcal{M}$, which is contained in

$$\begin{aligned} \mathcal{F} = \ker(\mathcal{E} = \mathcal{L}\oplus\mathcal{M} &\rightarrow \mathcal{Q}) \\ <\lambda,\mu> &\rightarrow \lambda(s)+\mu(s). \end{aligned} \tag{25}$$

We have a commutative diagram in $\mathfrak{K}(X)$

$$\begin{array}{ccc} \mathcal{E} & \longrightarrow & \mathcal{F}\oplus\mathcal{Q} \\ \downarrow & & \downarrow \\ \mathcal{L}\oplus\mathcal{M} \longrightarrow \mathcal{L}(-s)\oplus\mathcal{Q}\oplus\mathcal{M} & \longrightarrow & \mathcal{L}(-s)\oplus\mathcal{M}\oplus\mathcal{Q} \end{array} \tag{26}$$

The right vertical arrow is given by $0\rightarrow\mathcal{L}(-s)\rightarrow\mathcal{F}\rightarrow\mathcal{M}\rightarrow 0$, and the horizontal arrow by $0\rightarrow\mathcal{F}\rightarrow\mathcal{E}\rightarrow\mathcal{Q}\rightarrow 0$, with $\mathcal{E}\rightarrow\mathcal{Q}$ defined in (25). The non-vanishing global section σ defines trivialisations T_σ of $IC_2(\mathcal{E})$ and $IC_2(\mathcal{F})$. Now the top row of (26) defines isomorphisms

$$\begin{array}{ccc} 0 \xrightarrow{T_\sigma} IC_2(\mathcal{E}) & \longrightarrow & IC_2(\mathcal{F})\otimes IC_2(\mathcal{Q})\otimes <\det(\mathcal{F}),\mathcal{O}_X(s)> \\ & & \downarrow T_\sigma \\ & & IC_2(\mathcal{Q})\otimes <\det(\mathcal{F}),\mathcal{O}_X(s)>, \end{array}$$

where the second arrow uses the canonical isomorphism $det(\mathcal{Q})\cong\mathcal{O}_X(s)$.

In view of the canonical isomorphism $<\mathcal{L}\otimes\mathcal{M},\mathcal{O}_X(s)>\rightarrow\mathcal{O}_S$ defined by ℓ and m, this defines

$$I_\mathcal{E}: IC_2(\mathcal{Q})\rightarrow s^*\mathcal{O}_X(s).$$

We want to compare $I_\mathcal{E}$ and $I_\mathcal{L}$. (26) gives us a commutative diagram

$$
\begin{array}{ccc}
IC_2(\mathcal{E}) & \longrightarrow & IC_2(\mathcal{F}) \otimes IC_2(\mathcal{Q}) \otimes <\det(\mathcal{F}), \mathcal{O}_X(s)> \\
\Big\downarrow & & \Big\downarrow \\
\begin{array}{r} IC_2(\mathcal{L}) \otimes IC_2(\mathcal{M}) \otimes \\ \otimes <\mathcal{L}, \mathcal{M}> \end{array} & \longrightarrow & \begin{array}{r} IC_2(\mathcal{L}(-s)) \otimes IC_2(\mathcal{Q}) \otimes <\mathcal{L}(-s), \mathcal{O}_X(s)> \otimes \\ \otimes <\mathcal{L}(-s), \mathcal{M}> \otimes <\mathcal{M}, \mathcal{O}_X(s)> \otimes IC_2(\mathcal{M}). \end{array}
\end{array}
$$

By applying 3.1.(ii) to the exact sequence $0 \to \mathcal{L} \to \mathcal{E} \to \mathcal{M} \to 0$ and the global section σ of $\mathcal{E}$, we find that by the left vertical arrow the trivialisation T_σ of $IC_2(\mathcal{E})$ corresponds to the trivialisation of $<\mathcal{L}, \mathcal{M}>$ which maps

$$<\ell, -m> \quad \text{to} \quad (-1)^{\deg(\mathcal{M})} \tag{27}$$

(of course, multiply by the canonical trivialisations of $IC_2(\mathcal{L})$ and $IC_2(\mathcal{M})$. The sign in (27) comes in from the sign convention in (21)).

In a similar manner, applying 3.1. to the short exact sequence $0 \to \mathcal{L}(-s) \to \mathcal{F} \to \mathcal{M} \to 0$ and the global section σ of $\mathcal{F}$, we conclude that the trivialisation T_σ of $IC_2(\mathcal{F})$ corresponds by the right horizontal arrow to the trivialisation of $<\mathcal{L}(-s), \mathcal{M}>$ which maps

$$<\ell, -m> \quad \text{to} \quad (-1)^{\deg(\mathcal{M})}, \tag{28}$$

where ℓ is viewed as s rational section of $\mathcal{L}(-s)$ which is singular at s but regular at the divisor of m.

If we identify both $<\det(\mathcal{F}), \mathcal{O}_X(s)>$ and $<\mathcal{L}(-s), \mathcal{O}_X(s)>$ to $s^*\mathcal{O}_X(s)^{-1}$, then under the right vertical arrow these identifications differ by the trivialisation of $<\mathcal{M}, \mathcal{O}_X(s)>$ which maps

$$<-m, \text{“}\mathbf{1}\text{''}> \quad \text{to} \quad -1. \tag{29}$$

By (27), (28), and (29), we conclude that $I_\mathcal{E} = -I_\mathcal{L}$. In a similar manner, using the commutative diagram

$$\begin{array}{ccccccccc}
0 & \longrightarrow & \mathcal{M}(-s) & \longrightarrow & \mathcal{F} & \longrightarrow & \mathcal{L} & \longrightarrow & 0 \\
 & & \downarrow & & \downarrow & & \| & & \\
0 & \longrightarrow & \mathcal{M} & \longrightarrow & \mathcal{E} & \longrightarrow & \mathcal{L} & \longrightarrow & 0 \\
 & & \downarrow & & \downarrow & & & & \\
 & & \mathcal{Q} & = & \mathcal{Q} & & & &
\end{array}$$

of exact sequences, one proves $I_{\mathcal{M}} = -I_{\mathcal{E}}$. Consequently, $I_{\mathcal{L}} = I_{\mathcal{M}}$, completing the verification of [D,9.5.(iii)] and the proof of the proposition.

In a similar manner one can compare the integrals of our Chern functors over the fibres of a higher-dimensional morphism with Elkiks line bundles (in case both functors are defined). The starting point is the comparison for an integral of a product of first Chern classes of line bundles, where one has to verify that the integral of the product of $\mathfrak{c}_1$'s satisfies the descent condition used by Elkik. The extension of this isomorphism for $\mathfrak{c}_1$ to the general case is easier than the comparison with Deligne's IC_2 carried out in the above proposition, because the construction we used in 1.10 is also used by Elkik.

References

[D] P. Deligne, *Le déterminant de la cohomologie*, current trends in arithmetic algebraic geometry, Contemporary Mathematics **67** (1987), pp. 93–177.

[DM] . Deligne, J. Milne, *Tannakian Categories*, Lecture Notes in Mathematics **900**, pp. 101-228

[E] R. Elkik, *Intersections Relatives de fibrés en droites et intégrales de classes de Chern*, Preprint.

[F1] J. Franke, *Chow categories*, to appear in Compositio Math.

[F2] –, *Chow categories*, mimeographed notes, distributed in June 1988.

[F3] –, Talk at the Arbeitstagung 1988 of the Max-Planck-Institute, in the proceedings of that Arbeitstagung.

[FU] W. Fulton, *Intersection Theory*, Springer, 1984

[M] J. Milnor, *Introduction to algebraic K-theory*, Annals of Mathematics Studies **72**, Princeton 1971

[Q] D. Quillen, *Higher algebraic K-theory I*, Lecture Notes in Mathematics **341**, pp. 77–139

[W] F. Waldhausen, *Algebraic K-theory of generalized free products, Part 1*, Annals of Mathematics, **108**:1, pp. 135–204

Curves of genus 2 covering elliptic curves and an arithmetical application

GERHARD FREY and ERNST KANI

Introduction

The purpose of the following paper is to show how one can relate elliptic curves E with covering curves C of genus 2 and arithmetical properties of E and C in the case that the ground field is a global field.

In the center of our interest is the height conjecture for elliptic curves (cf. [5]) which would have nice implications for elliptic curves as well as for ternary diophantine equations; as it is well known it would imply an asymptotic version of Fermat's last theorem. It turns out that the height of E is closely connected with the self intersection number of the canonical sheaf of the arithmetical surface corresponding to C. This procedure should be compared with the very interesting approach of Parshin [14], and in fact a (stronger) version of a general inequality for self intersections of canonical sheaves conjectured by Parshin in [14] and known as Bogomolov-Miyaoka-Yau inequality in the geometric case would suffice to prove the height conjecture for E. We have to make the (hopefully only) technical restriction that the absolute values of the j-invariant of E are bounded. The difference to Parshin's approach is that we need the inequality "only" for curves of genus 2 (Corollary 4.5).

In the first section we give a short but systematic report on how to construct curves of genus 2 covering elliptic curves. It turns out that the choice of level-n-structures of pairs of elliptic curves is the essential datum for these coverings. For the arithmetic of C the Weierstraß points play an important role; in fact one has to know their configuration on the special fibres of the minimal model of C. In §2 we study this configuration in detail (cf. [13]).

In §3 we state height conjectures for curves and abelian varieties over global fields (all known to be true if the ground field is a function field), and in the last section we show, as announced above already, how the self intersection of canonical sheaves of curves of genus 2 is related to the height of elliptic curves covered by them. Here the essential tool is the relation between Néron-Tate heights on Jacobians and heights on curves which can be found in [20] and which is applied to Weierstraß points.

Acknowledgements: The authors are very thankful for helpful discussions with A. Brumer and L. Szpiro. The first author wants to thank Queen's University, Kingston, Canada, for its warm hospitality during a visit supported by NSERC and especially the organizers of the conference at Texel, Netherlands, who provided a generous and stimulating athmosphere which made the visit at Texel both pleasant and mathematically fruitful.

This research was supported in part by the Natural Science and Engeneering Research Council of Canada (NSERC) through a research grant hold by the second author.

1 Construction of coverings of genus 2 of elliptic curves

Our aim in this section is to construct curves of genus 2 covering two given elliptic curves. In order to get an idea what one has to expect we begin by looking at the situation that we have already a curve C of genus 2 defined over a field K and a K-morphism

$$\varphi_1 \colon C \longrightarrow E_1$$

of degree n prime to char(K) where E_1 is an elliptic curve defined over K. We assume that there is no non trivial unramified covering between E_1 and C. Let J be the Jacobian of C and

$$\varphi_1^* \colon E_1 \longrightarrow J \qquad \text{resp.}$$

$$\varphi_{1,*} \colon J \longrightarrow E_1$$

be the maps induced by φ_1.

The maximality of $(C \xrightarrow{\varphi_1} E_1)$ implies that φ_1 is injective and that $\ker(\varphi_{1,*})$ is an elliptic curve E_2^* (cf. Serre [16], p. 130).

Obviously we have

$$\varphi_{1*} \cdot \varphi_1^* = n \cdot \mathrm{id}_{E_1}$$

and hence $\varphi_1^*(E_1) =: E_1^*$ intersects E_2^* exactly in $E_1^*[n]$, the n-torsion points of E_1^*.

The projection $\varphi_{2,*}\colon J \longrightarrow J/E_1^* =: E_2$ induces a covering:

$$\varphi_2\colon C \longrightarrow E_2$$

of degree n, too, and the map

$$h = \varphi_1^* \times \varphi_2^*\colon E_1 \times E_2 \longrightarrow J$$

is an isogeny of degree n^2.

Let $H := \ker(h)$. Since $H \cap E_i = \{0_{E_i}\}$ we get a map

$$\alpha\colon E_1[n] \longrightarrow E_2[n]$$

with graph H which is defined over K. Let e_n be the Weil pairing on $(E_1 \times E_2)[n]$. Then H is isotropic with respect to e_n, i. e.:

$$e_n(\alpha(P), \alpha(Q)) = e_n(P, Q)^{-1}$$

for all P, $Q \in E_1[n]$.
We summarize: To

$$\varphi_1\colon C \longrightarrow E_1$$

we associated

$$\varphi_2\colon C \longrightarrow E_2$$

and a K-rational isomorphism

$$\alpha\colon E_1[n] \longrightarrow E_2[n]$$

whose graph H is isotropic in $(E_1 \times E_2)[n]$ such that $H = \ker(\varphi_1^* \times \varphi_2^*)$.

In fact α is the essential datum of C and this is what we want to describe in the case that n is odd. Although the following construction seems to be known in principle (cf. Serre [17], Ibukiyama, Katsura, Oort [7]) it does not to our knowledge appear explicitly in the literature and we therefore sketch the essential steps of the construction.

So begin with elliptic curves E_1 and E_2 defined over K with zero points 0_1 and 0_2. Let $n > 1$ be an odd number prime to $\mathrm{char}(k)$ and assume that

$$\alpha\colon E_1[n] \longrightarrow E_2[n]$$

is a K-rational isomorphism with $H_\alpha := \text{graph}(\alpha)$ isotropic in $(E_1 \times E_2)[n]$ (with respect to the e_n-pairing).

The divisor $\Theta := 0_1 \times E_2 + E_1 \times 0_2 \in \text{Div}(E_1 \times E_2)$ defines a principal polarization with corresponding map

$$\theta: E_1 \times E_2 \longrightarrow (E_1 \times E_2)^\wedge \ .$$

Let

$$h: E_1 \times E_2 \longrightarrow E_1 \times E_2/H_\alpha =: J_\alpha$$

be the canonical map.

By $\mathcal{D}_\alpha$ we denote the set of effective divisors D on $E_1 \times E_2 \otimes \bar{K}$ (where $\bar{K}$ = alg. closure of K) satisfying

i) $D \sim n\Theta$, and

ii) $T_x(D) = D$ for all $x \in H_\alpha \otimes \bar{K}$.

$\mathcal{C}_\alpha$ is the set of effective divisors C on $J_\alpha \otimes \bar{K}$ with

i) $h^*(C) \sim n\Theta$, and

ii) $(C \cdot C) = 2$.

Since h is etale we have: A divisor D on $(E_1 \times E_2) \otimes \bar{K}$ is invariant under translations by all $x \in H_\alpha \otimes \bar{K}$ if and only if $D \in h^*(\text{Div}(J_\alpha))$ and so h^* induces a bijection between $\mathcal{C}_\alpha$ and $\mathcal{D}_\alpha$.

The isotropy condition on H_α ensures (use [11], p. 231) that there is a divisor $\bar{D}$ on J_α with $h^*(\bar{D}) \sim n\Theta$ and so $h^0(\bar{D}) > 0$ and hence $\mathcal{C}_\alpha \neq 0$.

Choose $C \in \mathcal{C}_\alpha$ and define

$$\lambda_C =: \lambda: J_\alpha \longrightarrow J_\alpha^\wedge$$

by

$$\lambda(x) := [T_x(C) - C] \ .$$

Since $n\Theta$ is ample each $C \in \mathcal{C}_\alpha$ is ample and the Riemann-Roch theorem (cf. [11], p. 150) yields:

$$\deg(\lambda) = [1/2(C \cdot C)] = 1 \ .$$

Hence λ is an isomorphism.
The map

$$h' := \theta^{-1} \circ h^\wedge \circ \lambda$$

has a kernel which acts freely and transitively on $\mathcal{C}_\alpha$ and so λ and h' are defined over K and independent of the choice of C.

Moreover $\#\mathcal{C}_\alpha = \#\ker(h') = \#\ker(h) = n^2$. Using [11], p. 59 and 143 and [9], p. 131 one checks that we have the following commutative diagram

$$\begin{array}{ccc} E_1 \times E_2 & \stackrel{\theta}{\longrightarrow} & (E_1 \times E_2)^\wedge \\ \downarrow h & & \downarrow (h')^\wedge \\ J_\alpha & \stackrel{\lambda}{\longrightarrow} & J_\alpha^\wedge \\ \downarrow h' & & \downarrow h^\wedge \\ E_1 \times E_2 & \stackrel{\theta}{\longrightarrow} & (E_1 \times E_2)^\wedge \end{array}$$

with $h' \circ h = n \cdot \mathrm{id}$ and $h^\wedge \circ h'^\wedge = n \cdot \mathrm{id}$.

For $i = 1, 2$ denote by e_i the natural injection of E_i into $E_1 \times E_2$ and by p_i the projections.
The maps

$$h_i := h \circ e_i \qquad \text{and}$$
$$h'_i := p_i \circ h'$$

obviously satisfy

$$h'_i \circ h_j = n \cdot \delta_{ij} \circ \mathrm{id}_{E_i}$$

and the sequences

$$0 \longrightarrow E_j \stackrel{h_j}{\longrightarrow} J_\alpha \stackrel{h'_i}{\longrightarrow} E_i \longrightarrow 0$$

are exact for $i, j \in \{1, 2\}$, $i \neq j$.
Put $\iota = -\mathrm{id}_{J_\alpha}$.

Proposition 1.1: *There exists a unique curve $C \in \mathcal{C}_\alpha$ such that $h^*(C) \sim n\Theta$ and $\iota(C) = C$.*

Proof. For $C' \in \mathcal{C}_\alpha$ we have: $h^*(\iota C') = -\mathrm{id}_{E_1 \times E_2}(h^*(C')) \sim -\mathrm{id}_{E_1 \times E_2}(n\Theta) = n\Theta$, and so ι operates on $\mathcal{C}_\alpha$.

The order of $\mathcal{C}_\alpha$ is odd and so ι has to have a fixed element C in $\mathcal{C}_\alpha$. Let $C' \in \mathcal{C}_\alpha$ be another fixed element, $C' = T_x C$ with $x \in \ker(h')$.

So $C' = T_x C = T_{-x} C$ and so $2x = 0$ or: $x = 0$ since the kernel of h' has odd order. Hence $C = C'$. Q.E.D.

Corollary 1.2: *The curve $C \in \mathcal{C}_\alpha$ with $\iota C = C$ is defined over K and there exists a unique $D \in \mathcal{D}_\alpha$ with $-\mathrm{id}_{E_1 \times E_2}(D) = D$.*

Let $\rho: C \longrightarrow J_\alpha$ resp. $\tilde{\rho}: D \longrightarrow E_1 \times E_2$ be the canonical injections. The restriction of h' to $\rho(C)$ gives a morphism $\bar{\pi}: C \longrightarrow \bar{C} \subset E_1 \times E_2$, and the restriction of h to $\rho(D)$ defines a morphism $\pi: D \longrightarrow C$.

Put $\varphi_i := h'_i|_C$ and $\tilde{\varphi}_i := p_i|_D$. Then we get the commutative diagram

$$\begin{array}{ccccc}
E_1 & \stackrel{\tilde{\varphi}_1}{\longleftarrow} & D & \stackrel{\tilde{\varphi}_2}{\longrightarrow} & E_2 \\
\downarrow n \circ \mathrm{id}_{E_1} & & \downarrow \pi & & \downarrow n \circ \mathrm{id}_{E_2} \\
E_1 & \stackrel{\varphi_1}{\longleftarrow} & C & \stackrel{\varphi_2}{\longrightarrow} & E_2 \\
 & & \downarrow \bar{\pi} & & \\
 & & \bar{C} & &
\end{array}$$

and so we are in a situation as described in the beginning of the section. We would have finished our task if C would be an irreducible curve of genus 2. In fact this is not true in general:

Proposition 1.3: *The following statements are equivalent:*

(i) D is irreducible

(ii) D is a smooth connected curve of genus $n^2 + 1$

(iii) C is irreducible

(iv) C is a smooth connected curve of genus 2.

The proposition easily follows if one uses the adjunction formula on abelian surfaces:

$$p_a(D) = 1/2(D \cdot D) + 1$$

resp.

$$p_a(C) = 1/2(C \cdot C) + 1$$

"In general" C will be irreducible. To be more precise we state

Proposition 1.4 *Suppose that C is reducible. Then*

$$C = E + E' \qquad \textit{(as divisors)}$$

where E and E' are K-rational elliptic curves meeting transversally in a unique K-rational point P_0, and E, E', E_1 and E_2 are all isogenous.

Proof: Since $(C \cdot C) = 2$ the first assertion is a well known result of Weil ([23] Satz 2). Now assume that $(E \cdot h_i(E_i)) = 0$. This implies that $E = T_x(h_i(E_i))$ for some $x \in J(K)$. It follows that $(E \cdot h_j(E_j)) = \#H_\alpha = n^2$.

On the other side we have: $((E + E') \cdot h_j(E_j)) = (n\Theta \cdot e_j(E_j)) = n$, and so we get a contradiction.

Hence $(E \cdot h_i(E_i))$ and similarly $(E' \cdot h_i(E_i))$ are positive numbers, and thus E and E' are not translates of $h_1(E_1)$ and $h_2(E_2)$. Therefore E_1 and E_2 and hence also E and E' have to be all isogenous. Q.E.D.

Remark: It is easy to construct examples where C is reducible. For example, choose elliptic curves E_1, E_2 admitting an isogeny $f: E_1 \longrightarrow E_2$ of degree $n-1$. Then $\alpha = f|_{E_1[n]}$ satisfies the isotropy condition, but its associated curve C is always reducible.

We now assume that E_1 is an elliptic curve defined over K with an "anti-canonical" K-rational level-n-structure $\alpha_1: E_1[n] \longrightarrow \mathbb{Z}/n \times \mathbb{Z}/n$ with $e_n(\alpha_1^{-1}(1,0), \alpha_1^{-1}(0,1)) = \exp(\frac{-2\pi i}{n})$. Let M_n be the modular curve parametrizing elliptic curves with canonical level-n-structures. A consequence of our results is:

Proposition 1.5: *Let $x \in M_n(K)$ correspond to the pair (E_2, α_2), where α_2 is a canonical level-n-structure of the curve E_2. There is a unique projective K-rational curve C_x of (arithmetic) genus 2 covering E_1 and E_2 of degree n such that the induced morphism $\alpha: E_1[n] \longrightarrow E_2[n]$ is given by $\alpha_2^{-1} \circ \alpha_1$. If E_2 is not isogenous to E_1 then C_x is irreducible without singularities.*

One gets even more: Fixing (E_1, α_1) as above the map $(E_2, \alpha_2) \longrightarrow C_{\alpha_2^{-1} \circ \alpha_1}$ induces a morphism of M_n/K to the moduli space of stable curves of genus 2 over K which is birational onto its image.

Remark: In Proposition 1.5 we have fixed E_1 since we shall be interested in this situation later on. Of course, there is a version of 1.5 with E_1, E_2 both variable, too.

2 The configuration of Weierstraß points

1.) The generic situation

We now assume that C is a smooth irreducible curve of genus 2 covering E_1 and E_2 as described in the last section. Let $\varphi_i: C \longrightarrow E_i$ be the covering maps. We continue to assume that $\deg(\varphi_i)$ is prime to $\mathrm{char}(K)$ and odd. Moreover we assume that $\mathrm{char}(K)$ is odd or zero. By $\bar{K}$ we denote as before the algebraic closure of K.

Let w be the hyperelliptic involution on C. The fixed point set $W \subset C(\bar{K})$ of w consists of 6 points, the Weierstraß points of C.

We can choose zero points of E_1 and E_2 such that

$$\varphi_i \circ w = -\mathrm{id}_{E_i} \circ \varphi_i \quad \text{, and so}$$

$$\varphi_i(W) \subset E_i[2].$$

Lemma 2.1 *i)* $\varphi_i(W) = E_i[2]$

ii) There is exactly one point $P_{0,i} \in E_i[2]$ with $\#\varphi_i^{-1}(P_{0,i}) \cap W = 3$[1].

Proof: More generally we claim: For all $P \in E_i[2]$ we have: $\#\varphi_i^{-1}(P) \equiv 1 \bmod 2$.

Since w permutes $\varphi_i^{-1}(P)$ it has a fixed point inside $\varphi_i^{-1}(P)$ and hence the lemma follows by the pigeonhole principle. So look at the claim. If $\varphi_i^{-1}(P)$ contains no ramified point then $\#\varphi_i^{-1}(P) = n \equiv 1 \bmod 2$.

Next assume that there is a $Q \in \varphi_i^{-1}(P)$ with ramification $e_Q = 2$. The Hurwitz genus formula implies that there is exactly one point Q' different from Q on C which is ramified over E_1 of degree 2, too, and so $Q + Q'$ is a canonical divisor of C. Hence $(Q + Q')$ is linearly equivalent to $2 \cdot \tilde{Q}$ for all $\tilde{Q} \in W$ and so Q and Q' are not in W.

Thus $w(Q) = Q'$ and so $Q' \in \varphi_i^{-1}(P)$. But then all points in $\varphi_i^{-1}(P) \setminus \{Q, Q'\}$ are unramified, and hence $\varphi_i^{-1}(P)$ contains $(n - 4) + 2 = n - 2$ points.

The last case is that there is a point $Q \in \varphi_i^{-1}(P)$ with ramification index $e_Q = 3$. Then all other points of $\varphi_i^{-1}(P)$ are unramified, and so $\#\varphi_i^{-1}(P) = (n - 3) + 1 = n - 2 \equiv 1 \bmod 2$. Q.E.D.

Now define $W_i := \varphi_i^{-1}(P_{0,i})$.

Proposition 2.2 *We have: $W = W_1 \dot{\cup} W_2$ and $\varphi_j(W_i) = E_j[2] \setminus \{P_{0,j}\}$ for $i \neq j$. In particular, the zero point of J does not lie on C.*

Because of the uniqueness of $P_{0,i}$ it follows that this point is rational over any field over which C and φ_1, φ_2 are rational. In the following we choose $P_{0,i}$ as zero point of E_i.

Proof: Suppose first that $E_1[2](K) = \{P_{0,1}\}$. Suppose that there is a point $Q \in \varphi_i^{-1}(P_{0,1}) \cap \varphi_i^{-1}(P_{0,2}) \cap W$. Then $h'(Q) = (P_{0,1}, P_{0,2})$ and so $Q \in \ker(h')$ and $\text{ord}(Q) | \gcd(n, 2) = 1$. So $Q = \{0_J\}$ and hence $\#\{\varphi_1^{-1}(P_{0,1}) \cap \varphi_2^{-1}(P_{0,2}) \cap W\} = 1$. So $\#(W_1 \cup W_2) = 5$ and there is a point $P \in W$ with $\{P\} = W \setminus (W_1 \cup W_2)$. P has to be Galois invariant, hence $\varphi_1(P) \in E_1[2](K) = \{P_{0,1}\}$ and $P \in W_1$ which is a contradiction. This proves the first assertion of the proposition, and from lemma 2.1 the second part follows immediately.

Now we get rid of the condition that $E_1[2](K) = \{P_{0,1}\}$. We can assume that K is algebraically closed. We find an extension $F|K$ and an elliptic curve $\tilde{E}_1$ defined over F which specializes to E_1 such that $\tilde{E}_1[2](F) = \{\tilde{P}_{0,1}\}$. Clearly we can choose $\tilde{\alpha}\colon \tilde{E}_1[n] \longrightarrow E_2[n]$ which lifts the isomorphism $\alpha\colon E_1[n] \longrightarrow E_2[n]$ determined by C (cf. Proposition 1.5) and which is rational over a field F_1 with $\tilde{E}_1[2](F_1) = \{\tilde{P}_{0,1}\}$. Since the assertion of the proposition is true for the covering $\tilde{C}$ of $\tilde{E}_1$ and E_2 corresponding to $\tilde{\alpha}$ it follows also for C. Q.E.D.

---

[1] *To ease notation we interpret $\varphi_i^{-1}(P)$ as a subset of $C(\bar{K})$.*

Let's have a brief look at the "degenerated situation" in which the data (E_1, E_2, α) lead to a reducible curve $C = E \cup E'$. Since $wC = C$ it permutes E and E' and so it fixes $\{P_0\} = E \cap E'$. So $P_0 \in J[2](K)$. Now assume that $E_i[2](K) = \{P_{0,i}\}$. Then $J[2](K)$ consists only of the zero point and so P_0 is the zero point. Hence E and E' are abelian subvarieties of J and the set of fixed points of $w|_C$ has seven elements and is the union of the points of order 2 of E and E'.

Hence we get a nice "irreducibility criterion":

Corollary 2.3: *Let E_1, E_2 be elliptic curves over K without K-rational points of order 2 and $\alpha: E_1[n] \longrightarrow E_2[n]$ an isomorphism with isotropic graph H in $(E_1 \times E_2)[n]$, and let C be the corresponding covering curve embedded into $J = (E_1 \times E_2)/H$ according to 1.2. Then C is irreducible if and only if $0_J \notin C(K)$.*

2.) The local situation

We continue to assume that $C|K$ is an irreducible projective curve of genus 2 covering the elliptic curves E_1 and E_2 with covering maps φ_1 and φ_2 of odd degree n. For simplicity we assume that all points of order 2 of E_1 and E_2 (and hence of J) are K-rational. Let $\mathfrak{p}$ be a non-archimedean place of K with ring of integers $O_\mathfrak{p}$, residue field $k_\mathfrak{p}$ and normed valuation $v_\mathfrak{p}$. We assume that E_1 and E_2 (and hence J and C) have semi-stable reduction at $\mathfrak{p}$.

By $\mathcal{C}$, $\mathcal{E}_1$, $\mathcal{E}_2$ and $\mathcal{J}$ we denote the regular minimal (resp. Néron) models of C, E_1, E_2 and J at $\mathfrak{p}$. By $\mathcal{C}^*$ we denote the closure of C in $\mathcal{J}$. The maps $\varphi_{i,*}$ resp. φ_i^* extend to $\mathcal{J}$ resp. $\mathcal{E}_i$ in a unique way, and the same is true for the hyerelliptic involution w. These extensions are denoted by the same letters. Let $\mathcal{W}$ resp. $\mathcal{W}^*$ be the closure of W in $\mathcal{C}$ resp. $\mathcal{J}$. The index $\mathfrak{p}$ at objects mentioned above always means that we look at the special fibre of this object over $k_\mathfrak{p}$. By $\delta_\mathfrak{p}$ we denote the number of singular points on $\mathcal{C}_\mathfrak{p}$. We want to describe $\mathcal{C}_\mathfrak{p}$ and the configuration of $\mathcal{W}_\mathfrak{p}$ inside of $\mathcal{C}_\mathfrak{p}$.

For our purposes it is enough to do this in the case that E_1 has good reduction at $\mathfrak{p}$. Hence we distinguish two cases: 1.) E_2 has good reduction modulo $\mathfrak{p}$, and 2.) E_2 has multiplicative reduction at $\mathfrak{p}$. The results we need can be found in an article of Parshin (cf. [13], p. 80ff) based on work of Ogg [12]. Since the argumentation of Parshin is very short we sketch how to obtain them in the two most interesting cases.

α) The case of good reduction

In this case the special fibre $\mathcal{J}_\mathfrak{p}$ of $\mathcal{J}$ is an Abelian variety. If $\mathcal{C}_\mathfrak{p}$ is irreducible then $\mathcal{C}_\mathfrak{p}$ is a non-singular curve of genus 2 and $\delta_\mathfrak{p} = 0$. Now assume that $\delta_\mathfrak{p} > 0$ and so $\mathcal{C}_\mathfrak{p}^*$ is the union of two elliptic curves intersecting in the zero point of $\mathcal{J}_p$ (cf. the discussion in the last section). It follows that the reductions of points in $\mathcal{W}^*$ lie in the non-singular part of $\mathcal{C}_p^*$. From this fact we can conclude that each point $\mathcal{P}_p$ of $\mathcal{W}_p$ lies on a component of $\mathcal{C}_p$ on which w acts non-trivially: Let Q_0 be a regular point on $\mathcal{C}^*$. We take a desingularization

$$\tau: \tilde{\mathcal{C}} \longrightarrow \mathcal{C}^*$$

which is an isomorphism in a neighbourhood of Q_0 of $\mathcal{C}^*$. We can assume that w acts on this neighbourhood. By the universal property of $\mathcal{C}$ we find a map

$$\Pi: \tilde{\mathcal{C}} \longrightarrow \mathcal{C}$$

which continues the identity map on the generic fibre of $\mathcal{C}$ resp. $\mathcal{C}^*$ and which is an isomorphism near the preimage $\tilde{Q}_0$ of Q_0. Since w has only finitely many fixed points on $\mathcal{C}_p^*$ we see that the image under Π of $\tilde{Q}_0$ lies in a component on which the action of w is non-trivial.[1)]

Now look closer at $\mathcal{C}_p$: By [12] or [13] one knows that $\mathcal{C}_p$ has the following shape:

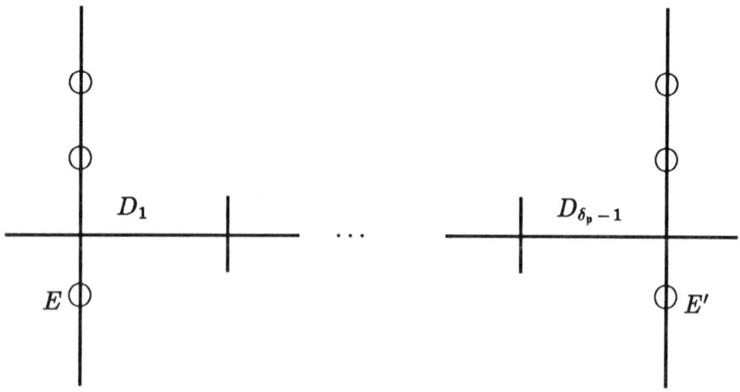

where E and E' are elliptic curves isogenous to $E_{1,p}$ and D_i is a projective line with self intersection -2 (on $\mathcal{C}$) for $i = 1$ to $\delta_p - 1$. w maps D_i into itself and fixes the singular points. If $P_p \in \mathcal{W}_p$ would lie on some D_i w

1) In the simple situation we are looking at our argument is unnecessarily involved. But since we'll have to use it in the next case we have preferrred to give it here already.

would have to act trivially on D_i and we would get a contradiction. Hence (as is to be expected) $\mathcal{W}_\mathfrak{p} \subset (E \cup E')$, as indicated by small circles in the picture.

Remarks: 1.) Of course for $v_\mathfrak{p}(n) = 0$ this result would follow immediately from the discussion of the degenerated situation above (corollary 2.3). 2.) For the arithmetical applications we'll discuss in §4 the degeneration described above causes the major trouble since neither the number $\delta_\mathfrak{p}$ nor the partition of $\mathcal{W}_{1,\mathfrak{p}}$ and $\mathcal{W}_{2,\mathfrak{p}}$ on E and E' can be controlled by help of $\mathcal{J}$.

β) The case of mixed reduction

We now assume that E_1 has good reduction and that E_2 has multiplicative reduction at $\mathfrak{p}$. We shall assume that the points of order n of E_1 (and hence of E_2, too) are K-rational. Let $\mathcal{J}_\mathfrak{p}^0$ be the connected component of the unity of $\mathcal{J}_\mathfrak{p}$. Let Z be the group of connected components of $\mathcal{J}_\mathfrak{p}$:

$$Z = \mathcal{J}_\mathfrak{p}/\mathcal{J}_\mathfrak{p}^0 .$$

We get

Lemma 2.4: *Z is a cyclic group of order dividing $v_\mathfrak{p}(j_2)$ where j_2 is the absolute invariant of E_2. If $\mathfrak{p} \not| \, n$ then $|Z| = \frac{-1}{n} v_\mathfrak{p}(j_2)$.*

Proof: The kernel of $\varphi_{1,*} \times \varphi_{2,*}$ is equal to $E_1[n]$ and so its reduction lies in $\mathcal{J}_\mathfrak{p}^0$. Hence $\varphi_{1,*} \times \varphi_{2,*}$ induces an injective map of Z into the group of components of $\mathcal{E}_{1,\mathfrak{p}} \times \mathcal{E}_{2,\mathfrak{p}}^0$ which is isomorphic to $\mathcal{E}_{2,\mathfrak{p}}/\mathcal{E}_{2,\mathfrak{p}}^0 \cong \mathbb{Z}/v_\mathfrak{p}(j_2)$.

Now assume that $v_\mathfrak{p}(n) = 0$. Let p be a prime and $p^s || n$. Let $\mathcal{J}_\mathfrak{p}^1$ be a component of $\mathcal{J}_\mathfrak{p}$ generating the p-primary part of Z.

Let Q_0 be a point of $J(K)$ with reduction in $\mathcal{J}_\mathfrak{p}^1$. If $p^k || \, |Z|$ we have: The reduction of $p^k Q_0$ lies in $\mathcal{J}_\mathfrak{p}^0$. We find a point $\tilde{Q}$ rational over a $\mathfrak{p}$-unramified extension K' of K with $p^k \tilde{Q} = p^k Q_0$ and $\tilde{Q}$ reducing to $\mathcal{J}_\mathfrak{p}^0$ mod $\mathfrak{p}$. Hence by replacing K by K' and Q_0 by $Q_0 - \tilde{Q} =: Q$ we can assume that Q has order p^k.

Take $Q' = (Q_1, Q_2) \in (E_1 \times E_2)(\bar{K})$ with $(\varphi_1^* \times \varphi_2^*)(Q') = Q$.

Claim: Q' is rational over a field extension K'' which is unramified at $\mathfrak{p}$.

For: Let $\mathfrak{P}$ be an extension of $\mathfrak{p}$ to K and $\tau \in G_\mathfrak{P}$, the inertia group of $\mathfrak{P}$. Q' is a point of finite order and this order is prime to $\mathfrak{p}$ and so $\tau Q_1 = Q_1$. So $\tau Q' - Q' = (0, \tau Q_2 - Q_2) \in H$, and so $\tau Q_2 = Q_2$.

It follows that $v_\mathfrak{p}(j_2) \cdot Q' \in (\mathcal{E}_{1,\mathfrak{p}} \times \mathcal{E}_{2,\mathfrak{p}})^0$. Let p^m be the maximal p-power dividing $v_\mathfrak{p}(j_2)$. Then $p^{m-s} Q'$ has a second component $p^{m-s} Q_2$ which differs from a second component of a point in H only by a point of order prime to p, and so $(\varphi_1^* \times \varphi_2^*)p^{m-s}Q' = p^{m-s}$. Q is a point which differs from a point with reduction in $\mathcal{J}_\mathfrak{p}^0$ only by a point of order prime to p and hence $p^{m-s}Q$ lies in $\mathcal{J}_\mathfrak{p}^0$ mod $\mathfrak{p}$. Hence $k \leq m - s$.

The converse estimate is clear since the group of connected components of $(\mathcal{E}_1 \times \mathcal{E}_2)_\mathfrak{p}$ is cyclic of order $-v_\mathfrak{p}(j_2)$ and H mod $(\mathcal{E}_1 \times \mathcal{E}_2)_\mathfrak{p}^0$ is cyclic of order n.

A consequence of lemma 2.3 is that we have a fairly natural way to compactify $\mathcal{J}$: First recall that $\mathcal{E}_2$ is constructed as open part (i.e. the regular part) of the projective minimal model $\tilde{\mathcal{E}}_2$ of E_2 at $\mathfrak{p}$. Next, the extension $\mathcal{H}$ of H to $\mathcal{E}_1 \times \mathcal{E}_2$ is a finite group scheme whose actions extends to $\tilde{\mathcal{P}} := \mathcal{E}_1 \times \tilde{\mathcal{E}}_2$ (on the second component it maps the Néron polygon into itself).

Then $\mathcal{P} := \tilde{\mathcal{P}}/\mathcal{H}$ is a proper $0_\mathfrak{p}$-scheme containing $\mathcal{J} \times 0_\mathfrak{p}$ for we claim:

Corollary 2.5: $(\varphi_1^* \times \varphi_2^*)(\mathcal{E}_1 \times \mathcal{E}_2) = \mathcal{J}$.

Proof: We have to show the surjectivity of the map in the special fibre. It is clear that $(\mathcal{E}_1 \times \mathcal{E}_2)_\mathfrak{p}^0$ is mapped surjectively to $\mathcal{J}_\mathfrak{p}^0$. Lemma 2.4 implies that $\varphi_1^* \times \varphi_2^*$ induces a surjection of the group of connected components, and hence the corollary follows.

The special fibre of $\mathcal{C}$

Since the abelian part of $\mathcal{J}_\mathfrak{p}$ has dimension 1 the components of $\mathcal{C}_\mathfrak{p}$ have genus ≤ 1, and there is exactly one component, which we call B, of genus 1. The other components are projective lines. Because of the semi-stability all singular points on $\mathcal{C}_\mathfrak{p}$ are ordinary double points.

In order to analyse $\mathcal{C}_\mathfrak{p}$, let us, for simplicity, work over $\bar{k}_\mathfrak{p}$ = alg. closure of $k_\mathfrak{p}$. We'll distinguish two cases.

β_1) First assume that divisors of the form $(\mathcal{P}_\mathfrak{p} - \mathcal{Q}_\mathfrak{p})$ with $\mathcal{P}_\mathfrak{p}, \mathcal{Q}_\mathfrak{p} \in B_{\text{reg}}(\bar{k}_\mathfrak{p})$ generate $\text{Pic}^0(\mathcal{C}_\mathfrak{p}) = \mathcal{J}_\mathfrak{p}^0$. Then $\mathcal{C}_\mathfrak{p}$ is equal to $B \cup C_1 \cup \ldots \cup C_{\delta_\mathfrak{p}-1}$ with C_i projective lines of self intersection -2, and we have the following configuration:

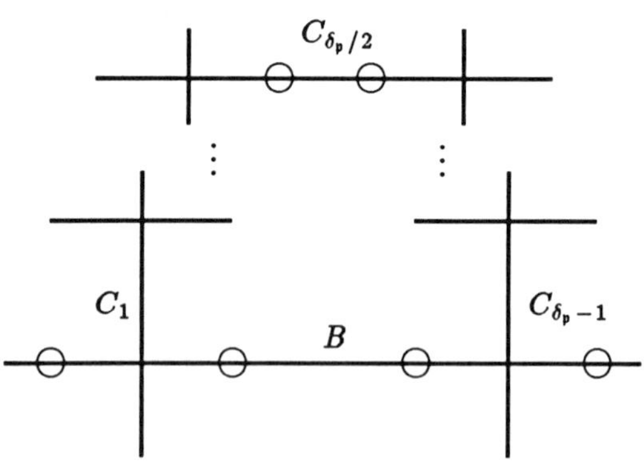

The number of rational components of $\mathcal{C}_\mathfrak{p}$ is linked to the number of connected components of $\mathcal{J}_\mathfrak{p}$ by results of Grothendieck [6](cf. also [15], p. 4ff). In our simple situation one gets by a short computation: $\delta_\mathfrak{p} = \#(\mathcal{J}_\mathfrak{p}/\mathcal{J}_\mathfrak{p}^0)$. Hence $\delta_\mathfrak{p} | v_\mathfrak{p}(j_2)$ and for $\mathfrak{p} \not| n$ we have: $\delta_\mathfrak{p} = -\frac{1}{n} v_\mathfrak{p}(j_2)$. The special fibres of Weierstraß points of $\mathcal{C}$ are located as indicated by the circles (cf. [13]) but since we don't need this we don't go into details of the proof.

β_2) Now assume that $\{(\mathcal{P}_\mathfrak{p} - \mathcal{Q}_\mathfrak{p}) : \mathcal{P}_\mathfrak{p}, \mathcal{Q}_\mathfrak{p} \in B_{\mathrm{reg}}(\bar{k}_\mathfrak{p})\}$ does not generate $\mathcal{J}_\mathfrak{p}^0$. Then $\mathcal{C}_\mathfrak{p}$ looks like this (circles denote special fibres of Weierstraß points again):

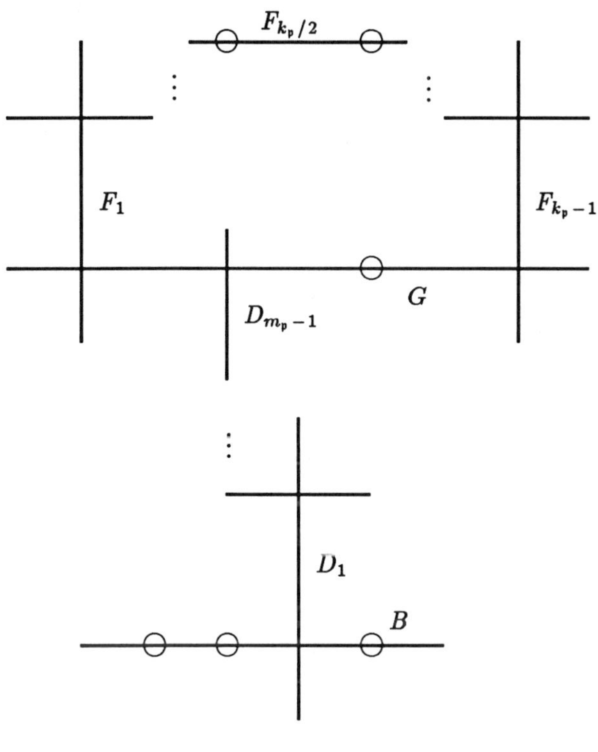

Obviously $\delta_\mathfrak{p} = m_\mathfrak{p} + k_\mathfrak{p}$ and again by Grothendieck's results [6] one computes:

$$k_\mathfrak{p} = \#(\mathcal{J}_\mathfrak{p}/\mathcal{J}_\mathfrak{p}^0).$$

Since for the arithmetical applications we have in mind it is essential where elements of $\mathcal{W}_\mathfrak{p}$ lie we'll give a sketch of the proof that the picture is correct. Let $\mathcal{C}^*$ be the closure of C in $\mathcal{J}$. We restrict the maps h_i' to $\mathcal{C}^*$. Each component of $\mathcal{C}_\mathfrak{p}^*$ is mapped surjectively either under h_1' or h_2' onto a component of $\mathcal{E}_{1,\mathfrak{p}}$ resp. $\mathcal{E}_{2,\mathfrak{p}}$ and is reduced. Hence it contains regular points and the discussion made in section α) above implies that these regular points correspond to points on $\mathcal{C}_\mathfrak{p}$ on which w acts non-trivial. In particular, it follows that there is one component B^* of genus 1 of $\mathcal{C}_\mathfrak{p}^*$ and on this component there is at most one singular point, namely the intersection of at most one rational component of $\mathcal{C}_\mathfrak{p}^*$ with B^*.

Rational components D^* of $\mathcal{C}_\mathfrak{p}^*$ are of the form $D^* = h(P \times D)$ with $P \in \mathcal{E}_{1,\mathfrak{p}}$ and D a component of $\mathcal{E}_{2,\mathfrak{p}}$. Looking at the action of the finite group scheme $\mathcal{H}$ (which is the extension of H to $\mathcal{E}_1 \times \mathcal{E}_2$) one sees that the points of order two of $\mathcal{J}_\mathfrak{p}$ which lie on $\mathcal{C}_\mathfrak{p}^* \setminus B^*$ are regular, and so it follows again by using the argument above that the special fibres of Weierstraß points lying in W_1 are situated on $F_{k_\mathfrak{p}/2}$ resp. G as indicated, and so the special fibres of points in W_2 are on $B \cup D_1 \cup \ldots \cup D_{m_\mathfrak{p}-1}$. Moreover, the intersection of rational components of $\mathcal{C}_\mathfrak{p}^*$ with B^* is mapped under h_1 to the zero point of $\mathcal{E}_{1,\mathfrak{p}}$, and since $2 \not| \, \mathfrak{p}$ this point is not the special fibre of the closure of a point in W_2 on $\mathcal{C}^*$ and hence W_2 is reduced into B (again use the action of w as above), and this concludes the discussion.

3 A digression about heights of curves

We now assume that K is a global field, i.e. either a finite extension of $\mathbb{Q}$ or a function field of one variable over a field of constants K_0 which we assume to be perfect and of characteristic not equal to 2.

$$m_k := \begin{cases} [K:\mathbb{Q}] & \text{if } K \supset \mathbb{Q} \\ 1 & \text{else.} \end{cases}$$

By Σ_K we denote the set of non-archimedean places of K, $S_{K,\infty}$ is the set of archimedean places. Take $\mathfrak{p} \in \Sigma_K$ and denote as above by $0_\mathfrak{p}$ its ring of integers and by $k_\mathfrak{p}$ its residue field. Define

$$\deg(\mathfrak{p}) := \begin{cases} [k_\mathfrak{p} : K_0] & \text{if } K \text{ is a function field,} \\ \log(\# k_\mathfrak{p}) & \text{if } K \supset \mathbb{Q}. \end{cases}$$

$v_\mathfrak{p} \in \mathfrak{p}$ is the normed valuation in $\mathfrak{p}$ and for the divisor $D = \prod_{\mathfrak{p} \in \Sigma_K} \mathfrak{p}^{z_\mathfrak{p}}$ define

$$\deg(D) = \sum_{\mathfrak{p}\in\Sigma_K} z_{\mathfrak{p}} \deg(\mathfrak{p}).$$

For $\mathfrak{p} \in S_{K,\infty}$ we choose an embedding σ of K into $\mathbb{C}$ and define for $x \in K$:

$$|x|_{\mathfrak{p}} = |\sigma x|,$$
$$v_{\mathfrak{p}}(x) = -\log(|\sigma x|) \text{ and}$$
$$\deg(\mathfrak{p}) = \begin{cases} 1 & \text{if } \sigma K \subset \mathbf{R}, \\ 2 & \text{if } \sigma K \not\subset \mathbf{R}. \end{cases}$$

By B we denote the arithmetical curve corresponding to K: In the function field case it is the regular irreducible projective curve over K_0 with function field K, in the number field case it corresponds to $\mathrm{Spec}(0_K) \cup S_{K,\infty}$ where 0_K is the ring of integers of K.
The genus of K is defined by

$$g(K) := \begin{cases} \text{genus of } B & \text{if } K \text{ is a function field} \\ \frac{1}{2}\log|\Delta_{K/\mathbf{Q}}| & \text{if } K \text{ is a number field with discriminant } \Delta_{K/\mathbf{Q}}. \end{cases}$$

Let x be an element of K^*. The height of x is defined by

$$h_K(x) := \sum_{\mathfrak{p}\in\Sigma_K\cup S_{K,\infty}} \max\{0, v_{\mathfrak{p}}(x) \deg \mathfrak{p}\}.$$

Clearly for $K' \supset K$ and $x \in K$ we get:

$$h_{K'}(x) = [K' : K] h_K(x).$$

Now take a projective irreducible smooth curve C of genus $g \geq 1$ defined over K. By $J(C)$ we denote its Jacobian.

Let $\mathcal{C}$ be the regular minimal (arithmetical) surface over B with generic fibre C and morphism

$$f: \mathcal{C} \longrightarrow B.$$

Let $\mathcal{J}(C)$ denote the Néron model of $J(C)$ over K.

For $\mathfrak{p} \in S_{K,\infty}$ and corresponding embedding $\sigma: K \longrightarrow \mathbb{C}$ let $\mathcal{C}_{\mathfrak{p}}$ resp. $\mathcal{J}(C)_{\mathfrak{p}}$ be equal to $\sigma C \otimes \mathbb{C}$ resp. $\sigma(J(C)) \otimes \mathbb{C}$ together with the Arakelov metric μ_σ on $\mathcal{C}_{\mathfrak{p}}$ which is induced by the flat metric on $\mathcal{J}(C)_{\mathfrak{p}}$.

We always assume that $\mathcal{C}_{\mathfrak{p}}$ is semi-stable for all $\mathfrak{p} \in \Sigma_K$. Divisors on $\mathcal{C}$ are Arakelov divisors and intersections of such divisors are taken in the sense of Arakelov ([1]). We are especially interested in the divisor

$$\omega := \omega_{\mathcal{C}/B},$$

the relative dualizing sheaf of $\mathcal{C}/B$.

The Faltings height of C is defined by

$$h_K(C) := \deg f_*\omega$$

(cf. Lang [10]).

Example (cf. [3], p. 29). If C is an elliptic curve E with absolute invariant $j_E \neq 0$ then

$$h_K(E) \sim \frac{1}{12} h_K(j_E).$$

Noether's formula. For $\mathfrak{p} \in \Sigma_K$ let $\delta_\mathfrak{p}$ denote the number of singular points in $\mathcal{C}_\mathfrak{p}$. For $\mathfrak{p} \in S_{K,\infty}$ Faltings (cf. [4]) has defined invariants $\delta_\mathfrak{p}$ of $\mathcal{C}_\mathfrak{p}$ such that

$$\omega^2 + \sum_{\mathfrak{p} \in \Sigma_K \cup S_{K,\infty}} \delta_\mathfrak{p} = 12 h_K(C).$$

We want to state some conjectures related with the height of curves. In the center of our interest are elliptic curves.

So assume that E/K is a semi-stable elliptic curve with absolute invariant j_E and conductor

$$\mathcal{N}_E = \sum_{\substack{v_\mathfrak{p}(j_E)<0 \\ \mathfrak{p} \in \Sigma_K}} \mathfrak{p}.$$

Conjecture (H_E): *There are constants $c_1(K)$ and $d_1(K)$ such that for all semi-stable elliptic curves E/K one has:*

$$h_K(E) \leq c_1(K) \deg \mathcal{N}_E + d_1(K).$$

It is clear that (H_E) is a stronger version of

Szpiro's conjecture(SZ): *Let Δ_E be the minimal discriminant divisor of E. Then*

$$\deg(\Delta_E) \leq c_2(K) \deg \mathcal{N}_E + d_2(K)$$

with $c_2(K)$, $d_2(K)$ depending only on K.

Obviously (H_E) is equivalent to (SZ) if we restrict ourselves to elliptic curves whose absolute invariants are bounded for all $\mathfrak{p} \in S_{K,\infty}$. Consequences of (H_E) or (SZ) can be found in [5].

The height conjecture (H_E) for elliptic curves can be generalized in two obvious ways: First we state the height conjecture for abelian varieties:

(H_A): *Fix $n \in \mathbb{N}$. There exist constants $c_n(K)$ and $d_n(K)$ such that for all semi-stable abelian varieties A of dimension n defined over K one has:*

$$h_K(A) \leq c_n(K) \deg \mathcal{N}_A + d_n(K)$$

where $h_K(A)$ is the Faltings height of A over K and $\mathcal{N}_A = \prod_{\substack{\mathfrak{p} \in \Sigma_K \\ A \text{ has bad reduction} \\ \text{modulo } \mathfrak{p}}} \mathfrak{p}$.

Secondly:

Conjecture (H_C). *Fix $n \in \mathbb{N}$. There exist constants $c'_n(K)$ and $d'_n(K)$ such that for all arithmetical semi-stable surfaces C/B whose generic fibres have genus n one has*

$$h_K(C) \leq c'_n(K) \deg \mathcal{N}_C + d'_n(k) \quad \text{with}$$
$$\mathcal{N}_C = \prod_{\substack{\mathfrak{p} \in \Sigma_K \\ \delta_{\mathfrak{p}} > 0}} \mathfrak{p}$$

For applications one would like to have even stronger versions:

$c_n(K)$ and $c'_n(K)$ depend on n only, and $d_n(K)$ resp. $d'_n(K)$ depend on n, linearly on $g(K)$ and polynomially on m_K.

These conjectures are known to be true in the function field case: If K is a function field over K_0 then Szpiro ([18], p. 55) has shown that for $n \geq 2$ and curves with non-trivial Kodaira-Spencer class one can take $c'_n(K) = \frac{1}{3}n(n-1)$ and $d'_n(K) = \frac{2}{3}n(n-1)(g(K)-1)$. For $n = 1$ (cf. [5]) one can take $c_1(K) = 1/2(g(K)-1)$. The height conjecture for abelian varieties over K has been proven by Griffiths. For $K_0 = \mathbb{C}$ one can take $c_n(K) = 4\pi \cdot n$ and $d_n(K) = 8\pi \cdot n(g(K)-1)$ (cf. [19], p. 87).

There is an important case in which (H_A) and (H_C) can be compared: Assume that A is the Jacobian $J(C)$ of a curve C/K. Then $h_K(A) = h_K(C)$ and $\mathcal{N}_A$ divides $\mathcal{N}_C$.

Hence the height conjecture (H_A) is stronger than the height conjecture for curves.

Now we can use Noether's formula to get weaker version of (H_C) resp. (H_A) for Jacobians.

Conjecture (S_J): *There exist constants $c(n)$ and $d(g(K),n)$ such that for each arithmetical semi-stable surface $\mathcal{C}/B$ whose generic fibre has genus n one has:*

$$\omega^2 \leq c(n) \deg \mathcal{N}_{J(C)} + d(g(K),n)$$

with $c(n)$ and $d(g(K),n)$ depending polynomially on n, $d(g(K),n)$ depending linearly on $g(K)$ and polynomially on m_K.

To get the version (S_C) corresponding to conjecture (H_C) one has to replace $\mathcal{N}_{J(C)}$ by $\mathcal{N}_C$.

In the function field case (S_C) follows from the Bogomolov-Miyaoka-Yau inequality for Chern classes (cf. [14]). In the same paper Parshin shows that (S_C) implies the height conjecture for elliptic curves. We thus have the following implications:

$$\begin{array}{ccccccc} (H_A) & \Rightarrow & (H_C) & \Rightarrow & (H_E) & \Rightarrow & (SZ) \\ \Downarrow & & \Downarrow & \nearrow \text{Parshin} & & & \\ (S_J) & \Rightarrow & (S_C) & & & & \end{array}$$

The purpose of the next section of this paper is to show that the conjecture (S_J) for curves of genus 2 implies the height conjecture at least for families of elliptic curves E with bounded absolute values of the invariants j_E. We hope that the condition about the boundedness of j_E is not necessary.

4 Application to elliptic curves

Again we assume that K is a global field with $\mathrm{char}(K) \nmid 2n$, and that E_1 and E_2 are non-isogenous semi-stable elliptic curves over K with all points of order $2n$ in $E_i(K)$. Let H be isotropic in $(E_1 \times E_2)[n]$ and C the corresponding covering curve which is assumed to be semi-stable over K with regular minimal model $\mathcal{C}$. For $P \in C(K)$ let $\mathcal{P}$ be the closure of P in $\mathcal{C}$ and $X_P := (\omega_{\mathcal{C}/B} - 2\mathcal{P}) \in \mathcal{J}$.

There is a close relation between the Néron-Tate height on J and the intersection form on $\mathcal{C}$ (cf. e.g. [20] or [8]). To state this relation it is useful to introduce a divisor $\Phi(\mathcal{P})$ whose support lies in the fibres of bad reduction of $\mathcal{C}$ with the following properties: For all $\mathfrak{p} \in \Sigma_K$ we have:

i) $\mathcal{P}_\mathfrak{p} \notin \Phi(\mathcal{P})$, and

ii) for all divisors D with $\mathrm{support}(D) \subset \mathcal{C}_\mathfrak{p}$ one has: $\Phi(\mathcal{P})_\mathfrak{p} \cdot D = 0$.

$\Phi(\mathcal{P})$ exists and is uniquely determined (cf. [20] again). In our situation we'll be able to determine $\mathcal{P}$ explicitely. But before doing so we state the mentioned relation in the special case that P, $P' \in C(K)$ are two different points whose image in $J(K)$ has finite order:

$$\tfrac{1}{2}\omega^2_{C/B} = \sum_{p\in\Sigma_K} \Delta(\mathcal{P},\mathcal{P}')_{\mathfrak{p}} - 4(\mathcal{P}\cdot\mathcal{P}').$$

where

$$\Delta(\mathcal{P},\mathcal{P}')_{\mathfrak{p}} := -\frac{1}{4}\left(\Phi(\mathcal{P})^2_{\mathfrak{p}} + \Phi(\mathcal{P}')^2_{\mathfrak{p}}\right) + \Phi(\mathcal{P})_{\mathfrak{p}}\cdot\Phi(\mathcal{P}')_{\mathfrak{p}}.$$

Now we determine $\Phi(\mathcal{P})_{\mathfrak{p}}$ and $\Delta(\mathcal{P},\mathcal{P}')_{\mathfrak{p}}$ for P, $P'\in W$ and $\mathfrak{p}\not| 2$. We use the notation from §2.

In the case of good reduction we get: $\Delta(\mathcal{P},\mathcal{P}')_{\mathfrak{p}} = 0$.

Now assume that $\mathfrak{p}|\mathcal{N}_C$.

Case α: $\mathcal{C}_{\mathfrak{p}}$ has two elliptic components E and E' connected by rational components $D_1, \ldots, D_{\delta_{\mathfrak{p}}-1}$. The intersection numbers are: $E^2 = E'^2 = -1$ and $D_i^2 = -2$, $(\omega_{C/B}\cdot E) = (\omega_{C/B}\cdot E') = 1$ and $(\omega_{C/B}\cdot D_i) = 0$.

Hence one easily checks:

$$\Phi(\mathcal{P}_{\mathfrak{p}} = \begin{cases} D_1 + 2D_2 + \ldots + (\delta_{\mathfrak{p}}-1)D_{\delta_{\mathfrak{p}}-1} + \delta_{\mathfrak{p}}E' & \text{if } \mathcal{P}_{\mathfrak{p}}\in E, \\ D_{\delta_{\mathfrak{p}}-1} + 2D_{\delta_{\mathfrak{p}}-2} + \ldots + (\delta_{\mathfrak{p}}-1)D_1 + \delta_{\mathfrak{p}}E & \text{if } \mathcal{P}_{\mathfrak{p}}\in E'. \end{cases}$$

and $\Phi(\mathcal{P})^2_{\mathfrak{p}} = -\delta_{\mathfrak{p}}$.

If $P'\in W$ is a different point then

$$\Phi(\mathcal{P})_{\mathfrak{p}}\cdot\Phi(\mathcal{P}')_{\mathfrak{p}} = \begin{cases} -\delta_{\mathfrak{p}} & \text{if } P_{\mathfrak{p}} \text{ and } P'_{\mathfrak{p}} \text{ are on the same elliptic component} \\ \delta_{\mathfrak{p}} & \text{else.} \end{cases}$$

So:

$$\Delta(\mathcal{P},\mathcal{P}')_{\mathfrak{p}} = \begin{cases} -\tfrac{1}{2}\delta_{\mathfrak{p}} & \text{if } \mathcal{P}_{\mathfrak{p}} \text{ and } \mathcal{P}'_{\mathfrak{p}} \text{ are on the same elliptic component} \\ \frac{3}{2}\delta_{\mathfrak{p}} & \text{else.} \end{cases}$$

We come to the case of mixed reduction:

Case β_1: $\mathcal{C}_{\mathfrak{p}}$ consists of $\frac{1}{n}v_{\mathfrak{p}}(j_2)$ components, if $\mathfrak{p}\not| n$, and of at most $v_{\mathfrak{p}}(j_2)$ components if $\mathfrak{p}|n$. Hence there exists an absolute constant c (independent of n) such that

$$|\Delta(\mathcal{P},\mathcal{P}')_{\mathfrak{p}}| \le \begin{cases} c\cdot\frac{1}{n}|v_{\mathfrak{p}}(j_2)| & \text{if } \mathfrak{p}\not| n \\ c\cdot|v_{\mathfrak{p}}(j_2)| & \text{if } \mathfrak{p}|n \end{cases}$$

Case β_2: Again we use the notation introduced in §2. In addition we number the points in W as follows: P_0, P_1 and P_2 are the points in W_2 and P_3, P_4 and P_5 are the points in W_1. We assume that $P_{3,\mathfrak{p}}\in G$.

Since $D_i^2 = F_j^2 = -2$, $G^2 = -3$, $\omega_{C/B} \cdot B = \omega_{C/B} \cdot G = 1$, $\omega_{C/B} \cdot D_i = \omega_{C/B} \cdot F_j = 0$, one easily checks: $\Phi(\mathcal{P}_i)_\mathfrak{p} = D_1 + \ldots + (m_\mathfrak{p} - 1)D_{m_\mathfrak{p}-1} + m_\mathfrak{p}(G + F_1 + \ldots + F_{k_\mathfrak{p}-1})$ for $i = 0,\ 1,\ 2$.
$\Phi(\mathcal{P}_3)_\mathfrak{p} = D_{m_\mathfrak{p}-1} + \ldots + (m_\mathfrak{p} - 1)D_1 + m_\mathfrak{p} \cdot B$ and

$$\Phi(\mathcal{P}_4)_\mathfrak{p} = \Phi(\mathcal{P}_5)_\mathfrak{p} = \sum_{i=1}^{\frac{k_\mathfrak{p}}{2}-1} i\left(F_{\frac{k_\mathfrak{p}}{2}-i} + F_{\frac{k_\mathfrak{p}}{2}+i}\right) + \frac{k_\mathfrak{p}}{2} \cdot H$$
$$+ \left(\frac{k_\mathfrak{p}}{2} + 1\right) D_{m_\mathfrak{p}-1} + \ldots + \left(\frac{k_\mathfrak{p}}{2} + m_\mathfrak{p}\right) \cdot B\,.$$

Hence

$$\Phi(\mathcal{P}_i)_\mathfrak{p}^2 = \begin{cases} -m_\mathfrak{p} & \text{for } 0 \le i \le 3 \\ -(k_\mathfrak{p} + m_\mathfrak{p}) & \text{for } i = 4,\ 5 \end{cases}$$

and

$$\Phi(\mathcal{P}_0)_\mathfrak{p} \cdot \Phi(\mathcal{P}_j)_\mathfrak{p} = \begin{cases} -m_\mathfrak{p} & \text{for } j = 1,\ 2 \\ m_\mathfrak{p} & \text{for } j = 3,\ 4,\ 5 \end{cases}$$

Hence

$$\Delta(\mathcal{P}_0, \mathcal{P}_j)_\mathfrak{p} = \begin{cases} -\frac{m_\mathfrak{p}}{2} & \text{for } j = 1,\ 2 \\ \frac{3}{2}m_\mathfrak{p} & \text{for } j = 3 \\ \frac{3}{2}m_\mathfrak{p} + \frac{1}{4}k_\mathfrak{p} & \text{for } j = 4,\ 5 \end{cases}$$

We continue to assume that $\mathfrak{p} \not| 2$ and that E_1 has good reduction modulo $\mathfrak{p}$. Then $\mathcal{P}_{j,\mathfrak{p}} \neq \mathcal{P}_{0,\mathfrak{p}}$ for $j = 1, \ldots, 5$ and hence $(\mathcal{P}_0 \cdot \mathcal{P}_j)_\mathfrak{p} = 0$.

A remark concerning the contribution of divisors of 2: Let $\mathfrak{p}|2$. We assume that E_1 has good reduction at $\mathfrak{p}$ and that E_1 is not isogenous to E_2 modulo $\mathfrak{p}$ if $\mathfrak{p} \not| \mathcal{N}_{E_2}$ and that E_1 is not supersingular modulo $\mathfrak{p}$ if $\mathfrak{p}|\mathcal{N}_{E_2}$. Then $(\mathcal{P}_0 \cdot \mathcal{P}_j)_\mathfrak{p}$ as well as $\Delta(\mathcal{P}_0, \mathcal{P}_j)_\mathfrak{p}$ can be estimated by $0(e_\mathfrak{p})$ where $e_\mathfrak{p}$ is the absolute ramification of $\mathfrak{p}$ in $K/\mathbb{Q}$.

By putting things together and summing up we get:

Proposition 4.1: *Let E_1 be an elliptic curve defined over K with good reduction at all non-archimedean places of K and E_2 a semi-stable elliptic curve over K which is not isogenous to E_1. If $\mathfrak{p}|2$ we assume that either E_1 is not supersingular modulo $\mathfrak{p}$ or that $\mathfrak{p} \not| \mathcal{N}_{E_2}$ and E_1 is not isogenous to E_2 modulo $\mathfrak{p}$. Let $\alpha: E_1[n] \longrightarrow E_2[n]$ be a K-isomorphism such that the graph of α is isotropic in $E_1[n] \times E_2[n]$.*

Let C be the corresponding curve of genus 2 with Weierstraß points P_0, P_1, $P_2 \in W_2$ and P_3, P_4, $P_5 \in W_1$. Then

$$\frac{5}{2}\omega^2_{C/B} = \frac{7}{2}\sum_{\mathfrak{p}\in\Sigma_K}\delta_{\mathfrak{p}} + O(h_{K,n}(E_2)) + O(\frac{1}{n}h_K(E_2)) + O(m_K) + s_\infty$$

with

$$s_\infty = -4\sum_{\mathfrak{p}\in S_{K,\infty}}\left(\sum_{j=1}^{5}(\mathcal{P}_0\cdot\mathcal{P}_i)_{\mathfrak{p}}\right)$$

and

$$h_{K,n}(E_2) = \sum_{\substack{\mathfrak{p}|n\\ \mathfrak{p}\in\Sigma_K}}\max\{0, -v_{\mathfrak{p}}(j_2)\cdot\deg\mathfrak{p}\}$$

and the term $O(m_K)$ estimates the contribution of primes $\mathfrak{p}|2$.

Using Noether's formula we get

Corollary 4.2 *The height of C over K is given by*

$$h_K(C) = \frac{1}{7}\omega^2_{C/B} + O(k_{K,n}(E_2)) + O(\frac{1}{n}h_K(E_2)) + O(m_K) + O(s_\infty) + \sum_{\mathfrak{p}\in S_{K,\infty}}\delta_{\mathfrak{p}}$$

On the other hand the Jacobian of C is isogenous to $E_1\times E_2$ under an isogeny of degree n^2, and using the behaviour of heights under isogenies (cf. [3]) one gets

Corollary 4.3:

$$h_K(E_2) = \frac{1}{7}\omega^2_{C/B} + O(h_{K,n}(E_2)) + O(\frac{1}{n}h_K(E_2)) + O(m_K) + O(g(K)\cdot\log n) + O(s_\infty) + \sum_{\mathfrak{p}\in S_{K,\infty}}\delta_{\mathfrak{p}}\,.$$

One could hope that the "∞-part" on the right hand side can be estimated using results of Bost ([2]) and the explicit construction of C. At least the following is true (use [2] and Proposition 1.5):

Assume that $M\in\mathbb{R}$ such that for all $\mathfrak{p}\in S_{K,\infty}$ the absolute value $|j_2|_{\mathfrak{p}}$ is bounded by M. Then $O(s_\infty)+\sum_{\mathfrak{p}\in S_{K,\infty}}\delta_{\mathfrak{p}}$ is bounded by $O(M, m_K)$.

Now we want to exploit corollary 4.3. We fix a global field K. We take a constant c such that the terms $O(h_{K,n}(E_2))$ and $O(\frac{1}{n}h_K(E_2))$ are bounded by $c \cdot h_{K,n}(E_2)$ resp. $c \cdot \frac{1}{n}h_K(E_2)$. There is a number N such that for all elliptic curves E/K we find an odd number $n_E \leq N$ with $h_{K,n_E}(E) \leq \frac{1}{2c}h_K(E)$ and $n_E > 3c$.

We find finitely many elliptic curves $\{E_{1,i}\}$ with integral j-invariants such that for all semi-stable curves $\tilde{E}/K$ there is one $E_1 \in \{E_{1,i}\}$ which satisfies: If $\mathfrak{p}|2$ and $\mathfrak{p} \not| \mathcal{N}_E$ then E_1 is not isogenous to $\tilde{E}$, and if $\mathfrak{p}|2$ and $\mathfrak{p}|\mathcal{N}_E$ then E_1 is not supersingular modulo $\mathfrak{p}$.

The necessary number of these curves $\{E_{1,i}\}$ depends on m_K only. There is a finite extension K' of K such that all curves $E_{1,i}$ have good reduction at all places of K' and such that the genus of K' and $m_{K'}$ depend only on K.

Now take a semi-stable curve E/K and assume that for all $\mathfrak{p} \in S_{K,\infty}$ we have: $|j_E|_{\mathfrak{p}} \leq M$.

Choose $n := n_E \leq N$ such that $h_{K,n}(E) \leq \frac{1}{2c}h_K(E)$ and take $E_1 \in \{E_{1,i}\}$ suitable for E. Define $K'' := K'(E_1[2n], E[2n])$.

The pair (E_1, E) has over K'' all the properties which are necessary to choose $\alpha\colon E_1[2n] \longrightarrow E[2n]$ and to construct a corresponding curve C such that the corollary 4.3 can be applied:

$$h_{K''}(E) \leq \frac{1}{7}\omega^2_{C/B} + \frac{5}{6}h_{K''}(E) + c_1 \cdot M_{K''} + c_2 g(K'')\log N \quad \text{or :}$$
$$\frac{1}{6}h_{K''}(E) \leq \frac{1}{7}\omega^2_{C/B} + c_1 \cdot M_{K''} + c_2 g(K'')\log N\,.$$

Now $m_{K''}$ is bounded by $O(m_{K'} \cdot N^7)$ and $g(K'')$ is bounded by a linear function of $\deg(\mathcal{N}_E)$ (depending on $g(K)$, m_K and n_E), and hence we get

Theorem 4.4: *Let $E|K$ be a semi-stable elliptic curve. Then there are constants c_1, c_2 depending on K only, and r_∞ depending on K and $\max_{\mathfrak{p}\in S_{K,\infty}}(|j|_{\mathfrak{p}})$ such that*

$$h_K(E) \leq c_1\omega^2_{C/B} + c_2 \deg \mathcal{N}_E + r_\infty$$

where C is the arithmetical surface belonging to a curve of genus 2 covering E and an elliptic curve E_1 with potentially good reduction everywhere.

Corollary 4.5: *Assume that the conjecture (S_J) is true for curves of genus 2 with split Jacobians. Then $\omega^2_{C/B} \leq c(K)\deg \mathcal{N}_E + d(K)$, and hence the height conjecture is true for elliptic curves for which the absolute values of the j-invariants are bounded.*

REFERENCES

[1] S. Ju. Arakelov: Intersection theory on an arithmetic surface (Russian). Izv. Akad. Nauk. SSSR 38 (1974) = Math. USSR Izv. 8 (1974), 1167–1180.

[2] J. B. Bost: Fonctions de Green-Arakelov, fonctions thêta et courbes de genre 2. C. R. Acad. Sci. Paris 306 (1987), 643–646.

[3] P. Deligne: Preuve des conjectures de Tate et Shafarevich [d'après G. Faltings]. Sem. Bourbaki (1983/84, no. 616, Astérisque 121/122 (1985), 25–41.

[4] G. Faltings: Calculus on arithmetic surfaces. Ann. Math. 119 (1984), 387–424.

[5] G. Frey: Links between solutions of $A - B = C$ and elliptic curves. Number Theory, Ulm 1987, Springer Lecture Notes in Math. 1380 (1989).

[6] A. Grothendieck: Modèles de Néron et monodromie (SGA 7, X). Springer Lecture Notes in Math. 288 (1972), 313–523.

[7] T. Ibukiyama, T. Katsura, F. Oort: Supersingular curves of genus two and class numbers. Comp. Math. 57 (1986), 127–152.

[8] E. Kani: Weil heights, Néron pairings and v-metrics on curves. Rocky Mt. J. 15 (1985), 417–449.

[9] S. Lang: Abelian Varieties. Interscience Publ., New York, 1959.

[10] S. Lang: Introduction to Arakelov Theory. Springer Verlag, New York, 1988.

[11] D. Mumford: Abelian Varieties. Oxford U. Press, London, 1970.

[12] A. Ogg: On pencils of curves of genus two. Topology 5 (1966), 355–362.

[13] A. N. Parshin: Minimal models of curves of genus 2. Izv. Akad. nauk. SSSR 36 (1972) = Math. USSR Izv. 6 (1972), 65–108.

[14] A. N. Parshin: The Bogomolov-Miyaoka-Yau inequality for arithmetical surfaces and its applications. Sem. Théorie des Nombres Paris 1986/87 (C. Goldstein, ed.) Progress in Math. 75, Birkhäuser, Boston (1989), 299–312.

[15] K. Ribet: On modular representations of Gal($\bar{\mathbb{Q}}/\mathbb{Q}$) arising from modular forms. Preprint MSRI 06420-87, June 1987 (68 pp.).

[16] J.-P. Serre: Groupes algébriques et corps de classes. Hermann, Paris, 1959.

[17] J.-P. Serre: Nombres des points des courbes algébriques sur $\mathbb{F}_q$. Sem. Théorie des Nombres Bordeaux 1982/83, no. 22 = Oeuvres/Collected Papers III, 664-668.

[18] L. Szpiro: Proprietés numériques du faisceaux dualisant relatif. In: Sem. sur les pinceaux des courbes de genre au moins deux (L. Szpiro,

ed.) Astérisque 86 (1981), 44–78.
[19] L. Szpiro: La Conjecture de Mordell [d'après G. Faltings]. Sém. Bourbaki 1983/84, no. 619 Astérisque 121/122 (1985), 83–103.
[20] L. Szpiro: Un peu d'effectivité. in: Sém. sur les pinceaux arithmétiques: La Conj. de Mordell (L. Szpiro, ed.), Astérisque 127 (1985), 275–287.
[21] L. Szpiro: Sur les propriétés numériques du dualisant relatif d'une surface arithmétique. To appear.
[22] P. Vojta: Diophantine inequalities and Arakelov theory. Appendix to [10].
[23] A. Weil: Zum Beweis des Torellischen Satzes. Göttinger Nachrichten 2957, Nr. 2 = Oeuvres Scientifiques/Collected Papers II, pp. 302–327.

Jacobians with complex multiplication

JOHAN DE JONG *   RUTGER NOOT

1 Introduction

1.1 In this paper we will say that a simple abelian variety X is *of CM type* if there is a number field K with $[K : \mathbf{Q}] = 2\dim(X)$ such that $K \subset \mathrm{End}^0(X)$. If X is any abelian variety, then we will say that X is of CM type if all its simple factors are. Equivalently, X is of CM type if there are number fields K_i such that $\sum[K_i : \mathbf{Q}] = 2\dim(X)$ and $\oplus K_i \subset \mathrm{End}^0(X)$.

1.2 In [Co] R. Coleman formulated the following conjecture (Conjecture 6):

Conjecture (Coleman). *Fix an integer $g \geq 4$. Then there are only finitely many complex algebraic curves C of genus g such that* Jac(C) *is of CM type.*

In this paper we give examples of families of curves that contradict this conjecture. This will be accomplished by giving two proofs of the following:

1.3 Claim. *Consider the following families of curves over* $\mathbf{C}$:

$$y^3 = x(x-1)(x-\lambda)(x-\mu)(x-\nu) \tag{1.3.1}$$

$$y^5 = x(x-1)(x-\lambda) \tag{1.3.2}$$

$$y^7 = x(x-1)(x-\lambda) \tag{1.3.3}$$

The curves given by 1.3.1 and 1.3.2 have genus 4, *those given by 1.3.3 have genus* 6. *In each of these families there are infinitely many fibres whose Jacobian is of CM type.*

*Supported by the Netherlands Organisation for Scientific Research (N.W.O.)

1.4 In our proof of this claim it is essential that the number of moduli of abelian varieties with certain endomorphisms is small. It seems very difficult to find similar examples for higher genera. We show that the hypergeometric families

$$y^N = x(x-1)(x-\lambda)$$

for N at least 8 and prime to 3 behave in a totally different way with respect to the questions raised above.

1.5 The authors wish to thank F. Oort for his help with the preparation of this paper. In particular, the ideas used in 5.8 are due to him. We thank B. van Geemen for his great help with Section 2.

2 Hodge theoretical proof of Claim 1.3

In this section we will use the notations of [De].

2.1 Fix a $\lambda_0 \in \mathbf{C}\backslash\{0,1\}$ and consider the corresponding fibre of 1.3.2:

$$C_{\lambda_0} : y^5 = x(x-1)(x-\lambda_0).$$

It has an automorphism of order 5 given by $(x,y) \mapsto (x, \zeta y)$ where $\zeta = \exp\left(\frac{2\pi i}{5}\right)$. This automorphism acts on the space of holomorphic differentials with eigenvalues ζ, ζ, ζ^2 and ζ^3. Let $L = \mathbf{Q}[\zeta]$, $V = H_1(\mathrm{Jac}(C_{\lambda_0}), \mathbf{Q}) = H_1(C_{\lambda_0}, \mathbf{Q})$ and let $\psi : V \times V \to \mathbf{Q}$ be the intersection form on $H_1(C_{\lambda_0}, \mathbf{Q})$. Now L acts on V via the automorphism of C_{λ_0} and this action has the property that $\psi(\ell x, y) = \psi(x, \bar{\ell} y)$ for all $\ell \in L$ and all $x, y \in V$. We are therefore in the situation of [De, 4.9].
Let $G \subset GL_L(V)$ be the algebraic group over $\mathbf{Q}$ defined by Deligne. The group $G(\mathbf{Q})$ is the set of $g \in GL_L(V)$ such that there is a $\mu(g) \in \mathbf{Q}^*$ satisfying:

$$\psi(gx, gy) = \mu(g)\psi(x,y) \; \forall x, y \in V.$$

Because V is the cohomology of an algebraic curve, it is equipped with a Hodge structure for which ψ is a polarisation. The elements of L act as endomorphisms of this Hodge structure. It follows that the map

$$h_0 : S \to GL_L(V)_{\mathbf{R}}$$

defining this Hodge structure factors through $G_{\mathbf{R}}$. Therefore we have a map

$$h_0 : S \to G_{\mathbf{R}}.$$

2.2 Lemma. *There is a compact open subgroup $K \subset G(\mathbf{A}^f)$ such that there is a non-constant holomorphic map*

$$j : \mathbf{C}\backslash\{0,1\} \to {}_K M_{\mathbf{C}}(G, h_0)$$

that maps $\lambda \in \mathbf{C}\backslash\{0,1\}$ to the isomorphism class of the principally polarised abelian variety $\mathrm{Jac}(C_\lambda)$ *together with the given action of* $\mathbf{Z}[\zeta]$ *and a level* 1 *structure (cf. [De, 4.12]).*

2.3 Proof. By horizontal continuation of id : $V \to V$ we get a non-canonical isomorphism

$$i_\lambda : V \to H_1(C_\lambda, \mathbf{Q})$$

for every $\lambda \in \mathbf{C}\backslash\{0,1\}$. This map commutes with the action of L and is an isometry for the intersection forms. This gives us a map

$$h_\lambda : S \to G_{\mathbf{R}}$$

for all $\lambda \in \mathbf{C}\backslash\{0,1\}$. All these maps are conjugate to h_0 because the set of conjugacy classes of maps h with the desired properties is finite ([Mu1], footnote on page 345) and we have a continuous family connecting h_0 and h_λ. Because i_λ respects integral lattices we can take

$$K = G(\mathbf{A}^f) \cap GL_{\hat{\mathbf{Z}}}(V \otimes \hat{\mathbf{Z}}).$$

Finally j is not constant because the Kodaira Spencer map corresponding to our family is not equal to zero (cf. 5.5). □

2.4 Lemma. *For all K such that there is a map j as in Lemma 2.2 we have:*

$$\dim {}_K M_{\mathbf{C}}(G, h_0) = 1.$$

2.5 Proof. In view of the above it suffices to show that the space of maps h that give a Hodge structure of the required kind has dimension 1. Such a map h is completely determined by the Hodge structure it defines on V. This Hodge structure consists of a $H^{-1,0} \subset V \otimes \mathbf{C}$ which is a 4-dimensional L-invariant linear subspace, on which $\zeta \in L$ acts with eigenvalues $\zeta^2, \zeta^3, \zeta^4, \zeta^4$ and which satisfies the Riemann bilinear relations:

1. $H^{-1,0}$ is isotropic for $\psi \otimes \mathbf{C}$.

2. $\forall\, \omega \in H^{-1,0}$, $\omega \neq 0$ we have: $i\psi(\omega, \overline{\omega}) > 0$.

If we write $V(\zeta^m)$ for the eigenvalue ζ^m eigenspace of $\zeta \in L$ in $V \otimes \mathbf{C}$, then we can decompose:

$$V \otimes \mathbf{C} = V(\zeta) \oplus V(\zeta^2) \oplus V(\zeta^3) \oplus V(\zeta^4).$$

We now see that

$$H^{-1,0} = V_2 \oplus V_3 \oplus V(\zeta^4) \subset V \otimes \mathbf{C}$$

with $V_i \subset V(\zeta^i)$ one dimensional. However, using 2.5.1 we see that the choice of V_2 determines V_3. It follows that the possible choices of $H^{-1,0}$ are parametrised by an open subset of $\mathbf{P}(V(\zeta^2))$, this open set being determined by 2.5.2. As this space has dimension one, we have now proved the lemma. □

2.6 By [Mu1, §2], the points of ${}_K M_{\mathbf{C}}(G, h_0)$ that correspond to abelian varieties that are of CM type are exactly the points that are in the image of a map

$$M_{\mathbf{C}}(H, h') \to M_{\mathbf{C}}(G, h_0) \to {}_K M_{\mathbf{C}}(G, h_0) \qquad (2.6.1)$$

for some torus $u : H \hookrightarrow G$ and some map h' as in [De, 3.13]. Combining [De] 5.1 and 5.2 we see that the set of points obtained in this way is dense. Using our map j we see that the set of $\lambda \in \mathbf{C}\backslash\{0, 1\}$ for which $\mathrm{Jac}(C_\lambda)$ is of CM type is dense. This finishes our proof of Claim 1.3 in the case of Family 1.3.2. The same arguments (mutatis mutandis) give a proof for the other two families.

This concludes the proof of Claim 1.3. The following proposition shows that we can say a bit more.

2.7 Proposition. *In the case of Family 1.3.2 the set*

$$\{\mathrm{End}^0(\mathrm{Jac}(C_\lambda)) \mid \lambda \in \mathbf{C}\backslash\{0,1\}\}$$

contains infinitely many CM fields of degree 8.

The proof will depend on some general facts that we state below. We then show why we can apply them.

2.8 Lemma. *Let X be a simple abelian variety of CM type, with CM type (K, Φ) and let*

$$h : S \to GL(H_1(X, \mathbf{Q}))_{\mathbf{R}}$$

be the map defining the Hodge structure on $H_1(X, \mathbf{Q})$. Then the field of definition of the map

$$h \circ r : \mathbf{G}_m \to GL(H_1(X, \mathbf{Q}))_{\mathbf{C}}$$

is the reflex field of (K, Φ).

2.9 Corollary. *Let $(H, h', u : H \hookrightarrow G)$ be as in 2.6 and let $E(H, h')$ be defined as in [De, 3.7]. If X is an abelian variety corresponding to a point in the image of the map 2.6.1, then $E(H, h')$ is the compositum of the reflex fields of the simple factors of X.*

2.10 Proof of 2.8 and 2.9. We have

$$H_1(X, \mathbf{Q}) \otimes \mathbf{C} = \bigoplus_{\phi : K \to \mathbf{C}} \mathbf{C}$$

and h is the map defining the subspace

$$H^{-1,0} = \bigoplus_{\phi \in \Phi} \mathbf{C} \subset H_1(X, \mathbf{Q}) \otimes \mathbf{C}.$$

Let M be the field of definition of the composite

$$\mathbf{G}_m \to S_{\mathbf{C}} \to GL(H_1(X, \mathbf{Q}))_{\mathbf{C}}.$$

Then $\sigma \in \mathrm{Aut}(\mathbf{C})$ fixes M exactly if it leaves $H^{-1,0} \subset H_1(X, \mathbf{Q}) \otimes \mathbf{C}$ invariant. Since this is the case if and only if $\sigma \circ \Phi = \Phi$ it follows from the definition of reflex field that M is the reflex field of (K,Φ). This proves 2.8. If X is of CM type and not simple it easily follows from 2.8, that the map $h \circ r$ is defined over the compositum of the reflex fields of the isogeny factors of X. Corollary 2.9 now follows immediately from Lemma 2.8, the definition of $E(H,h')$ in [De, 3.7] and the fact that H is commutative. □

2.11 Theorem. *(cf. [De, 5.1]). Suppose we are in the situation of [De, 5.1] and that in addition to the conditions of that theorem, there is a $(H,h',H \hookrightarrow G)$ (as in loc.cit.) such that $[E(H,h') : E(G,h)] = d > 1$. Then for all finite extensions F of $E(G,h)$ there exists $(T,t',T \hookrightarrow G)$ such that $[E(T,t') : E(G,h)] \geq d$ and $E(T,t')$ is linearly disjoint from F. In this case the set of fields occurring as the reflex of the CM type of the abelian variety corresponding to a point of this Shimura variety is infinite.*

2.12 Proof. This can be derived from Deligne's proof. He considers the following diagram:

$$\begin{array}{ccc} W & \xrightarrow{f} & V \\ \downarrow & & \downarrow \\ \mathrm{Spec}(E(G,h)) & \rightarrow & \mathrm{Spec}(\mathbf{Q}) \end{array} \qquad (2.12.1)$$

The reader is referred to [De] for the meaning of this. His arguments show that to find such T it is sufficient to find $v \in V(\mathbf{Q})$ such that $f^{-1}(v)$ is the spectrum of an extension of $E(G,h)$ of degree greater then d that is linearly disjoint from F. By looking at $W' = W \otimes F$ and $g = f \circ \mathrm{pr}_1$ we see that it is enough to find a $v \in V(\mathbf{Q})$ such that $g^{-1}(v)$ is the spectrum of an extension of F of degree greater then d. By [De, 5.1.2] $W/\mathrm{Spec}(E(G,h))$ is absolutely irreducible. Because by our assumptions the degree of g is greater then $d[E(G,h) : \mathbf{Q}]$ the existence of the v we look for follows from Hilbert's irreducibility theorem because V is an open subset of an affine space. The last assertion of the theorem is an easy consequence of the first one and 2.9. □

2.13 Lemma. *For the group G and the map h_0 defined in 2.1 we have: $E(G,h_0) = \mathbf{Q}[\zeta]$.*

2.14 Proof. From the definition of $E(G, h_0)$ in [De, 3.7] and [De, §1] it is clear that $E(G, h_0)$ is the intersection of the fields of definition of the spaces $H^{-1,0} \subset V \otimes \mathbf{C}$ of 2.5. By 2.5 such $H^{-1,0}$ are parametrised by an open subset of $\mathbf{P}(V(\zeta^2))$. Since this space has points that are rational over $\mathbf{Q}[\zeta]$ there exist $H^{-1,0}$ that are defined over $\mathbf{Q}[\zeta]$. It is clear that no $H^{-1,0}$ can be defined over a smaller field. □

2.15 Lemma. *If for Family 1.3.2 we have an* H *as above such that* $E(H, h') \neq E(G, h_0)$*, then the abelian varieties arising from* H *via the map 2.6.1 are simple.*

2.16 Proof. If $X = \text{Jac}(C_\lambda)$ is of CM type but not simple for some λ, then we have the following possibilities:

- $X \sim E^4$ for some elliptic curve E. Because $\mathbf{Q}[\zeta]$ does not have a proper CM subfield it follows that $\text{End}^0(X) = M_4(M)$, where M is linearly disjoint from $\mathbf{Q}[\zeta]$. Therefore $\zeta \in \mathbf{Q}[\zeta]$ acts on the space of holomorphic differentials of X with eigenvalues $\zeta, \zeta^2, \zeta^3, \zeta^4$. This is a contradiction so this case can not occur.

- $X \sim Y^2$ where Y is a simple abelian surface. In this case $E(H, h')$ is the reflex of the CM type of Y by 2.9. Let $M = \text{End}^0(Y)$, then this reflex is contained in the Galois closure M' of M. Because M is a CM field it has a subfield of degree 2, so $[M' : \mathbf{Q}] \leq 8$. Since $E(G, h) \subset E(H, h')$ we have $\mathbf{Q}[\zeta] \subset M'$. Suppose that $\text{Gal}(M'/\mathbf{Q})$ has order 8, then the subgroup corresponding to $\mathbf{Q}[\zeta]$ is a normal subgroup of order 2 such that the quotient is cyclic of order 4. But then $\text{Gal}(M'/\mathbf{Q})$ is abelian so M is Galois. Because this contradicts the assumption that $[M' : \mathbf{Q}] = 8$ we must have that $M = M' = \mathbf{Q}[\zeta]$. We therefore have $E(H, h') = \mathbf{Q}[\zeta] = E(G, h)$.

- $X \sim Y_1 \oplus Y_2$ with $\text{Hom}(Y_1, Y_2) = \{0\}$. Because the tangent space at the origin of X has no linear subspaces on which $\mathbf{Q}[\zeta]$ acts trivially, we have $\dim Y_i \geq 2$ for $i = 1, 2$, so $\dim Y_1 = \dim Y_2 = 2$ and $\mathbf{Q}[\zeta]$ acts on both the Y_i. Since ζ can not act on the space of holomorphic differentials with only one eigenvalue (this doesn't give a CM type) we can assume that ζ acts with eigenvalues ζ, ζ^2 and ζ, ζ^3 on the spaces of holomorphic differentials of Y_1 and Y_2 respectively. Because these actions differ only by the automorphism $\zeta \mapsto \zeta^2$ of $\mathbf{Q}[\zeta]$ it follows from [S,T], §6.1, corollary of Theorem 2 that $Y_1 \sim Y_2$, contrary to the assumption. □

2.17 Proof of Proposition 2.7. If we take λ a primitive sixth root of unity, then the curve C_λ is isomorphic to the curve given by $y^5 = x^3 - 1$. The Jacobian of this curve has CM with $\mathbf{Q}[\zeta_{15}]$ and the reflex of its CM type is $\mathbf{Q}[\zeta_{15}] \neq \mathbf{Q}[\zeta]$. So by 2.13 and 2.9 the conditions of 2.11 are satisfied. By 2.15 the CM points that 2.11 gives us correspond to simple abelian varieties. This concludes the proof of Proposition 2.7. □

3 Quasi canonical liftings and complex multiplication

3.1 Let k be an algebraically closed field of characteristic p and let X_0/k be an ordinary abelian variety. Then, as explained in [Ka, Theorem 2.1] the formal moduli space $\mathcal{M}_{X_0/k}$ of liftings of X_0 is canonically isomorphic to

$$\mathrm{Hom}_{\mathbf{Z}_p}(T_p X_0(k) \otimes T_p X_0^t(k), \widehat{\mathbf{G}}_m).$$

If $\alpha \in T_p X_0(k)$ and $\alpha_t \in T_p X_0^t(k)$ we shall denote by $q(-, \alpha, \alpha_t)$ the corresponding function on $\mathcal{M}$. If Y_0 is another ordinary abelian variety over k and if R is a complete local ring with residue field k, then suppose that X/R and Y/R are abelian schemes lifting X_0 and Y_0 respectively. In this case a homomorphism $f_0 \in \mathrm{Hom}(X_0, Y_0)$ can be lifted to an element $f \in \mathrm{Hom}(X, Y)$ if and only if

$$q(Y/R, f_0(\alpha), \alpha_t) = q(X/R, \alpha, f_0^t(\alpha_t)) \tag{3.1.1}$$

for all $\alpha \in T_p X_0(k)$ and $\alpha_t \in T_p Y_0^t(k)$. It immediately follows from this that $f_0 \in \mathrm{End}(X_0)$ can be lifted to an element $f \in \mathrm{End}(X)$ if and only if

$$q(X/R, f_0(\alpha), \alpha_t) = q(X/R, \alpha, f_0^t(\alpha_t)) \tag{3.1.2}$$

for all $\alpha \in T_p X_0(k)$ and $\alpha_t \in T_p X_0^t(k)$.

3.2 From now on we will work with the field $\overline{\mathbf{F}}_p$ and R will be a finite extension of $W(\overline{\mathbf{F}}_p)$, the ring of infinite Witt-vectors over $\overline{\mathbf{F}}_p$. The ring R is a local ring with maximal ideal m. The *canonical lifting* of X_0 to R is the abelian scheme X/R for which $q(X/R,-,-) \equiv 1$. Note that this lifting is defined over $W(\overline{\mathbf{F}}_p)$. Using 3.1.2, we see that $\mathrm{End}(X) = \mathrm{End}(X_0)$. We will call a lifting X/R of X_0 for which $q(X/R,-,-)^M \equiv 1$ for some $M \in \mathbf{Z}$ a *quasi canonical lifting* of X_0. Note that, since $q(X/R,-,-) \equiv 1 \bmod m$, we can assume that M is a power of p. By 3.1.1 and some linear algebra it is easily checked that 'being a quasi canonical lifting' is an isogeny invariant of abelian varieties. It trivially follows from the above that if X/R is a quasi canonical lifting of X_0, then $\mathrm{End}^0(X) = \mathrm{End}^0(X_0)$.

3.3 Lemma (Messing). *Let X/R be a lifting of X_0 and let $\pi \in$ $\mathrm{End}(X_0)$ be the Frobenius element, as in [Ta]. Then X is the canonical lifting of X_0 if and only if π lifts to an element of* $\mathrm{End}(X)$.

3.4 Proof. See also [Me, Appendix 1]. If

$$q(X/R, \pi(\alpha), \alpha_t) = q(X/R, \alpha, \pi^t(\alpha_t))$$

for all $\alpha \in T_pX_0(\overline{\mathbf{F}}_p)$ and $\alpha_t \in T_pX_0^t(\overline{\mathbf{F}}_p)$ then it follows from the fact that π is invertible on $T_pX_0(\overline{\mathbf{F}}_p)$ and π^t is divisible by p on $T_pX_0^t(\overline{\mathbf{F}}_p)$, that $q(X/R,\alpha,\alpha_t) = q(X/R, \pi^{-1}(\alpha), \pi^t(\alpha_t)/p)^p$ for all α, α_t. By induction we get that $q(X/R,\alpha,\alpha_t)$ is a p^m-th power in R for all m, so $q(x/R,-,-) \equiv 1$. □

3.5 Proposition. *Let $X_0/\overline{\mathbf{F}}_p$ be an ordinary abelian variety. Let R be a finite extension of $W(\overline{\mathbf{F}}_p)$ and suppose that X/R is any lifting. Then X is of CM type if and only if it is a quasi canonical lifting of X_0.*

3.6 Proof. In view of the above and [Ta, Theorem 2] it suffices to show that if X is of CM type then it is a quasi canonical lifting of X_0. We immediately reduce to the case that X is simple by considering the isogeny factors of X separately. Then there is a field K with $K \subset \mathrm{End}^0(X)$ and $[K : \mathbf{Q}] = 2\dim(X)$. Since X_0 is ordinary we have $\mathrm{End}^0(X_0) = \bigoplus_{i=1}^n M_{m_i}(K_i)$ with the K_i number fields. Because $\mathrm{End}^0(X) \subset \mathrm{End}^0(X_0)$ we have $n = 1$, as follows immediately from the dimensions and from the fact that if $K \subset$

$M_m(K')$ then $[K:\mathbf{Q}] \le m[K':\mathbf{Q}]$. Since $\dim(X) = \dim(X_0)$ we conclude that $K \subset M_{m_1}(K_1)$ is maximal commutative, so it contains the center $\mathbf{Q}(\pi)$ of $\mathrm{End}^0(X_0)$. Here π is the Frobenius element of $\mathrm{End}(X_0)$ as in 3.3. It follows that some integral multiple $M\pi$ of π lifts to an element of $\mathrm{End}(X)$. Now consider the lifting X'/R of X_0 with $q(X'/R,-,-) = q(X/R,-,-)^M$. It is obvious from Equation 3.1.2 that π lifts to an endomorphism of X'. It now follows from 3.3 that $q(X'/R,-,-) \equiv 1$. □

3.7 If X_0 is a principally polarised abelian variety, with polarisation $\mu : X_0 \to X_0^t$, then we can use μ to identify $T_pX_0(\overline{\mathbf{F}}_p)$ an $T_pX_0^t(\overline{\mathbf{F}}_p)$. If X/R lifts X_0 we can define a new q-parameter as follows:

$$q_\mu(X/R, \alpha_1, \alpha_2) := q(X/R, \alpha_1, \mu(\alpha_2)) \tag{3.7.1}$$

Using 3.1.2 it is easy to verify that μ lifts to a principal polarisation of a lifting X/R of X_0 if and only if $q_\mu(X/R,-,-)$ is symmetric. We can reformulate 3.1.2 as follows:
An endomorphism $f_0 \in \mathrm{End}(X_0)$ can be lifted to $\mathrm{End}(X)$ if and only if

$$q_\mu(X/R, f_0(\alpha_1), \alpha_2) = q_\mu(X/R, \alpha_1, f_0'(\alpha_2)) \tag{3.7.2}$$

for all $\alpha_1, \alpha_2 \in T_pX_0(\overline{\mathbf{F}}_p)$. Here $'$ is the Rosati-involution on $\mathrm{End}^0(X_0)$ determined by the polarisation μ.

4 Second proof of Claim 1.3

4.1 We now give the second proof of Claim 1.3. We give it in detail only for the family 1.3.3. Take $p \equiv 1 \pmod{21}$ and prime. Take $S = \mathrm{Spec}(\overline{\mathbf{F}}_p[\lambda, (\lambda(1-\lambda))^{-1}])$. We first prove that the generic member of the family $C \to S$ given by

$$y^7 = x(x-1)(x-\lambda)$$

is ordinary: If we take $\lambda \in \overline{\mathbf{F}}_p$ a primitive sixth root of unity, then C_λ is isomorphic to the curve given by $y^7 = x^3 - 1$ so $\mathrm{Jac}(C_\lambda)$ has CM with $\mathbf{Z}[\zeta_{21}]$. As p splits completely in $\mathbf{Z}[\zeta_{21}]$, $\mathrm{Jac}(C_\lambda)$ is ordinary and therefore the same holds for the generic member of $C \to S$.

4.2 We now fix λ_0 in $\overline{\mathbf{F}}_p$ such that the Jacobian of $C_0 = C_{\lambda_0}$ is ordinary. Because $\zeta_7 \in \mathbf{Z}_p$ we can diagonalise the action of ζ_7 on $T_p\mathrm{Jac}(C_0)(\overline{\mathbf{F}}_p)$: it is $\mathrm{diag}(\zeta_7^3, \zeta_7^4, \zeta_7^5, \zeta_7^5, \zeta_7^6, \zeta_7^6)$ (use [Ka, 4.2]). Say this diagonal form is assumed using the base $\alpha_1, \ldots, \alpha_6$ of $T_p\mathrm{Jac}(C_0)(\overline{\mathbf{F}}_p)$. Now $q(-,-,-)$ will be as in 3.7.1 for the canonical polarisation on $\mathrm{Jac}(C_0)$. We will write $\mathcal{M}$ for the formal moduli space of liftings of $\mathrm{Jac}(C_0)$ and $\mathcal{M}'$ for the moduli space of principally polarised liftings X of $\mathrm{Jac}(C_0)$ with $\mathbf{Z}[\zeta_7] \subset \mathrm{End}(X)$. If we take for f_0 the automorphism determined by ζ, Equation 3.7.2 shows that $\mathcal{M}'$ is the subscheme of $\mathcal{M}$ given by the equations:

$$q(-, \alpha_i, \alpha_j) = q(-, \alpha_j, \alpha_i)$$

$$q(-, \alpha_i, \alpha_j) \equiv 1 \qquad \text{if } \{i, j\} \neq \{1, 2\}.$$

Therefore $\mathcal{M}' \subset \mathcal{M}$ is a (relatively) one dimensional formal subtorus and $q(-, \alpha_1, \alpha_2)$ is a coordinate on $\mathcal{M}'$. It follows that the Torelli map corresponding to the family 1.3.3 over $\mathrm{Spf}(W(\overline{\mathbf{F}}_p)[[\lambda - \widetilde{\lambda_0}]])$:

$$\mathrm{Spf}(W(\overline{\mathbf{F}}_p)[[\lambda - \widetilde{\lambda_0}]]) \to \mathcal{M}'$$

is surjective. Here $\widetilde{\lambda_0} \in W(\overline{\mathbf{F}}_p)$ lifts λ_0. Every torsion point of $\mathcal{M}'$ is also a torsion point of $\mathcal{M}$, so it corresponds to a quasi canonical lifting of $\mathrm{Jac}(C_0)$. Moreover, it is the image of a point of $\mathrm{Spf}(W(\overline{\mathbf{F}}_p)[[\lambda - \widetilde{\lambda_0}]])$. Therefore the fibre of C over this point is a curve for which the Jacobian of the generic fibre is of CM type. Because there are infinitely many torsion points on a formal torus, this shows that for every choice of $\lambda_0 \in \overline{\mathbf{F}}_p$ for which $\mathrm{Jac}(C_{\lambda_0})$ is ordinary, there are infinitely many liftings of C_{λ_0} to characteristic zero with Jacobian of CM type.

4.3 Again similar arguments work for the families 1.3.1 and 1.3.2 if we take $p \equiv 1 \pmod{15}$.

5 Similar families with different behaviour

5.1 In this section we examine the hypergeometric families C of smooth projective curves over $S = \mathrm{Spec}(\mathbf{Z}[\lambda, (\lambda(\lambda-1))^{-1}])$ given by

$$y^N = x(x-1)(x-\lambda) \tag{5.1.1}$$

with N at least 8 and prime to 3. A computation of the Hasse-Witt matrix as in 5.6 shows that the choice of a prime p with $p \equiv 1 \pmod{N}$ will ensure that in characteristic p the generic member of $C \to S$ is ordinary. Choose a $\lambda_0 \in \overline{\mathbf{F}}_p$ such that $\mathrm{Jac}(C_{\lambda_0})$ is ordinary. The calculation of 4.2 shows that for this family the deformation space $\mathcal{M}'$ of $\mathrm{Jac}(C_{\lambda_0})$ with the automorphism of order N is a formal subtorus of $\mathcal{M}$ of relative dimension 2 or more. Therefore the argument we applied in 4.2 to the Families 1.3 to show that the image of the Torelli mapping is a subtorus of $\mathcal{M}$ can not be applied to this family. We can even show, using the concept of pre-W_2-canonicalness, that for generic λ_0 the image of the Torelli mapping

$$\mathrm{Spf}(W(\overline{\mathbf{F}}_p)[[\lambda - \widetilde{\lambda_0}]]) \to \mathcal{M}'$$

is *not* a formal subtorus. Here, again, $\widetilde{\lambda_0}$ lifts λ_0. First we give a

5.2 Definition. An ordinary curve C defined over a perfect field k of characteristic p is called *pre-W_2-canonical* if the canonical lifting of $\mathrm{Jac}(C)$ to $W_2(k) = W(k)/(p^2)$ is a Jacobian variety (necessarily of a lifting of C to $W_2(k)$).

5.3 If the image of the Torelli mapping is a formal subtorus then the canonical lifting of the curve C_{λ_0} is is a Jacobian, so C_{λ_0} is pre-W_2-canonical. Therefore the following lemma implies our contention.

Lemma. *For $N \geq 8$ and prime to 3 and $p \equiv 1 \pmod{N}$ the generic member of the family $C \to S$ is not pre-W_2-canonical.*

5.4 Proof. We introduce some more terminology from [D,O]. We shall denote by $F_S : S \to S$ the absolute Frobenius endomorphism on S. The map

$$\mu_C : \mathrm{Sym}^2\Gamma(C, \Omega_C) \to \Gamma(C, \Omega_C^{\otimes 2})$$

will be the natural one and K the kernel of μ. The map

$$\gamma : F^*\Gamma(C, \Omega_C) \to \Gamma(C, \Omega_C)$$

is the inverse transpose of Hasse-Witt and finally

$$\mathrm{Kod} : \Theta_{S/\overline{\mathbf{F}}_p} \to H^1(C, \Theta_{C/S})$$

is the Kodaira-Spencer mapping. It follows from [D,O, Proposition 3.2] that the generic member of $C \to S$ is pre-W_2-canonical if and only if the following map vanishes identically:

$$\begin{aligned} F_S^* K &\xrightarrow{\text{inc}} F_S^* \mathrm{Sym}^2 \Gamma(C, \Omega_C) \xrightarrow{\mathrm{Sym}^2(\gamma)} \mathrm{Sym}^2 \Gamma(C, \Omega_C) \longrightarrow \\ &\xrightarrow{\mu_{C/S}} \Gamma(C, \Omega_C^{\otimes 2}) \xrightarrow{\mathrm{Kod}^\vee} \Omega_{S/\overline{\mathbf{F}}_p} \end{aligned} \tag{5.4.1}$$

5.5 We first calculate $\mathrm{Kod} : \Theta_{S/\overline{\mathbf{F}}_p} \to H^1(C, \Theta_{C/S})$. To do this we let $U_1 = C \backslash \{x = \lambda\}$ and try to find some U_2 with $U_1 \cup U_2 = C$ such that we can extend ∂_λ to derivations D_1 and D_2 on $U_1 = C \backslash \{x = \lambda\}$ and U_2 respectively. The cocycle $D_1 - D_2$ on $U_1 \cap U_2$ is $\mathrm{Kod}(\partial_\lambda)$. To define D_1 we set $D_1(x) = 0$ and from Equation 5.1.1 it follows that

$$D_1(y) = \frac{x - x^2}{N y^{N-1}}$$

which is defined on U_1. We set $D_2(y) = 0$ and find

$$D_2(x) = \frac{x^2 - x}{3x^2 - 2(\lambda + 1)x + \lambda}$$

and this is defined in the point $x = \lambda$. This means that we can find the looked for U_2. So $\mathrm{Kod}(\partial_\lambda)$ is the cocycle for $\Theta_{C/S}$ on $U_1 \cap U_2$ for which

$$\mathrm{Kod}(\partial_\lambda)(\mathrm{d}x) = \frac{x - x^2}{3x^2 - 2(\lambda + 1)x + \lambda}. \tag{5.5.1}$$

5.6 Let $\frac{N+1}{3} \leq \ell_1 < \ell_2 \leq \frac{2N-1}{3}$ such that $N \neq \ell_1 + \ell_2$ This is possible since $N \geq 8$. Obviously

$$0 \neq \omega = \frac{\mathrm{d}x}{y^{\ell_1}} \frac{\mathrm{d}x}{y^{N-\ell_1}} - \frac{\mathrm{d}x}{y^{\ell_2}} \frac{\mathrm{d}x}{y^{N-\ell_2}} \in K$$

so we can calculate the image of it under the mapping 5.4.1. To do this we have to calculate the image of the differential form

$$\frac{\mathrm{d}x}{y^{N-\ell}}$$

under γ for $\ell = \ell_i, N - \ell_i$, $i = 1, 2$. Because γ commutes with the $\mathbf{Z}[\zeta_N]$ action on $\Gamma(C, \Omega_C)$ this calculation can be done on the different weight spaces separately. Because of our choice of the ℓ_i, these all have dimension one, so this can be done as in [Mu2, p215–216], see also [Ma, 3.2.8]. The result is that

$$\gamma : \frac{\mathrm{d}x}{y^{N-\ell}} \mapsto \tau_p(N-\ell)^{-1}\frac{\mathrm{d}x}{y^{N-\ell}}. \tag{5.6.1}$$

Here $\tau_p(a)$ is the coefficient of X^{p-1} in $(x(x-1)(x-\lambda))^{\frac{p-1}{N}a}$. Combining these results we see that the image of ω under the map of 5.4.1 is

$$\mathrm{res}_{\{x=\lambda\}}\mathrm{Kod}(\partial_\lambda)\left(((\tau_p(\ell_1)\tau_p(N-\ell_1))^{-1} - (\tau_p(\ell_2)\tau_p(N-\ell_2))^{-1})\frac{(\mathrm{d}x)^2}{y^N}\right)$$

which is equal to

$$\frac{N}{\lambda(1-\lambda)}((\tau_p(\ell_1)\tau_p(N-\ell_1))^{-1} - (\tau_p(\ell_2)\tau_p(N-\ell_2))^{-1})\mathrm{d}x \tag{5.6.2}$$

It is straightforward to verify that this is not zero. □

Using the above we can prove an even stronger result:

5.7 Proposition. *Let p be a prime and $\lambda_0 \in \overline{\mathbf{F}}_p$. Suppose that* $\mathrm{Jac}(C_{\lambda_0})$ *is ordinary. Let $R_p = \cup W(\overline{\mathbf{F}}_p)[p^{1/n}]$. Then there are only finitely many liftings $\lambda \in R_p$ of λ_0 such that* $\mathrm{Jac}(C_\lambda)$ *is of CM type.*

5.8 Proof. Suppose that there are infinitely many $\lambda \in R_p$ that lift λ_0, such that $\mathrm{Jac}(C_\lambda)$ is of CM type. If we let $A = \mathrm{End}^0(\mathrm{Jac}(C_{\lambda_0}))$ then it follows from 3.5 that all our infinitely many liftings have endomorphism algebra A. In characteristic zero, abelian varieties of CM type are defined over a number field. It follows that there is an infinite subset $\Lambda \subset \overline{\mathbf{Q}}$ such that $\mathrm{Jac}(C_\lambda)$ is has CM with A for every $\lambda \in \Lambda$. If we let F be the set of conductors of the rings $\mathrm{End}(\mathrm{Jac}(C_\lambda))$ for $\lambda \in \Lambda$, then $f_1, f_2 \in F$ differ only by a power of p. Since every $\lambda \in \Lambda$ gives a torsion point of $\mathcal{M}$ and because a formal torus has only finitely many p^m torsion points for fixed m, we see that F is infinite. We wish to find a prime $q \equiv 1 \pmod N$ such that q splits completely in the CM field of every simple factor of $\mathrm{Jac}(C_{\lambda_0})$ and such that q does not divide any $f \in F$. The compositum of these CM fields and $\mathbf{Q}[\zeta_N]$ is a finite extension of $\mathbf{Q}$, so by the Tchebotarev density theorem there exists a prime q that spits completely. This q splits in $\mathbf{Q}[\zeta_N]$, so $q \equiv 1 \pmod{N}$. It also splits in all our CM fields, so it is the prime that we where looking for. Because we have $\mathrm{End}^0(\mathrm{Jac}(C_\lambda)) = A$ for all $\lambda \in \Lambda$, q splits in the CM field of every simple component of $\mathrm{Jac}(C_\lambda)$ for all $\lambda \in \Lambda$. Since $\Lambda \subset \overline{\mathbf{Q}}$, each $\lambda \in \Lambda$ gives us a curve C_λ over a finite extension R_λ of $W(\overline{\mathbf{F}}_p)$. We consider the reduction modulo q of these curves. Because every $\mathrm{Jac}(C_\lambda)$ is of CM type, there is a finite extension R'_λ of R_λ such that $\mathrm{Jac}(C_\lambda)/R'_\lambda$ has good reduction modulo q. From the splitting behaviour of q it follows that for every $\lambda \in \Lambda$, the reduction modulo q of $\mathrm{Jac}(C_\lambda)$ we thus obtain is ordinary. It follows from Proposition 3.5 that every one of these Jacobians is a quasi canonical lifting of its reduction. Since under reduction modulo q the conductor of the endomorphism ring can only change by a power of q, the endomorphism ring does not change at all under reduction modulo q. So by Lemma 3.3 they are even the canonical lifting, therefore our reductions modulo q are all pre-W_2-canonical. The endomorphism rings of these reductions are different, because their conductors are different, so it follows that there are in fact infinitely many different reductions. We arrive at a contradiction with Lemma 5.3. □

References

[Co] R. Coleman: *Torsion points on curves.* In Galois representations and arithmetic algebraic geometry (Y. Ihara ed.), Advanced studies in pure mathematics 12 (1987), 235–247.

[De] P. Deligne: *Travaux de Shimura.* Séminaire Bourbaki, exposé 389, Février 1971. Springer LNM 244, 123–165.

[D,O] B. Dwork and A. Ogus: *Canonical liftings of Jacobians.* Compos. Math. 58 (1986), 111–131.

[Ka] N. Katz: *Serre-Tate local moduli.* In Surfaces algébriques (édité par J. Giraud, L. Illusie et M. Raynaud), Springer LNM 868 (1981); Exposé 5bis, 138–202.

[Ma] L. van der Marel: *Formal groups and differential equations.* Thesis, Groningen 1989.

[Me] W. Messing: *The crystals associated to Barsotti-Tate groups: with applications to abelian schemes.* Springer LNM 264 (1972).

[Mu1] D. Mumford: *A note on Shimura's paper "Discontinuous groups and abelian varieties"*. Math. Ann. 181 (1969), 345–351.

[Mu2] D. Mumford: *Abelian varieties.* Oxford university press (1970).

[S,T] G. Shimura and Y. Taniyama: *Complex multiplication of abelian varieties and its applications to number theory.* Publ. math. soc. Japan 6 (1961).

[Ta] J. Tate: *Endomorphisms of abelian varieties over finite fields.* Invent. Math. 2 (1966), 134–144.

Familles de courbes hyperelliptiques à multiplications réelles

J.-F. Mestre

Pour tout entier n, notons G_n le polynôme

$$G_n(T) = \prod_{k=1}^{[n/2]} (T - 2\cos\frac{2k\pi}{n}),$$

où $[x]$ est la partie entière de x. Disons qu'une courbe C de genre $[n/2]$, définie sur un corps k, est à multiplications réelles par G_n s'il existe une correspondance $\mathcal{C}$ sur C telle que G_n soit le polynôme caractéristique de l'endomorphisme induit par $\mathcal{C}$ sur les différentielles de première espèce de C.

L'anneau des endomorphismes de la jacobienne Jac C de C contient alors un sous-anneau isomorphe à $\mathbf{Z}[X]/(G_n(X))$, dont les éléments sont invariants par l'involution de Rosati. En particulier, si n est un nombre premier > 2, Jac$C$ est à multiplications réelles par $\mathbf{Z}[2\cos\frac{2\pi}{n}]$, dans la terminologie usuelle (*cf.* [9], par exemple.)

Dans cet article, nous construisons, pour tout entier $n \geq 4$, une famille de dimension 2, définie sur $\mathbf{C}$, de courbes hyperelliptiques de genre $[n/2]$ à multiplications réelles par G_n.

Plus précisément, à toute courbe elliptique E définie sur un corps k de caractéristique nulle, munie d'un sous-groupe cyclique G d'ordre n rationnel sur k, nous associons une famille à un paramètre de courbes hyperelliptiques de genre $[n/2]$ définies sur k à multiplications réelles par G_n. Si G est engendré par un point rationnel sur k, la correspondance associée est définie sur k.

Dans le cas où $n = 5$, on retrouve ainsi une construction connue, due à Humbert, (*cf.* par exemple [5], p. 374, [10], p. 220, ou encore [2]), et que nous rappelons: soient X une courbe de genre 2 dont la jacobienne est à multiplications réelles par $[\frac{1+\sqrt{5}}{2}]$, et w son involution hyperelliptique.

Soient C une conique du plan, et $f: X/w \to C$ un isomorphisme. Si P est l'image sur C d'un point de Weierstrass de X, il existe une numérotation $P_1, \ldots, P_5$ des images sur C des points de Weierstrass de X distincts de P telle qu'il existe une conique passant par P et inscrite au pentagone formé par $P_1, \ldots, P_5$ (i.e. tangente aux droites P_1P_2, P_2P_3, ..., P_5P_1.) Si l'on rapproche cet énoncé du théorème de Poncelet, et de son interprétation par la théorie des courbes elliptiques, on voit que la donnée de X est équivalente à la donnée d'une courbe elliptique E munie d'un point d'ordre 5, d'un revêtement ϕ à 2 feuillets de E sur une courbe de genre 0, et d'un point quelconque de cette dernière, distinct des 4 points de ramification de ϕ.

Dans le §1, nous construisons la famille de courbes hyperelliptiques ci-dessus.

Plus généralement, à toute isogénie $f: E_1 \to E_2$ de courbes elliptiques, définie sur un corps k, nous associons une courbe hyperelliptique C_f sur $k(T)$, où T est une indéterminée; à tout élément $R \in \mathrm{Ker} f$ est associée une correspondance $\mathcal{C}_R$ sur C_f, telle que le polynôme caractéristique de l'endomorphisme induit par $\mathcal{C}_R$ sur les formes de première espèce de C_f est un produit de polynômes G_m convenables.

Cette construction permet par exemple d'obtenir une famille à 2 paramètres, définie sur $\mathbf{Q}$, de courbes hyperelliptiques de genre 19 dont la jacobienne est isogène au produit de 19 courbes elliptiques.

Dans le §2, nous donnons quelques exemples, associés à des isogénies à noyau cyclique.

Pour $n = 5, 7, 9$, la courbe $X_1(n)$, classifiant les courbes elliptiques munies d'un point d'ordre n, est $\mathbf{Q}$-isomorphe à la droite projective. On obtient alors une famille à deux paramètres, définie sur $\mathbf{Q}$, de courbes de genre 2 (resp. 3, resp. 4) à multiplications réelles par G_5 (resp. G_7, resp. G_9). Nous en donnons une description explicite, et examinons également le cas où $N = 13$: on trouve alors une famille à deux paramètres, définie sur $\mathbf{Q}$, de courbes hyperelliptiques de genre 6, dont la jacobienne est à multiplications réelles par G_{13}, mais les endomorphismes correspondants ne sont pas en général définis sur $\mathbf{Q}$.

Dans le paragraphe suivant, nous examinons le cas des courbes C_f associées à des isogénies f de degré pair. Le fait que $X_1(8)$ (resp. $X_1(12)$) est $\mathbf{Q}$-isomorphe à $\mathbf{P}_1$ entraîne l'existence d'une famille à 2 paramètres, définie sur $\mathbf{Q}$, de surfaces abéliennes à multiplications réelles par $\mathbf{Z}[\sqrt{2}]$ (resp. $\mathbf{Z}[\sqrt{3}]$.)

Dans le dernier paragraphe, nous montrons que les constructions précédentes permettent d'obtenir, pour tout nombre premier $p \equiv \pm 2 \bmod 5$, une extension régulière de $\mathbf{Q}(T)$ de groupe de Galois $\mathrm{PSL}_2(\mathbf{F}_{p^2})$.

Je tiens à remercier J.-P. Serre pour l'aide précieuse qu'il a bien voulu m'accorder tout au long de ce travail.

1. Les courbes C_f

Soient E_1 et E_2 deux courbes elliptiques définies sur un corps k de caractéristique 0, x_1 (resp. x_2) une fonction de E_1 (resp. de E_2) de degré 2 ayant un pôle double en l'origine, $f : E_1 \to E_2$ une isogénie de degré n définie sur k, et G le noyau de f.

Soit enfin u la fonction, de degré n, rendant commutatif le diagramme

$$\begin{array}{ccc} E_1 & \to & E_2 \\ x_1 \downarrow & & \downarrow x_2 \\ \mathbf{P}_1 & \xrightarrow{u} & \mathbf{P}_2 . \end{array}$$

On dit que u est "l'équation aux abscisses" de f.

On note C_f la courbe hyperelliptique définie sur le corps $K = k(T)$ d'équation affine

$$y^2 = u(x) - T,$$

où T est une indéterminée.

Si P_T est un point de $E_1(\overline{K})$ tel que $x_2(f(P_T)) = T$, la courbe C_f est donc un revêtement à deux feuillets de $\mathbf{P}_1$ ramifié en les points $x_1(P_T + R)$, où R parcourt G, et en les points $x_1(S)$, où S parcourt de G tels que $2S = 0$.

Par suite,

PROPOSITION 1 .- *Le genre de la courbe hyperelliptique C_f est égal à $(n + m - 1)/2$, où n (resp. m) est le cardinal (resp. le nombre de points d'ordre égal à 2) de G.*

1.1. Revêtement associé à la composition de deux isogénies

Soient E_1, E_2 et E_3 trois courbes elliptiques, x_1, x_2 et x_3 trois fonctions d'ordre 2 sur E_1, E_2 et E_3 respectivement, ayant un pôle double en l'origine. Si f_1 (resp. f_2) est une isogénie de E_1 sur E_2 (resp. de E_2 sur E_3) de degré n_1 (resp. n_2), on note f l'isogénie $f_2 \circ f_1$ de E_1 sur E_3 et u (resp. u_1, resp. u_2) l'équation aux abscisses de f (resp. f_1, resp. f_2). On a donc $u = u_2 \circ u_1$.

L'application

$$(x,y) \mapsto (u_1(x), y)$$

définit un revêtement de degré n_1 de la courbe C_f d'équation $y^2 = u(x) - T$ sur la courbe C_{f_2} d'équation $y^2 = u_2(x) - T$.

Ceci permet de ramener partiellement l'étude des courbes C_f à l'étude des diverses courbes C_g, où g est une isogénie factorisant f.

EXEMPLE.- Soit E une courbe elliptique, et $f: E \to E$ la multiplication par 6. Le genre de C_f est égal à 19, et il existe 19 isogénies $g: F \to E$ de noyau cyclique telle qu'il existe $h: E \to F$, avec $6 = g \circ h$:

- Trois d'entre elles sont de degré 2; les courbes C_g associées sont de genre 1; notons-les E_i, $i = 1, \ldots, 3$.
- Quatre sont de degré 3; les courbes associées sont aussi de genre 1; notons-les F_i, $i = 1, \ldots, 4$.
- Enfin, les douze dernières sont de degré 6; les courbes associées sont de genre 3, et chacune est un revêtement d'une courbe correspondant à une isogénie de degré 2, ainsi que d'une courbe correspondant à une isogénie de degré 3, donc leur jacobienne est isogène à un produit de trois courbes elliptiques, une courbe de type E_i, une de type F_i, et une nouvelle courbe, que nous notons G_i ($i = 1, \ldots, 12$).

On obtient ainsi un homomorphisme

$$\mathrm{Jac}\, C_f \to \prod E_i \times \prod F_i \times \prod G_i,$$

défini sur $k'(T)$, où k' est l'extension de k obtenue par adjonction des points d'ordre 6 de k. Cet homomorphisme est une isogénie; on peut par exemple le prouver en utilisant les correspondances sur C_f définies en 1.2 et 1.3.

Par suite:

THÉORÈME 1 .- *Soient E une courbe elliptique définie sur un corps k de caractéristique nulle, et x une fonction de degré 2 de E ayant un pôle double en 0. Soit u la fonction rationnelle de degré 36 telle que $x(6P) = u(x(P))$ pour tout point P de E. Alors la courbe hyperelliptique d'équation affine*

$$Y^2 = u(X) - T,$$

où T est une indéterminée, est de genre 19, et sa jacobienne est isogène à un produit de 19 courbes elliptiques.

Les courbes elliptiques étant paramétrées par la droite affine, on en déduit l'existence *d'une famille à deux paramètres, définie sur* $\mathbf{Q}$*, de courbes hyperelliptiques de genre* 19 *dont la jacobienne est isogène à un produit de* 19 *courbes elliptiques.*

REMARQUE.- Le corps k' contient les racines sixièmes de l'unité, et ne peut donc être égal à $\mathbf{Q}$. Par contre, la courbe modulaire classifiant les courbes elliptiques munies d'un sous-groupe isomorphe à $\mathbf{Z}/6\mathbf{Z} \times \mathbf{Z}/2\mathbf{Z}$ est $\mathbf{Q}$-isomorphe à la droite projective. Le genre des courbes C_f associées est égal à 7; on en déduit donc l'existence d'une famille à deux paramètres s et t, définie sur $\mathbf{Q}$, de courbes hyperelliptiques de genre 7 dont la jacobienne est $\mathbf{Q}(s,t)$-isogène à un produit de 7 courbes elliptiques.

1.2. Involutions de C_f associées aux points d'ordre 2 de G

Supposons dans ce paragraphe que l'ordre n de G soit pair. Soit alors $R \in G$ un point d'ordre 2 de la courbe E_1. L'involution de E_1 donnée par $P \mapsto P + R$ commute à l'involution $P \mapsto -P$, donc $x_1(P+R)$ est une fonction rationnelle de $x_1(P)$, et est une involution: il existe a, b, c tels que

$$x_1(P+R) = \frac{ax_1(P)+b}{cx_1(P)-a}.$$

Soit donc $\mathcal{C}_R : C_f \to C_f$ l'involution donnée par

$$(x,y) \mapsto (\frac{ax+b}{cx-a}, y).$$

Si l'on note F la courbe quotient de E_1 par le sous-groupe engendré par R, et h le morphisme canonique de E_1 sur F, on a $f = g \circ h$, où g est une isogénie de F sur E_2. La courbe quotient $C_f/\mathcal{C}_R$ est alors isomorphe à la courbe C_g.

Soient x_3 une fonction de degré 2 de F, ayant un pôle double en 0, et u l'équation aux abscisses de g.

La courbe C_g a alors comme équation

$$y^2 = u(x) - T.$$

Soient S un point de E_1 tel que $2S = R$, et Q un point d'ordre 2 de E_1 distinct de R. La courbe $C_f/w \circ \mathcal{C}_R$, où w est l'involution hyperelliptique de C_f, a alors comme équation

$$y^2 = (u(x) - T)(x - x_3(\rho(S)))(x - x_3(\rho(S+Q))).$$

Soit g le genre de C_f. Si g est pair, les genres de C_f et $C_f/(w \circ \mathcal{C}_R)$ sont égaux; sinon, ils sont respectivement égaux à $(g-1)/2$ et $(g+1)/2$.

1.3. Correspondances sur C_f associées aux points d'ordre > 2 de G

Soit $f : E_1 \to E_2$ une isogénie de noyau G et de degré n (non nécessairement pair), et u l'équation aux abscisses de f. Pour tout point P de E_1 et tout point R de G, on a

$$u(x_1(P+R)) = u(x_1(P)).$$

Par ailleurs, les fonctions $P \mapsto x_1(P+R) + x_1(P-R)$ et $P \mapsto x_1(P+R)x_1(P-R)$ sont des fonctions invariantes par l'involution $P \mapsto -P$, donc des fonctions rationnelles de x_1, définies sur le corps $k(x_1(R))$. Notons-les s et p. Si Z est une indéterminée, on a

$$(Z - x_1(P+R))(Z - x_1(P-R)) = Z^2 - s(x_1(P))Z + p(x_1(P)).$$

Soit à présent R un point de G d'ordre > 2. L'équation ci-dessus permet de lui associer la $2-2$ correspondance symétrique $\mathcal{C}_R \subset C_f \times C_f$ définie sur $k(x_1(R))(T)$, d'équations

$$y^2 = u(x) - T, \quad Y^2 = u(X) - T, \quad X^2 - s(x)X + p(x) = 0, \quad Y = y. \quad (1)$$

Soit $P = (x, y)$ un point de C_f; si $Q \in E_1$ est tel que $x = x_1(Q)$, l'image du diviseur (P) par l'endomorphisme de $\mathrm{Pic}(C_f)$ associé à la correspondance $\mathcal{C}_R$ est le diviseur $((x_1(Q+R), y)) + ((x_1(Q-R), y))$.

1.4. Action de la correspondance $\mathcal{C}_R$ sur $\Omega^1(C_f)$.

Pour tout $R \in G$, notons ω_R la forme différentielle régulière de C_f définie par

$$\omega_R = \frac{1}{x - x_1(R)} \frac{dx}{y}.$$

(Par convention, on pose $\omega_0 = 0$.)

On a $\omega_S = \omega_R$ si et seulement si $R = \pm S$. L'ensemble des formes ω_R, $R \in G - \{0\}$, est une base de $\Omega^1(C_f)$.

Pour étudier l'action des correspondances $\mathcal{C}_R$ sur $\Omega^1(C_f)$, nous aurons besoin du lemme suivant:

LEMME 1 .- *La fonction F qui à trois points P, Q, R de E_1 associe*

$$F(P,Q,R) = (x_1(P) - x_1(Q-R))(x_1(P) - x_1(Q+R))(x_1(Q) - x_1(R))^2$$

est symétrique en P, Q, R.

Il est clair que la permutation $Q \leftrightarrow R$ ne change pas l'expression ci-dessus. Il en est de même lorsqu'on permute P et Q. En effet, Q et R étant fixés, les fonctions f et g définies respectivement par $f(P) = (x_1(P)-x_1(Q-R))(x_1(P)-x_1(Q+R))(x_1(Q)-x_1(R))^2$ et $g(P) = (x_1(Q)-x_1(P-R))(x_1(Q)-x_1(P+R))(x_1(P)-x_1(R))^2$ ont même diviseur, à savoir $(Q-R)+(Q+R)+(-Q-R)+(-Q+R)$, donc f et g sont proportionnelles. En faisant tendre P vers 0, on en déduit qu'elles sont égales.

Lorsque E_1 est définie sur $\mathbf{C}$, ce lemme est conséquence de la formule $\wp(u) - \wp(v) = \sigma(u+v)\sigma(u-v)\sigma^{-2}(u)\sigma^{-2}(v)$, et du fait que la fonction σ est impaire. En effet,

$$\begin{aligned}
&(\wp(u) - \wp(v-w))(\wp(u) - \wp(v+w))(\wp(v) - \wp(w))^2 \\
&= \sigma(u+v+w)\sigma(u+v-w)\sigma(-u+v-w)\sigma(-u+v+w) \\
&\qquad \sigma^{-4}(u)\sigma^{-4}(v)\sigma^{-4}(w) \\
&= -\sigma(u+v+w)\sigma(u+v-w)\sigma(u-v+w)\sigma(v-u+w)\sigma^{-4}(u) \\
&\qquad \sigma^{-4}(v)\sigma^{-4}(w),
\end{aligned}$$

expression symétrique en u, v, w. Par le principe de prolongement des identités algébriques, on pourrait d'ailleurs en déduire le cas où k est quelconque.

Rappelons que l'endomorphisme T_R de $\Omega^1(C_f)$ associé à la correspondance $\mathcal{C}_R$ est égal à $\mathrm{Tr} \circ p_1^*$, où p_1 est la première projection de $\mathcal{C}_R$ sur C_f et Tr la fonction trace de $\Omega^1(\mathcal{C}_R)$ sur $\Omega^1(C_f)$ associée à la seconde projection.

Notons $z = x_1(P)$, $z_1 = x_1(P-R)$ et $z_2 = x_1(P+R)$. Pour tout couple de points (Q, R) de E_1, on a donc:

$$(z-x_1(Q-R))(z-x_1(Q+R))(x_1(Q)-x_1(R))^2 = (z_1-x_1(Q))(z_2-x_1(Q))(z-x_1(R))^2.$$

En dérivant logarithmiquement cette expression, on obtient:

$$\frac{dz_1}{z_1 - x_1(Q)} + \frac{dz_2}{z_2 - x_1(Q)} = \frac{dz}{z - x_1(Q-R)} + \frac{dz}{z - x_1(Q+R)} - 2\frac{dz}{z - x_1(R)}.$$

Comme $T_R(\omega_Q) = T_R(\frac{1}{z - x_1(Q)}\frac{dz}{y}) = \frac{1}{z_1 - x_1(Q)}\frac{dz_1}{y} + \frac{1}{z_2 - x_1(Q)}\frac{dz_2}{y}$, on a donc

$$T_R(\omega_Q) = \omega_{Q-R} + \omega_{Q+R} - 2\omega_R,$$

(toujours avec la convention $\omega_0 = 0$.).

PROPOSITION 2 *L'action de la correspondance $\mathcal{C}_R$ sur l'espace des formes différentielles de première espèce de C_f est donnée par les formules:*

$$\omega_Q \mapsto \omega_{Q-R} + \omega_{Q+R} - 2\omega_R,$$

les formes ω_Q, $Q \in G$, étant définies comme plus haut.

1.5. Le cas où G est cyclique d'ordre n

Supposons maintenant que G est un groupe cyclique d'ordre n, et soit R un générateur de G.

Pour tout $S \in G$, posons $v_S = \omega_{S-R} - \omega_{S+R}$. Notons que $v_{-S} = -v_S$, donc $v_S = 0$ si et seulement si $2S = 0$.

Le sous-espace Ω' de $\Omega^1(C_f)$ engendré par les v_S, $S \in G$, est stable par T_R. Plus précisément, on a

$$T_R(v_S) = v_{S+R} + v_{S-R}.$$

Si n est impair, on vérifie facilement que Ω' est égal à $\Omega^1(C_f)$ tout entier. Il est alors clair que l'endomorphisme T_R de $\Omega^1(C_R)$ a pour polynôme caractéristique

$$G_n(X) = \prod_{k=1}^{(n-1)/2} (X - 2\cos\frac{2k\pi}{n}).$$

Si n est pair, l'espace Ω' est de dimension $(n-2)/2$. Posons alors

$$w = \omega_{\frac{n}{2}R} + 2(-1)^{n/2}(\sum_{i=1} (n-2)/2)(-1)^i \omega_{iR}.$$

On vérifie immédiatement que $T_R(w) = -2w$, et que $\Omega^1(C_f)$ est somme directe de Ω' et de la droite engendrée par w. Le polynôme caractéristique de T_R agissant sur Ω' est égal à

$$\prod_{k=1}^{(n-2)/2} (X - 2\cos\frac{2k\pi}{n}),$$

donc le polynôme caractéristique de T_R agissant sur $\Omega^1(C_f)$ est égal à

$$\prod_{k=1}^{n/2} (X - 2\cos\frac{2k\pi}{n}).$$

PROPOSITION 3 .- *Soit* $f : E_1 \to E_2$ *une isogénie de noyau* G *cyclique d'ordre* n. *Le polynôme caractéristique de la correspondance* $\mathcal{C}_R$ *agissant sur* $\Omega^1(C_f)$ *est égal à*

$$G_n(X) = \prod_{k=1}^{[n/2]} (X - 2\cos\frac{2k\pi}{n}).$$

REMARQUE.- Soit Z le normalisé du produit fibré de E_1 et C_f relativement aux revêtements $x_1 : E_1 \to \mathbf{P}_1$ et $x : C_f \to \mathbf{P}_1$.

Si E_1 a comme équation $z^2 = h(x)$, où h est de degré 3, un système d'équations affines de Z est par exemple donné dans $\mathbf{P}_3$ par

$$z^2 = h(x), \quad y^2 = u(x).$$

À tout point $R \in G$, on peut associer un automorphisme ϕ_R de Z d'ordre égal à l'ordre de R, en posant $\phi(x, y, z) = (x(P + R), y, z(P + R))$, P étant l'image de (x, y, z) par la projection de Z sur E_1.

Par ailleurs, soit v l'involution de Z donnée par $(x, y, z) \mapsto (x, y, -z)$, et soit G' le groupe d'automorphismes de Z engendré par G et v. La courbe C_f est la courbe quotient Z/v, et l'on a

$$v \circ \phi_R \circ v = \phi_{-R}.$$

Les correspondances $\mathcal{C}_R$ ne sont autres que les images par $Z \to C_f$ de la correspondance graphe de ϕ_R dans $Z \times Z$. Si G est cyclique d'ordre n, G' est le groupe diédral D_n, et on retrouve le fait que le polynôme caractéristique de $\mathcal{C}_R$ agissant sur $\Omega^1(C_f)$ est le polynôme G_n.

Ce point de vue avait déjà été développé par A. Brumer [3].

2. Exemples

On trouve dans Kubert ([6], p. 217) une description des courbes modulaires $X_1(n)$ de genre 0, classifiant les couples (E, R) formés d'une courbe elliptique munie d'un point S d'ordre n, et une paramétrisation explicite de tels couples.

D'après la section précédente, à un tel couple est associée une famille à un paramètre de courbes hyperelliptiques à multiplications réelles par G_n.

Par ailleurs, si E_1 est une courbe elliptique définie sur un corps k et G un sous-groupe fini de $E(\overline{k})$, les formules permettant d'obtenir explicitement la courbe quotient $E_2 = E_1/G$ et une isogénie $f : E_1 \to E_2$ de noyau G ont été établies par Vélu ([11]).

2.1. Exemples avec n impair

- LE CAS $n = 5$

La courbe modulaire $X_1(5)$ est $\mathbf{Q}$-isomorphe à $\mathbf{P}_1$. Si E_1 a comme équation $y^2 + (1-U)xy - Uy = x^3 - Ux^2$, le point $R = (0,0)$ de E_1 est d'ordre 5. Les formules donnant l'isogénie f et la courbe quotient $E_2 = E_1/f$. figurent par exemple dans [7].

On trouve alors la famille de courbes hyperelliptiques $C_5(U,T)$

$$Y^2 = (1-Z)^3 + UZ((1-Z)^3 + UZ^2 - Z^3(1-Z)) - TZ^2(Z-1)^2.$$

- LE CAS $n = 7$

Dans le cas de $X_1(7)$, des calculs analogues conduisent à une famille de courbes $C_7(U,T)$ à multiplications réelles par $G_7(X)$:

$$\begin{aligned}
Y^2 = {} & U^4 - (3U^3 + 3U^4 - U^5)Z + U^2(3 + 8U + 2U^2 - 3U^3)Z^2 \\
& -U(1 + 7U + 8U^2 - 4U^3 - U^4 - U^5)Z^3 \\
& +U(1 + 13U - 12U^2 + 9U^3 - 6U^4)Z^4 \\
& +(1 - 7U + 5U^2 - 3U^3 + 2U^4 + U^5)Z^5 \\
& +2U(1 - U^2)Z^6 + U(U-1)Z^7 - TZ^2(Z-U)^2(Z-1)^2,
\end{aligned}$$

où U est le paramètre de $X_1(7)$ adopté dans [7].

- LE CAS $n = 9$

On trouve dans [6], p. 217, une paramétrisation des courbes elliptiques munies d'un point d'ordre 9: le point $(0,0)$ est d'ordre 9 sur la courbe elliptique d'équation

$$y^2 - (U^3 - U^2 - 1)xy - U^2(U-1)(U^2-U+1)y = x^3 - U^2(U-1)(U^2-U+1)x^2,$$

où U paramètre $X_1(9)$.

Les formules de Vélu donnent alors comme équation des courbes hyperelliptiques $C_9(U,T)$ de genre 4 associées

$$
\begin{aligned}
Y^2 &= U^4(U-1)(U^2-U+1)^3Z^9 - 2U^3(U-1)(U^2-U+1)^2 \\
&(U^3+U+1)Z^8 + U(U^2-U+1) \\
&(1-10U+23U^2-44U^3+50U^4-39U^5+23U^6-7U^7+U^8+U^9)Z^7 \\
&+(-1+32U-107U^2+243U^3-353U^4+369U^5 \\
&-279U^6+154U^7-67U^8+22U^9-6U^{10})Z^6 \\
&+(-1-5U-32U^2+101U^3-237U^4+323U^5 \\
&-312U^6+209U^7-91U^8+25U^9-2U^{10}+U^{11})Z^5 \\
&+(5+11U+17U^2-29U^3+83U^4-97U^5+83U^6-51U^7+19U^8-6U^9)Z^4 \\
&+(-10-8U-14U^2+2U^3-12U^4+5U^5+U^6+U^8)Z^3 \\
&+(10+2U+11U^2-4U^3+5U^4-3U^5)Z^2+(-5-3U^2+U^3)Z+1 \\
&-TZ^2(Z-1)^2((U^2-U+1)Z-1)^2(UZ-1)^2.
\end{aligned}
$$

On a $G_9(X) = (X+1)(X^3-3X+1)$, donc, pour toute courbe $C_9(U,T)$ de cette famille, la jacobienne de $C_9(U,T)$ contient une variété abélienne de dimension 3 à multiplications réelles par $\mathbf{Z}[2\cos\frac{2\pi}{9}]$.

- Le cas $n = 13$

La courbe $X_0(13)$, classifiant les courbes elliptiques munies d'un groupe cyclique d'ordre 13, est $\mathbf{Q}$-isomorphe à $\mathbf{P}_1$. À tout point de $P_1(\mathbf{Q}) \simeq X_0(13)(\mathbf{Q})$ distinct des pointes est associée une courbe elliptique E_1 munie d'un sous-groupe cyclique G d'ordre 13 rationnel sur $\mathbf{Q}$, donc une isogénie $f: E_1 \to E_2 = E_1/G$ définie sur $\mathbf{Q}$; si l'équation aux abscisses de f est $X = p(x)/q^2(x)$, on en déduit comme précédemment que la courbe hyperelliptique de genre 6 d'équation $z^2 = p(x) - Tq^2(x)$, où T est un paramètre, est à multiplications réelles par G_{13}. Si $R \in G$ est rationnel sur une extension k de $\mathbf{Q}$, la correspondance $\mathcal{C}_R$, et donc l'endomorphisme induit sur la jacobienne, est défini sur k. Mais la courbe $X_1(13)$ n'a pas de point rationnel sur $\mathbf{Q}$ distinct des pointes, et la correspondance $\mathcal{C}_R$ n'est jamais définie sur $\mathbf{Q}$.

2.2. Exemples avec n pair

Soit $f: E_1 \to E_2$ une isogénie de degré n et à noyau cyclique, et soit R un générateur de Ker$f$. Posons $R_2 = (n/2)R$, $E_3 = E_1/ < R_2 >$, et $g: E_3 \to E_2$ l'isogénie de degré $n/2$ déduite de f comme en 1.2. On a vu que, si $s = \mathcal{C}_{R_2}$, construite comme en 1.2, la courbe C_f/s n'est autre que la courbe C_g. Plus précisément, soit x_3 une fonction de E_3 de degré 2 ayant un pôle double en 0, et u l'équation aux abscisses de g. La courbe C_g a alors comme équation

$$y^2 = u(x) - T.$$

De même, la courbe $C' = C_f / w \circ s$, où w est l'involution hyperelliptique de C_f, a comme équation

$$y^2 = (u(x) - T)(x - a)(x - b),$$

où a et b sont les abscisses de points d'ordre 2 convenables de E_3 (*cf* 1.2).

- $n = 8$. Dans ce cas, la courbe C_f est de genre 4, et les courbes C_g et C' sont de genre 2. Le polynôme caractéristique de C_R est le polynôme $X(X+2)(X^2-2)$. L'isogénie g se factorise en un produit de 2 isogénies de degré 2, donc la jacobienne de la courbe C_g est isogène à un produit de 2 courbes elliptiques, alors que la jacobienne de la courbe C' est à multiplications réelles par $\mathbf{Z}[\sqrt{2}]$.

La courbe $X_1(8)$ est $\mathbf{Q}$-isomorphe à $\mathbf{P}_1$. Par suite, *il existe une famille à deux paramètres, définie sur* $\mathbf{Q}$, *de surfaces abéliennes à multiplications réelles par* $\mathbf{Z}[\sqrt{2}]$.

De façon explicite, une famille $C_4(U,T)$ à 2 paramètres U et T de courbes de genre 2 dont la jacobienne est à multiplications réelles par $\mathbf{Z}[\sqrt{2}]$ est donnée par

$$\begin{aligned}Y^2 = &((U^2+1)^2X + U + 1)((U-1)^2(U+1)X + 1)\\ &(U^2(U-1)^2(U^2+1)^2X - T + \frac{(U^2+1)^2}{X} + \frac{U+1}{X^2}\\ &+\frac{U^2(U^2-1)^2}{(U^2+1)^2X + U + 1}).\end{aligned}$$

Remarque.- On retrouve ainsi un autre résultat de Humbert (*loc. cit.*, p. 379): soient X une courbe de genre 2, v son involution hyperelliptique, C une conique non dégénérée, et $\phi : X/v \to C$ un isomorphisme. Notons P_i, $i = 1, \dots, 6$ les images par ϕ des points de Weierstrass de X. La jacobienne de X est à multiplications réelles par $\mathbf{Z}[\sqrt{2}]$ si et seulement s'il existe une conique passant par P_1 et P_2 et inscrite à l'un des quadrilatères formés par les points P_i, $i = 3, \dots, 6$.

Via l'interprétation du théorème de Poncelet à l'aide de la théorie des courbes elliptiques, une telle configuration équivaut à la donnée d'une courbe elliptique muni d'un point R d'ordre 4 et d'un point d'ordre 2 distinct de $2R$, donc à la donnée d'une courbe de type C'.

- $n = 12$. Soit $f : E_1 \to E_2$ une isogénie de degré 12 à noyau cyclique, et soit R un générateur de Ker$f$. La courbe C_f est de genre 6, le polynôme caractéristique de $\mathcal{C}_R$ agissant sur les formes de première espèce de C_f est égal à $X(X+2)(X-1)(X+1)(X^2-3)$.

Si ϕ est l'endomorphisme de Jac C_f induit par la correspondance $\mathcal{C}_R$, la variété abélienne A_f égale à l'image de Jac C_f par $\phi\circ(\phi+2)\circ(\phi^2-1)$ est à multiplications réelles par $\mathbf{Z}[\sqrt{3}]$.

La courbe $X_1(12)$ est $\mathbf{Q}$-isomorphe à $\mathbf{P}_1$. Par suite, *il existe une famille à deux paramètres, définie sur* $\mathbf{Q}$, *de surfaces abéliennes à multiplications réelles par* $\mathbf{Z}[\sqrt{3}]$.

Ici encore, on pourrait expliciter la famille à deux paramètres correspondantes, en utilisant d'une part les formules de Kubert donnant une paramétrisation de $X_1(12)$ et d'autre part les formules de Vélu. Nous nous contentons ici de donner un exemple, la formule générale nous semblant un peu fastidieuse à écrire:

Soit E la courbe elliptique, intitulée $90G$ dans les tables de [1], p. 92, dont une équation est

$$y^2 + xy + x = x^3 - x^2 - 122x + 1721.$$

Son groupe de Mordell-Weil est cyclique d'ordre 12, engendré par le point de coordonnées $(-9, 49)$.

En utilisant les formules de Vélu, on trouve alors comme équation de la courbe hyperelliptique $C_3(T)$ correspondante

$$\begin{aligned} Y^2 = {} & (X+2)(93312 + 279936X + 134784X^2 - 304560X^3 \\ & +2532888X^4 + 5715072X^5 + 2899412X^6 - 779631X^7 - 448605X^8 \\ & -308497X^9 + 118326X^{10} - 2988X^{11} + 432X^{12}) \\ & -TX^2(X+2)^2(X-6)^2(3X+2)^2(2X+3)^2(X-1)^2, \end{aligned}$$

où T est un paramètre.

On note $A_3(T)$ la sous-variété abélienne de Jac$(C_3(T))$ à multiplications réelles par $\mathbf{Z}[\sqrt{3}]$.

REMARQUE.- Soit $f\, E_1 \to E_2$ une isogénie de degré 12 à noyau cyclique. Les courbes C_g et C' construites par la méthode indiquée au début de ce paragraphe sont de genre 3. L'isogénie g est ici de degré 6, donc la jacobienne de C_g est isogène à un produit de 2 courbes elliptiques. La jacobienne de C' est isogène au produit d'une courbe elliptique et d'une surface abélienne à multiplications réelles par $\mathbf{Z}[\sqrt{3}]$.

Réciproquement, toute surface abélienne à multiplications réelles par $\mathbf{Z}[\sqrt{3}]$ peut être obtenue par la construction précédente, à partir d'une courbe elliptique munie d'un point R d'ordre 6 et d'un point d'ordre 2 distinct de $3R$.

3. Application à l'obtention d'extensions régulières de $\mathbf{Q}(T)$ à groupe de Galois $\mathrm{PSL}_2(\mathbf{F}_{p^2})$.

Soit A une surface abélienne définie sur $\mathbf{Q}(T)$, non constante (i.e. de modules non constants), et dont l'anneau des $\mathbf{Q}(T)$-endomorphismes contient un sous-anneau isomorphe à l'anneau des entiers d'un corps quadratique réel M. Notons $A[p]$ le sous-groupe des points de p-torsion de A, et G le groupe de Galois de l'extension $L/\mathbf{Q}(T)$, où $L = \mathbf{Q}(T)(A[p])$. Si p est inerte dans M, $A[p]$ est un $\mathbf{F}_{p^2}$-espace vectoriel de dimension 2, et G est isomorphe à un sous-groupe du sous-groupe $\mathrm{GL}'_2(\mathbf{F}_{p^2})$ de $\mathrm{GL}_2(\mathbf{F}_{p^2})$ formé des matrices dont le déterminant est dans $\mathbf{F}_p^*$. On voit aisément que l'image de $\mathrm{GL}'_2(\mathbf{F}_{p^2})$ dans $\mathrm{PGL}_2(\mathbf{F}_{p^2})$ est égale à $\mathrm{PSL}_2(\mathbf{F}_{p^2})$.

Par suite, si $G = \mathrm{GL}'_2(\mathbf{F}_{p^2})$, le sous-corps M de L fixé par les matrices scalaires de G est une extension non constante, donc régulière, de $\mathbf{Q}(T)$. Or, pour que ceci soit vrai, il suffit que, pour une spécialisation $t \in \mathbf{Q}$ de T, les points de p-division de la spécialisation correspondante de A engendrent une extension de $\mathbf{Q}$ de groupe de Galois $\mathrm{GL}'_2(\mathbf{F}_{p^2})$.

Certaines des familles de courbes hyperelliptiques à multiplications réelles décrites dans les sections précédentes permettent de construire de telles extensions. Considérons par exemple la famille $C_5(U,T)$ ci-dessus. Nous avons montré dans [8] que, pour tout $p \equiv \pm 2 \bmod 5, p > 2$, le groupe de Galois des points de p-division de la jacobienne de $C_5(-17/4, 1)$ est égal à $\mathrm{GL}'_2(\mathbf{F}_{p^2})$, d'où le théorème:

THÉORÈME 2 .- *Pour tout nombre premier* $p \equiv \pm 2 \bmod 5$, *il existe une extension régulière de* $\mathbf{Q}(T)$ *de groupe de Galois* $\mathrm{PSL}_2(\mathbf{F}_{p^2})$.

REMARQUE 1.- W. Feit, dans [4], avait déjà donné une démonstration de ce théorème, à ceci près qu'il restait à prouver qu'une certaine courbe de genre 0 a un point rationnel, ce que J-P. Serre a récemment démontré. La méthode utilisée par Feit est différente de celle présentée ici.

REMARQUE 2.- Par une méthode analogue, on peut prouver que, *pour tout nombre premier p assez grand,* $p \not\equiv \pm 1 \bmod 24$, *il existe une extension régulière de* $\mathbf{Q}(T)$ *de groupe de Galois* $\mathrm{PSL}_2(\mathbf{F}_{p^2})$. D'après un théorème de Ribet ([9], p. 801, th. 5.5.2.), il suffit en effet d'exhiber *une* courbe dans chacune des deux familles de courbes $C_4(U,T)$ et $C_3(T)$ de la section précédente qui n'a pas partout bonne réduction potentielle. Par exemple, choisissons dans la famille $C_4(U,T)$ ci-dessus la courbe $C_4(2,12)$. Elle est donnée par l'équation

$$Y^2 = 45 + 177X + 215X^2 + 75X^3 + 20X^4 + 12X^5.$$

Le discriminant du membre de gauche est égal à $2^{12}3^4 1201^3$, et sa réduction mod 1201 est la courbe d'équation

$$Y^2 = 12(X - 1125)(X - 239)^2(X - 799)^2.$$

Par suite, la courbe $C_4(2,12)$ n'a pas bonne réduction potentielle en 1201, et on peut appliquer le théorème de Ribet cité plus haut: pour tout p assez grand, $p \equiv \pm 3 \bmod 8$, le groupe de Galois de l'extension de $\mathbf{Q}$ obtenue par adjonction des points de p-torsion de la jacobienne de $C_4(2,12)$ est égal à $\mathrm{GL}_2'(\mathbf{F}_{p^2})$.

On procède de même avec la courbe $C_3(T)$: pour tout nombre rationnel t, et tout nombre premier $l > 5$ divisant strictement le dénominateur de t, la réduction de la courbe $C_3(t)$ est stable en l, et est même complétement torique. Prenons par exemple $C_3(1/7)$. Toujours d'après le théorème de Ribet, pour tout p assez grand, $p = \pm 5 \bmod 12$, le groupe de Galois de l'extension de $\mathbf{Q}$ obtenue par adjonction des points de p torsion de la variété abélienne $A_3(1/7)$ est égal à $\mathrm{GL}_2'(\mathbf{F}_{p^2})$.

En fait, il est probable que, pour *tout* $p \equiv \pm 3 \bmod 8$, (resp. $p \equiv \pm 5 \bmod 12$) le groupe de Galois des points d'ordre p de Jac $C_4(2,12)$ (resp. de $A_3(1/7)$) est égal à $\mathrm{GL}_2'(\mathbf{F}_{p^2})$. Pour le démontrer, il faudrait faire une étude détaillée de ces courbes $C_4(2,12)$ et $C_3(1/7)$, analogue à celles de [8], section 2.

Bibliographie

[1] Numerical tables on elliptic curves, in: Modular Functions of One Variable IV, *Lecture Notes in Math.* **476** (1975), 74–144.

[2] W. BARTH et R. MOORE. Geometry in the space of Horrocks-Mumford surface, *Topology.* **28** (1989), 231–245.

[3] A. BRUMER. Courbes à automorphismes et courbes à multiplications réelles, *non publié.*

[4] W. FEIT. Rigidity of Aut($_2(p^2)$), $p \equiv \pm 2 \pmod 5$, $p \neq 2$, in: proc. rutgers group theory year, 1983-1984, *Cambridge Univ. Press,* 1984 351–356.

[5] G. HUMBERT. *Oeuvres.* Volume II, Gauthier-Villars, (1936).

[6] D.S. KUBERT. Universal bounds on the torsion of elliptic curves, *Proc. London Math. Soc.* (3). **33** (1976), 193–237.

[7] J.-F. MESTRE. Courbes elliptiques et groupes de classes d'idéaux de certains corps quadratiques, *J. reine angew. Math.* **343** (1983), 23–35.

[8] J.-F. MESTRE. Courbes hyperelliptiques à multiplications réelles, *C. R. Ac. Sc. Paris.* **307** (1988), 721–724.

[9] K. RIBET. Galois action on division points of abelian varieties with real multiplications, *Amer. J. of Math.* **98** (1976), 751–804.

[10] G. VAN DER GEER. *Hilbert Modular Surfaces.* Springer-Verlag, (1988).

[11] J. VÉLU. Isogénies entre courbes elliptiques, *C. R. Acad. Sc. Paris.* **273** (1971), 238–241.

SÉRIES DE KRONECKER ET FONCTIONS L DES PUISSANCES SYMÉTRIQUES DE COURBES ELLIPTIQUES SUR $\mathbf{Q}$

Jean-François Mestre & Norbert Schappacher

Table des Matières

0. Introduction

La formulation, aujourd'hui très générale, des *conjectures de Beilinson* a été inaugurée par Bloch, et influencée en particulier par une étude numérique du cas des courbes elliptiques E définies sur $\mathbf{Q}$ faite en 1981 par Bloch et Grayson, [Bloch, Grayson 1986]. Nous reprenons ici le travail de Bloch et Grayson et présentons ensuite une étude analogue du carré symétrique d'une courbe elliptique. Notre travail repose sur une version explicite des conjectures de Beilinson relatives aux valeurs $L(Sym^k E, k+1)$, $k \geq 2$, donnée récemment par Deninger: [Deninger 1988], [Deninger 1989]. Nous vérifions numériquement les relations conjecturées par Beilinson et Deninger dans le cas $k = 2$, pour des courbes elliptiques pour lesquelles cette conjecture explicite donne un énoncé non trivial. Il s'avère que celles d'entre ces courbes qui admettent un point de torsion rationnel sont *en nombre fini*. *Voir* plus précisément, 3.4.1 et 3.4.6.

Les conjectures de Beilinson expriment la valeur $L^{(r)}(s)$ de la première dérivée non nulle en tout entier s, de la fonction L d'un motif sur $\mathbf{Q}$, en termes de *régulateurs* définis à moyen de la K-théorie (ou *cohomologie motivique*) du motif. Si E est une courbe elliptique sur $\mathbf{Q}$, la valeur spéciale $L(E, 2)$ est ainsi conjecturalement liée au groupe $K_2(E)$. Une façon de se donner des éléments de $K_2(E)$ utilise des diviseurs *définis sur* $\mathbf{Q}$ à support dans les points de torsion de E. Les régulateurs de ces éléments de $K_2(E)$ s'expriment en termes de séries de Kronecker attachées à E; $L(E, 2)$ s'exprime donc conjecturalement comme combinaison linéaire à coefficients rationnels de certaines séries de Kronecker. Dans le cas d'une courbe *à multiplications complexes*, la fonction analytique $L(E, s)$ est déjà combinaison linéaire de séries de Kronecker — *cf.* [Rohrlich 1987], [Deninger, Wingberg 1988]. Pour *certaines* courbes sans multiplication complexe, Bloch et Grayson [Bloch, Grayson 1986] ont mis en évidence l'existence de telles combinaisons linéaires au point $s = 2$.

Le phénomène le plus important découvert dans [Bloch, Grayson 1986] est la nécessité d'introduire une *condition d'intégralité* dans la conjecture. Si $\mathcal{E}$ est un modèle minimal régulier sur $\mathbf{Z}$ de la courbe E, on conjecture que seuls les éléments du sous-groupe $K_2(\mathcal{E})$ de $K_2(E)$ ont des régulateurs liés à la valeur $L(E, 2)$. Bloch et Grayson ont calculé explicitement cette restriction.

Dans la première partie de notre article, nous reprenons le travail de Bloch et Grayson. Nous insistons sur le fait que l'existence de suffisamment de diviseurs rationnels à support dans $E(\overline{\mathbf{Q}})_{tors}$ semble nécessaire à l'obtention de telles relations linéaires — *voir* 1.3.2. En général, on ne peut donc s'attendre à une telle relation entre séries de Kronecker et la valeur $L(E, 2)$.

Au §1.6 nous comparons les éléments de $K_2(E)$ obtenus à ceux donnés par la méthode modulaire de Beilinson [Schappacher, Scholl 1988] si E est paramétrée par une courbe $X_o(N)$.

Bloch, dans [Bloch, Grayson 1986], se demandait s'il n'existe pas de relations linéaires entre valeur L et séries de Kronecker pour s entier ≥ 3. Tout au début de notre travail, nous nous sommes posés la même question. Mais nos essais dans cette direction nous ont inspiré la conviction que ce n'était pas le cas. C'est Deninger qui a trouvé une raison théorique à cela en 1987 :— Dans le cas des courbes elliptiques E *à multiplications complexes* il pouvait établir un lien entre les valeurs spéciales $L(E, s)$, pour s entier > 2, et des régulateurs de Beilinson correspondants — *voir* [Deninger 1988], [Deninger 1989]. Sa méthode consiste à considérer d'abord certaines puissances symétriques de E et à revenir au niveau de la courbe, au moyen des multiplications complexes. Or, pour toute puissance symétrique d'une courbe E *quelconque*, Deninger obtient toujours des éléments d'un certain groupe de cohomologie motivique attachée à cette puissance. Mais, les multiplications complexes faisant défaut, il est impossible de ramener cette version explicite des conjectures de Beilinson à la courbe elle-même.

Nous donnons, dans la deuxième partie de cet article, un resumé de la partie du travail de Deninger qui s'applique aux courbes elliptiques définies sur $\mathbf{Q}$. C'est la seule généralisation à d'autres valeurs spéciales de fonctions L des conjectures explicites du §1 que nous connaissons actuellement. Mais ici, encore plus que dans le cas de Bloch et Grayson, l'absence de suffisamment de diviseurs rationnels à support dans $E(\overline{\mathbf{Q}})_{tors}$, pour une courbe E "générique", restreint la portée de ces conjectures explicites — *voir* 2.4.2. Par exemple, pour toute courbe il existe un k_0 — généralement assez petit; voir plus loin le cas $k = 2$ — tel que, pour tout $k \geq k_0$, la méthode de Deninger ne suffit pas à fabriquer un régulateur non-nul de $Sym^k E$.

Le §2.5 contient la généralisation aux puissances symétriques, due à Scholl, de la formule explicite donnant la *condition d'intégralité*, c'est-à-dire la généralisation de la formule trouvée par Bloch et Grayson qui caractérise le sous-groupe $K_2(\mathcal{E})$ de $K_2(E)$.

La troisième partie de ce travail est consacrée aux vérifications numériques des conjectures de Beilinson-Deninger pour le carré symétrique de certaines courbes elliptiques sur $\mathbf{Q}$, ainsi qu'à la classification de familles de courbes pour lesquelles ces conjectures de Beilinson-Deninger donnent un énoncé non-trivial. On procède de façon différente suivant que la condition d'intégralité intervient ou non.

Si elle n'intervient pas — c'est à dire, si la courbe a potentiellement bonne réduction partout (mais n'admet pas de multiplication complexe ...) — on peut se servir de diviseurs contenant soit tous les conjugués d'un point de torsion donné, soit tous les éléments d'un sous-groupe fini rationnel. Pour les sous-groupes rationnels cycliques nous arrivons à traiter complètement le cas de toutes les courbes admettant un régulateur non-nul: 3.2, 3.3.

Si au contraire il y a obstruction d'intégralité — c'est à dire, si la courbe a réduction (potentiellement) multiplicative quelque part —, alors les diviseurs contenant tous les conjugués d'un point de torsion non rationnels, ou tous les éléments d'un sous-groupe fini rationnel, sont exclus. Nous montrons au numéro 3.4 qu'il n'y a qu'un nombre fini de telles courbes avec régulateur non-nul, si on se restreint aux diviseurs formés de multiples d'un seul point de torsion rationnel. Au dernier numéro 3.5, nous en présentons des exemples numériques, mettant en évidence l'analogue, pour le carré symétrique, des "relations exotiques" entre séries de Kronecker constatées par Bloch et Grayson au niveau de la courbe elle-même (*cf.* 1.5.3).

1. La courbe

Dans ce numéro, nous reprenons la situation étudiée auparavant dans [Bloch, Grayson 1986].

1.1 Notations

Soient E une courbe elliptique sur $\mathbf{Q}$ et $\mathcal{E}$ son modèle minimal régulier sur $\mathbf{Z}$.* Fixons une différentielle non nulle ω de E définie sur $\mathbf{R}$. Soit $\omega_1 = |\int_{E(\mathbf{R})^0} \omega|$ la période réelle de (E, ω). Choisissons ω_2 tel que $\Lambda = \omega_1 \mathbf{Z} + \omega_2 \mathbf{Z} \subset \mathbf{C}$ soit le réseau rendant $(\mathbf{C}/\Lambda, dz)$ isomorphe à (E, ω) sur $\mathbf{C}$ et tel que $\tau = \frac{\omega_2}{\omega_1}$ ait une partie imaginaire $y = \Im(\tau)$ positive. On emploie les notations (légèrement modifiées) de [Weil 1976, *chap. VIII*] : $A(\Lambda) = \frac{{\omega_1}^2 y}{\pi}$, $\langle \gamma, x \rangle = exp(\frac{\gamma \bar{x} - x \bar{\gamma}}{A(\Lambda)})$. Pour un diviseur $\mathbf{a} = \sum_{x \in \mathbf{C}/\Lambda} a_x(x)$ sur $E(\mathbf{C})$ et un entier $\nu \geq 0$, on pose—si $\Re(s) > \frac{\nu}{2} + 1$—,

$$K_\nu(0, \mathbf{a}, s, \Lambda) = \sum_{x \in \mathbf{C}/\Lambda} a_x \sum_{\gamma \in \Lambda}{}' \langle \gamma, x \rangle \frac{\bar{\gamma}^\nu}{(\gamma \bar{\gamma})^s}.$$

Pour $s \in \mathbf{C}$ quelconque, les valeurs de ces *doubles séries de Kronecker* sont définies par prolongement analytique. Si le réseau Λ de K_ν n'est pas précisé, on convient que $\Lambda = \mathbf{Z} + \mathbf{Z}\tau$ — ce qui est indépendant du choix de ω.

Pour un entier $d > 1$, on écrit $\mathbf{Q}[E_d]^o$ le $\mathbf{Q}$-espace vectoriel des diviseurs à coefficients rationnels sur E qui sont *définis sur* $\mathbf{Q}$ *en tant que diviseurs,* de degré zéro et à support dans les points de d-torsion de E. Sur $\mathbf{C}$ les éléments de $\mathbf{Q}[E_d]^o$ s'identifient à des diviseurs $\mathbf{a}$ comme avant sur $E(\mathbf{C}) = \mathbf{C}/\Lambda$.

1.2 Cohomologie motivique

Soit X une variété quasi-projective lisse sur $\mathbf{Q}$. Une façon d'introduire les groupes de *cohomologie motivique* de X est de définir, pour tout $j \geq 0$, $i \leq 2j$,

$$H^i_{\mathcal{M}}(X, \mathbf{Q}(j)) = Gr^j_\gamma(K_{2j-i}(X) \otimes \mathbf{Q}),$$

où le gradué est relatif à la filtration γ sur le K-groupe de Quillen tensorisé par $\mathbf{Q}$. — *Cf.* [Beilinson 1985], [Schneider 1988, §3]. On sait démontrer

* Si besoin est, on supposera que E est paramétrée par une courbe modulaire $X_o(N)$. De même, dans nos exemples nous citons parfois des courbes par le nom qui leur est donné dans la "Table 1" de [Modular Functions ... IV].

que l'image de $K_{2j-i}(\mathcal{X}) \otimes \mathbf{Q}$ dans $H^i_{\mathcal{M}}(X, \mathbf{Q}(j))$ ne dépend pas du choix d'un modèle *régulier* propre $\mathcal{X}$ de X sur $\mathbf{Z}$—si un tel modèle existe. Elle est notée $H^i_{\mathcal{M}}(X, \mathbf{Q}(j))_{\mathbf{Z}}$.

Considérons la suite exacte suivante, extraite de la suite exacte de localisation de $\mathcal{E}$ par rapport à sa fibre générique E — *cf.* [Bloch, Grayson 1986], [Deninger, Wingberg 1988], [Jannsen 1990].

$$\text{(1.2.0)}\quad H^2_{\mathcal{M}}(E, \mathbf{Q}(2))_{\mathbf{Z}} \hookrightarrow H^2_{\mathcal{M}}(E, \mathbf{Q}(2)) \xrightarrow{\partial=\coprod \partial_p} \coprod_p H^3_{\mathcal{M},\mathcal{E}(p)}(E, \mathbf{Q}(2))_{\mathbf{Z}} \to H^3_{\mathcal{M}}(E, \mathbf{Q}(2))_{\mathbf{Z}}$$

La somme est étendue à tous les nombres premiers p.

Beilinson conjecture que:

$$\text{(1.2.1 ?)}\quad dim_{\mathbf{Q}} H^2_{\mathcal{M}}(E, \mathbf{Q}(2))_{\mathbf{Z}} = ord_{s=0} L(E, s) = 1,$$
$$\text{(1.2.2 ?)}\quad H^3_{\mathcal{M}}(E, \mathbf{Q}(2))_{\mathbf{Z}} = 0.$$

On sait [Jannsen 1990, §12] que l'"homologie motivique"

$$H^3_{\mathcal{M},\mathcal{E}(p)}(E, \mathbf{Q}(2))_{\mathbf{Z}} = K'_1(\mathcal{E}(p)) \otimes_{\mathbf{Z}} \mathbf{Q}$$

est non nulle si et seulement si le facteur eulérien $L_p(E, s)^{-1}$ en p de $L(E, s) = \prod_p L_p(E, s)$ s'annule en $s = 0$, donc si et seulement si E a mauvaise réduction multiplicative déployée en p. Cette homologie motivique est alors de dimension 1.

Ceci implique la conjecture :

$$\text{(1.2.3 ?)}\quad dim_{\mathbf{Q}} H^2_{\mathcal{M}}(E, \mathbf{Q}(2)) = ord_{s=0} L_S(E, s) = 1 + \#\{p \mid L_p(E, 0)^{-1} = 0\},$$

où S est l'ensemble des nombres premiers de mauvaise réduction pour E, et $L_S = \prod_{p \notin S} L_p$ est la fonction L sans facteurs eulériens en S. L'assertion $1.2.3_?$ est un cas particulier de la première partie d'une *S-conjecture de Beilinson*, proposée déjà dans une lettre de Deligne à Soulé [Deligne 1985], et développée par Ramakrishnan [Ramakrishnan 1989], Jannsen et Scholl [non publiés].

Notons que la finitude de la dimension de $H^2_{\mathcal{M}}(E, \mathbf{Q}(2))$ — ou de $H^2_{\mathcal{M}}(E, \mathbf{Q}(2))_{\mathbf{Z}}$ — n'est toujours pas établie. Remarquons d'autre part qu'on a beaucoup de mal à construire explicitement des éléments non nuls de ces groupes. Toutefois on sait, grâce à la méthode des courbes modulaires développée par Beilinson—*voir* 1.6; *cf.* [Beilinson 1985], [Beilinson 1986], [Schappacher, Scholl 1988]—, que $H^2_{\mathcal{M}}(E, \mathbf{Q}(2))_{\mathbf{Z}} \neq 0$ pour toute courbe E paramétrée par une courbe modulaire—donc conjecturalement pour toute courbe elliptique sur $\mathbf{Q}$.

1.3 Comment construire des éléments de $H^2_{\mathcal{M}}(E, \mathbf{Q}(2))$

Notons r l'application du *régulateur de Bloch-Beilinson* — *cf.* [Schneider 1988], [Rohrlich 1987], [Deninger, Wingberg 1988];

$$r : H^2_{\mathcal{M}}(E, \mathbf{Q}(2)) \longrightarrow H^2_{\mathcal{D}}(E_{/\mathbf{R}}, \mathbf{R}(2)) = H^1(E(\mathbf{C}), \mathbf{R}(1))^{\overline{F_\infty}}.$$

Ici, $\overline{F_\infty}$ est l'involution correspondant à la structure réelle $H^1_{dR}(E_{/\mathbf{R}}) \otimes \mathbf{R}(1)$, où $\mathbf{R}(1) = 2\pi i\mathbf{R}$.

Bloch conjecture [Bloch, Grayson 1986, p. 84, note (*)] que l'application r est injective sur $H^2_{\mathcal{M}}(E, \mathbf{Q}(2))$, et les résultats de [Bloch, Grayson 1986] sont pleinement en accord avec cette conjecture — *voir* 1.5.3, *cf. aussi* [Schappacher, Scholl 1988, 1.1.3 (iii)]. Dans le cadre général de Beilinson, on est amené à conjecturer au moins l'injectivité de la restriction de r à $H^2_{\mathcal{M}}(E, \mathbf{Q}(2))_{\mathbf{Z}}$. Si la conjecture plus forte de Bloch est juste, alors l'application ξ définie ci-après est unique.

1.3.1 Lemme. *Pour tout $d > 1$, il existe une application $\mathbf{Q}$-linéaire*

$$\xi = \xi_d : \mathbf{Q}[E_d]^o \longrightarrow H^2_{\mathcal{M}}(E, \mathbf{Q}(2))$$

tel que, pour tout $\mathbf{a} = \sum_{d.x=0} a_x(x) \in \mathbf{Q}[E_d]^o$, on ait

$$[r(\xi(\mathbf{a})), \omega] = A(\Lambda)^2 K_1(0, \mathbf{a}, 2, \Lambda),$$

où $[\alpha, \beta] \overset{\text{déf}}{=} \frac{1}{2\pi i}\int_{E(\mathbf{C})} \alpha \wedge \beta$, pour deux 1-formes α, β sur E. De plus, si $e > 1$ et $i_e : \mathbf{Q}[E_d]^o \hookrightarrow \mathbf{Q}[E_{de}]^o$, alors $\xi_d = \xi_{de} \circ i_e$.

L'application ξ peut se construire facilement en termes de symboles $\{f, g\} \in K_2(\mathbf{Q}(E))$ relevables dans $H^2_{\mathcal{M}}(E, \mathbf{Q}(2))$: Etant donné $\mathbf{a}$, appelons P le sous-groupe de E_d engendré par les points qui apparaissent dans $\mathbf{a}$ avec un coefficient non nul, et posons $d' = \#(P)$. Comme $\mathbf{a}$ est défini sur $\mathbf{Q}$, il en est de même du sous-groupe P. Il existe donc des fonctions rationnelles sur $\mathbf{Q}$, $f, g \in \mathbf{Q}(E)$ ayant comme diviseurs $div(f) = d'\mathbf{a}$, $div(g) = 2[d'(0) - \sum_{y \in P}(y)]$. Par un lemme de Bloch [Bloch 1980,

p. 8.6f], [Deninger, Wingberg 1988, Lemma 5.2], quitte à modifier le symbole $\{f, g\} \in K_2(\mathbf{Q}(E))$ par des symboles du type $\{h, c\}$, $h \in \mathbf{Q}(E)^*, c \in \mathbf{Q}^*$, on tombe dans $H^2_{\mathcal{M}}(E, \mathbf{Q}(2)) \hookrightarrow K_2(\mathbf{Q}(E)) \otimes \mathbf{Q}$. Normalisons le régulateur de $\{\phi, \psi\} \in K_2(\mathbf{Q}(E))$, avec $div(\phi) = \sum p_x(x)$, $div(\psi) = \sum q_x(x)$, comme étant $A(\Lambda)^2 \sum_{x,y} p_x q_y \; K_1(0, x-y, 2, \Lambda)$—c'est la normalisation qu'on trouve dans [Bloch, Grayson 1986] (avec $\Lambda = \mathbf{Z} + \mathbf{Z}\tau$); dans [Deninger, Wingberg 1988] et [Deninger 1988], [Deninger 1989], il y a un facteur correctif de $-\frac{1}{2}$. Alors on pose $\xi_d(\mathbf{a}) = \frac{1}{2d'^2}\{f, g\}$, et la formule donnant $r(\xi(\mathbf{a}))$ dans 1.3.1 découle du fait que $K_1(0, x, s)$ est une fonction impaire de la variable x. La construction donnée de ξ_d ne dépend pas du d tel que $\mathbf{a} \in \mathbf{Q}[E_d]^o$, ce qui démontre la dernière assertion de 1.3.1.

Mais en fait, *la construction qu'on vient de donner ne dépend pas du sous-groupe rationnel P de E_{tors} utilisé tel que* **a** *soit à support dans P.* Ceci découle aussitôt de la construction de ξ employée par Deninger—*voir* [Deninger 1988]; *cf.* [Schappacher, Scholl 19**]—; noter pourtant que la normalisation adoptée par Deninger est différente de la nôtre ce qui fait que son application est sensible au changement du sous-groupe P.

1.3.2 *L'action du groupe de Galois* $\mathrm{Gal}(\overline{\mathbf{Q}}/\mathbf{Q})$

Voici la *contrainte essentielle*, dans le cas d'une courbe elliptique, de cette façon de se donner des éléments en cohomologie motivique (nous en verrons une autre en 2.4.2, qui s'applique à toutes les puissances symétriques et devient prédominante surtout pour les puissances paires):

Si l'image de $\mathrm{Gal}(\overline{\mathbf{Q}}/\mathbf{Q})$ *dans* $\mathrm{Aut}(E_d) = GL_2(\mathbf{Z}/d\mathbf{Z})$ *contient* $\begin{pmatrix} -1 & 0 \\ 0 & -1 \end{pmatrix}$, *alors le régulateur* $r(\xi(\mathbf{Q}[E_d]^o))$ *vaut* 0 — *donc conjecturalement aussi* $\xi(\mathbf{Q}[E_d]^o) = 0$.

C'est à cause du fait, déjà utilisé dans la démonstration précédente, que $K_\nu(0, x, s)$, comme fonction de la variable x, a la parité de ν.

L'hypothèse concernant l'image de l'action de Galois est souvent satisfaite. Elle vaut systématiquement pour tout $d > 1$, si l'image de $\mathrm{Gal}(\overline{\mathbf{Q}}/\mathbf{Q})$ dans $\mathrm{Aut}(E_{tors}) = GL_2(\hat{\mathbf{Z}})$ est aussi grand que possible, c'est-à-dire d'indice deux — par exemple, si E est la courbe *37A*, discutée dans [Serre 1972, *exemple 5.5.6*].

1.3.3 *Remarque.* Dans [Ramakrishnan 1990, *Question 8*], on trouve un exemple, dû à R. Ross, d'un symbole $\{\phi, \psi\} \in K_2(\mathbf{Q}(E))$, sur la courbe $E : y^2 = x^3 + x$, tel que $div(\phi)$ contienne un point d'ordre *infini* de E et tel que $6.\{\phi, \psi\}$ soit dans le noyau du symbole modéré $\mathcal{T}$ de la suite exacte de localisation de E relatif à $\mathbf{Q}(E)$. Ici, le point clé de la démonstration réside dans le fait que les valeurs $\mathcal{T}_Q(\{\phi, \psi\})$, $Q \in E(\overline{\mathbf{Q}})$, sont des racines d'unité. D. Grayson a vérifié numériquement que le régulateur de $6.\{\phi, \psi\}$ semble lié à $L(E,2)$ comme le prédit la conjecture de Beilinson — *voir* 1.4.

Récemment J. Nekovář a obtenu, par une voie entamée par l'un de nous (N.S.), l'exemple suivant (entre autres). Soit E la courbe $y^2 = x^3 - 4x + 4$. Posons $f = y - (2x-2)$; $g = y$. Alors le diviseur de f contient des points rationnels *d'ordre infini* de E, mais on trouve que $\{f, g\}$ définit bien un élément de $K_2(E)$. D. Grayson a vérifié numériquement que son régulateur est non nul et semble lié à la valeur $L(E,2)$ selon la conjecture de Beilinson.

1.4 Conjecture de Beilinson

Elle établit un lien entre le premier terme du développement de $L(E,s)$ en $s = 0$ et le régulateur de $H^2_{\mathcal{M}}(E, \mathbf{Q}(2))_{\mathbf{Z}}$. Elle dit en particulier que, pour $\mathbf{a} \in \mathbf{Q}[E_d]^o$, on a

$$(1.4.0\,?) \qquad \omega_1^{-1}\,[r(\xi(\mathbf{a})), \omega] \in L'(E,0).\mathbf{Q},$$

pourvu que $\xi(\mathbf{a})$ *appartienne à* $H^2_{\mathcal{M}}(E, \mathbf{Q}(2))_{\mathbf{Z}}$.

Compte tenu de l'équation fonctionnelle de $L(E,s)$ et des propriétés élémentaires des séries de Kronecker, ceci équivaut à dire que:

$$(1.4.1\,?) \qquad \partial(\xi(\mathbf{a})) = 0 \implies y^2 K_1(0, \mathbf{a}, 2) \in L(E,2).\mathbf{Q}.$$

(Rappelons qu'un réseau non écrit dans K_ν est toujours $\mathbf{Z} + \mathbf{Z}\tau$.)

1.4.2 *Invariance par isogénie*

L'existence d'un diviseur $\mathbf{a} \in \mathbf{Q}[E_{tors}]^o$ tel que

$$\frac{y^2 K_1(0, \mathbf{a}, 2)}{L(E,2)}$$

soit un rationnel non nul est une notion invariante par isogénies rationnelles. En fait, si $\phi : E' \to E$ est une isogénie définie sur $\mathbf{Q}$ et $\mathbf{a}$ un diviseur convenable sur E, alors $y'^2 K_1(0, \phi^*\mathbf{a}, 2, \mathbf{Z} + \mathbf{Z}\tau')$ ne diffère de $y^2 K_1(0, \mathbf{a}, 2)$ que par un facteur rationnel non nul. — En fait, par un calcul direct on se réduit au cas où l'isogénie est induite, sur $\mathbf{C}$, par l'inclusion $\Lambda' \hookrightarrow \Lambda$ de deux réseaux. Alors on note que, par rapport à l'accouplement $\langle .,. \rangle_{\Lambda'}$, on a l'identité $([\Lambda : \Lambda'].\Lambda)^\perp = \Lambda/\Lambda' \subseteq \mathbf{C}/\Lambda'$. D'où la relation

$$K_\nu(0, \mathbf{a}, s, \Lambda) = [\Lambda : \Lambda']^{2s-\nu-2} K_\nu(0, \phi^*\mathbf{a}, s, \Lambda').$$

1.5 La formule et les calculs de Bloch et Grayson

L'article [Bloch, Grayson 1986] a mené à la formulation des conjectures de Beilinson que nous connaissons aujourd'hui: il a mis en évidence la nécessité de se restreindre au sous-groupe $H^2_{\mathcal{M}}(E, \mathbf{Q}(2))_{\mathbf{Z}}$ de $H^2_{\mathcal{M}}(E, \mathbf{Q}(2))$. La caractérisation explicite de ce sous-groupe par le lemme ci-après est à la base de l'article [Bloch, Grayson 1986].

Soit p un nombre premier où E a réduction multiplicative déployée. La fibre spéciale en p de $\mathcal{E}$ est alors un polygone de Néron I_ν. Fixons une numérotation des ses ν côtés telle que la composante numéro 0 contienne l'origine de E. Identifions $H^3_{\mathcal{M},\mathcal{E}(p)}(\mathcal{E}, \mathbf{Q}(2)) = K'_1(\mathcal{E}(p)) \otimes_{\mathbf{Z}} \mathbf{Q}$ à $\mathbf{Q}$ en choisissant une base—*cf.* 1.2 plus haut. Ceci nous permet d'écrire les images de ∂_p comme des nombres rationnels déterminés à un facteur près.

1.5.1 Lemme. *(i) Si* $\nu = 1$, *alors* $\partial_p = 0$.

(ii) Soit $x \in E(\mathbf{Q})_\nu$ *tel que* $x \bmod p$ *appartienne à la première composante. Alors, pour* $j = 1, ..., \nu$,

$$\partial_p(\xi((jx) - (0))) = c\ B_3(\frac{j}{\nu}),$$

où $c \neq 0$ *ne dépend que de* ν, *et* $B_3(X) = X^3 - \frac{3}{2}X^2 + \frac{1}{2}X$ *est le troisième polynôme de Bernoulli.*

Cette formulation du résultat de Bloch et Grayson nous a été signalée par A.J. Scholl. Il permet en fait de calculer $\partial_p(\xi(\mathbf{a}))$, pour tout $\mathbf{a} \in \mathbf{Q}[E_{tors}]^o$. Une formule légèrement plus générale est utilisée dans [Bloch, Grayson 1986] sans démonstration. D'autre part, 1.5.1 est un cas particulier du lemme 2.5.1 plus loin.

1.5.2 *Remarque.* On voit que si, par exemple, tous les nombres premiers p de réduction multiplicative déployée de E sont tels que la fibre spéciale en p n'ait qu'un seul côté, alors les éléments $\xi((jx) - (0))$ de Bloch et Grayson ne sont jamais dans $H^2_{\mathcal{M}}(E, \mathbf{Q}(2)) - H^2_{\mathcal{M}}(E, \mathbf{Q}(2))_{\mathbf{Z}}$, qui devrait pourtant être non vide. Comme exemple, on peut citer la courbe *11A*, c'est-à dire, $X_1(11)$.

1.5.3 L'article [Bloch, Grayson 1986] peut se voir d'au moins trois points de vue différents:

a) Bloch et Grayson y vérifient numériquement la conjecture $1.4.1_?$, pour 39 courbes elliptiques parmi une collection de 41 dont la torsion rationnelle n'est pas réduite à des points d'ordre 2, et pour des diviseurs $\mathbf{a} \in \mathbf{Q}[E(\mathbf{Q})_{tors}]^o$ convenables.

b) Bloch et Grayson y vérifient sur les valeurs approchées des séries de Kronecker que $dim_{\mathbf{Q}} r(\xi(\mathbf{Q}[E(\mathbf{Q})_{tors}]^0))$ semble être bornée par la dimension de $H^2_{\mathcal{M}}(E, \mathbf{Q}(2))$, selon 1.2.3?. (En fait, $dim_{\mathbf{Q}} H^2_{\mathcal{M}}(E, \mathbf{Q}(2))$ n'est pas écrite dans [Bloch, Grayson 1986]; il faut, pour chaque courbe, déterminer quels sont les nombres premiers de réduction multiplicative *déployée, resp. non déployée.*) De plus, les auteurs vérifient numériquement que les séries de Kronecker semblent respecter les restrictions imposées par ∂. Parfois, il arrive que les $\xi(\mathbf{a})$ qu'on construit tombent automatiquement dans le noyau de ∂_p, pour un premier p tel que $L_p(E,1) = 0$ (*cf.* 1.5.2.). Ceci se traduit par plus de relations linéaires entre les séries de Kronecker $[r(\xi(\mathbf{a})), \omega]$ que la seule dimension de l'espace ambiant $H^2_{\mathcal{M}}(E, \mathbf{Q}(2))$ ne laisserait attendre. Ces relations ainsi constatées numériquement sont parfois appelées *"relations exotiques"*.

c) Bien que ce ne soit pas rédigé dans [Bloch, Grayson 1986], Bloch et Grayson ont certainement dû constater que, si $\partial(\xi(\mathbf{a})) \neq 0$, alors $y^2 K_1(0, \mathbf{a}, 2)$ ne semble pas être rationnellement lié à $L(E, 2)$.

1.6 Le côté modulaire

Si la courbe E est paramétrée par une courbe modulaire — *i.e.*, $X_0(N) \to E$, ou plus généralement $\overline{M} \to E$, pour n'importe quelle courbe modulaire $\overline{M}$ — alors le théorème de Beilinson relatif à la valeur en $s = 2$ des fonctions L de courbes modulaires — *voir* [Schappacher, Scholl 1988]; *cf.* [Beilinson 1985], [Beilinson 1986] — permet de construire des éléments non nuls dans $H^2_{\mathcal{M}}(E, \mathbf{Q}(2))_{\mathbf{Z}} \subseteq H^2_{\mathcal{M}}(E, \mathbf{Q}(2))$ dont les régulateurs satisfont à la conjecture de Beilinson, *i.e.*, sont liés à la valeur $L(E, 2)$. Malheureusement, il semble être impossible, *en général*, de les situer par rapport au sous-espace vectoriel $\Xi \overset{\text{déf}}{=} \xi(\mathbf{Q}[E_{tors}]^o)$ de $H^2_{\mathcal{M}}(E, \mathbf{Q}(2))$ — bien que conjecturalement $H^2_{\mathcal{M}}(E, \mathbf{Q}(2))_{\mathbf{Z}}$ soit de dimension 1. Toutefois, les quelques *cas particuliers* où on arrive à comparer des symboles de $H^2_{\mathcal{M}}(\overline{M}, \mathbf{Q}(2))$ considérés par Beilinson à Ξ *fournissent les seuls exemples actuellement connus de courbes elliptiques E sans multiplication complexe pour lesquelles on sait démontrer des relations entre $L(E, 2)$ et des séries de Kronecker du type envisagé numériquement par Bloch et Grayson.* C'est pourquoi nous en parlons ici.

1.6.1 Soit donc une application non constante $\psi : \overline{M} \to E$, pour une courbe modulaire $\overline{M}$ sur $\mathbf{Q}$. Supposons qu'il y ait un sous-groupe rationnel fini $P \subseteq E_{tors}$ tel que $\psi^{-1}(P)$ soit entièrement contenu dans l'ensemble des pointes $\overline{M} - M$ de $\overline{M}$. Alors, étant donné un diviseur $\mathbf{a} \in \mathbf{Q}[P]^o$, le régulateur $[r(\xi(\mathbf{a})), \omega]$ calculé "sur la courbe elliptique" est le même que le régulateur d'un symbole d'unités modulaires dans $\{\mathcal{O}^*(M), \mathcal{O}^*(M)\}$ relevé dans $H^2_{\mathcal{M}}(\overline{M}, \mathbf{Q}(2))$—et en fait dans $H^2_{\mathcal{M}}(\overline{M}_{\mathbf{Z}}, \mathbf{Q}(2))$—, calculé "sur la courbe modulaire". — *Cf.* [Schappacher, Scholl 1988, 1.1] pour la construction et l'énoncé des propriétés des symboles construites à partir des unités modulaires. Vue la construction de $\xi(\mathbf{a})$ donnée dans la démonstration du lemme 1.3.1, l'égalité des régulateurs découle immédiatement de la formule $[r_E(\sigma), \omega] = [r_{\overline{M}}(\psi^*(\sigma)), \psi^*\omega]$ qui est elle-même une conséquence immédiate de l'identité, valable sur toute courbe algébrique lisse X,

$$[r_X(\{f, g\}), \omega] = \int_{X(\mathbf{C})} \log|f| \, \overline{d \log g} \wedge \omega.$$

Voir [Deninger, Wingberg 1988, 1.10], et [Schappacher, Scholl 1988, 1.3].

Or, les régulateurs provenant d'unités modulaires de $\overline{M}$ satisfont à la conjecture de Beilinson relative à $L(\overline{M}, s)$ en $s = 0$ — c'est le théorème de Beilinson, *voir* [Schappacher, Scholl 1988, 1.1.2]. En passant à la composante propre correspondant à E sous l'action de l'algèbre de Hecke, on en déduit que $y^2 K_1(0, \mathbf{a}, 2) \in L(E, 2).\mathbf{Q}$, et que en fait $\partial(\xi(\mathbf{a})) = 0$, comme l'exige la conjecture $1.4.1_?$.

Reste donc à vérifier, chaque fois que la situation discutée ici se présente, si $r(\xi(\mathbf{Q}[P]^o))$ est non nul. Pour une fois, c'est un énoncé qui se *démontre* sur ordinateur.

1.6.2 Voici un cadre qui couvre tous les exemples que nous connaissons actuellement où l'on peut démontrer des relations entre séries de Kronecker et la valeur $L(E, 2)$ par la méthode esquissée en 1.6.1.

Soit le diagramme cartésien

$$\begin{array}{ccc} \overline{M} & \longrightarrow & X_\circ(N) \\ \psi \Big\downarrow & & \Big\downarrow \varphi \\ E & \xrightarrow{\lambda} & E' \end{array}$$

où $\varphi : X_\circ(N) \to E'$ est une "courbe de Weil forte involutoire", *i.e.*, **1.** il n'y a pas de courbe elliptique E_1 non isomorphe à E' qui s'insère dans l'application $\varphi : X_\circ(N) \to E_1 \to E'$, et **2.** E' est le quotient $X_\circ(N)/U$ par un sous-groupe U du groupe W engendré dans $\mathrm{Aut}(X_\circ(N))$ par les involutions *d'Atkin-Lehner* w_p, pour les diviseurs premiers p de N. Ensuite, λ est une isogénie définie sur $\mathbf{Q}$, de degré $d > 2$, et $\overline{M}$ est le produit fibré $E \times_{E'} X_\circ(N)$. Posons $P \overset{\text{déf}}{=} \ker\lambda$.

Supposons maintenant que $\overline{M}$ est une courbe modulaire et que l'application $\overline{M} \longrightarrow X_\circ(N)$ est l'application naturelle entre courbes modulaires. Alors, comme E' est involutoire, $\varphi^{-1}(0_{E'})$ est contenue dans l'ensemble des pointes de $X_o(N)$. Par conséquent, $\psi^{-1}(P)$ ne consiste que de pointes de $\overline{M}$. Alors, *si* $r(\xi(\mathbf{a})) \neq 0$, *pour un* $\mathbf{a} \in \mathbf{Q}[P]^o$, on trouve que $y^2 K_1(0, \mathbf{a}, 2) \in L(E,2).\mathbf{Q}^*$.

1.6.3 *Exemples*

a) $N = 11$, $E' = X_o(11)$, $E = X_1(11) = \overline{M}$, $U = \{1\}$, P engendré par $x = (\infty) - (0)$. Le fait que $K_1(0, 2x, 2) \neq 0$ a été vérifié par Grayson: *voir* la table de [Bloch, Grayson 1986], courbe *11A*.

b) $N = 20$, $E' = E = X_o(20) = \overline{M}$, $U = \{1\}$, $P = E(\mathbf{Q})_{tors} = \{pointes\}$, $\#(P) = 6$. Alors [Bloch, Grayson 1986], table, *20B*, montre que $r(\xi(\mathbf{Q}[P]^o)) \neq 0$.

c) $N = 24$, $E' = E = X_o(24) = \overline{M}$, *i.e.*, la courbe *24B*, $U = \{1\}$, $P = E(\mathbf{Q})_{tors} = \{pointes\} \cong \mathbf{Z}/4\mathbf{Z} \times \mathbf{Z}/2\mathbf{Z}$. Nous avons vérifié sur ordinateur que $K_1(0, x, 2) \neq 0$, pour $x \in \mathbf{C}/\mathbf{Z} + \mathbf{Z}\tau$ correspondant à un point rationnel de E d'ordre 4. Donc $r(\xi(\mathbf{Q}[P]^o)) \neq 0$.

d)? $N = 26$, $E' =$"26B", $E =$"26A", $U = \{1, w_2\}$, $\#(P) = 3$. Nous ignorons pour le moment si $\overline{M}$ est modulaire.

2. Les puissances symétriques

Ce numéro est consacré à l'énoncé des conjectures explicites générales relatives aux puissances symétriques d'une courbe elliptique sur $\mathbf{Q}$, analogues aux conjectures figurant en §1.

2.1 Les motifs

Afin de pouvoir parler de la cohomologie motivique d'un motif, il faut travailler avec la catégorie $\mathcal{M}_\mathbf{Q}$ des motifs sur $\mathbf{Q}$ construite à partir de la catégorie d'objets les $h(X)$, pour toutes les variétés projectives lisses X sur $\mathbf{Q}$, avec morphismes (si Y est équidimensionnelle)

$$Hom(h(X), h(Y)) = CH^{dim Y}(X \times Y) \otimes \mathbf{Q},$$

des cycles algébriques *à équivalence rationnelle près* (et non pas à équivalence homologique près).

On dispose dans cette catégorie d'un motif $h_1(E)$, et un motif $Sym^k E$ peut y être construit par un des procédés équivalents suivants.

(2.1.1) On peut suivre la méthode de Beilinson dans [Beilinson 1986, §4], et considérer E^k comme la sous-variété de E^{k+1} sur laquelle la somme des composantes vaut zéro — elle sera notée ici $E^{k'}$. Alors $Sym^k E$ est le sous-motif de $h(E^{k'})$ sur lequel le groupe des permutations S_{k+1} agit par le caractère signe.

(2.1.2) Deninger [Deninger 1989, §8] montre que ceci équivaut à prendre la partie de $h_1(E)^{\otimes k}$—c'est un sous-motif de $h(E^k)$ muni d'une action de $\mathcal{S}_k$—sur laquelle $\mathcal{S}_k$ agit par le caractère signe.

(2.1.3) Finalement — *cf.* [Scholl 1990, §1] — $Sym^k E$ peut être vu comme la partie de $h(E)^{\otimes k}$ sur laquelle

- $\mathcal{S}_k$ agit par le caractère signe;
- en chaque composante, $h(-1)\colon h(E) \to h(E)$ agit par -1;

Alors, pour toute théorie de cohomologie $H^\bullet$ compatible avec nos motifs, qui admet une formule de Künneth et qui satisfait $H(h_1(\text{point})) = H^1(\text{point})$, on a

$$H(Sym^k(E)) = Sym^k H^1(E).$$

2.2 Cohomologie motivique

La cohomologie motivique se factorise par la catégorie des motifs que nous considérons. On peut donc parler de $H^i_{\mathcal{M}}(Sym^k E, \mathbf{Q}(j))$. La "partie intégrale" $H^i_{\mathcal{M}}(Sym^k E, \mathbf{Q}(j))_{\mathbf{Z}}$ peut être définie comme l'image dans $H^i_{\mathcal{M}}(Sym^k E, \mathbf{Q}(j))$ de $H^i_{\mathcal{M}}(\tilde{\mathcal{E}^k}, \mathbf{Q}(j))$, où $\tilde{\mathcal{E}^k}$ est la normalisation de la puissance de $\mathcal{E}$ sur $\mathbf{Z}$. De même, dans la troisième construction de $Sym^k E$ (2.1.3), la généralisation de l'"homologie motivique" utilisée en 1.2 s'écrit comme la partie propre par rapport aux deux opérations décrites en (2.1.3) de l'espace $H^i_{\mathcal{M},\tilde{\mathcal{E}^k}(p)}(\tilde{\mathcal{E}^k}, \mathbf{Q}(j))$. Nous le notons $H^i_{\mathcal{M},p}(Sym^k E, \mathbf{Q}(j))$.

On obtient alors la suite exacte suivante, analogue de 1.2.0.

$$(2.2.0)\quad H^{k+1}_{\mathcal{M}}(Sym^k E, \mathbf{Q}(k+1))_{\mathbf{Z}} \hookrightarrow H^{k+1}_{\mathcal{M}}(Sym^k E, \mathbf{Q}(k+1)) \xrightarrow{\partial=\coprod \partial_p}$$

$$\xrightarrow{\partial=\coprod \partial_p} \coprod_p H^{k+2}_{\mathcal{M},p}(Sym^k E, \mathbf{Q}(k+1)) \longrightarrow H^{k+2}_{\mathcal{M}}(Sym^k E, \mathbf{Q}(k+1))_{\mathbf{Z}}.$$

Soit $L(Sym^k E, s) = \prod_p L_p(Sym^k E, s)$ la fonction L du motif $Sym^k E$. Elle est définie par rapport aux *Frobenius géométriques* agissant sur les réalisations l-adiques du motif en question. Par conséquent, $L(Sym^1 E, s) = L(E, s)$, la fonction L habituelle de la courbe elliptique E; *cf.* [Schappacher 1988, I.1.8].

Voici trois conséquences des conjectures de Beilinson, analogues aux énoncés correspondants de 1.2.

(2.2.1 ?) $dim_{\mathbf{Q}} H^{k+1}_{\mathcal{M}}(Sym^k E, \mathbf{Q}(k+1))_{\mathbf{Z}} = ord_{s=0} L(Sym^k E, s) = [\frac{k}{2}]+1$

(2.2.2 ?) $H^{k+2}_{\mathcal{M}}(Sym^k E, \mathbf{Q}(k+1))_{\mathbf{Z}} = 0$

(*2.2.3 ?*)

$$dim_{\mathbf{Q}} H^{k+1}_{\mathcal{M}}(Sym^k E, \mathbf{Q}(k+1)) = [\frac{k+2}{2}] + \#\{p \mid L_p(Sym^k E, 0)^- 1 = 0\}$$

2.3 L'application de Deninger

Imitant les ***symboles d'Eisenstein*** de Beilinson [Beilinson 1986], Deninger [Deninger 1988], [Deninger 1989, §8], construit, pour tout entier $d > 1$, une application

$$\xi : \mathbf{Q}[E_d]^\circ \longrightarrow H^{k+1}_{\mathcal{M}}(Sym^k E, \mathbf{Q}(k+1))$$

qui, pour $k = 1$, coincide avec l'application ξ de 1.3.1.

2.3.0 *Remarque.* En fait, la construction de Deninger donne parfois un peu plus, et nous allons en profiter. Ainsi, pour k pair, on peut construire une application qui associe un élément de $H^{k+1}_{\mathcal{M}}(Sym^k E, \mathbf{Q}(k+1))$ à chaque diviseur de degré zéro à support dans E_d sur lequel $\mathrm{Gal}(\overline{\mathbf{Q}}/\mathbf{Q})$ agit par ± 1. — En effet, la première étape de la construction de Deninger est de passer d'un diviseur donné β au produit $\beta \otimes \alpha \otimes ... \otimes \alpha$, avec k exemplaires d'un même diviseur α, et sur l'image ultérieure de ce produit le groupe symétrique S_{k+1} agit par le caractère signe — *voir* [Deninger 1989, 8.5]. Cette image est donc invariante par $\beta \mapsto -\beta$.

Comme pour $k = 1$, écrivons r l'application du régulateur de Beilinson;

$$r : H^{k+1}_{\mathcal{M}}(Sym^k E, \mathbf{Q}(k+1)) \longrightarrow H^{k+1}_{\mathcal{D}}(Sym^k E_{/\mathbf{R}}, \mathbf{R}(k+1)).$$

Ce groupe de *cohomologie de Deligne* est la partie de $Sym^k H^1(E(\mathbf{C}), \mathbf{C})$, obtenue par la projection $\mathbf{C} \longrightarrow \mathbf{R}(k)$, qui est invariante sous l'involution de de Rham $\overline{F_\infty}$. Il est donc de dimension $[\frac{k}{2}]+1$ sur $\mathbf{R}$. Si ω, η est une base de $H^1_{dR}(E_{/\mathbf{Q}})$, alors une base de $H^{k+1}_{\mathcal{D}}(Sym^k E_{/\mathbf{R}}, \mathbf{R}(k+1))$ est donnée par les projections par $\mathbf{C} \longrightarrow \mathbf{R}(k)$ des différentielles

$$\omega^{(k)}, \omega^{(k-1)}\eta, ..., \omega^{(k-[\frac{k}{2}])}\eta^{([\frac{k}{2}])}$$

comme $\mathbf{R}$-base [Deligne 1979, 7.7 (b)]. La projection sera notée par une tilde.

Pour deux k-formes α, β sur E^k, soit $[\alpha, \beta] = \frac{1}{(2\pi i)^k} \int_{E^k(\mathbf{C})} \alpha \wedge \beta$. Alors, ξ est caractérisé (conjecturalement de façon unique) par les relations suivantes—*cf.* 1.3.1. Si $\mathbf{a} = \sum_{d.x=0} a_x(x) \in \mathbf{Q}[E_d]^\circ$, on a, pour $j = 0, ..., [\frac{k}{2}]$,

$$(2.3.1) \qquad [r(\xi(\mathbf{a})), \tilde{\omega}^{(k-j)}\tilde{\eta}^{(j)}] = A(\Lambda)^{k-j+1} K_{k-2j}(0, \mathbf{a}, k-j+1).$$

2.3.2 *Remarque.* Remarquons que les deux membres de 2.3.1 sont homogènes de poids $k-2j$ par rapport à ω: le changement $\omega \mapsto \lambda\omega, \lambda \in \mathbf{R}$, multiplie le côté droit par λ^{k-2j}. Ceci est en accord avec le fait que, à gauche, $\omega \mapsto \lambda\omega$ implique que $\eta \mapsto \lambda^{-1}\eta$ — se rappeler par exemple la relation de Legendre.

2.4 Conjecture de Beilinson

Ici encore, le formalisme est tout à fait analogue au cas $k = 1$—*voir* 1.4. Pour trouver la période qui généralise le ω_1 dans 1.4 on remarque que, si ω est définie sur $\mathbf{Q}$ alors $\omega_1 = c^+(H^1(E)(1))$ dans le formalisme de Deligne [Deligne 1979] et on fait appel à [Deligne 1979, 7.7], avec les données $c^-(H^1(E)(1)) = \Im(\omega_2)i = \omega_1 yi$ et $\delta(H^1(E)(1)) = 2\pi i$. [Deligne 1979, 7.7] permet alors de calculer la période $c^+(Sym^k E(k))$ dans la conjecture suivante.

2.4.0 Conjecture. *Soient* $\mathbf{a}_0, ..., \mathbf{a}_{[\frac{k}{2}]} \in \mathbf{Q}[E_d]^\circ$ *tels que leurs images par* ξ *appartiennent à* $H_{\mathcal{M}}^{k+1}(Sym^k E, \mathbf{Q}(2))_{\mathbf{Z}}$. *Alors*

$$c^+(Sym^k E(k))^{-1} \text{dét}([r(\xi(\mathbf{a}_i)), \tilde{\omega}^{(k-j)}\tilde{\eta}^{(j)}])_{i,j} \in L^{([\frac{k}{2}]+1)}(Sym^k E, 0).\mathbf{Q}.$$

Nous allons donner la reformulation analogue à 1.4.1? de cette conjecture dans le cas $k = 2$ au §3.

2.4.1 *Remarques.*

(i) Comme en 2.3.2, on vérifie que la conjecture ne dépend pas du choix de ω qui affecte le côté gauche de la relation. En fait, l'homogénéité de $c^+(Sym^k E(k))$ est de degré $\frac{k}{2}(\frac{k}{2}+1)$ en λ, si k est pair, et $([\frac{k}{2}]+1)^2$, si k est impair—ce sont précisément les degrés qu'on obtient par la règle donnée en 2.3.2 pour le déterminant.

(ii) La conjecture 2.4.0 est *invariante par isogénie*, dans le sens suivant. Soit une isogénie $\phi : E' \to E$ définie sur $\mathbf{Q}$. Alors en prenant des diviseurs $\mathbf{a}'_i = \phi^*(\mathbf{a}_i)$, on ne change que par un facteur rationnel le déterminant de 2.4.0. Ceci se voit sur 2.3.1 par les calculs esquissés en 1.4.2.

(iii) Les puissances symétriques *paires* d'une courbe elliptique E sont *invariantes par torsion quadratique*. Plus précisément si E est la tordue de E' par $\sqrt{D}$, alors $\sqrt{D}$ définit sur $\mathbf{Q}(\sqrt{D})$ une isogénie $\phi : E' \to E$ telle que $\text{Gal}(\overline{\mathbf{Q}}/\mathbf{Q})$ agisse sur $\mathbf{a}'_i = \phi^*(\mathbf{a}_i)$ par ± 1. Grâce à 2.3.0, ceci permet de définir $\xi(\mathbf{a}'_i)$, et de toutes les séries de Kronecker dans son régulateur — qui ont toutes la même parité que l'ordre de la puissance symétrique — $\sqrt{D}$ sort à une puissance paire. Donc finalement on est dans la même situation qu'en (ii): à un facteur rationnel près, les régulateurs pour les puissances symétriques paires restent inchangés par torsion quadratique. — Il en est de même, bien sûr, des fonctions L des puissances symétriques paires.

Vues les propriétés bien connues de l'action de Galois sur les points de torsion d'une courbe elliptique sans multiplication complexe [Serre 1972], le lemme suivant implique que, pour k assez grand, il est impossible de trouver des diviseurs $\mathbf{a}_i$ rationnels sur $\mathbf{Q}$ tels que le déterminant dans 2.4.0 soit non nul. En fait, les calculs de 3.3 et 3.5 indiquent bien que ces k "assez grands" ont tendance à être plutôt petits pour la plupart des courbes elliptiques....

2.4.2 Lemme. *Soient $\nu \geq 0, d > 1$, et $s \in \mathbf{C}$. Alors pour tout $w \in \mathbf{C}/\Lambda$, on a*

$$\sum_{d.x=w} K_\nu(0,x,s,\Lambda) = d^{\nu+2-2s} K_\nu(0,w,s,\Lambda).$$

En particulier, $\sum_{d.x=0} K_\nu(0,x,s,\Lambda)$ ne diffère de $K_\nu(0,0,s,\Lambda)$ que par un facteur qui ne dépend que de $\nu - 2s$.

Ce facteur est donc partout le même quand on écrit sous la forme de 2.3.1 les coefficients de la matrice dont on prend le déterminant en 2.4.0. Noter de plus que $K_\nu(0,0,s,\Lambda) = 0$ si ν est impair.

La démonstration du lemme est une simple application des relations d'orthogonalité des caractères.

2.5 Calcul de ∂

Voici la généralisation annoncée plus haut du lemme 1.5.1. Nous quittons, pour une fois, le corps des nombres rationnels: si F est une extension de $\mathbf{Q}$, nous notons par E_F, $\mathcal{E}_F$, etc. les extensions des scalaires de E, $\mathcal{E}$, etc. à F, voir $\mathcal{O}_F$. Le lemme suivant fait référence à la suite exacte analogue à 2.2.0 sur F.

2.5.1 Lemme. *Soit $\mathbf{a} = \sum_{d.x=0} a_x(x) \in \mathbf{Q}[E_d]^\circ$. Soit F un corps de nombres tel que tous les points de d-torsion de E soient rationnels sur F. Soit v une place de F telle que la fibre spéciale $\mathcal{E}_F(v)$ soit un polygone de Néron I_ν, dont nous numérotons les côtés comme en 1.5. Notons $d(\mathbf{a}, i)$, pour $i = 0, ..., \nu - 1$, le degré de la réduction de $\mathbf{a}$ restreinte à la i-ième composante. Alors on a l'équivalence*

$$\partial_v(\xi(\mathbf{a})) = 0 \iff \sum_{i=0}^{\nu-1} d(\mathbf{a}, i) B_{k+2}\left(\frac{i}{\nu}\right) = 0,$$

où $B_n(X)$ est le n-ième polynôme de Bernoulli.

Ce lemme sera démontré dans [Schappacher, Scholl 19**]. Disons simplement qu'il y a deux démonstrations possibles de ce lemme. L'une s'appuie sur les constructions dans [Beilinson 1986, 3.1.7, et avant]. En effet, la restriction aux pointes correspond par spécialisation à la réduction en une place multiplicative — *cf.* 3.4.3 plus loin. L'autre démonstration consiste à généraliser la formule 1.5.1 de Bloch et Grayson par la théorie des invariants des groupes symétriques $\mathcal{S}_k$ ....

2.5.2 Rappelons que les polynômes de Bernoulli sont définis par l'identité

$$\frac{te^{tX}}{e^t - 1} = \sum_{k=0}^{\infty} B_k(X)\frac{t^k}{k!},$$

de sorte qu'on ait:

$$B_0(X) = 1, \qquad B_1(X) = X - \frac{1}{2}, \qquad B_2(X) = X^2 - X + \frac{1}{6}$$

$$B_3(X) = X^3 - \frac{3}{2}X^2 + \frac{1}{2}X, \qquad B_4(X) = X^4 - 2X^3 + X^2 - \frac{1}{30}$$

$$B_5(X) = X^5 - \frac{5}{2}X^4 + \frac{5}{3}X^3 - \frac{1}{6}X, \qquad B_6(X) = X^6 - 3X^5 + \frac{5}{2}X^4 - \frac{1}{2}X^2 + \frac{1}{42}$$

3. Le carré symétrique

Le lemme 2.4.2 indique combien il est difficile de se procurer suffisamment d'éléments indépendants de $H^{k+1}_{\mathcal{M}}(Sym^k E, \mathbf{Q}(k+1))_{\mathbf{Z}}$. Pour disposer facilement de courbes avec un déterminant régulateur non nul, il est commode de se borner à des espaces de dimension 2, c'est-à-dire à $k \leq 3$. Or, entre 2 et 3, $k = 2$ semble préférable, car les séries K_ν figurant dans le régulateur ont la parité de k. C'est la raison pour laquelle nous nous sommes d'abord occupés du carré symétrique.

3.1 La fonction L

Nous tirons du travail de Coates et Schmidt [Coates, Schmidt 1987] les renseignements suivants.

3.1.1 Le facteur eulérien en p, $L_p(Sym^2 E, s)^{-1}$ s'annule en $s = 0$ si et seulement si la courbe E a mauvaise réduction multiplicative en p.

Selon $2.2.3_?$, ceci prédit que $dim_{\mathbf{Q}} H^3_{\mathcal{M}}(Sym^2 E, \mathbf{Q}(3)) - 2$ serait le nombre des places de réduction multiplicative.

3.1.2 *Equation fonctionnelle*

Soit C le conducteur du système de représentations l-adiques $H_l(Sym^2 E)$. On pose $\Lambda(Sym^2 E, s) = C^{s/2}\pi^{-s/2}\Gamma(\frac{s}{2})(2\pi)^{-s}\Gamma(s)L(Sym^2 E, s)$. Alors

$$\Lambda(Sym^2 E, 3 - s) = \Lambda(Sym^2 E, s).$$

Ceci, joint à 2.3.1 et à la formule $c^+(Sym^2 E(2)) = \pi {\omega_1}^2 y$ tirée de [Deligne 1979, 7.7], nous permet de traduire 2.4.0, pour $k = 2$, en un énoncé en $s = 3$:

3.1.3 Conjecture. *Soient* $\mathbf{a}, \mathbf{b} \in \mathbf{Q}[E_d]^\circ$ *tels que leurs images par* ξ *appartiennent à* $H^3_{\mathcal{M}}(Sym^2 E, \mathbf{Q}(3))_{\mathbf{Z}}$ *— autrement dit, tels que* $\partial(\xi(\mathbf{a})) = \partial(\xi(\mathbf{b})) = 0$. *Alors on a l'identité suivante à un facteur* $\varrho \in \mathbf{Q}$ *près.*

$$(3.1.3_?) \quad \frac{y^4}{\pi^2} \text{ dét} \begin{pmatrix} K_2(0,\mathbf{a},3) & K_0(0,\mathbf{a},2) \\ K_2(0,\mathbf{b},3) & K_0(0,\mathbf{b},2) \end{pmatrix} = \varrho\, C^{3/2} L(Sym^2 E, 3).$$

Par la suite, on posera (rappelons encore une fois que K_ν est relative au réseau $\mathbf{Z} + \mathbf{Z}\tau$):

$$R(\mathbf{a},\mathbf{b}) = \frac{y^4}{\pi^2} \text{ dét} \begin{pmatrix} K_2(0,\mathbf{a},3) & K_0(0,\mathbf{a},2) \\ K_2(0,\mathbf{b},3) & K_0(0,\mathbf{b},2) \end{pmatrix}.$$

3.2 Courbes ayant bonne réduction potentielle

Nous examinons la conjecture 3.1.3 en deux étapes : d'abord, dans ce numéro et le suivant, en exigeant que la courbe ait bonne réduction potentielle partout, ce qui permet d'éviter complètement l'obstruction d'intégralité; puis, aux §§ 3.4 et 3.5, pour des courbes à réduction semi-stable en une place au moins, en tenant compte des restrictions imposées dans ce cas par le lemme 2.5.1.

Même si la courbe E a bonne réduction potentielle partout, une action du groupe de Galois trop grande sur les points de torsion de E nous empêchera de trouver deux diviseurs rationnels $\mathbf{a}, \mathbf{b} \in \mathbf{Q}[E_{tors}]^{\circ}$ tels que $R(\mathbf{a}, \mathbf{b})$ soit non-nul. Si par exemple $\mathrm{Gal}(\overline{\mathbf{Q}}/\mathbf{Q})$ s'envoie surjectivement sur $\mathrm{Aut}(E_d)$, on déduit ausitôt de 2.4.2 qu'il est impossible de trouver $\mathbf{a}, \mathbf{b} \in \mathbf{Q}[E_d]^{\circ}$ avec $R(\mathbf{a}, \mathbf{b}) \neq 0$. Il faut donc une hypothèse restrictive sur l'action du groupe de Galois, et nous allons bientôt supposer *qu'il y existe un sous-groupe cyclique fini $P \subset E_{tors}$ défini sur $\mathbf{Q}$*. Toutefois, avant d'imposer ceci, force est de constater qu'*il y a des cas intermédiaires* dont la paramétrisation modulaire n'est pas aussi bien comprise que les courbes $X_0(N), X_1(N)$ ce qui fait que *nous ignorons pour l'instant si leur nombre est fini*. Ce sont les courbes elliptiques sur $\mathbf{Q}$ sans multiplication complexe dont l'image du groupe de Galois dans $\mathrm{Aut}(E_p)$, pour un nombre premier p, est le normalisateur d'un sous-groupe de Cartan non déployé. (Rappelons que les sous-groupes de Cartan non déployés n'apparaissent pas comme image de Galois *sur* $\mathbf{Q}$. — *Voir* [Serre 1972, 5.2.iv].)

3.2.0 *Exemple*

Soit $E : y^2 = x^3 - 6x + 8$ la courbe elliptique d'invariant $j = -1728$, conducteur $N = 6912 = 2^8 3^3$, discriminant $\Delta = -2^9 3^3$. Le conducteur du carré symétrique vaut $C = 2^6 3^4$. Les facteurs eulériens du carré symétrique aux mauvaises places sont $L_2(Sym^2 E, s) = (1 + 2^{1-s})^{-1}$ et $L_3(Sym^2 E, s) \equiv 1$. Dans la table 4 de [Modular Functions ...., IV, p. 132], E s'appelle "25* − 1". E n'admet pas de multiplication complexe; l'image de Galois dans $\mathrm{Aut}(E_3)$ est le normalisateur d'un sous-groupe de Cartan déployé. En fait, cette image peut s'interpréter de deux façons différentes comme un tel normalisateur : les corps quadratiques associés aux deux quotients de l'image par les sous-groupes de Cartan sont respectivement $\mathbf{Q}(\sqrt{6})$ et $\mathbf{Q}(\sqrt{-2})$. Il n'y a donc pas de point rationnel d'ordre 3 sur E, mais les diviseurs $\mathbf{a}, \mathbf{b}$ suivants sont définis sur $\mathbf{Q}$. Soit P le point de E dont le paramètre analytique dans $\mathbf{C}/\mathbf{Z} + \mathbf{Z}\tau$ vaut $1/3$. L'orbite de P sous $\mathrm{Gal}(\overline{\mathbf{Q}}/\mathbf{Q})$ est $O(P) = \{\pm\frac{1}{3}, \pm\frac{1\pm\tau}{3}\}$. On pose $\mathbf{a} = 4.(0) - \sum_{Q \in O(P)}(Q)$ et $\mathbf{b} = 4.(0) - \sum_{2.Q=0}(Q)$. Alors on trouve que numériquement $R(\mathbf{a}, \mathbf{b}) \neq 0$, et en fait

$$\frac{C^{3/2}\ L(Sym^2 E,3)}{R(\mathbf{a},\mathbf{b})} \approx -2^8 3^4 = -20736.$$

Ceci est en accord avec la conjecture 3.1.3.

Passons maintenant au cas où l'image du groupe de Galois est contenue dans un sous-groupe de Cartan déployé : À $\mathbf{Q}$-isogénie et à torsion par un caractère quadratique près — ce sont les deux opérations qui n'affectent pas le carré symétrique de la courbe; *cf* 2.4.1, (ii), (iii) —, il y a vingt-quatre courbes elliptiques sur $\mathbf{Q}$ sans multiplication complexe qui ont potentiellement bonne réduction partout — autrement dit, dont l'invariant j est entier —, et qui *possèdent un sous-groupe cyclique fini P défini sur $\mathbf{Q}$.*

En fait, on possède grâce aux travaux de Mazur [Mazur 1978] une liste complète des groupes cycliques d'ordre premier qui apparaissent comme sous-groupe de $E(\mathbf{Q})$, pour E une courbe elliptique sur $\mathbf{Q}$. Si N décrit les ordres possibles de ces sous-groupes cycliques, alors notre liste de courbes découle de la recherche systématique des points rationnels et d'invariant j entier sur les courbes modulaires $X_0(N)$. Nous supprimons ici les détails des calculs.

Pour chacune des 24 courbes ainsi trouvées, il n'y a en fait qu'un seul sous-groupe cyclique fini P qui est rationnel sur $\mathbf{Q}$. Ecrivons d son ordre (d est donc toujours premier), et posons

$$\mathbf{a} = d(0) - \sum_{x\in P}(x).$$

L'hypothèse de bonne réduction potentielle entraîne que

$$H^3_{\mathcal{M}}(Sym^2 E, \mathbf{Q}(3))_{\mathbf{Z}} = H^3_{\mathcal{M}}(Sym^2 E, \mathbf{Q}(3)).$$

La conjecture 3.1.3 prédit donc en particulier pour ces 24 courbes, que, pour $\mathbf{b}_n = n^2(0) - \sum_{nx=0}(x)$, le quotient

$$\textit{(3.2.1)} \qquad \varrho_n = \frac{R(\mathbf{a},\mathbf{b}_n)}{C^{3/2}\ L(Sym^2 E,3)} \qquad (n \geq 2),$$

est un nombre rationnel.

Noter que, d'après le lemme 2.4.2, on a

$$\varrho_n = \varrho_2\ \frac{4}{15}\ \frac{n^4-1}{n^2}.$$

Donc, afin de ne pas introduire des facteurs parasites, on peut calculer soit ϱ_d, soit ϱ_2. Il s'avère en fait que les résultats sont plus simples quand on calcule ϱ_2. C'est donc ce que nous donnons comme résultats dans la table suivante. — Bien entendu, les calculs ne déterminent qu'approximativement ces nombres; dans aucun cas on ne sait *démontrer* que ce soit des nombres rationnels. Toutefois, les résultats cohérents présentés dans la table ci-après semblent convaincants.

3.2.2 *Les résultats*

Dans la table suivante, nous notons ϱ_2^{-1} au lieu de ϱ_2; car, pour chacune des vingt-quatre courbes, on trouve numériquement que

$$\varrho_2^{-1} \approx \pm m\, 2^\alpha\, 3^\beta\, d^\gamma,$$

avec $3 \leq \alpha \leq 14$, $0 \leq \beta \leq 2$, $1 \leq \gamma \leq 3$ — en fait, $\gamma \leq 2$, sauf si $d = 5$ —, et m *vaut toujours* 1, *sauf pour quatre courbes:*

n° 5 $m = 13$; *n*° 7 $m = 1/5$; *n*° 15 $m = 1/17$; *n*° 24 $m = 1/21$.

Ces *numérateurs* (!) de ϱ_2 dans les trois derniers cas devraient indiquer que le réseau engendré par les images $\xi(\mathbf{a})$, $\xi(\mathbf{b})$ n'est pas le réseau maximal pour le raffinement "sur **Z**" de la conjecture de Beilinson.

D'autre part, le *dénominateur* 13 de ϱ_2 obtenu pour la courbe *n*° 5 semble bien plus intéressant. Pourtant nous en ignorons à présent une interprétation dans le cadre de la conjecture de Bloch et Kato [Bloch, Kato 1989]. — Aussi faut-il dire que c'est la seule courbe pour laquelle nous avons trouvé un dénominateur étrange — tous les facteurs bizarres pour les courbes semi-stables du numéro 3.5 semblent s''expliquer' par des "relations exotiques".

NB. Remarquons en passant,

— que dans le cas de la conjecture de Beilinson relatif à la valeur $L(E, 2)$, Bloch et Grayson n'ont pas trouvé de facteur inattendu comme dans notre calcul de la courbe *n*° 5. Peut-être faudrait-il pousser plus loin les calculs de Bloch et Grayson ?

— que Rohrlich [Rohrlich 1987] a très bien su contrôler les facteurs rationnels intervenant dans le cas d'une *courbe à multiplication complexe*, pour la conjecture de Beilinson relatif à la valeur $L(E, 2)$.

3.2.3 *Remarques sur les calculs*

Les séries K_ν de Kronecker ont été calculées comme des intégrales de fonctions thêta d'après [Weil 1976, p.79–80].

Pour la valeur $L(Sym^2 E, 3)$, une méthode de calcul satisfaisante découle de [Friedman 1988]. Toutefois, nous ne l'avons pas implémentée, car le calcul naïf, certes peu précis, du produit eulérien semble suffisant pour nos besoins. La précision des calculs est d'environ huit chiffres significatifs.

3.3 Table numérique

Pour chacune des 24 courbes elliptiques sans multiplication complexe avec j entier qui admettent un sous-groupe cyclique fini non trivial P défini sur $\mathbf{Q}$, nous donnons ici les informations suivantes.

ligne 1: numéro, (éventuellement, sigle relatif à [Modular Functions .. IV],) conducteur N de la courbe et signe de son discriminant Δ.

ligne 2: a_1, a_2, a_3, a_4, a_6; $|\Delta|$, C, $L_p(Sym^2 E, s)$, d, (a, b), ϱ_2^{-1}, où :

les a_i sont les coefficients de l'équation de Weierstrass généralisée;

$|\Delta|$ est la valeur absolue de son discriminant;

C est le conducteur de $Sym^2 E$;

$|\Delta|$ et C sont donnés par leurs exposants des diviseurs premiers p de N;

les facteurs eulériens aux mauvaises places* p sont codés comme ceci:

1 : $L_p(Sym^2 E, s)^{-1} = 1$
$+$: $L_p(Sym^2 E, s)^{-1} = 1 + p^{1-s}$
$-$: $L_p(Sym^2 E, s)^{-1} = 1 - p^{1-s}$;

d est l'ordre du sous-groupe P et $\frac{a+b\tau}{d}$ (mod $\mathbf{Z} + \mathbf{Z}\tau$) est un générateur de
$P(\mathbf{C}) \subset \mathbf{C}/\mathbf{Z} + \mathbf{Z}\tau$;

finalement, on donne le nombre rationnel suggéré par notre calcul du quotient $\varrho_2^{-1} = \frac{C^{\frac{3}{2}} L(Sym^2 E, 3)}{R(\mathbf{a}, \mathbf{b}_2)}$, les notations étant celles de 3.2.

Par exemple, la cinquième courbe est donnée par l'équation $E : y^2 = x^3 + x^2 - 72x - 380$. Le conducteur de E est $N = 15376 = 2^4 \,.\, 31^2$, son discriminant $\Delta = -30505984 = -2^{10} \,.\, 31^3$. (Le réseau Λ attaché au modèle donné de E est donc un losange.) Le carré symétrique de E a conducteur $C = 3844 = 2^2 \,.\, 31^2$. Pour les deux diviseurs premiers de N, $p = 2, 31$, le facteur eulérien de $L(Sym^2(E), s)$ en p est $(1+p^{1-s})^{-1}$. La courbe possède un sous-groupe — et donc en l'occurence un point rationnel — d'ordre 2 sur $\mathbf{Q}$ qui, sur $\mathbf{C}$, est engendré par $\frac{1}{2}$ (mod $\mathbf{Z} + \mathbf{Z}\tau$) $\in \mathbf{C}/\mathbf{Z} + \mathbf{Z}\tau$. Enfin, le calcul donne approximativement $\varrho_2 \approx \frac{1}{106496} = \frac{1}{2^{13} \,.\, 13}$.

---

* Nous remercions G. Henniart qui nous a indiqué des méthodes efficaces pour déterminer les facteurs eulériens de $L(Sym^2 E, s)$ aux mauvaises places, ainsi que le conducteur C.

1. *128A* **N = 128 = 2^7** $+$
0 -1 0   1   -1   7   *8*   1   2 *1 0* 2^9

2. *200C* **N = 200 = $2^3 5^2$** $+$
0 1 0   -3   -2   *4 3*   *4 2*   1 -   2 *0 1* -2^9

3. **N = 1152 = $2^7 3^2$** $+$
0 0 0   -6   4   *8 3*   *8 2*   1 +   2 *1 0* $2^{11}\,3$

4. **N = 4225 = $5^2 13^2$** $+$
1 1 1   -23   -44   *3 3*   *2 2*   - -   2 *1 0* $2^{12}\,3$

5. **N = 15376 = $2^4 31^2$** $-$
0 1 0   -72   -380   *10 3*   *2 2*   + +   2 *1 0* $2^{12}\,13$

6. **N = 8712 = $2^3 3^2 11^2$** $+$
0 0 0   -99   242   *10 3 3*   *4 2 2*   1 + +   2 *0 1* $-2^{15}\,3$

7. **N = 900 = $2^2 3^2 5^2$** $+$
0 0 0   -15   -50   *8 3 3*   *2 2 2*   + + -   2 *1 0* $\frac{2^{11}\,3}{5}$

8. **N = 2312 = $2^3 17^2$** $+$
0 1 0   -28   12   *8 3*   *4 2*   1 -   2 *0 1* -2^{14}

9. **N = 1568 = $2^5 7^2$** $-$
0 1 0   -2   -8   *6 3*   *6 2*   1 +   2 *1 0* $2^{12}\,3$

10.a *196C* **N = 196 = $2^2 7^2$** $+$
0 1 0 -114 -127   *4 8*   *2 2*   + -   3 *1 0* $2^3\,3^2$

10.b *196C tordu par* **$\mathbf{Q}(\sqrt{7})$** **N = 784 = $2^4 7^2$** $+$
0 1 0   -2   -1   *4 2*   *2 2*   + -   3 *1 0* $2^4\,3^2$

11. **N = 2704 = $2^4 13^2$** $-$
0 1 0   -4   -12   *8 2*   *2 2*   + -   3 *1 0* $2^7\,3^2$

12. **N = 2700 = $2^2 3^3 5^2$** $+$
0 0 0   -15   10   *8 3 2*   *2 4 2*   + 1 +   3 *1 0* $2^9\,3^4$

13. **N = 1296 = $2^4 3^4$** $+$
0 0 0   -9   9   *4 6*   *2 4*   + -   3 *1 0* $2^4\,3^3$

14. $\mathbf{N = 1296 = 2^4 3^4}$ $+$
0 0 0 9 -18 *8 6* *2 4* + - 3 *1 0* $2^5\ 3^3$

15. $\mathbf{N = 1369 = 37^2}$ $+$
0 1 1 -12 -17 *3* *2* - 5 *0 1* $-\frac{2^8\ 3\ 5^3}{17}$

16. $\mathbf{N = 43264 = 2^8 13^2}$ $-$
0 -1 0 9 -53 *9 3* *6 2* 1 - 5 *1 0* $2^7\ 3^2\ 5^3$

17. $\mathbf{N = 4225 = 5^2 13^2}$ $+$
0 1 1 -108 394 *4 3* *2 2* + - 5 *1 0* $2^8\ 5^3$

18. $\mathbf{N = 6400 = 2^8 5^2}$ $-$
0 1 0 17 13 *9 4* *6 2* 1 + 5 *1 0* $2^6\ 3\ 5^3$

19. $\mathbf{N = 3969 = 3^4 7^2}$ $+$
1 -1 0 -9 -8 *4 2* *4 2* - - 7 *0 1* $2^7\ 3\ 7^2$

20. $\mathbf{N = 1369 = 37^2}$ $-$
1 -1 1 2 -2 *2* *2* - 7 *1 0* $2^7\ 7^2$

21. *121I* $\mathbf{N = 121 = 11^2}$ $-$
1 1 1 -305 7888 *10* *2* + 11 *1 9* $2^6\ 11$

22. $\mathbf{N = 20736 = 2^8 3^4}$ $-$
0 0 0 6 8 *9 4* *6 4* 1 - 13 *1 0* $2^6\ 3^2\ 13^2$

23. $\mathbf{N = 9025 = 5^2 19^2}$ $+$
0 0 1 -95 356 *3 2* *2 2* - - 13 *1 0* $2^8\ 13^2$

24. $\mathbf{N = 1225 = 5^2 7^2}$ $-$
1 1 1 -8 6 *3 2* *2 2* - - 37 *1 0* $\frac{2^7\ 37^2}{3\ 7}$

3.4 Des courbes qui admettent un régulateur non-nul

Considérons des couples (E, T), où E est une courbe elliptique sur $\mathbf{Q}$ sans multiplication complexe, et $T \in E(\overline{\mathbf{Q}}_{tors})$ d'ordre $d \geq 2$ tel que, pour tout $\sigma \in \mathrm{Gal}(\overline{\mathbf{Q}}/\mathbf{Q})$, on ait $\sigma(T) = \pm T$. On écrira P le sous-groupe engendré par T et L la plus petite extension de $\mathbf{Q}$ telle que $P \subset E(L)$. Donc $[L : \mathbf{Q}] \leq 2$. L'ensemble de tous les couples (E, T) est stable par $\mathbf{Q}$-isogénie et aussi par torsion quadratique. D'après 2.3.0, on dispose d'une application $\xi : L[P]^{\circ,\pm} \longrightarrow H^{k+1}_{\mathcal{M}}(Sym^2 E, \mathbf{Q}(3))$. Rappelons de 3.1.3 la notation $R(\mathbf{a}, \mathbf{b})$ pour le régulateur de deux diviseurs $\mathbf{a}, \mathbf{b} \in L[P]^{\circ,\pm}$.

3.4.1 Théorème. *À $\mathbf{Q}$-isogénie et à torsion quadratique près, il n'y a qu'un nombre fini de couples (E, T) comme ci-dessus tels qu'il existe de diviseurs $\mathbf{a}, \mathbf{b} \in L[P]^{\circ,\pm}$ avec $\partial(\xi(\mathbf{a})) = \partial(\xi(\mathbf{b})) = 0$ et tels que $R(\mathbf{a}, \mathbf{b})$ soit non-nul.*

Démonstration. Le cas où E a partout bonne réduction potentielle ayant été traité en 3.2, nous supposons que l'ensemble fini S des nombres premiers p tel que E ait mauvaise réduction potentiellement multiplicative en p, est *non-vide.* (Par exemple, si le conducteur N de la courbe est sans facteur carré, toute mauvaise réduction est multiplicative.) Quitte à remplacer (E, T) par une tordue quadratique, on suppose que $L = \mathbf{Q}$, *i.e.*, $T \in E(\mathbf{Q})$. Pour F un corps de nombres soit $\mathcal{S} = \{v \mid v \text{ place de } F,\ v|p, \text{ pour un } p \in S\}$. Choisissons F tel que $E_d \subset E(F)$ et tel que E/F_v soit une courbe de Tate pour tout $v \in \mathcal{S}$. Alors la fibre spéciale $\mathcal{E}_F(v)$ est de type I_ν pour un ν avec $d|\nu$. Pour tout $v \in \mathcal{S}$, nous fixons une numérotation cyclique du polygone I_d inscrit dans $\mathcal{E}_F(v)$, en donnant, bien sûr, le numéro 0 à la composante neutre. Pour calculer l'obstruction imposée par ∂_v à un diviseur $\mathbf{a} \in \mathbf{Q}[P]^\circ$ il suffira, d'après 2.5.1, de connaître le numéro i_v de la composante de I_d sur laquelle T s'envoie par réduction modulo v. Quitte à renumêroter nos polygones de Néron, nous supposons toujours que $0 \leq i_v \leq [\frac{d}{2}]$. (De même, comme $B_4(\frac{d-m}{d}) = B_4(\frac{m}{d})$ et que les séries K_0 et K_2 sont paires, il suffit de travailler avec des diviseurs dans $\mathbf{Q}[P]^\circ$de la forme $\mathbf{a} = \sum_{k=0}^{[d/2]} a_k(k.T)$.) Avec ces notations, on a le

3.4.2 Lemme. *Pour qu'il existe deux diviseurs $\mathbf{a}, \mathbf{b} \in \mathbf{Q}[P]^\circ$ avec $\partial(\xi(\mathbf{a})) = \partial(\xi(\mathbf{b})) = 0$ et tels que $R(\mathbf{a}, \mathbf{b})$ soit non-nul, il faut qu'il existe j_1, j_2, $0 \leq j_1 < j_2 \leq [\frac{d}{2}]$ tels que, pour tout $v \in \mathcal{S}$, $j_1 \neq i_v \neq j_2$.*

Démonstration du lemme. Le représentant entre 0 et 1 d'un nombre réel modulo $\mathbf{Z}$ est noté par $\langle . \rangle$. Vu le lemme 2.5.1, il s'agit de montrer que

$$\mathcal{D} \stackrel{\text{déf}}{=} \text{dét}\big(B_4(\langle \frac{k.i}{d} \rangle)\big)_{0 \leq k,i \leq [\frac{d}{2}]} \neq 0.$$

D'après [Mazur 1978], on a $d \leq 12$ et $d \neq 11$. On démontre donc le lemme en calculant le déterminant pour ces 10 valeurs de d .... — Toutefois, donnons aussi une preuve théorique, mais seulement dans le cas où d est un nombre premier impair :

Dans ce cas le déterminant $\mathcal{D}' \overset{\text{déf}}{=} \text{dét}\big(B_4(\langle \frac{k.i}{d} \rangle)\big)_{1 \leq k,i \leq \frac{d-1}{2}}$ est un déterminant du groupe $\mathcal{G} = (\mathbf{Z}/d\mathbf{Z})^*/\{\pm 1\}$ et est donc — d'après la formule de Dedekind [Lang 1978, chap. 3, §6] — au signe près égal au produit suivant sur les caractères pairs modulo p :

$$\mathcal{D}' = \pm \prod_{\chi} \sum_{\mathcal{G}} \chi(k) B_4(\langle \frac{k}{d} \rangle) = \pm \prod_{\chi} B_{4,\chi} = \mp \prod_{\chi} L(-3, \chi) = \mp \zeta_{\mathbf{Q}(\mu_p)^+}(-3)$$

— un nombre non nul; *cf.* [Washington 1982, Thm. 4.2].
Or, on a $\mathcal{D} = -\frac{1}{30}\mathcal{D}^*$, où $\mathcal{D}^* = \text{dét}\big(B_4^*(\langle \frac{k.i}{d} \rangle)\big)_{1 \leq k,i \leq \frac{d-1}{2}}$, avec $B_4^*(X) = B_4(X) + \frac{1}{30} = X^4 - 2X^3 + X^2$. Par la formule de Dedekind et les relations d'orthogonalité on a

$$\mathcal{D}^* = \prod_{\chi \neq 1} \sum_{\mathcal{G}} \chi(k) B_4(\langle \frac{k}{d} \rangle) \cdot \sum_{\mathcal{G}} B_4^*(\langle \frac{k}{d} \rangle)$$

Les termes de la dernière somme sont positifs et les autres facteurs sont non-nuls parce que $\mathcal{D}' \neq 0$. Donc $\mathcal{D} \neq 0$, si d est premier.

q.e.d.

Exploitons le lemme du point de vue modulaire : Le couple (E, T) décrit un point rationnel de la courbe $X_1(d)$ et i_v apporte des renseignements sur les pointes de $X_1(d)$ qui, modulo v, se confondent avec le point (E, T). Plus précisément, dans l'interprétation modulaire des points de $X_1(d)$, une pointe correspond à un polygone de Néron I_ν, $\nu | d$, avec, sur sa composante numéro 1 (à isomorphisme près on peut normaliser n'importe quel générateur du groupe des composantes pour être sur la première composante), un point ζ^n, *i.e.*, une certaine puissance du générateur fixé ζ de $\mu_d \subset \mathbf{G}_m$. La pointe que nous venons de décrire est souvent notée $[\frac{n}{m}]$, où $m = \frac{d}{\nu}$. Comme plus haut nous inscrivons tous ces polygones I_ν dans I_d. Alors on voit que, en chaque place $v \in \mathcal{S}$, le point $(E, T) \in X_1(d)(F)$ est congru modulo v à une pointe de la forme $[\frac{n}{d/i_v}]$. Notons aussi que, en dehors des places de $\mathcal{S}$, le point (E, T) ne peut être congru à une pointe. Le lemme 3.4.2 se traduit donc comme ceci.

3.4.3 Lemme. *Pour qu'il existe deux diviseurs* $\mathrm{a}, \mathrm{b} \in \mathbf{Q}[P]^\circ$ *avec* $\partial(\xi(\mathrm{a})) = \partial(\xi(\mathrm{b})) = 0$ *et tels que* $R(\mathrm{a}, \mathrm{b})$ *soit non-nul, il faut qu'il existe deux* *pointes* $[\frac{n_1}{m_1}], [\frac{n_2}{m_2}]$ *de* $X_1(d)$ *avec* $0 \leq \frac{d}{m_1} < \frac{d}{m_2} \leq [\frac{d}{2}]$ *et telles que, sur* Spec $\mathbf{Z}$, *la section du point* (E,T) *de* $X_1(d)$ *ne rencontre nulle part celles de* $[\frac{n_1}{m_1}]$, $[\frac{n_2}{m_2}]$.

Ce lemme nous ramène à vérifier l'énoncé de finitude suivant.

3.4.4 Proposition. *Soit* C *une courbe affine de la forme*

$$X_1(d) - \{[\frac{n_1}{m_1}], [\frac{n_2}{m_2}]\},$$

les notations étant comme plus haut. Alors C *n'a qu'un nombre fini de points entiers.*

Cette proposition est en fait un cas particulier du théorème de Siegel sur les points entiers, à savoir:

3.4.5 Théorème. [Siegel 1929, *Zweiter Teil*, en particulier §1] *Soit une courbe projective irréductible* $\overline{C}$ *sur* $\mathbf{Q}$ *et soit* $C = \overline{C} - \{P_1, P_2\}$ *une courbe affine avec deux "points à l'infini". Alors* C *ne possède une infinité de points entiers que si* $\overline{C}$ *est de genre* 0 *et* P_1, P_2 *sont quadratiques conjugués.*

Appliquons ceci à la situation de 3.4.4 : D'après [Ogg 1973], une pointe $[\frac{n}{m}]$ **de** $X_1(d)$ **ne peut être quadratique sur** $\mathbf{Q}$ **que si** $m = \frac{d}{2} \in \mathbf{Z}$. **Or, dans ce cas, sa conjuguée est aussi de la forme** $[\frac{n'}{m}]$. **Donc 3.4.5 implique 3.4.4, et nous avons démontré le théorème 3.4.1.**

q.e.d.

3.4.6 *Remarques.*

(i) Notons explicitement que nous n'avons pas vraiment utilisé dans la démonstration la liberté offerte par le carré symétrique de passer d'une courbe E à une tordue quadratique, disons E'. En fait, E et E' ne peuvent pas en même temps définir un point rationnel de $X_1(d)$. Autrement dit nous avons démontré l'énoncé suivant:

À $\mathbf{Q}$*-isomorphisme près, il n'y a qu'un nombre fini de courbes elliptiques* E *sur* $\mathbf{Q}$ *d'invariant* j *non-entier, munies d'un point rationnel* T *d'ordre fini, telles qu'il existe des diviseurs* $\mathbf{a}, \mathbf{b} \in \mathbf{Q}[\langle T \rangle]^\circ$ avec $\partial(\xi(\mathbf{a})) = \partial(\xi(\mathbf{b})) = 0$ *et tels que* $R(\mathbf{a}, \mathbf{b})$ *soit non-nul.*

(ii) Nous n'avons pas insisté sur l'analyse des régulateurs R non-nuls dans les cas où les points rationnels de la courbe elliptique E forment un groupe du type $\mathbf{Z}/2\mathbf{Z} \times \mathbf{Z}/2\nu\mathbf{Z}$, $1 \leq \nu \leq 4$ au lieu d'un groupe cyclique P .... *Cf.* toutefois l'exemple de la courbe 24B.

3.5 Exemples numériques

Nous présentons ici une petite liste d'exemples de courbes admettant un régulateur non-nul. L'intérêt est de voir explicitement les obstructions d'intégralité et des "relations exotiques" analogues à celles décrites en 1.5.3.b, provenant de conditions d'intégralité trivialement satisfaites par les diviseurs utilisés .

Les huit premiers exemples qui suivent sont — à isogénie près — toutes les courbes de conducteur inférieur à 30 qui admettent un régulateur $R(\mathbf{a}, \mathbf{b}) \neq 0$ de diviseurs satisfaisant $\partial(\xi(\mathbf{a})) = \partial(\xi(\mathbf{b})) = 0$.

La courbe 11A

Les coefficients a_1, a_2, a_3, a_4, a_6 sont respectivement $0, -1, 1, 0, 0$. C'est une courbe semi-stable de conducteur 11, donc le conducteur du carré symétrique vaut $C = 121$. Son groupe de Mordell-Weil sur $\mathbf{Q}$ est engendré par un *point P d'ordre 5*. Le discriminant vaut $\Delta = -11$. Donc, dans $\mathbf{C}/\mathbf{Z} + \mathbf{Z}\tau$, P (choisi convenablement) est donné par $\frac{1}{5}$. La fibre spéciale de $\mathcal{E}$ en $p = 11$ est de type I_1. Donc les diviseurs composés de points rationnels ne rencontrent pas d'obstruction d'intégralité. Une $\mathbf{Q}$-base de l'espace des $[K_2(0, \mathbf{a}, 3), K_0(0, \mathbf{a}, 2)]$ avec $\mathbf{a} \in \mathbf{Q}[E(\mathbf{Q})_{tors}]^\circ$ est donnée par les diviseurs $\mathbf{a}_1 = (0) - (P)$; $\mathbf{a}_2 = (0) - (2.P)$ — se rappeller que K_2 et K_0 sont des fonctions paires; donc $(3.P), (4.P)$ donnent les mêmes valeurs que $(2.P), (P)$.

Alors on trouve numériquement que

$$\frac{C^{3/2} L(Sym^2 E, 3)}{R(\mathbf{a}_1, \mathbf{a}_2)} \approx -500 = -2^2\ 5^3$$

Ici le fait qu'il n'y a pas d'obstruction à l'intégralité suffit pour assurer l'existence d'un régulateur non-trivial; mais ne se traduit pas par une "relation exotique".

La courbe 14A

Les coefficients sont $1, 0, 1, -1, 0$. C'est une courbe semi-stable de conducteur 14, donc le conducteur du carré symétrique vaut $C = 14^2$. Son groupe de Mordell-Weil sur $\mathbf{Q}$ est engendré par *un point P d'ordre 6*. Dans les coordonnées du modèle de Weierstrass, prenons $P = (1, 0)$. Comme le discriminant $\Delta = -28$ est négatif, l'image de P (convenablement choisi) dans $\mathbf{C}/\mathbf{Z} + \mathbf{Z}\tau$ est donnée par $\frac{1}{6}$. La fibre spéciale de $\mathcal{E}$ en $p = 2$ est de type I_2 où P s'envoie sur la composante non-triviale. En $p = 7$, on trouve le type I_1, de sorte qu'ici, il n'y a pas d'obstruction d'intégralité pour les diviseurs composés de points rationnels. Une $\mathbf{Q}$-base de l'espace

des $[K_2(0, \mathbf{a}, 3), K_0(0, \mathbf{a}, 2)]$ avec $\mathbf{a} \in \mathbf{Q}[E(\mathbf{Q})_{tors}]^\circ$ tel que $\partial(\xi(\mathbf{a})) = 0$ est donnée par les diviseurs $\mathbf{a}_1 = (P) - (3.P)$, $\mathbf{a}_2 = (0) - (2.P)$.
On trouve numériquement que

$$\frac{C^{3/2} L(Sym^2 E, 3)}{R(\mathbf{a}_1, \mathbf{a}_2)} \approx -648 = -2^3\ 3^4$$

Ici encore, les conditions d'intégralité permettent simplement d'obtenir un régulateur non-trivial.

La courbe 15A

Les coefficients sont $1, 1, 1, 0, 0$. C'est une courbe semi-stable de conducteur 15, donc le conducteur du carré symétrique vaut $C = 15^2$. Son groupe de Mordell-Weil sur $\mathbf{Q}$ est engendré par *un point P d'ordre 4*. Comme le discriminant $\Delta = -15$ est négatif, l'image de P (convenablement choisi) dans $\mathbf{C}/\mathbf{Z} + \mathbf{Z}\tau$ est donnée par $\frac{1}{4}$. En $p = 3$ aussi bien qu'en $p = 5$, on trouve le type I_1 comme fibre spéciale de $\mathcal{E}$, de sorte qu'il n'y a pas d'obstruction d'intégralité du tout pour les diviseurs composés de points rationnels. Une $\mathbf{Q}$-base de l'espace des $[K_2(0, \mathbf{a}, 3), K_0(0, \mathbf{a}, 2)]$ avec $\mathbf{a} \in \mathbf{Q}[E(\mathbf{Q})_{tors}]^\circ$ est donnée par les diviseurs $\mathbf{a}_1 = (0) - (P)$; $\mathbf{a}_2 = (0) - (2.P)$.
On trouve numériquement que

$$\frac{C^{3/2} L(Sym^2 E, 3)}{R(\mathbf{a}_1, \mathbf{a}_2)} \approx -1024 = -2^{10}$$

P étant d'ordre 4, les conditions triviales d'intégralité laissent juste de la place pour un régulateur non-trivial.

La courbe 17A

Les coefficients sont $1, -1, 1, -1, 0$. C'est une courbe semi-stable de conducteur 17, donc le conducteur du carré symétrique vaut $C = 17^2$. Son groupe de Mordell-Weil sur $\mathbf{Q}$ est engendré par *un point P d'ordre 4*. Le discriminant $\Delta = +15$ est positif, et on vérifie que $\frac{1}{4} + \frac{\tau}{2}$ dans $\mathbf{C}/\mathbf{Z} + \mathbf{Z}\tau$ correspond à un point rationnel P. En $p = 17$, la fibre spéciale de $\mathcal{E}$ est de type I_1. Donc on n'a pas d'obstruction d'intégralité pour les diviseurs à support dans les points rationnels. Une $\mathbf{Q}$-base de l'espace des $[K_2(0, \mathbf{a}, 3), K_0(0, \mathbf{a}, 2)]$ avec $\mathbf{a} \in \mathbf{Q}[E(\mathbf{Q})_{tors}]^\circ$ est, comme dans l'exemple précédent, donnée par les diviseurs $\mathbf{a}_1 = (0) - (P)$; $\mathbf{a}_2 = (0) - (2.P)$.
Et comme dans le calcul précédent, on trouve numériquement que

$$\frac{C^{3/2} L(Sym^2 E, 3)}{R(\mathbf{a}_1, \mathbf{a}_2)} \approx -1024 = -2^{10}$$

Ici encore, la condition triviale d'intégralité permet simplement d'obtenir un régulateur non-trivial.

La courbe 20A

Les coefficients sont $0, 1, 0, -1, 0$. En $p = 2$, la courbe a bonne réduction potentielle, tandis que la réduction est semi-stable en $p = 5$, avec fibre spéciale de type I_1. Le conducteur est 20; le discriminant $\Delta = +80$; le conducteur du carré symétrique vaut $C = 2^2.5^2 = 100$. Le facteur eulérien en $p = 2$ de $L(Sym^2 E, s)$ est $(1+p^{1-s})^{-1}$. Le groupe de Mordell-Weil sur $\mathbf{Q}$ est engendré par *un point P d'ordre 6*. On vérifie que $\frac{1}{6}+\frac{\tau}{2}$ dans $\mathbf{C}/\mathbf{Z}+\mathbf{Z}\tau$ correspond à un point rationnel P. Comme le I_1 en $p = 5$ n'impose pas d'obstruction d'intégralité pour les diviseurs à support dans $E(\mathbf{Q})$, on se retrouve, pour la première fois dans cette liste d'exemples, avec plus de deux diviseurs de degré zéro linéairement indépendants satisfaisant aux conditions d'intégralité. Autrement dit, on devrait trouver une "relation exotique" entre leurs régulateurs. Soient $\mathbf{a}_1 = (0) - (P)$; $\mathbf{a}_2 = (0) - (2.P)$; $\mathbf{a}_3 = (0) - (3.P)$.
Alors on trouve numériquement que

$$\frac{C^{3/2}L(Sym^2 E, 3)}{R(\mathbf{a}_1, \mathbf{a}_2)} \approx 216 = 2^3.3^3 \qquad \frac{C^{3/2}L(Sym^2 E, 3)}{R(\mathbf{a}_3, \mathbf{a}_2)} \approx 432 = 2^4.3^3$$

Ceci suggère — noter que les déterminants se retrouvent au dénominateur dans notre façon d'afficher les résultats — que $\xi(2\mathbf{a}_3 - \mathbf{a}_1) \in \xi(\mathbf{a}_2).\mathbf{Q}$. En fait, on vérifie aussitôt numériquement la "relation exotique" suivante entre vecteurs de séries de Kronecker.

$$[K_2(0, 2\mathbf{a}_3 - \mathbf{a}_1, 3), K_0(0, 2\mathbf{a}_3 - \mathbf{a}_1, 2)] \approx \frac{29}{16} [K_2(0, \mathbf{a}_2, 3), K_0(0, \mathbf{a}_2, 2)]$$

Donc on trouve par exemple :

$$\frac{C^{3/2}L(Sym^2 E, 3)}{R(\mathbf{a}_1, \mathbf{a}_3)} \approx \frac{2^8.3^3}{29}.$$

La courbe 21C

Les coefficients sont $1, 0, 0, -39, 90$. C'est une courbe semi-stable de conducteur 21, donc le conducteur du carré symétrique vaut $C = 21^2$. Les points rationnels sont engendrés par *un point P d'ordre 8* rationnel sur $\mathbf{Q}$. L'image de P dans $\mathbf{C}/\mathbf{Z} + \mathbf{Z}\tau$ est donnée par $\frac{1}{8} + \frac{\tau}{2}$. La fibre spéciale de $\mathcal{E}$ en $p = 3$ est de type I_8 où P s'envoie sur la composante numéro 3, pour un sens de numérotation convenablement choisi. En $p = 7$, on trouve le type

I_1, de sorte qu'il n'y a pas d'obstruction d'intégralité pour les diviseurs composés de points rationnels.

Posons $\mathbf{b}_i = (0) - (i.P)$ et $\eta_i = [K_2(0, \mathbf{b}_i, 3), K_0(0, \mathbf{b}_i, 2)]$. Commençons cette fois-ci par la "relation exotique" :

$$7\eta_1 - 78\eta_2 + 9\eta_3 + 36\eta_4 \approx 0.$$

Convenons de l'utiliser afin d'éliminer η_3 de la discussion qui suit. Quant à l'obstruction d'intégralité (en $p = 7$), on a $-\partial(\xi(\mathbf{b}_i)) = -\partial_7(\xi(\mathbf{b}_i)) = B_4(\langle \frac{3i}{8} \rangle) + \frac{1}{30} = (\langle \frac{3i}{8} \rangle)^2(\langle \frac{3i}{8} \rangle - 1)^2$. Toutes les solutions de $\partial(\xi(\mathbf{a})) = 0$ sont engendrées — modulo la "relation exotique" ci-dessus, et modulo le noyau conjecturalement trivial de $\xi(\mathbf{a}) \mapsto [K_2(0, \mathbf{a}, 3), K_0(0, \mathbf{a}, 2)]$ — par

$$\mathbf{a}_1 = 225\mathbf{b}_4 - 256\mathbf{b}_1; \quad \mathbf{a}_2 = 144\mathbf{b}_4 - 256\mathbf{b}_2.$$

On trouve numériquement que

$$\frac{C^{3/2} L(Sym^2 E, 3)}{R(\mathbf{a}_1, \mathbf{a}_2)} \approx -\frac{2}{3^2}$$

C'est la combinaison linéaire des $\mathbf{b}_i$ qui donne le résultat le plus simple. Les autres s'en déduisent par la relation exotique. Par exemple, on trouve des résultats comme :

$$\frac{C^{3/2} L(Sym^2 E, 3)}{R(25\mathbf{b}_2 - 16\mathbf{b}_1, 49\mathbf{b}_4 - 256\mathbf{b}_3)} \approx -\frac{2^9}{29 \, . \, 37}$$

La courbe 24B

Les coefficients sont $0, -1, 0, -4, 4$. En $p = 2$, la courbe a bonne réduction potentielle, tandis que la réduction est semi-stable en $p = 3$, avec fibre spéciale de type I_2. Le conducteur est 24; le discriminant $\Delta = +2^8.3^2$; le conducteur du carré symétrique vaut $C = 2^4.3^2 = 144$. Le facteur eulérien en $p = 2$ de $L(Sym^2 E, s)$ est 1. Le groupe de Mordell-Weil sur $\mathbf{Q}$ est isomorphe à $\mathbf{Z}/4\mathbf{Z} \times \mathbf{Z}/2\mathbf{Z}$. Soit $P = (0, 2)$ *un point rationnel d'ordre 4*. Dans $\mathbf{C}/\mathbf{Z} + \mathbf{Z}\tau$, P correspond à $\frac{1}{4} + \frac{\tau}{2}$. P se réduit sur la composante neutre en $p = 3$. (L'autre facteur du groupe des points rationnels passe par tous les côtés du 2-gone de Néron en $p = 3$.) Donc il n'y a pas d'obstruction d'intégralité pour les diviseurs composés de multiples de P. Posons $\mathbf{a}_1 = (0) - (P)$; $\mathbf{a}_2 = (0) - (2.P)$.
On trouve numériquement que

$$\frac{C^{3/2} L(Sym^2 E, 3)}{R(\mathbf{a}_1, \mathbf{a}_2)} \approx \frac{2^{12}}{3^2}$$

La courbe 26D

Les coefficients sont $1, -1, 1, -3, 3$. La courbe est semi-stable de conducteur 26, donc le conducteur du carré symétrique vaut 26^2. Les points rationnels sont engendrés par *un point P d'ordre 7*. Comme le discriminant $\Delta = -2^7.13$ est négatif, l'image de P (convenablement choisi) dans $\mathbf{C}/\mathbf{Z}+\mathbf{Z}\tau$ est donnée par $\frac{1}{7}$. La fibre spéciale de $\mathcal{E}$ en $p = 2$ est de type I_7 où il s'avère que P s'envoie sur la composante numéro 2, pour un sens de numérotation convenablement choisi. En $p = 13$, on trouve le type I_1 de sorte que là, il n'y a pas d'obstruction d'intégralité pour les diviseurs composés de points rationnels. Une $\mathbf{Q}$-base de l'espace des $[K_2(0, \mathbf{a}, 3), K_0(0, \mathbf{a}, 2)]$ avec $\mathbf{a} \in \mathbf{Q}[E(\mathbf{Q})_{tors}]^\circ$ tel que $\partial(\xi(\mathbf{a})) = 0$ est donnée par

$$\mathbf{a}_1 = 4(0) - 4(P) - 9(2.P) + 9(3.P); \quad \mathbf{a}_2 = -3(0) - (P) + 4(2.P).$$

On trouve numériquement que

$$\frac{C^{3/2}L(Sym^2 E, 3)}{R(\mathbf{a}_1, \mathbf{a}_2)} \approx -196 = -2^2.7^2$$

La courbe 66I

Les coefficients sont $1, 0, 0, -45, 81$. C'est une courbe semi-stable de conducteur 66, donc le conducteur du carré symétrique vaut 66^2. Les points rationnels sont engendrés par *un point P d'ordre 10 rationnel* sur $\mathbf{Q}$. Dans les coordonnées du modèle de Weierstrass, $P = (-6, 15)$; l'image de P dans $\mathbf{C}/\mathbf{Z}+\mathbf{Z}\tau$ est donnée par $\frac{1}{10} + \frac{\tau}{2}$. La fibre spéciale de $\mathcal{E}$ en $p = 2$ est de type I_{10} où P s'envoie sur la composante numéro 3, pour un sens de numérotation convenablement choisi. En $p = 3$, on trouve le type I_5 avec la réduction de P sur la première composante. En $p = 11$, on trouve une fibre spéciale du type I_1, de sorte qu'il n'y a pas d'obstruction d'intégralité pour les diviseurs composés de points rationnels. Une $\mathbf{Q}$-base de l'espace des $[K_2(0, \mathbf{a}, 3), K_0(0, \mathbf{a}, 2)]$ avec $\mathbf{a} \subset \mathbf{Q}[E(\mathbf{Q})_{tors}]^\circ$ tel que $\partial(\xi(\mathbf{a})) = 0$ est donnée par les diviseurs $\mathbf{a}_i = \sum_{j=0}^{5} a_{ij}(i.P)$; $i = 1, 2, 3$, où: $\mathbf{a}_1 = (a_{1j}) = (-1, 10, -5, 5, -10, 1)$; $\mathbf{a}_2 = (a_{2j}) = (10, -1, 13, -13, 1, -10)$; $\mathbf{a}_3 = (a_{3j}) = (-5, 0, -4, 0, 9, 0)$.

On trouve numériquement que

$$\frac{C^{3/2}L(Sym^2 E, 3)}{R(\mathbf{a}_1, \mathbf{a}_3)} \approx -800 = -2^5.5^2 \qquad \frac{C^{3/2}L(Sym^2 E, 3)}{R(\mathbf{a}_2, \mathbf{a}_3)} \approx 800 = 2^5.5^2$$

On constate donc numériquement, comme il convient, l'existence d'une dépendance linéaire "exotique" entre les régulateurs de $\xi(\mathbf{a}_3)$ et de $\xi(\mathbf{a}_1 +$

$\mathbf{a}_2$). En fait, on vérifie numériquement pour $\eta_i = [K_2(0, \mathbf{a}_i, 3), K_0(0, \mathbf{a}_i, 2)]$ que

$$2\eta_1 + 2\eta_2 + 5\eta_3 \approx 0.$$

Cette relation devrait déjà exister dans l'image de ξ ...

La courbe 90G

Les coefficients sont $1, -1, 1, -122, 1721$. C'est une courbe semi-stable de conducteur 90, donc le conducteur du carré symétrique vaut 90^2. Les points rationnels sont engendrés par *un point P d'ordre 12*. Comme le discriminant $\Delta = -2^{12}.3^7.5^3$ est négatif, l'image de P (convenablement choisi) dans $\mathbf{C}/\mathbf{Z} + \mathbf{Z}\tau$ est donnée par $\frac{1}{12}$. La fibre spéciale de $\mathcal{E}$ en $p = 2$ est de type I_{12} où P s'envoie sur la composante numéro 5, pour un sens de numérotation convenablement choisi. En $p = 3$, on trouve le type I_1^*. Le groupe des composantes connexes de $\mathcal{E}_3$ est cyclique d'ordre 4 et P s'envoie dans une composante d'ordre 4. Mais après extension des scalaires à l'extension ramifiée $\mathbf{Q}(\sqrt{-3})$, la courbe acquiert une fibre spéciale de type I_2 en l'unique place au-dessus de 3. En fait, elle est isomorphe sur ce corps à la courbe 30C, dont le I_1 sur $\mathbf{Q}$ se déploie en un I_2 sur $\mathbf{Q}(\sqrt{-3})$. P s'envoie dans la composante non-triviale de I_2. Finalement, en $p = 5$, on trouve une fibre spéciale du type I_3, où P s'envoie sur une composante non-triviale, disons le numéro 1. Nous avons donc trois conditions d'intégralité et une par le degré des diviseurs, imposées aux 7 multiples $(i.P)$, $i = 0, ..., 6$. On s'attend donc à trouver une "relation exotique" entre les régulateurs de 3 éléments $\xi(\mathbf{a})$ tels que $\mathbf{a} \in \mathbf{Q}[E(\mathbf{Q})_{tors}]^\circ$, $\partial(\xi(\mathbf{a})) = 0$. Voici de tels $\mathbf{a}$: $\mathbf{a}_1 = (1, 2, 2, 1, -1, -3, -2)$; $\mathbf{a}_2 = (8, 10, -7, -15, -8, 5, 7)$; $\mathbf{a}_3 = (0, -2, -3, 1, 4, 1, -1)$.
On trouve numériquement que

$$\frac{C^{3/2} L(Sym^2 E, 3)}{R(\mathbf{a}_1, \mathbf{a}_2)} \approx -2^8.3^2 \qquad \frac{C^{3/2} L(Sym^2 E, 3)}{R(\mathbf{a}_1, \mathbf{a}_3)} \approx 2^8.3^2.13.$$

Ceci suggère déjà que $\xi(\mathbf{a}_2 + 13\mathbf{a}_3) \in \xi(\mathbf{a}_1).\mathbf{Q}$. En fait, on trouve numériquement la "relation exotique" entre les $\eta_i = [K_2(0, \mathbf{a}_i, 3), K_0(0, \mathbf{a}_i, 2)]$:

$$6\eta_1 + \eta_2 + 13\eta_3 \approx 0.$$

Une courbe de conducteur 2730

Soit finalement E la courbe elliptique donnée par l'équation $y^2 + 43xy - 210y = x^3 - 210x^2$, de conducteur $N = 2.3.5.7.13 = 2730$ et discriminant $\Delta = +2^{12}.3^6.5^3.7^4.13$. Elle possède *un point rationnel* $P = (0, 210)$

d'ordre 12, qui, dans $\mathbf{C}/\mathbf{Z} + \mathbf{Z}\tau$ est donné par $\frac{1}{12} + \frac{\tau}{2}$. E est semi-stable; donc $C = N^2$. Les fibres spéciales de $\mathcal{E}$ aux mauvaises places sont respectivement: de type I_{12} en $p = 2$; de type I_6 en $p = 3$; de type I_3 en $p = 5$; de type I_4 en $p = 7$; de type I_1 en $p = 13$. Pour tout p, notre point P s'envoie sur un générateur du groupe des composantes de I_ν. En dehors de $p = 2, 13$, on peut le normaliser comme étant le côté numéro 1. En $p = 2$, P s'envoie sur le cinquième côté du I_{12}. Ces quatre contraintes d'intégralité nous permettent donc seulement de trouver deux diviseur intégraux dont le régulateur est non-nul: $\mathbf{a}_1 = (20, -42, 39, -26, 15, -12, 6)$; $\mathbf{a}_2 = (-3, -2, 16, -26, 38, -52, 29)$. Alors on trouve numériquement que

$$\frac{C^{3/2}L(Sym^2 E, 3)}{R(\mathbf{a}_1, \mathbf{a}_2)} \approx -2^{14}.3^4$$

Bibliographie

A. Beilinson (1985), Higher regulators and values of L-functions, *J. Soviet Math.* **30** (1985), 2036–2070.

A. Beilinson (1986), Higher regulators of modular curves; *in: Contemporary Mathematics* **55** (1986), Part I, 1–34

S. Bloch (1980), Lectures on algebraic cycles; *Duke Univ. Math. Series* **IV** 1980

S. Bloch, D. Grayson (1986), K_2 and the L-functions of elliptic curves. Computer calculations; *in: Contemporary Mathematics* **55** (1986), Part I, 79–88

S. Bloch, K. Kato (1989), L–functions and Tamagawa numbers of motives, *preprint*

J. Coates, C.-G. Schmidt (1987), Iwasawa theory for the symmetric square of an elliptic curve; *J. reine angewandte Math.* **375/376** (1987), 104–156

P. Deligne (1979), Valeurs de fonctions L et périodes d'intégrales; *in: Proc. Symp. Pure Math.* **33** (1979), 313–346

P. Deligne (1985), lettre à C. Soulé, 20/1/1985.

C. Deninger (1988), Higher regulators of elliptic curves with complex multiplication; *in: Sém. Théorie de Nombres Paris* 1986/87, C. Goldstein (ed.), Birkhäuser 1988

C. Deninger (1989), Higher regulators and Hecke L-series of imaginary quadratic fields I, *Inventiones Math.* **96** (1989), 1–69; II, *à paraître*

C. Deninger, K. Wingberg (1988), On the Beilinson conjectures for elliptic curves with complex multiplication; *in:* Rapoport, Schappacher, Schneider (ed.s), Beilinson's Conjectures on Special Values of L-Functions, *Perspectives in Math.* **4**, Acad. Press 1988, 249–272

E. Friedman (1988), Hecke's integral formula, *preprint Max-Planck-Institut für Mathematik, Bonn*, 1988

U. Jannsen (1990), Mixed motives and algebraic K-theory; *Lect. Notes Mathematics (Springer)* **1400**

D. Kubert (1977), Universal bounds on the torsion of elliptic curves, *Proc. London Math. Soc.* **33** (1977), 193–237

S. Lang (1978), Cyclotomic fields; *GTM* **59**, Springer 1978

B. Mazur (1978), Rational Isogenies of prime degree, *Inventiones Math.* **44** (1978), 129–162

Modular Functions of One Variable, IV, Antwerp Proceedings 1972; *Lect. Notes Mathematics (Springer)* **476**, 1975

A.P. Ogg (1973), Rational points on certain elliptic modular curves, Proc. Symp. Pure Math. **24**, AMS 1973, 221–231

D. Ramakrishnan (1989), Regulators, Algebraic Cycles, and Values of L-functions, *Contemporary Mathematics* **83**, 183–310

D. Ramakrishnan (1990), Problems arising from the Tate and Beilinson conjectures in the context of Shimura varieties; *in :* Automorphic forms, Shimura varieties, and L-functions, part II, (Clozel, Milne, ed.s), *Perspectives in Math.* **11**, Acad. Press 1990, 227–252

D.E. Rohrlich (1987), Elliptic curves and values of L- functions; *in:* *Can. Math. Soc. Conf. Proc.* **7** (1987), 371–387

N. Schappacher (1988), Periods of Hecke Characters, *Lecture Notes in Mathematics (Springer)* **1301**

N. Schappacher, A.J. Scholl (1988), Beilinson's Theorem on Modular Curves; *in:* Rapoport, Schappacher, Schneider (ed.s), Beilinson's Conjectures on Special Values of L- Functions, *Perspectives in Math.* **4**, Acad. Press 1988, 273–304

N. Schappacher, A.J. Scholl (19**), On the Tame symbol in Symmetric Powers, *en préparation.*

A.J. Scholl (1990), Motives for modular forms; *Inventiones Math., en presse*

P. Schneider (1988), Introduction to the Beilinson conjectures; *in:* Rapoport, Schappacher, Schneider (ed.s), Beilinson's Conjectures on Special Values of L-Functions, *Perspectives in Math.* **4**, Acad. Press 1988, 1–35

J.-P. Serre (1972), Propriétés galoisiennes des points d'ordre fini des courbes elliptiques, *Inventiones Math.* **15** (1972), 259–331; *Œuvres*, III: no. 94

C.L. Siegel (1929), Über einige Anwendungen diophantischer Approximationen; *Abhandlungen der Preußischen Akademie der Wissenschaften. Physikalisch-Mathematische Klasse 1929, Nr. 1* — Gesammelte Abhandlungen **I**, 209–266.

L. Washington (1982), Introduction to cyclotomic fields; *GTM* **83**, Springer 1982

A. Weil (1976), Elliptic functions according to Eisenstein and Kronecker; *Ergebnisse* **88**, Springer 1976.

Hyperelliptic supersingular curves

Frans Oort

Consider algebraic curves of genus 3 in positive characteristic which are hyperelliptic, and whose Jacobian is a supersingular abelian variety. The main result of this paper is: every component of this locus in the moduli space has dimension one (cf. Theorem (1.12)).

This is a particular case of a general expectation: consider two subsets of a moduli space both given by properties of the objects in consideration. If these properties are "of a different nature", "do not imply each other", then one could try to show that the intersection of those subsets is as transversal as possible. In the case we consider, $g = 3$, the moduli space of curves, and of polarized abelian varieties has dimension $3g - 3 = \frac{1}{2}g(g+1) = 6$, the hyperelliptic locus $\mathcal{B}$ has dimension $2g + 2 - 3 = 5$, and the supersingular locus $\mathcal{S}$ has dimension $[g^2/4] = 2$, and we expect $\dim(\mathcal{B} \cap \mathcal{S}) = 5+2-6 = 1$.

Computations in [18], pp. 194-197 seemed to indicate that components of dimension 2 could be possible in this intersection. Those attempts by Koblitz stimulated this research.

For some time progress seemed difficult: it is easy to define when a curve C is hyperelliptic (a double cover of a rational curve), but how do recognize whether its Jacobian J is supersingular? The Hasse-Witt matrix of C tells us the p-rank of J, but "the p rank is zero" does not imply "supersingular" for $g \geq 3$. Starting from the other side, we can construct all supersingular abelian threefolds (cf. [16]), but how do we detect whether the good curve obtained from the polarization (cf. [28]) is hyperelliptic?

We have not been able to solve the problem directly. But we could obtain enough information on $\mathcal{H}$ and on $\mathcal{B}$ (almost formally smooth in characteristic $p \neq 2$, cf. (4.5)) and (4.6), and on $\mathcal{S}$ (every component has one very singular point, cf. (2.9)), in order to be able to conclude the expected transversality. In characteristic $p = 2$ we finish by two explicit computations ([12] and (3.1)).

Several arguments and methods in this paper should be generalized and improved: is there a direct approach?, how can we define "admissible coverings" of algebraic curves in positive characteristics?, how do we generalize to other strata and to higher dimensions? We hope that these results stimulate the search for such improvements.

Notation: The dual of an abelian variety X is denoted by X^t, its formal group by $\hat{X}$. Notations and facts about existence of moduli spaces are as in [22], [23], [2]. If T is an algebraic set, we write $\dim T = m$ if every component of T has dimension exactly m.

On topics in this paper I had several discussions and correspondences with: B. van Geemen, T. Katsura, K.Z. Li, R. Noot, J-P. Serre. I thank them for their stimulating suggestions and questions.

1 Preliminaries

1.1 Definition: *A* good *curve is (either a non-singular elliptic curve or) a stable curve (of genus $g \geq 2$) such that* Jac (C) *is an abelian variety. Note that a stable curve is good if and only if each component is non-singular, and the graph of the components is a tree.*

1.2 Good curves in $g = 3$:

Type (3) a non-singular curve,

Type (2) $E \cup C_1$ with $g(C_1) = 2$, and $g(E) = 1$,

Type (1) $E_1 \cup E_2 \cup E_3$, elliptic tails E_1, E_3 attaches to E_2 at points $P, Q \in E_2$, with $P \neq Q$,

Type (0) elliptic tails E_1, E_2, E_3 attaches at three different points to $\mathbb{P}^1$ (all components are non-singular, all singularities are normal crossings).

1.3 We denote by $\mathcal{M}_g^{\sim}$ (cf. [23]) the coarse moduli space of *good* curves of genus g. Note that the space of curves of genus $g = 3$ of Type (i) has relative dimension $i + 3$ over Spec($\mathbb{Z}$). In fact the moduli for curves of Type (2) are: moduli for C_1, for E, and for $P \in C_1$; for curves of Type (1) we have: the moduli for the curves E_i, $1 \leq i \leq 3$, ánd for the class $[P - Q] \in Pic(E_2)$.

1.4 The **Torelli-morphism**

$$j : \mathcal{M}_g^{\sim} \to \mathcal{A}_{g,1}$$

from the moduli space of good curves to the moduli space of principally polarized abelian varieties is defined, and we have:

(1.4.1) j *is proper,*
(1.4.2) j *is not finite if* $g \geq 3$.

Note that (1.4.1) follows from[28], Lemma 7, by the valuative criterion for properness, cf. [11], II, Th. (4.7). The fact (1.4.2) was already long ago observed by Griffiths: take $C = C_1 \cup C_2$ with $g(C_1) = 2$, then j "forgets" $P \in C_1$.

1.5 Definition: *A hyperelliptic non-singular curve is a non-singular, complete, irreducible curve of genus* $g \geq 2$ *which admits a double covering to a rational curve; we denote by* $\mathcal{H}_g^0 \subset \mathcal{M}_g, g \geq 2$, *the moduli space of such curves. We denote by* $\mathcal{H}_g \subset \mathcal{M}_g^{\sim}$ *the Zariski-closure of* $\mathcal{H}_g^0$. *A curve is called a* good hyperelliptic *curve if its moduli point is in* $\mathcal{H}_g$ *(i.e. if it is a specialization of a smooth hyperelliptic curve). We denote by* $\mathcal{B}_g \subset \mathcal{A}_{g,1}$ *its image:*

$$j(\mathcal{H}_g) = \mathcal{B}_g \subset \mathcal{A}_{g,1},$$

and the points in $\mathcal{B}_g$ *correspond to* "hyperelliptic Jacobians".

1.6 Definition: *Let* C *be a good curve,* $g \geq 2$, *suppose* E *is an elliptic component of* C, *let* P *be one of the intersection points of* E *with another component; the* quasi-Weierstrass *points on* E *in* C *are the points* Q *so that* $[P - Q] \in E[2]$ *(i.e. this class is killed by* 2).

1.7 Lemma: *If* C *is a good hyperelliptic curve, then its components are of genus at most one, or hyperelliptic. Every component of genus at least one intersects the other components in (quasi-) Weierstrass points.*

1.8 Corollary:

$$j : \mathcal{H}_3 \to \mathcal{B}_3 \text{ is finite.}$$

1.9 Remark: $\mathcal{H}_g \otimes \mathbb{Z}[\frac{1}{2}] \to \mathcal{B}_g \otimes \mathbb{Z}[\frac{1}{2}]$ is finite for all g at least two, but probably $\mathcal{H}_g \to \mathcal{B}_g$ is not finite for g at least 4 at $p = 2$.

Proof of (1.7): The generic point of $\mathcal{H}_g$ corresponds to a smooth curve C which admits an involution i with rational quotient. By [2], Th. (1.11) this involution extends to every stable specialization of C. Moreover

$$(i \in \text{Aut } (\text{Jac}(C))^* = -id_{\text{Jac}(C)},$$

and this remains true under specialization. If C' is a good hyperelliptic curve with components $C'_1, \ldots, C'_t, C'_{t+1}, \ldots, C'_n$ such that C'_j is non-rational for $j \leq t$, and rational for $j > t$, then $i' \in \text{Aut } (C')$ gives

$$-id = (i')^* \text{ on } \text{Jac}(C) = \Pi_{j=1}^{t} \text{Jac}(C'_j).$$

This implies that i' is not the identity on the components C'_j for $j \leq t$, and because the genus of $C'_j / < i' >$ is zero this implies that these curves C'_j are (hyper)elliptic. Moreover i' maps each C'_j for $j \leq t$ to itself, thus the intersection points with the other components are fixed points of i' (note that the graph of components is a tree, and there are no tails consisting of rational curves only). Hence these points are (quasi-)Weierstrass on that component. This shows (1.7).

Proof of (1.8): If a curve of genus three is of Type (2), and it is hyperelliptic, the point P is a Weierstrass point on C_1. In case of hyperelliptic Type (3) we have $[P - Q] \in E_2[2]$. Finally note that j is injective (hence finite) at smooth curves. Thus every fibre of $\mathcal{H}_3 \to \mathcal{B}_3$ is finite, thus the morphism is quasi-finite, hence finite by (1.4.1). This shows (1.8).

After fixing a prime number p, we write $\mathcal{S} \subset \mathcal{A}_{3,1} \otimes \mathsf{F}_p$ for the supersingular locus (cf. Section 2), and we write $\mathcal{S}' \subset \mathcal{M}_3^{\sim} \otimes \mathsf{F}_p$ for the set of good supersingular curves of genus 3. In this paper we show:

1.10 Proposition: *No component of $\mathcal{S} \subset \mathcal{A}_{3,1} \otimes \mathsf{F}_p$ is contained in $\mathcal{B}_3 \otimes \mathsf{F}_p$ (for the proof we refer to Section 3 and to (4.7)).*

In (4.1) and (4.2) we show:

1.11 Lemma: *Let K be a field, C_1 and E proper non-singular curves over K, with $g(C_1) = 2$ and $g(E) = 1$. Let $P \in C_1(K)$ be a Weierstrass point of C_1, and let C be the good curve obtained by attaching the elliptic tail to C_1 at P. Then C is hyperelliptic.*

From (1.10) and (1.11) we deduce:

1.12 Theorem: *Let* char $(k) = p > 0$. *Then*

$$\dim(\mathcal{S} \cap \mathcal{B}_3 \otimes \mathsf{F}_p) = 1 = \dim(\mathcal{S}' \cap \mathcal{H}_3 \otimes \mathsf{F}_p)$$

(i.e. every component of these intersections has dimension exactly one).

Proof: Take C_1 and E supersingular, and P as in (1.11). Then the good curve C obtained in this way is in $\mathcal{S}' \subset \mathcal{H}_3 \otimes \mathsf{F}_p$ by (1.11), hence both intersections are non-empty.

Clearly $\dim(\mathcal{B}_3 \otimes \mathsf{F}_p) = 5$, we know (cf. Section 2) that $\dim(\mathcal{S}) = 2$ in $g = 3$, and $\dim(\mathcal{A}_{3,1} \otimes \mathsf{F}_p) = 6$. Moreover, going to some finite covering of $\mathcal{A}_{3,1}$ (by a fine moduli scheme) we work in a 6-dimensional non-singular variety. Hence (1.10) implies that all components of $\mathcal{S} \cap \mathcal{B}_3 \otimes \mathsf{F}_p$ have dimension one. The second statement then follows using (1.8).

1.13 Remark: For an algebraically closed field k we have: Pic $(\mathcal{A}^*_{3,1}) \otimes \mathbb{Q}$ has rank one (the star denote the Satake compactification, or the minimal compactification, in the sense of [8], Section 6). This can be seen as in [7], page 138. Thus the closure of the divisor $\mathcal{B}_3 \otimes k$ is ample on $\mathcal{A}^*_{3,1}$. Note that $\mathcal{S} \subset \mathcal{A}_{3,1} \otimes k$ is complete (p-rank zero abelian varieties do not degenerate, cf. [30], p. 99. Thus we see that $\mathcal{S} \cap \mathcal{B}_3 \otimes \mathsf{F}_p \neq \emptyset$ (without using 1.11).

2 Supersingular abelian varieties

In this section we recall definitions and properties, and we refine some earlier results on the moduli space of supersingular abelian varieties of dimension 2 and 3.

For an abelian variety X (sometimes abbreviated by AV) we denote by X^t its dual, and by $\hat{X}$ its formal group.

We fix a prime number p, and by k we denote in this section an algebraically closed field, char $(k) = p > 0$.

An elliptic curve E is called *supersingular* if E has no geometric points of order exactly p (there are many other characterizations ...). We denote by H the formal group $H = G_{1,1}$ as in [19], p.35, which is defined over $\mathbf{F}_p$. Note that H, and its Serre-dual H^t are p-divisible formal groups of dimension one. An elliptic curve E over an algebraically closed field in characteristic p is supersingular if and only if $\hat{E} = H$. We recall:

2.0 Theorem / Definition: *Let $g \geq 2$. Let X be an abelian variety over k of dimension g. Fix a supersingular elliptic curve E over k. Then:*

$$\hat{X} \sim H^g \Leftrightarrow X \sim E^g$$

(*here "$\sim$" stands for isogeny;* definition*: if the above property is satisfied the abelian variety X is called* supersingular*). Furthermore:*

$$\hat{X} \simeq H^g \Leftrightarrow X \simeq E^g$$

(definition*: and such an abelian variety is called* superspecial*).*

These facts result from work by Eichler [5], work by Deligne published by Shioda [34], and furthermore see [30], Th. 4.2, and [31], Th. 2. Note that the definition above of "superspecial" differs from that in [30]. Note that "p-rank is zero" is weaker than "supersingular" for $g \geq 3$, "supersingular" is weaker than "superspecial" for $g \geq 2$.

We denote by $\mathcal{S} = \mathcal{S}_g \subset \mathcal{A}_{g,1} \otimes \mathbf{F}_p$ the supersingular locus.

There are precise conjectures about the dimension of $\mathcal{S}_g$, it should be equal to $[g^2/4]$, cf. [26], p.616, about the number of components, and about its structure. These have been proven for $g \leq 4$. We need the descriptions of $\mathcal{S}_3$ and of $\mathcal{S}_3$, which we take from [16].

We write α_p for the local group scheme of rank p in characteristic p such that its dual is also local (cf. [29], p. I.2-11). For a group scheme or a formal group G in characteristic p we write

$$a(G) = \dim_k \text{ Hom } (\alpha_p, G).$$

Now we fix $g = 3$, and we describe $\mathcal{S} = \mathcal{S}_3$.

Let τ be the (unique) principal quasi-polarization on H, and let μ be the quasi-polarization on H^3 obtained by pulling back

$$\text{diag}_3(\tau) = (\tau, \tau, \tau) : H^3 \to (H^t)^3 \text{ by } F : H^3 \to H^3.$$

Clearly

$$\ker(\mu) = (H[F^2])^3 = (H[p])^3$$

(we write $H[F]$ for the kernel of $F : H \to H$, etc.). We note that the above property characterizes μ. We denote by $\mathcal{F} = \mathcal{F}(\mu)$ (analogous to [16]), the moduli space of polarized flag type quotients (pftq) starting at (H^3, μ). The structure of $\mathcal{F}$ is described as follows. Take the Fermat curve of degree $p+1$:

$$C := \mathcal{Z}(X^{p+1} + Y^{p+1} + Z^{p+1}) \subset \mathbf{P}^2,$$

and

$$\pi : \mathcal{F} \to C$$

is a $\mathbf{P}^1$-fibration, actually $\mathcal{F} = \text{Proj}(\mathcal{O}_C \oplus \mathcal{O}_C(1))$. This fibration has a section

$$\begin{array}{ccc} \mathcal{F} & \stackrel{\pi}{\to} & C \\ \cup & & \\ T & \nearrow\sim & \end{array}$$

which is constructed as follows. A point of $\mathcal{F}$ is a sequence of formal groups with quasi-polarizations:

$$(H^3, \mu) = (Y_2, \mu_2) \to (Y_1, \mu_1) \to (Y_0, \mu_0),$$

with

$$\text{Ker } (\varphi_2 : Y_2 \to Y_1) = \alpha_p^2,$$

$$\text{Ker } (\varphi_1 : Y_1 \to Y_0) = \alpha_p.$$

To the point $\varphi_2 \in C$ we associate

$$(\varphi_1, \varphi_2) \in T \subset \mathcal{F}$$

as follows:

$$\begin{array}{rccc} H^3 = & Y_2 & \overset{\varphi_2}{\to} & Y_1 \\ & \cup & & \uparrow \\ \alpha_p^3 = & Y_2[F] & \to & N = \alpha_p, \end{array}$$

the restriction of φ_2 to $Y_2[F]$ gives

$$\varphi_2(Y_2[F]) = N = \alpha_p,$$

and we *choose*

$$\varphi_1 : Y_1 \to Y_0 = Y_1/N.$$

Clearly

$$(Y_2 \to Y_1 \to Y_0) = (F : Y_2 = H^3 \to H^3).$$

We conclude that for every $(\varphi_1, \varphi_2) \in T$ the end of the pftq has $a(Y_0) = 3$.

2.1 Lemma: *Let* $C := \mathcal{Z}(X^{p+1} + Y^{p+1} + Z^{p+1})$, *let* $a, b, c \in \mathbf{F}_{p^2}$ *not all zero, and let* $R := (x : y : z) \in \mathbf{P}^2$ *be a point in the intersection of* C *and of the line* $\mathcal{Z}(aX + bY + cZ)$. *Then*

$$R \in \mathbf{P}^2(\mathbf{F}_{p^2}).$$

Proof: (Thanks to Schloss Ringberg, and to H.W. Lenstra we can give the following elegant proof): Let k be an algebraic closure of $\mathbf{F}_p$, and let W_1 and W_2 be the linear spaces

$$W_1 = k.(a, b, c) + k.(x^p, y^p, z^p) \subset k^3,$$

$$W_2 = k.(x, y, z) + k.(x^{p^2}, y^{p^2}, z^{p^2}) \subset k^3.$$

Because

$$x^{p+1} + y^{p+1} + z^{p+1} = 0 = ax + by + cz,$$

and

$$a, b, c \in \mathbf{F}_{p^2},$$

these spaces are perpendicular

$$W_1 \perp W_2$$

under the usual inner product. Hence

$$\dim W_1 \leq 1 \quad \text{or} \quad \dim W_2 \leq 1.$$

Both cases imply $(x : y : z) \in \mathbf{P}^2(\mathbf{F}_{p^2})$, which proves the lemma.

There exists a principal quasi-polarization, which we call τ, on H. Note that $p^s.\tau : H \to H^t$ is a quasi-polarization on H with $\deg(p^s\tau) = p^{2s}$. Let

$$0 \to \alpha_p \to H^2 \overset{\psi}{\to} H^2 \to 0$$

be an exact sequence given by a "very good direction" (cf. [21], pp. 138/139, cf. [15], p.119), i.e. if $\tau_2 = \operatorname{diag}_2(\tau)$ on H^2, and

$$\rho := \psi^*(\tau_2), \text{ then } \operatorname{Ker}(\rho) = \alpha_p \oplus \alpha_p = H^2[F].$$

2.2 Lemma (K.Li): *Let μ be a quasi-polarization on the superspecial p-divisible group $G = H^g$. Then there exists integers $s_1, \ldots, s_a, m_1, \ldots, m_b$, with $a + 2b = g$, and an isomorphism*

$$G \cong H^a \oplus (H^2)^b$$

so that μ is given by

$$\mu \cong (p^{s_1}\tau, \ldots, p^{s_a}\tau, p^{m_1}\rho, \ldots, p^{m_b}\rho).$$

(A proof will be published soon).

2.3 Proposition: *Let $\pi : \mathcal{F} \to C$ be as above, let $\mathcal{Y} = \mathcal{Y}_0 \to \mathcal{F}$ be the universal family, then:*

(1) $x \in T \Leftrightarrow a(\mathcal{Y}_x) = 3$,

(2) $\pi(x) \in C(\mathbf{F}_{p^2}) \Leftrightarrow a(\mathcal{Y}_x) \geq 2$,

(3) $x \notin T, \quad \pi(x) \notin C(\mathbf{F}_{p^2}) \Leftrightarrow a(\mathcal{Y}_x) = 1$,

(4) $\forall R \in C(\mathbf{F}_{p^2}), \quad \#\{x \in \mathcal{F} \mid \pi(x) = R, \quad a(\mathcal{Y}_x) = 3\} - p^2 + 1.$

Proof: We have seen that (1) holds by the construction of T. Starting with a pftq with $\varphi_{g-1} : Y_{g-1} \to Y_{g-2}$, we know it can end with $a(Y_0) = 1$ if and only if $R = \varphi_{g-1} \in \mathbf{P}^{g-1}$ is on a hyperplane defined by an equation with coefficients in $\mathbf{F}_{p^2}$ (cf. [6a] pp. 29/30, this corrects the statement on [26], p. 609 in case of Flag_{g-2}). By Lemma (2.1) this is the same as $R \in \mathbf{P}(\mathbf{F}_{p^2})$! This shows (2), and it proves that $a(\mathcal{Y}_x) = 1$ implies $\pi(x) \notin C(\mathbf{F}_{p^2})$. By [26], p. 609 for such a point $S = \pi(x)$, with $S \notin C(\mathbf{F}_{p^2})$ there is exactly one $(\varphi_2, \varphi_1) = x \in \pi^{-1}(S)$ with $a(\mathcal{Y}_x) = 3$ (and this is $x \in T$). This shows (3). Now consider $R \in C(\mathbf{F}_{p^2})$. Then we can choose coordinates such that

$$(Y_1, \mu_1) = (H^2, \rho) \times (H, \tau),$$

with

$$\mathrm{Ker}\ (\rho) = (E[F])^2, \text{ and where } \tau \text{ is an } \cong .$$

Then

$$\varphi_1 : Y_1 \to Y_0$$

gives $a(Y_0) = 3$ iff $\mathrm{Ker}\ (\varphi_1) \subset \mathrm{Ker}\ (\rho)$ and

$$\varphi_1 \in \mathbf{P}^1(\mathbf{F}_{p^2})$$

(i.e. this is a "good direction"). This finishes the proof of the proposition.

Description of the supersingular locus $\mathcal{S} = \mathcal{S}_3 \subset \mathcal{A}_{3,1} \otimes \mathbf{F}_p$ taken from [16], Section 6:

2.4 Proposition: *We fix a supersingular elliptic curve E.*
(1) For each principal polarization λ on E^3 there is a finite group $G_\lambda \subset \mathrm{Aut}(\mathcal{F})$, an irreducible component $W_\lambda \subset \mathcal{S}$ of $\mathcal{S}$, and morphisms

$$\mathcal{F} \xrightarrow{\beta} \mathcal{F}/G_\lambda \xrightarrow{\psi} W_\lambda$$

such that ψ is birational, $\beta(T)$ is the only curve which is blown down under ψ and its image is the point:

$$\psi(\beta(T)) = [(E^3, \lambda)].$$

(2) Every component of $\mathcal{S}$ has dimension equal to two, it is obtained in this way, and the number of components of $\mathcal{S}$ equals the class number $H_3(1,p)$ (as studied by Hashimoto in [12]).

Proof: Let μ' be the polarization $F^*(\lambda)$ on $Y_2 = (E^{(1/p)})^3 \cong E^3$, and let μ be the quasi-polarization induced on the formal group H^3 of Y_2. By Lemma (2.2) this is in the form used in the construction of $\mathcal{F}$ recorded above. Thus we use that construction for pftq starting from (H^3, μ'). We write $G_\lambda = \mathrm{Aut}(Y_2, \mu')$. Let $T' \subset \mathcal{F}$ be an irreducible curve, $T' \neq T$. If $\pi(T') = C$, by (2.3.3) there is a point $x \in T'$ with $a(\mathcal{Y}_x) = 1$, hence under the map into $\mathcal{A}_{3,1} \otimes \mathsf{F}_p$ it is not blown down. If T' is a fibre of $\pi : \mathcal{F} \to C$ the moduli of $\mathcal{Y}_x$ vary if x varies in T', hence T' is not blown down. For the rest all statements of the proposition are contained in [16].

2.5 Remark:

$$\#(C(\mathsf{F}_{p^2})) = 3(p+1) + (p^2-1)(p+1).$$

2.6 Although we do not need it in the sequel of this paper, we describe the way in which components of $\mathcal{S}_3$ fit together.

We write $\mathcal{S}_g(\geq i)$ for the locus of supersingular abelian varieties with $a(X) \geq i$. We remark that for any $g \geq 2$ the locus $\mathcal{S}_g(\geq (g-1))$ is connected (T. Ekedahl, unpublished). Note that the stratum $\mathcal{S}_3(\geq 2)$ is one-dimensional. It consist of images of fibres in $\mathcal{F}$ above points in $C(\mathcal{F}_{p^2})$. We know that components of $\mathcal{S} = \mathcal{S}_3$ do not intersect at points with $a(X) = 1$ (this holds for any g: at such points the pftq is unique, or apply deformation theory). Thus components of $\mathcal{S}$ intersect in a subset of $\mathcal{S}(\geq 2)$. It can be seen (in the situation of a fine moduli scheme), that if we move along a component of $\mathcal{S}(\geq 2)$, which is a P^1, we meet p^2+1 singular points which are of the form (E^3, λ), which determine each a component of $\mathcal{S}$. These components (in the fine situation) all contain this P^1. Now take a component W_λ (still in the fine situation), take two different components of $\mathcal{S}(\geq 2)$ (i.e. two of such fibres in $\mathcal{F}$), mapped into W_λ , hence intersecting at the singular point of W_λ. On each of these lines we choose a singular point of other components, say of W_γ and of W_δ. In this way it may happen that two different components intersect at a point (and it seems reasonable to expect that such cases also occur in the coarse situation). In conclusion: the intersection of two different components of $\mathcal{S}$ is a union of components of $\mathcal{S}(\geq 2)$ and of points with $a(X) = 3$.

Now we express Moret-Bailly families (as in [21], pp. 138/139) from the terminology of "very good directions" (as in [15], p.199) explicitly in the infinitesimal deformation theory for $g = 2$.

2.7 Proposition: *Let $\tau_2 = (\tau, \tau)$ be the principal quasi-polarization of $H^2 = H \times H$. The supersingular locus in the formal deformation space of $(G = H \times H, \tau_2)$ consists of all directions*

$$\begin{pmatrix} -\zeta^p & 1 \\ 1 & \zeta \end{pmatrix} \quad \textit{with } \zeta^{p+1} = -1.$$

Proof: The local deformation space of (G, τ_2) is three-dimensional. With the help of the Dieudonné-module theory (in the form of [25], Lemma 0.2), we see that (G, τ_2) corresponds to:

$$\begin{aligned} F\xi_1 &= \eta_1 \\ F\xi_2 &= \eta_2 \\ \eta_1 &= V\xi_1 \\ \eta_2 &= V\xi_2, \end{aligned}$$

where the quasi-polarization is given by $<\xi_1, \eta_1>= \varepsilon =<\xi_2, \eta_2>$, with $\varepsilon \neq 0$. Then a deformation is given by [25], p. 242

$$\begin{aligned} FX_1 &= Y_1 + T_{11}X_1 + T_{12}X_2 \\ FX_2 &= Y_2 + T_{21}X_1 + T_{22}X_2 \\ Y_1 &= VX_1 \\ Y_2 &= VX_2 \end{aligned}$$

with $T_{12} = T_{21}$. This is the universal deformation space of (H^2, τ_2) (because that is also smooth on 3 parameters). We see that the Hasse-Witt matrix is:

$$A = \begin{pmatrix} T_{11} & T_{12} \\ T_{12} & T_{22} \end{pmatrix}.$$

A solution of $A^{(p)}.A = 0$ is of the form

$$T_{1,2} = \begin{array}{rl} T_{11} = & -\zeta^p.t \\ T_{21} = & t \\ T_{22} = & \zeta.t \end{array} \qquad \text{with } \zeta^{p+1} = -1.$$

For $g = 2$ this gives the supersingular locus, because in that case "p-rank is zero" implies supersingular. Thus the $p + 1$ branches of the supersingular moduli space for $g = 2$ are given in this way.

2.8 Corollary: *In the formal deformation space of (H^2, τ_2) the tangent lines to the $p + 1$ components of the supersingular locus span the tangent space of the deformation space.*

Proof: In case $p = 2$ the α can be taken from $\mathbb{F}_4^* = \{1, \alpha, \alpha^2\}$. The matrix

$$\begin{pmatrix} 1 & 1 & 1 \\ \alpha^2 & 1 & \alpha \\ \alpha & 1 & \alpha^2 \end{pmatrix}$$

has determinant $\alpha^2 + \alpha = 1$, hence the row vectors are linearly independent.

In case $p \geq 3$, we choose ζ so that $\zeta^{p+1} = -1$ and $\zeta^{p-3} \neq -1$. The matrix

$$\begin{pmatrix} 1 & \zeta & -\zeta^p \\ 1 & \zeta^p & -\zeta \\ 1 & -\zeta & \zeta^p \end{pmatrix}$$

has determinant $-2\zeta^2(\zeta^{p-3} + 1) \neq 0$, and the corollary is proved.

2.9 Corollary: *Let E be a supersingular curve, let n be a large integer not divisible by p, let λ be a principal polarization on $X = E^3$, and let α be a level-n-structure on X. Let*

$$\mathcal{S} = \mathcal{S}_3^{(n)} \subset \mathcal{A}_{3,1}^{(n)} \otimes \mathbb{F}_p$$

be the supersingular locus in this fine moduli scheme, and let $W = W_{\lambda,\alpha}$ be the component of the supersingular locus which has a singular point at $[(X, \lambda, \alpha)]$. Note that the map $\mathcal{F} \to W$ is surjective, and it blows down the curve T to that singular point. The tangent space of W at this point equals the tangent space, of dimension 6, to $\mathcal{A}_{3,1}^{(n)} \otimes \mathbb{F}_p$ at that point.

Proof: By (2.2) we can write

$$(\hat{X}, \hat{\lambda}) = (H^3, \tau_3).$$

We note that the components of $\mathcal{S}(\geq 2)$ inside W through the point $x = (X, \lambda, \alpha)$ come from fibres of $\pi : \mathcal{F} \to C$. We start with $H^3 = (H_1 \times H_2) \times H_3$. The supersingular deformations of $(H^2 = H_1 \times H_2, \tau_2)$ span by (2.8) the 3-dimensional tangent space. In the deformation space this gives the directions

$$\begin{pmatrix} -\zeta^p & 1 & 0 \\ 1 & \zeta & 0 \\ 0 & 0 & 0 \end{pmatrix} \qquad (1,2).$$

In the same way we can take the deformation space of $(H^2 = H_1 \times H_3, \tau_2)$, and obtain the directions

$$\begin{pmatrix} -\zeta^p & - & 1 \\ 0 & 0 & 0 \\ 1 & 0 & \zeta \end{pmatrix} \qquad (1,3),$$

and by taking $H^2 = H_2 \times H_3$ we obtain the vectors

$$\begin{pmatrix} 0 & 0 & 0 \\ 0 & -\zeta^p & 1 \\ 0 & 1 & \zeta \end{pmatrix} \qquad (2,3).$$

In the deformation space with coordinates $T_{11}, T_{12}, T_{13}, T_{22}, T_{23}, T_{33}$ the tangents in (1,2) span T_{11}, T_{12}, T_{22}. In the same way we see that (1,3) span T_{11}, T_{13}, T_{33}, and (2,3) span T_{22}, T_{23}, T_{33}. Thus these $3(p+1)$ fibres on $\mathcal{F}$ gives curves on W which give tangent vectors to W which together span the total tangent space of the formal moduli space of (H^3, τ_3). This proves the proposition.

3 The proof in case $p = 2$.

3.1 Example: *Let* char $(k) = 2$, *and let* C *be the plane quartic curve*

$$\mathcal{Z}(T^4 + TU^3 + X^3U) =: C \subset \mathbf{P}^2.$$

This curve C *is:*
non-singular
non-hyperelliptic, and
supersingular.

Proof: The zeros of the partial derivatives have no common points if char $(k) \neq 3$, hence in our case we have a non-singular plane curve.

Any non-singular plane quartic is non-hyperelliptic, because the canonical series is given by the intersections with lines in the plane, hence it is non-compound.

There are many ways to see that C is supersingular in characteristic 2. We indicate one. Let $G := \mathbf{F}_4^+$. Clearly

$$G \hookrightarrow \mathrm{Aut}(C), \qquad \delta \mapsto \varphi_\delta$$

by:

$$\left.\begin{array}{l} T \mapsto T + \delta U \\ X \mapsto X \\ U \mapsto U \end{array}\right\} \text{ for } \delta \in \mathbf{F}_4.$$

If moreover $\delta \neq 0$ this is an involution on C with quotient

$$C \to C/ < \varphi_\delta >= E_\delta := \mathcal{Z}(Y^2Z + \delta^2YZ^2 + X^3)$$

$$(x, t) \mapsto (x, y = t^2 + \delta t) \text{ in affine coordinates.}$$

In other words: the involution $(t \mapsto t + \delta, x \mapsto x)$ on $K = k(t, u)$, with $t^4 + t + x^3 = 0$ has as fixed field

$$k(t, x) \supset L_\delta := k(y_\delta = t^2 + \delta t, x)$$

with $y = y_\delta$ and $y^2 + \delta^2 y + x^3 = 0$. Let $\mathbf{F}_4^* = \{\beta, \gamma, \delta\}$, then $\varphi_\delta : K \to K$ is the identity on L_δ, and

$$\varphi_\delta(t^2 + \beta t) = t^2 + \beta t + (\beta + \delta)\delta,$$

hence φ_δ maps L_β and L_γ both to itself, and

$$\varphi_\delta \mid L_\beta \qquad \text{and} \qquad \varphi_\delta \mid L_\gamma$$

are non-trivial involutions. This shows that the images of

$$X := \mathrm{Jac}(C) \leftarrow E_\beta, \qquad X \leftarrow E_\gamma, \qquad X \leftarrow E_\delta$$

are elliptic curves in X, each not contained in the sum of the other two. Thus

$$X \twoheadleftarrow E_\beta \times E_\gamma \times E_\delta$$

is surjective. Moreover each of the elliptic curves is supersingular (and they are isomorphic), which proves the claim.

3.2 Remark: On the curve E there is a regular differential $\omega = dx$; thus the tangent spaces of the three elliptic curves have the same image in X, thus $a(X) = 1$.

3.3 Remark: The proof of the fact that C is supersingular can also be given as follows: it is easy to see that $C \subset \mathbf{P}^2$ has exactly one bitangent, namely the line $\mathcal{Z}(U)$ (and the point $(x = 1 : t = 0 : u = 0)$ is the hyperflex). From this one deduces that the 2-rank of $X = \mathrm{Jac}(C)$ is zero. The non-trivial morphism $X \to E$ shows that the formal isogeny type $(1,2) + (2,1)$ is excluded, hence X is supersingular. The fact that the 2-rank of X is zero can also be seen using [35], Th. 4.1, for the covering $C \to E$.

One can also proceed as follows. The curve C is defined over $\mathbf{F}_2$, and the number of points of C over $\mathbf{F}_q$ equals:

$$N_1 = 3, \quad N_2 = 5, \quad N_3 = 9 \text{ with } \quad N_i = \#C(\mathbf{F}_{2^i}).$$

By

$$N_i = 1 + 2^i - \sum_{j=1}^{6} \lambda_j^i,$$

we see that the characteristic polynomial of Frobenius is

$$P = \lambda^6 - 8 = (\lambda^2 - 2)(\lambda^4 + 2\lambda^2 + 4),$$

and we see that $\mathrm{Jac}(C)$ up to isogeny over $\mathbf{F}_4$ splits as a product of three supersingular elliptic curves.

3.4 Proof of (1.10) $\mathrm{char}(k) = 2$: By a class number computation by

Hashimoto we know that $\mathcal{S}$ is irreducible in characteristic 2 (cf. [12], cf. [16], Corollary 6.8). By the previous example we see that $\mathcal{S} \not\subset \mathcal{B}$ in characteristic two, hence we have proved (1.10), $p = 2$.

3.5 Remark: In (5.6.1) we shall see it is not surprising that we have in (3.1) a curve with a V_4-action with three elliptic quotients.

4 Good hyperelliptic curves

4.1 Lemma: *Let k (be algebraically closed) and*

$$\mathrm{char}(k) \neq 2.$$

If C is a singular good hyperelliptic curve, $g(C) = 3$, then all components of C are non-rational, i.e. the Type (0) as in (1.2) is excluded in char $\neq 2$, *and all singular points are (quasi-)Weierstrass on every component.*

4.2 Lemma: *Let*

$$\mathrm{char}(k) = 2.$$

If C is a singular good hyperelliptic curve, $g(C) = 3$, then all singular points are (quasi-)Weierstrass on every component, in particular in case of Type (1) the curve E_2 is ordinary, and, conversely, every good curve of Type (2), (1) or (0) with this property for the singular points is hyperelliptic.

Proof of (4.1) and (4.2): If char$(k) = 0$, by [10] we have a characterization of all specializations of smooth hyperelliptic curves of genus 3: take the covering degree $= 2$, take the number of branch points $= 8$. Thus the curves as in (4.1) are precisely the good singular hyperelliptic curves in characteristic zero.

Note that Type (0) cannot appear as a specialization in char $\neq 2$ because the specialization σ_s of the hyperelliptic involution would have 9 non-singular fixed points.

Suppose that C_0 is a curve over a field of positive characteristic of Type (2) or (1) as indicated in the lemmas. Lift the curve and the (quasi-) Weierstrass points to characteristic zero, there we obtain a singular curve of the same type, which is hyperelliptic by the first part of the proof. Hence its specialization C_0 is hyperelliptic.

Finally, suppose C_0 is of Type (0) in char $= 2$. Choose liftings E'_1, E'_2, E'_3 in characteristic zero of these elliptic tails, i.e.

$$k \overset{\rho}{\leftarrow} R \hookrightarrow \mathbb{C},$$

and elliptic curves E_i' with $E_i' \otimes k = E_i$, points $P, Q \in E_2'(R)$, with $P \neq Q$ and $[P - Q] \in \mathrm{Pic}(E_2')[2]$, such that $\rho(P) = \rho(Q)$. The stable curve C over $K =$ Fract(R) obtained by glueing E_1' and E_3' to E_2' at P and Q is hyperelliptic, hence specializes to a hyperelliptic curve by definition. Moreover Pic(C) has good reduction, hence C specializes to a good hyperelliptic curve C_0. Because P and Q coincide after specialization, Type (1) is excluded, hence C_0 is of Type (0). Note that we can construct this example with E_2 either ordinary or supersingular. This ends the proof of the lemmas.

4.3 Remark: In [4], Appendix 2, Weierstrass points on stable curves in characteristic zero are studied.

4.4 Remark: We should like to have a more conceptual proof of (4.1). We need a generalization of the concept "admissable covering". In [1], 0.3, we find a notion of "pseudo-revêtement" (the case of unramified double coverings in char $\neq 2$). This was generalized in [10], §4, to the notion of admissible coverings of curves in characteristic zero. We see in [17], p. 179, Th. 2.7, that the notion of "n-pointed curve of genus g", for $n \geq 0, g \geq 0$ and $2g - 2 + n > 0$, works in all characteristics. It seems plausible that Theorem 4 of [10] should hold, with the same definition of admissible covering over $\mathbf{Z}[1/(k!)]$. If so, this would give another proof of (4.1).

However, it is not so clear how to generalize the notion of coverings of (stable) curves to the case of curves over fields of characteristic p, where p is not bigger than the degree of the covering.

For a good curve C, a level-n-structure, with $(n, \mathrm{char}(k)) = 1$, is an isomorphism

$$\mathrm{Jac}(C)[n] \cong (\mathbf{Z}/n)^{2g}$$

(we do not need symplectic structures here). We denote by $\mathcal{M}_g^{\sim(n)}$ the moduli scheme over $\mathbf{Z}[1/n]$ of good curves of genus g in characteristics prime to n with a level-n-structure. By $\mathcal{H}_g^{(n)}$ the same for good hyperelliptic curves.

4.5 Proposition: *For $n \geq 3$ the morphism*

$$T := \mathcal{H}_g^{(n)} \otimes \mathbf{Z}[1/2n] \to S := \mathrm{Spec}(\mathbf{Z}[1/2n])$$

is smooth.

Proof: Because $n \geq 3$ we work with fine moduli schemes, and over T we have a universal curve $\mathcal{C} \to T$, and an involution $\sigma \in \mathrm{Aut}(\mathcal{C}.T)$. By Lemma (4.1), and because char $\neq 2$, this involution σ has fixed points exactly in a divisor $W' \subset \mathcal{C}$ which is étale and of degree 8 over T, and in the double points of the fibres. Let $W \to W'$ be the normalization in the Galois closure of $W' \to T$. Then we have a morphism

$$\psi : W \to \mathcal{M}_{0,8} \otimes_{\mathbf{Z}} \mathbf{Z}[1/2n],$$

where $\mathcal{M}_{0,8}$ is the moduli space of stable curves of arithmetic genus zero with 8 points in the sense of [17]. Note that $\mathcal{M}_{0,8}$ is smooth over $\mathbf{Z}$ (cf. [17], Th. 2.7). A geometric point of $\mathcal{M}_{0,8} \otimes_{\mathbf{Z}} \mathbf{Z}[1/2n]$ corresponds to a curve D with components isomorphic to $\mathbf{P}^1$ and marked points as in the Types (0), (1), (2) of (1.2) and of (4.1). There is exactly one curve C with a double covering $C \to D$ ramifying above the marked points and in the singular points. Thus $\psi(W) := U$ is dense in $\mathcal{M}_{0,8}$, thus U is smooth over $\mathbf{Z}$. For any $u \in U$, and for any formal neighbourhood $U' \subset U$ of u in U, there exists a covering $\mathcal{C}' \to \mathcal{D}'$ of the desired type, as follows by the construction as in [10], pp. 58-60. We conclude that the morphisms

$$T \leftarrow W' \leftarrow W \to \mathcal{M}_{0,8}$$

are étale, hence formal isomorphisms at each point, thus T is smooth over S. This proves the proposition.

4.6 Proposition: *Let $n \geq 3$, consider the morphism*

$$\psi : \mathcal{H}_3^{(n)} \otimes \mathbf{Z}[1/2n] \to \mathcal{B}_3^{(n)} \otimes \mathbf{Z}[1/2n].$$

1) For any geometric point y of $\mathcal{H}_3^{(n)} \otimes \mathbf{Z}[1/2n]$ the differential $d\psi$ is injective at y.
2) For any geometric point x of $\mathcal{B}_3^{(n)} \otimes \mathbf{Z}[1/2n] =: B$, with $\mathrm{char}(x) = p$, the formal completion of B at x consists of a finite number of branches each formally smooth over $\mathbf{Z}_p$.

Proof: A fiber of $\mathcal{M}_3^{(n)} \to \mathcal{A}_{3,1}^{(n)}$ is locally isomorphic with a smooth curve of genus 2 in case of Type (2), or a curve of genus 1+ base point in case of Type (1), and in case of Type (3) it is locally an isomorphism. The subscheme of Weierstrass points on a hyperelliptic curve is reduced in characteristic $\neq 2$, and the same holds for $E[2]$ in case of an elliptic curve in characteristic $\neq 2$. Thus $\mathcal{H}_3^{(n)}$ is transversal to the mentioned fibers and the first statement is proved. The second follows from the first. This finishes the proof of the proposition.

4.7 Corollary: *(2.9) and (4.7.2) give a proof for (1.10) in case* $p \geq 3$.

Indeed, the tangent space to $W_\lambda \subset \mathcal{A}_{3,1}^{(n)}$ has dimension 6 at the point $x = [X, \lambda, \alpha] \in W_\lambda$ and each component of the formal completion at x of $\mathcal{B}_3^{(n)} \otimes \mathbf{F}_p$ with $p \geq 3$ is formally smooth, hence has a tangent space of dimension 5. Thus

$$W_\lambda \not\subset \mathcal{B}_3^{(n)} \otimes \mathbf{F}_p.$$

This finishes the proof of (1.10). Thus the proof of Theorem (1.12) is complete.

5 Existence of some supersingular curves: curves with involution.

In this section we show that for $p = 2$ there do not exist (non-singular) hyperelliptic supersingular curves of genus $g = 3$, while for every $p \geq 3$ such curves do exist.

For a curve C in characteristic p we write

$$a(C) = a(\mathrm{Jac}(C)) = \dim \mathrm{Hom}(\alpha_p, \mathrm{Jac}(C)).$$

First we show that from a result of Ekedahl it follows that for every $p \geq 3$ there exists a non-singular hyperelliptic supersingular curve with $g = 3$ (and in the rest of this section we shall make this result more precise). In fact, if $n = g = 3$, and $p > 2$, by [6], Prop. 7.5 (iii), page 163, there exists an indecomposable principally polarized supersingular abelian variety $(X = E^3, \lambda)$. Let C be the good curve with $\mathrm{Jac}(C) = (X, \lambda)$ (cf. [28]). As the polarization is indecomposable, this curve is irreducible (and non-singular). We see that $a(C) = 3$. If C is hyperelliptic, we are done. If C is not hyperelliptic, consider the component $W_\lambda = V \subset \mathcal{S}$ defined by (X, λ) as in (2.4).

5.1 Proposition: *Let V be an irreducible closed set in $\mathcal{A}_{3,1} \otimes \mathsf{F}_p$ contained in the* (p-rank=0)-*locus with* $\dim V > 0$. *Then*

$$V \cap \mathcal{B} \neq \emptyset$$

(to be on the safe side, we claim and show this only for $p > 7$).

Proof: If $q > 7$, then the prime q does not divide the order of an automorphism of a good curve of genus 3 (cf. [13], Th. 1). Thus the Chow-ring with $\mathbb{Q}$-coefficients in characteristic $p > 7$ of $\mathcal{M}_3^{\sim}$ and of $\bar{\mathcal{M}}_3$ can be described as in the case of characteristic zero, and we see there exists integers $c > 0, d, e$ such that $c\bar{H} + d\delta_0 + e\delta_1$ is ample on $\bar{\mathcal{M}}_3$ (cf. [4], p.67; cf. [7], p.31; cf. [24], last section, or [10], p.24). Because the (p-rank zero)-locus does not meet the boundary of $\mathcal{A}_{g,1} \otimes \mathsf{F}_p$ (cf. [30], pp. 98/99), it follows that V meets $\mathcal{B}$.

For $3 \leq p \leq 7$ we know there exist such hyperelliptic curves (cf. [18], p.196, "Type 5"), thus we have proved:

5.2 Theorem: *For every $p \geq 3$ there exists an irreducible non-singular hyperelliptic curve of genus 3 with p-rank equal to zero.*

In order to derive more precise statements we study curves with involutions.

5.3 Lemma: *Let $\tau \in \operatorname{Aut}(C)$, consider the Jacobian $X = \operatorname{Pic}(C)$; let π be the natural projection*

$$\pi : C \to D := C/ < \tau >,$$

and let

$$\tau^* =: t \in \operatorname{Aut}(X), \quad \pi_* : X \to Y := \operatorname{Pic}(D), \quad \pi^* : Y \to X$$

be the induced actions. Then

$$g(D) = \dim(Tr(t)(X)) \quad \text{with} \begin{cases} n = \text{ order } (\tau) \\ Tr(t) = t^{n-1} + \ldots + t + 1. \end{cases}$$

Proof: Clearly

$$\pi_* \pi^* = \deg(\pi).1_y = n,$$

and

$$\pi^* \pi_* = Tr(t) \text{ on } X,$$

and the result follows.

5.4 Proposition: *Let $\tau \in \operatorname{Aut}(C)$, $g(C) = 3$, and suppose τ is an involution*

$$\tau^2 = id \neq \tau.$$

(i) *If τ has no fixed points on C, then C is hyperelliptic.*

(ii) *If C is hyperelliptic, with $C/ < \sigma >= \mathbf{P}^1$ and $\tau \neq \sigma$ then either τ or $\sigma\tau$ has no fixed points on C.*

Proof: If τ has no fixed points on C, then $D := C/ < \tau >$ has genus 2 by the Zeuthen-Hurwitz formula, hence D is hyperelliptic. Moreover the covering $C \to D$ is given by class field theory:

$$\begin{array}{ccc} C & \hookrightarrow & Z \\ \downarrow & & \downarrow \\ C/<\tau>=D & \hookrightarrow & Y=\mathrm{Jac}(D) \end{array} \quad , \xi=[Z\to Y]\in \mathrm{Ext}(Y,\mathbf{Z}/2),$$

(cf. [33], VI.8). The hyperelliptic involution β' on D gives -1_y on $Y = \mathrm{Jac}(D)$, hence it leaves this class $\xi \in \mathrm{Ext}(Y, \mathbf{Z}/2)$ invariant, thus there exists an involution $\beta \in \mathrm{Aut}(C)$ which lifts β'. Note that

$$\mathrm{Jac}(C) = X \sim \mathrm{Ker}\,(X \to Y) \oplus \mathrm{Jac}(D),$$

further τ^* is -1_E on the connected component E of Ker $(X \to Y)$, and σ^* is -1_y on $Y = \mathrm{Jac}(D)$. Because $g(C) = 3$ we have $\dim(E) = 1$, and hence either σ^* or $(\sigma\tau)^*$ is -1_E on E. This proves (i).

Suppose $C/ < \sigma >= \mathbf{P}^1$, and $\tau \neq \delta$. If τ has a fixed point then $g(C/ < \tau >) = 1$, by the previous lemma we see that $g < C/ < \tau\delta >) = 2$, and we are done.

For further reference we state:

5.5 Let C be an algebraic curve (complete, non-singular) over a field K, and let (X, λ) be the canonically polarized Jacobian variety. We obtain:

$$\mathrm{Aut}(C) \to \mathrm{Aut}(X, \lambda).$$

(5.5.h) If C is hyperelliptic $\mathrm{Aut}(C) \cong \mathrm{Aut}(X, \lambda)$;
(5.5.nh) if C is non-hyperelliptic $\mathrm{Aut}(C) \times \{\pm 1\} \cong \mathrm{Aut}(X, \lambda)$
(cf. [20], pp.790, 792).

5.6 Theorem ($p = 2$): *Suppose* $\mathrm{char}(k) = 2$.
(5.6.1) *If (X, λ) is a principally polarized supersingular abelian variety of dimension* 3, *then*

$$(\mathbf{Z}/2)^3 \hookrightarrow \mathrm{Aut}(X, \lambda).$$

(5.6.2) *There does* not *exist a supersingular, irreducible hyperelliptic curve of genus* 3 *in characteristic* $p = 2$.

Proof: Let (E, ρ) be a supersingular principally polarized elliptic curve, and $(Y, \mu) = (E, \rho)^3$. Note that the automorphisms $(\pm 1, \pm 1, \pm 1)$ act on $(E, \rho)^3$. In $\mathrm{char}(k) = p = 2$ over an algebraically closed field there is only one principal polarization on E^3 (cf. [6], Prop. 7.5(iii); cf. [12], p.493). For every principally polarized supersingular abelian threefold (X, λ) there exists a morphism

$$(Y = E^3, \mu) \to (X, \lambda)$$

(cf. Prop. 2.4) with

$$\mathrm{Ker}\,(E^3 \to X) \subset E^3[F^2] = E^3[p].$$

In case $p = 2$, we see that -1 acts trivial on $E[2]$, and we conclude:

$$(\mathbb{Z}/2)^3 \hookrightarrow \mathrm{Aut}(X, \lambda),$$

this shows the first claim.

Suppose (X, λ) would be the Jacobian of an irreducible hyperelliptic curve C. Then $(\mathbb{Z}/2)^3 \hookrightarrow \mathrm{Aut}(C)$ by (5.5.h), and by (5.4) we conclude there exists an automorphism $\tau \in \mathrm{Aut}(C)$ without fixed points on C. We would obtain

$$C \to C/ < \tau >= D, \text{ genus}(D) = 2,$$

an unramified covering of degree $p = 2$. However $X = \mathrm{Jac}(C)$ maps onto $Y = \mathrm{Jac}(D)$, hence the p-rank (Y), with $p = 2$, is zero, thus Y has no unramified covering of degree 2,

$$\mathrm{Ext}(Y, \mathbb{Z}/2) = 0,$$

and by class field theory we conclude that D has no unramified (irreducible) coverings of degree 2. This contradiction proves our claim.

5.7 Lemma: *Let (X, λ) be a principally polarized abelian variety defined over a field K, which over some extension of K is the Jacobian of a complete, irreducible, non-singular non-hyperelliptic curve. Then there exists a separable extension $K \subset L$ of degree at most 2 and an algebraic curve C over L such that $(X, \lambda) \otimes_K L = \mathrm{Jac}(C)$.*

Proof: Using an appropriate level-n-structure, and using the Torelli map

$$j^{(n)} : \mathcal{M}_g^{(n)} \to \mathcal{A}_{g,1}^{(n)},$$

and using [27], Coroll. 2.8, we conclude there exists a finite (separable) Galois extension $K \subset M$, an algebraic curve D defined over M and an isomorphism

$$\mathrm{Jac}(D) \cong_M P := (X, \lambda) \otimes_K M.$$

The descent data for P from M to K consist of a homomorphism

$$\mathrm{Gal}(M/K) =: G \to \mathrm{Aut}(P, M).$$

Let $H \subset G$ be the biggest subgroup of G factoring through $\mathrm{Aut}(D, M)$:

$$\mathrm{Gal}(G/M) \supset H \to \mathrm{Aut}(D, M) \supset \mathrm{Aut}(P, M).$$

By (5.5.nh) the index of H in G is at most 2, we write $L := M^H$ for the field of H-invariant elements in M. The descent data tell us that D is already defined over L, i.e. there exists C over L with $C \otimes L \cong D$, with moreover $\mathrm{Jac}(C) \cong (X, \lambda) \otimes_K L$. This proves the lemma.

5.8 It is a well-known result by A. Weil that for a curve C of genus $(C) = g$ over a finite field $K = \mathbf{F}_q$ the number of rational points is bounded by

$$\#C(K) \leq q + 1 + 2g\sqrt{q} \qquad (W).$$

Let $\pi_i, \bar{\pi}_i$, with $1 \leq i \leq g$ be the eigenvalues of the Frobenius endomorphism of C/K. Then

$$C(K) = q + 1 - \sum(\pi_i + \bar{\pi}_i).$$

Thus we see that equality in (W) implies that ($q = p^n$ with n even and that) $\mathrm{Jac}(C) = X$ is supersingular. (In (5.6.2) we have seen that hyperelliptic curves with $g = 3$ and equality in (W) do not exist in characteristic 2. Professor Serre kindly communicated to me a construction of supersingular non-hyperelliptic curves with equality in (W), with the help of $V_4 \subset \mathrm{Aut}(C)$ over certain finite fields of characteristic 2, together with the question whether all supersingular curves of genus three in characteristic two could be obtained in this way. We answer that question in (5.6) in a positive way.

5.9 Proposition ($p = 2$): *a) (Serre) Choose an* even *integer* $n = 2m$ *with* m odd, *and* $m \geq 3$; *we write* $K_n = \mathbf{F}_q$ *with* $q = 2^n$. *There exists a (complete, non-singular, non-hyperelliptic) curve* C *with*

$$g(C) = 3, \text{ and } C(K_n) = q + 1 + 6\sqrt{q}.$$

b) Such pairs give a dense subset of $\mathcal{S}_3 \subset \mathcal{A}_{3,1} \otimes \mathcal{F}_2$.

Proof: Note that $\Gamma := \mathcal{Z}(X^3 + Y^3 + Z^3)$ (here $3 = p + 1$) has a point

$$\gamma \in \Gamma(K_m), \quad \gamma \notin \Gamma(\mathbf{F}_2)$$

(choose $x, y \in K_m$, with $x^3+y^3 \neq \mathbf{F}_2$, because m is odd we have $(3, 2^m-1) = 1$, hence we can solve $Z^3 = x^3 + y^3$ in K_m). The fibre space $\pi : \mathcal{F} \to \Gamma$ (as in 2.4) has a section $T \subset \mathcal{F}$, hence the fibre over γ is isomorphic with $\mathbf{P}^1$ over K_m, and we can choose $\varphi \in \mathcal{F}(K_m)$, $\varphi \in T$, $\pi(\varphi) = \gamma$. Let $E/\mathbf{F}_2$ given by $E = \mathcal{Z}(Y^2 + Y + X^3)$. Then $E/\mathbf{F}_2$ has eigenvalues of Frobenius equal to $\pm\sqrt{-2}$. By the construction in (2.4) (there is only one component of $\mathcal{S}_3$ for $p = 2$), this φ gives (X, λ), the isogeny $\varphi : E^3 \to X$ is defined over K_m, thus the eigenvalues of Frobenius of X/K_m are all $\pm(-2)^{m/2}$. By [28] we know that (X, λ) is a Jacobian over some extension of K_m. Note that $a(X) = 1$ (use 2.3.3), thus this curve is irreducible, hence by (5.6.2) it is not hyperelliptic, thus by (5.7) we obtain:

$$\begin{gathered} \text{There exists a curve } C \text{ defined over } K_n, \\ \operatorname{Jac}(C) = (X, \lambda),\ X \sim_{K_n} E^3,\ a(X) = 1, \\ \text{the eigenvalues of Frobenius of } C/K_n \\ \text{are } (\pm\sqrt{-2})^n = -2, \text{ thus} \\ \#C(K_n) = q + 1 + 6\sqrt{q}, \text{ with } q = 2^n. \end{gathered}$$

For any odd m, with $m \geq 3$, for any $\gamma \in \Gamma(K_m)$, and for any $\varphi \in \mathcal{F}(K_m)$ as above we obtain such a curve C/K_n. These points φ are dense in $\mathcal{F}$, hence their images are dense in $\mathcal{S}$. This concludes the proof of the proposition.

5.10 Remark ($p = 2$): Let $\operatorname{char}(k) = 2$, let E be given by $E = \mathcal{Z}(Y^2+Y+X^3)$, and let $g \in \mathbf{Z}_{\geq 2}$. If g is not divisible by 3 there exists a hyperelliptic curve C such that:

$$\operatorname{genus}(C) = g, \quad 2-\operatorname{rank}(\operatorname{Jac}(C)) = 0, \quad \exists C \twoheadrightarrow E.$$

In fact, if $g = 1 + 3a$, choose some monic polynomial f of degree $2a + 1$, and consider

$$C = \mathcal{Z}(Y^2 + Y + f^3), \text{ and use [35], Th.4.1.}$$

If $g = 2 + 3b$, choose some monic polynomial f of degree $2b + 2$, consider $\eta = Y + X^{3b+3}$, and $h := f^3 + X^{6b+6}$. Note that $\deg(h) = 6b + 5 = 2g + 1$, and let

$$C = \mathcal{Z}(\eta^2 + \eta + h).$$

However if g is divisible by 3 we cannot find such a construction. Therefore we ask:

Question: Let g be divisible by 3. Does there exist a hyperelliptic curve C in characteristic $p = 2$ with $g(C) = g$, with 2-rank $(\mathrm{Jac}(C)) = 0$ and which admits a surjection $C \twoheadrightarrow E$? (We have seen that the answer is negative for $g = 3$).

For the rest of this section we consider $p \neq 2$, and we study some curves with

$$V_4 \subset \mathrm{Aut}(C), \text{ with } V_4 = \mathbf{Z}/2 \times \mathbf{Z}/2, \text{ and } g(C) = 3.$$

The three non-identity elements of V_4 give quotients of C, and these can have the following genera: $(1,1,1)$, $(0,1,2)$, $(1,2,2)$. Note further that $\mathrm{char}(k) \neq 2$ does not divide the order of V_4, hence the inertia group of the V_4-action is cyclic for every point of C, hence only $(\mathbf{Z}/2)$-inertia can appear. It is easy to describe such situations. We use the first one, which we describe for further reference:

5.11 (1, 1, 1):

$$\begin{array}{ccccc} & & C & & \\ & \swarrow & \downarrow & \searrow & \\ E_\alpha & & E_\beta & & E_\gamma \\ & \searrow & \downarrow & \swarrow & \\ & & \mathbf{P}^1 & = C/V_4, & \end{array}$$

$\delta(C/\mathbf{P}^1) = 12$, $\delta(C/E) = 4$ for $E \in \{E_\alpha, E_\beta, E_\gamma\}$ (here δ denotes the degree of the different of the covering), and ($p \neq 2$) such a situation can be obtained as follows:

choose mutually different points
$A_1, A_2, B_1, B_2, C_1, C_2 \in \mathbf{P}^1$, let
$\pi : E_\alpha \to \mathbf{P}^1$ ramify in B_1, B_2, C_1, C_2, and let
$C \to E_\alpha$ ramify in the points above A_1 and A_2,
$E/\mathbf{P}^1$ and C/E quadratic.

Note that C thus obtained is hyperelliptic if and only if there exists $s \in \mathrm{Aut}(\mathbf{P}^1)$, $s^2 = id$, having $\{A_1, A_2\}$, $\{B_1, B_2\}$, and $\{C_1, C_2\}$ as orbits.

5.12 Theorem: *Consider complete, non-singular curves C with $g(C) = 3$.*
(1: nh, $a = 3$): *There exists a non-hyperelliptic supersingular curve C in characteristic p with $a(C) = 3$ iff $p \geq 3$.*
(2: nh, $a = 1$): *There exists a non-hyperelliptic supersingular curve with $a(C) = 1$ for every p.*
(3: h, $a = 3$): *If $p \in \{2, 3, 5\}$ there does not exist a hyperelliptic supersingular curve with $a(C) = 3$ in characteristic p; if $p \geq 7$ and $p \equiv 3$ (mod4) such curves do exist.*
(4: h, $a = 1$): *There exists a hyperelliptic supersingular curve with $a(C) = 1$ in characteristics p iff $p > 2$.*
For a **proof** of (5.12) we start with some easy facts. Note that $\mathcal{S}_3 \subset \mathcal{A}_{3,1} \otimes \mathbb{F}_p$ has pure dimension 2, the subset where $a \geq 2$ and/or C is hyperelliptic has components of dimension at most one, thus every component of $\mathcal{S}_3$ has points corresponding to C non-hyperelliptic, and $a(C) = 1$ (hence C irreducible); this proves (2).

Note that the first statement in (3) is contained in [6], Th. 1.1(ii). For $p = 7$ there exists a hyperelliptic C with $a(C) = 1$: in [18], on page 196 we see there exist at least 22 isomorphism classes of hyperelliptic supersingular curves, and by [12], p. 493 at most 5 of these have $a(C) = 3$. We see that (1) implies (4): if $2 < p \leq 7$ there exist supersingular hyperelliptic curves, cf. [18], p. 196, and for $p = 3$, or $p = 5$ the case $a(C) = 3$ is excluded (for these cases, also see (5.15) below). For $p > 7$, granting (1) we can choose a non-hyperelliptic supersingular curve C' with $a(C') = 3$. Using (2.4), (1.12), (2.3.2) and (5.1) we conclude the proof of (4).

In order to finish the proof of (5.12) we give the following explicit constructions:

5.13 Let $p \geq 7$, let $\lambda \in \mathbb{P}^1, \lambda \notin \{0, 1, -1, \infty\}$, consider

$$A_1 = 0,\ A_2 = \infty,\ B_1 = 1,\ B_2 = \lambda,\ C_1 = -1,\ C_2 = -\lambda.$$

Let C be the curve constructed from these date by the Construction (5.11). Suppose that $Y^2 = X(X-1)(X-\lambda)$ defines a supersingular elliptic curve and suppose that $p \equiv 3$ (mod 4). Then C is hyperelliptic, supersingular, and $a(C) = 3$.

5.14 For every $p > 2$, there exist $\lambda, \gamma \in k$, such that (5.11) applied to

$$A_1 = 0,\ A_2 = \infty,\ B_1 = 1,\ B_2 = \lambda C_1 = \gamma,\ C_2 = \gamma\lambda$$

gives a non-hyperelliptic curve C with $a(C) = 3$.

Note that (5.13) and (5.14) together with the previous arguments finish the proof of (5.12).

Proof of (5.13) ($p \neq 2$)**:** As usual we write $h = h_p$ for the polynomial

$$h(T) := \sum_{i=0}^{k} \binom{k}{i}^2 T^i, \quad k := (p-1)/2.$$

Note that the curve defined by $Y^2 = X(X-1)(X-\lambda)$ is supersingular iff $h(\lambda) = 0$ (cf. [11], Coroll. 4.22). We choose

$$\lambda \in \mathbf{P}^1 \quad \text{with} \quad \lambda \notin \{0, 1, -1, \infty\}.$$

If $p \geq 5$ there exists such a λ with $h(\lambda) = 0$ (because the zeros of h are simple, and degree $(h) = k \geq 2$). Let $A_1, \ldots, B_2$ be as in (5.14) above. The transformation

$$x \mapsto \frac{\lambda}{X} \quad \text{on} \quad \mathbf{P}^1$$

interchanges 0 and ∞, 1 and λ, and it interchanges γ and $\gamma\lambda$ iff

$$\frac{\lambda}{\gamma} = \gamma\lambda, \quad \text{i.e.} \gamma^2 = 1.$$

We see: if $\gamma \neq \pm 1$, then Construction (5.11) gives a non-hyperelliptic curve; if $\gamma = -1$, it gives a hyperelliptic curve. Here we take $\gamma = -1$, $p \geq 7$, the points

$$\{0, \infty, 1, \lambda, -1, -\lambda\}$$

mutually different, and $h(\lambda) = 0$. Then

$$E_1 : Y^2 = X(X-1)(X-\lambda)$$

and

$$E_2 : Y^2 = X(X+1)(X+\lambda)$$

are supersingular, and

$$E_3 : Y^2 = (X^2 - 1)(X^2 - \lambda^2)$$

is isomorphic to the curve defined by $Y^2 = X^4 - 1$, which has CM by -1. Hence if $p \equiv 3 \pmod 4$ this curve is also supersinguar. Thus in that case C is supersingular. Moreover there is an isogeny from $X = \operatorname{Jac}(C)$ to $E_1 \times E_2 \times E_3$ of degree a power of 2 (and note that $\operatorname{char}(k) \neq 2$), hence $a(C) = 3$. This concludes the proof of (5.13).

Proof of (5.14): We are going to choose $\gamma \neq \pm 1$, we choose $\lambda \in \mathbf{P}^1$ with $h(\lambda) = 0$, and we consider the third curve

$$E_3 : Y^2 = (X-1)(X-\lambda)(X-\gamma)(X-\gamma\lambda).$$

The map

$$X \mapsto \frac{X-1}{X-\lambda} \cdot \frac{\gamma - \lambda}{\gamma - 1} \qquad \begin{cases} 1 \mapsto 0 \\ \lambda \mapsto \infty \\ \gamma \mapsto 1 \end{cases}$$

transforms E_3 into:

$$E_3 : \quad Y^2 = X(X-1)\left(X - \frac{(\gamma\lambda - 1)(\gamma - \lambda)}{\lambda(\gamma-1)^2}\right).$$

We try to find $\gamma \notin \{0, 1, \lambda, -1, \frac{1}{\lambda}\}$ so that E_3 is supersingular. This is the case if

$$h\left(\frac{(\gamma\lambda - 1)(\gamma - \lambda)}{\lambda(\gamma-1)^2}\right) = 0,$$

i.e. if

$$H(\lambda, \gamma) := \sum_{i=0}^{k} \binom{k}{i}^2 \{\lambda(\gamma-1)^2\}^{k-i}\{(\lambda\gamma - 1)(\gamma - \lambda)\}^i = 0.$$

Note that as a polynomial in γ, we have

$$H(\lambda, \gamma) = 2\lambda^k \gamma^{2k} + \text{(lower order terms in } \gamma).$$

Note that:

$$H(\lambda, 1) = (\lambda - 1)^k (1-\lambda)^k,$$

$$H(\lambda, \lambda) = (\lambda - 1)^{2k} \lambda^k,$$

$$H(\lambda, 1/\lambda) = (1-\lambda)^{2k} / \lambda^k.$$

These values are non-zero for $\lambda \notin \{0, 1, \infty\}$. Note that

$$H(\lambda, 0) = \sum \binom{k}{i}^2 \lambda^{k-i}\lambda^i,$$

moreover

$$\sum_{i=0}^{k} \binom{k}{i}^2 = \binom{2k}{k}$$

(expand $(X+Y)^k(Z+X)^k$ and put $Y = Z$), thus

$$H(\lambda, 0) = \binom{2k}{k} \lambda^k \neq 0.$$

If $p = 3$, then $h = T + 1$, we choose $\lambda = -1$, and solve

$$\frac{(-\gamma - 1)(\gamma + 1)}{-(\gamma - 1)^2} + 1 = 0 \quad \text{by} \quad \gamma = \sqrt{-1}.$$

Suppose $p > 3$. Note that for $k = (p-1)/2$ we have

$$\binom{2k}{k} \equiv (-1)^k \quad (modp)$$

(expand $(X^p - 1)/(X - 1) \equiv (X - 1)^{p-1}$), thus for $p > 3$ we have

$$\binom{2k}{k} \not\equiv 2 \quad (modp).$$

This shows that

$$H(\lambda, \gamma) \not\equiv 2\lambda^k(\gamma + 1)^{2k}$$

($p > 3$, substitute $\gamma = 0$ on both sides). We have seen that $0, 1, \lambda$ and $1/\lambda$ are not zeros of $H(\lambda, \gamma)$; we choose γ so that $\gamma \neq -1$ and $H(\lambda, \gamma) = 0$, and we are done. As above it follows that $a(C) = 3$. This concludes the proof of (5.14), and hence of Theorem (5.12).

5.15 Remark: Consider ($p \neq 2$) the curve defined by $Y^2 = X^7 - 1$. Its Jacobian $X = \text{Jac}(C)$ has CM by $\mathbf{Z}[\zeta_7]$. If $p \equiv 6 \pmod 7$, we see that $\mathbf{Z}[\zeta_7] \otimes \mathbf{Z}_p$ is the product of three rings, and accordingly the Barsotti-Tate group (p-divisible group) of X splits as a direct sum; in this way it follows that C is supersingular, with $a(C) = 3$. If $p \equiv 3$ or $5 \pmod 7$, the prime p

is inert in $\mathbb{Q}(\zeta_7)$, it follows that C is supersingular; as T. Katsura showed to me, a computation of the Cartier operator shows that $a(C) = 1$ in this case. I do not know whether there exist hyperelliptic curves C with $a(C) = 3$ for every $p \geq 7$.

5.16 Remark: We obtain a hyperelliptic Jacobian with $g = 3$ by taking the product of an elliptic curve and an abelian surface. Thus we see that the locus of hyperelliptic supersingular Jacobians has $H_1 \times H_2(1,p)$ components, where the moduli points correspond to reducible polarizations. In char$(k) = 2$ this number is one, and it is the only component of the hyperelliptic supersingular locus. For every $p \geq 3$ there is at least one other component, as follows by (5.12.4).

6 Some questions and remarks

6.1 (T. Ekedahl): For a given prime p, does there exist a smooth non-isotrivial family $\mathcal{C} \to \mathbf{P}^1$ of supersingular hyperelliptic curves of genus 3? If so, it should come from a fibre of $\pi : \mathcal{F} \to C$ (as in Section 2).

6.2 Let $g \geq 3$, $0 \leq f \leq g$, and $V_f \subset \mathcal{A}_{g,1} \otimes \mathbf{F}_p$ the subset of the moduli space consisting of those principally polarized abelian varieties (X, λ) with $\dim(X) = g$, and p-rank $f(X) \leq f$. Let $\mathcal{T}_g \subset \mathcal{A}_{g,1} \otimes \mathbf{F}_p$ denote the Torelli locus, i.e. the closure of the image of

$$j : \mathcal{M}_g \otimes \mathcal{F}_p \to \mathcal{A}_{g,1} \otimes \mathcal{F}_p.$$

We expect:

$$\dim(\mathcal{T}_g \cap V_f) = 3g - 3 - (g - f) \qquad (?).$$

Clearly any component of this intersection has at least this dimension (by [18], p. 163, Th. 7, or by [25], Th. 4.1), and it is easy to show, using deformation of $(E, \rho)^g$, that $\mathcal{T}_g \cap V_f$ has at least one component of this dimension.

6.3 Remark: It seems difficult to have a good guess on the dimensions of the components of

$$(\mathcal{T}_g \cap \mathcal{S}_g) \subset \mathcal{A}_{g,1} \otimes \mathbf{F}_p$$

for large g. Note that

$$(3g - 3) + [g^2/4] < \frac{1}{2} g(g+1)$$

for large g. On the one hand it seems reasonable to expect that there are components of $\mathcal{S}_g$ which do not meet $\mathcal{T}_g$. On the other hand, using reducible curves, it is easy to see that for $g \geq 3h$ there is at least one component of $\mathcal{T}_g \cap \mathcal{S}_g$ of dimension $2h$. Moreover $\mathcal{S}_g$ has many components (especially for large p). I do not see any reasonable simple guess for the dimension(s) of the components of $\mathcal{T}_g \cap \mathcal{S}_g$, or for $\mathcal{B}_g \cap \mathcal{S}_g$.

6.4 We use the notation for types as in (1.2). We have seen that the following cases do occur:

	hyperelliptic	supersingular
	$p = 2$	$p > 2$
Type (3)	no	yes
Type (2)	yes	yes
Type (1)	yes	yes
Type (0)	yes	no.

References

[1] A. Beauville - Variétés de Prym et Jacobiennes intermédiares. Ann. Sci. E. Norm. Sup. 10 (1977), 304-392.

[2] P. Deligne & D. Mumford - The irredicibility of the space of curves of given genus. Publ. Math. 36, IHES 1969 (Volume dédié au Prof. Zariski à l'occasion de son 70e anniversaire), pp. 75-109.

[3] M. Demazure - Lectures on p-divisible groups. Lect. N. Math. 303. Springer-Verlag 1972.

[4] S. Diaz - Exceptional Weierstrass points and the divisor on moduli spaces that they define. Mem. AMS 56, N. 327 (1985).

[5] M. Eichler - Über die Idealklassenzahl hyperkomplexe Systeme. Math. Zeitschr. 43 (1938), 481-494.

[6] T. Ekedahl - On supersingular curves and abelian varieties. Math. Scand. 60 (1987), 151-178. 6a: Prépublications 84 T 50, Univ. Paris-Sud 1984.

[7] C. Faber - Chow rings of moduli spaces of curves. Ph.D. Thesis, Amsterdam, 1988.

[8] G. Faltings - Arithmetische Kompaktifizierung des Modulraums der abelsche Varietäten. Arbeitstagung Bonn 1984. Lect. N. Math. 1111, Springer-Verlag 1985; pp. 321-383.

[9] W. Fulton - Hurwitz schemes and irreducibility of moduli of algebraic curves. Ann. Math. 90 (1969), 542-575.

[10] J. Harris & D. Mumford - On the Kodaira dimension of the moduli space of curves. Invent. Math. 67 (1982), 23-86.

[11] R. Hartshorne - Algebraic geometry. Grad. Texts Math. 52, Springer--Verlag 1977.

[12] K. Hashimoto - Class numbers of positive definite ternary quaternion hermitian forms. Proc. Japan Acad. 59 (1983), 490-493.

[13] M. Homma - Automorphisms of prime order of curves. Manusc. Math. 33 (1980), 99-109.

[14] T. Ibukiyama, T. Katsura & F. Oort - Supersingular curves of genus two and class numbers. Compos. Math. 57 (1986), 127-152.

[15] T. Katsura & F. Oort - Families of supersingular abelian surfaces. Compos. Math. 62 (1987), 107-167.

[16] T. Katsura & F. Oort - Supersingular abelian varieties of dimension two or three and class numbers. In: Algebraic Geometry, Sendai 1985, Ed. T. Oda. Adv. Stud. Pure Math. 10 (1987), Kinokuniya Cy & North Holl. Cy 1987; pp. 253-281.

[17] F. Knudsen - The projectivity of the moduli space of stable curves, II. Math. Scand. 52 (1983), 161-199.

[18] N. Koblitz - p-adic variation of the zeta-function over families of varieties defined over finite fields. Compos. Math. 31 (1975), 119-218.

[19] Yu. I. Manin - The theory of commutative formal groups over fields of finite characteristic. Usp. Math. Nauk 18 (1963), 3-90 [Russ. Math. Surveys 18 (1963), 1-80].

[20] T. Matsusaka - On a theorem of Torelli. Amer. Journ. Math. 80 (1985), 784-800.

[21] L. Moret-Bailly - Familles de courbes et des variétés abéliennes sur $\mathbb{P}^1$. In: Sém. sur les pinceaux de courbes de genre au moins deux. Ed. L. Szpiro. Astérisque 86, Soc. Math. France 1981; pp. 109-140.

[22] D. Mumford - Geometric invariant theory. Ergebnisse, Vol. 34, Springer-Verlag 1965.

[23] D. Mumford - The structure of the moduli space of curves and of abelian varieties. Actes Congrès Intern. Math. Nice 1970, Tome 1, Gauthiers--Villars, Paris 1971; pp. 457-465.

[24] D. Mumford - Stability of projective varieties. l'Ens. Math. 23 (1977), 39-110.

[25] P. Norman & F. Oort - Moduli of abelian varieties. Ann. Mat. 112 (1980), 413-439.

[26] T. Oda & F. Oort - Supersingular abelian varieties. Intl. Sympos. on Algebraic Geometry, Kyoto 1977 (Ed. M. Nagata), Kinokuniya Cy 1987; pp. 595-621.

[27] F. Oort & J. Steenbrink - The local Torelli problem for algebraic curves. In: Journ. géom. algébrique d'Angers 1979, Ed. A. Beauville. Sijthoff & Noordhoff 1980; pp. 157-204.

[28] F. Oort & K. Ueno - Principally polarized abelian varieties of dimension two or three are Jacobian varieties. Journ Fac. Sc., Univ. Tokyo, Sec. IA, 20 (1973), 377-381.

[29] F. Oort - Commutative group schemes. Lect. N. Math. 15, Springer--Verlag 1966.

[30] F. Oort - Subvarieties of moduli spaces. Invent. Math. 24 (1974), 95-119.

[31] F. Oort - Which abelian surfaces are products of elliptic curves? Math. Ann. 214 (1975), 35-47.

[32] F. Oort - Abelian varieties: moduli and lifting problems. Algebraic Geometry, Copenhagen 1978 (Ed. K. Lønsted). Lect. N. Math. 732, Springer-Verlag 1979; pp. 477-495.

[33] J-P. Serre - Groups algébriques et corps de classes. Act. sc. ind. 1264, Hermann 1959.

[34] T. Shioda - Supersingular K3 surfaces. Algebraic Geometry, Copenhagen 1978 (Ed. K. Lønsted). Lect. N. Math. 732, Springer-Verlag 1979; pp. 564-591.

[35] D. Subrao - The p-rank of Artin-Schreier curves. Manuscr. Math. 16 (1975), 169-193.

Letter to Don Zagier by A.N. Parshin. *

Bonn, July 3, 1988.

Dear Don,

Here is a more detailed explanation of what we talked about during our quick lunch. For your convenience I have divided this letter into several parts which may be of independent interest.

1. Some general remarks.

Let me explain again what I said in my Arbeitstagung talk. (I use the notations employed there,cf. also references at the end).

For a fibration $f: V \to B$ the inequality has the following form: $\omega^2_{V/B} \leq a_1\delta + a_2\chi_0 + a_3$. It implies both Conjectures of Shafarevich as they were stated in his Stockholm lecture. For the first Finiteness Conjecture the implication was shown by me in 1968, the boundedness of the height and the rigidity of families were proved by Arakelov in 1972. Also the second conjecture can be proved using the inequality. Let $d = \deg f_*\omega_{V/B}$. By Arakelov's Theorem $d \geq 0$, and $d = 0$ is equivalent to the isotriviality of $f: V \to B$. Now suppose that we have a smooth family $f: V \to B$ and that the inequality $\omega^2_{V/B} \leq a_1\delta + a_2\chi_0$ (with positive a_i) holds. By Noether's formula $\omega^2_{V/B} + \delta = 12d$ and consequently $0 < 12d \leq (a_1+1)\delta + a_2\chi_0$. The smoothness implies $\delta = 0$ and we get that $\chi_0 = (2g-2)(2g(B)-2) > 0$. It means that there are no non-isotrivial families with $g(B) \leq 1$. In the arithmetic case the inequality has the following form :

$$\omega^2_{V/B} \leq a_1\delta + a_2(2g-2)\log|D_{K/\mathbb{Q}}| + a_3[K:\mathbb{Q}] .$$

I forgot the degree $[K:\mathbb{Q}]$ in my talk and I am greatly indebted to Serge Lang for pointing this out to me. Unfortunately, in this situation it is impossible to use the inequality to get the Shafarevich Conjectures. In the first place, to bound δ we need some information about the archimedean fibres and so one can only prove a weaker result (when we fix not only the genus g , the ground field K and the set S of bad reductions, but also

---

* This letter is a slightly edited version of the original letter.

the distance between the corresponding point in the moduli scheme and the boundary). Also Arakelov's result $d \geq 0$ is not valid in the arithmetic case. (Deligne discovered elliptic curves with $d < 0$.) But if we look more carefully at there arguments used by Shafarevich for his conjectures we can consider their analogues for relative dimension 0, namely the finite extensions $L \supset K$ of global fields (of dimension 1). It is well known that Minkowski's inequality $D_{K/\mathbb{Q}} >$(some function of $[K:\mathbb{Q}]$) plays an essential role in the proofs of both the Hermite Theorem (there are finitely many $L \supset K$'s with given degree $[L:K]$ and ramification set S) and Minkowski's Theorem that there is no $L \supset \mathbb{Q}$ which is unramified ($S = \emptyset$) and of degree > 1. These results are also valid in the function field case and instead of Minkowski's inequality you can use the inequality $g \geq 0$. Like the inequality for the surfaces it has to do with the Chern classes. Take for example an algebraic curve consisting of s irreducible components. Then if we put $c_0 = s$ we have $c_1 \leq 2c_0$ for this curve.

There are two approaches to get Minkowski's inequality and $g \geq 0$ simultaneously. One is the geometry of numbers (apply it to the discrete subgroup $K \subset A_K$ of the adele group and use that

$$\mu(A_K/K) = \begin{cases} |D_{K/\mathbb{Q}}|^{1/2} & \text{in the arithmetic case} \\ q^{1-g} & \text{in the function field case.)} \end{cases}$$

The other is an application of the functional equations of the zeta functions.

2. Curves over $\mathbb{F}_q$ and the inequality $g \geq 0$.

Let B be a smooth geometrically irreducible curve over $\mathbb{F}_q$ of genus g and let $\zeta_B(s)$ be its zeta function. Then we know that

i) $|\zeta(s)|$ is bounded for $Re(s) > 1 + \epsilon$ (by the Euler product $\zeta_{B(s)} = \prod(1 - N_v^{-s})^{-1}$).

ii) $P(s) = (1 - q^{-s})(1 - q^{1-s})\zeta_B(s)$ is an entire function on the whole plane $\mathbb{C}$ and $|P(s)|$ is bounded in each strip $a_0 \leq \mathrm{Re}(s) \leq a_1$ (by the functional equation, see Lang, Algebraic Number Theory, Ch.17, Sec.2 for the (of course more complicated) number field case which we should have in mind.

iii) $P(s) \neq$ constant for $g \neq 0$.

These properties will imply the inequality $g \geq 0$ in the following way. Consider the functional equation for $P(s)$:

$$P(s) = q^g q^{-2gs} P(1-s) \qquad (*)$$

By i) and ii) $|P(s)|$ is bounded by C in $\mathrm{Re}(s) \geq \frac{1}{2}$. Assume that $g < 0$ and let $s \to -\infty$ (and $1 - s \to +\infty$). We get from $(*)$ that for $\mathrm{Re}(s) < \frac{1}{2}$ one has $|P(s)| \leq q^g q^{-2gs} C \leq C'$, which contradicts iii).

We can also proceed in a way more close to the usual proof using "explicit formulas" (see for example Poitou's talk in Seminaire Bourbaki,1976, exp. 479, no 4. Let $\phi(s) = q^{gs} P(s)$. Then we have the functional equation $\phi(s) = \phi(1-s)$ and we get an expression for

$$\phi(s) = q^{gs}(1 - q^{-s})(1 - q^{1-s})\zeta_B(s).$$

Taking the logarithmic derivative of that we get

$$g \log q = -\frac{q^{-s} \log q}{1 - q^{-s}} - \frac{q^{1-s} \log q}{1 - q^{1-s}} \underbrace{- \frac{\zeta'(s)}{\zeta(s)} + \frac{\phi'(s)}{\phi(s)}}_{positive}$$

[This is positive by the Euler product for the first term $-\zeta'(s)/\zeta(s)$ and by the decomposition $\frac{\phi'(s)}{\phi(s)} = \sum_\rho \frac{1}{s-\rho}$ over the zeroes of P(s).] Rewriting this as

$$g = -\frac{1}{q^s - 1} - \frac{1}{q^{s-1} - 1} + (>0) \qquad (**)$$

and letting $s \to +\infty$ we see that $g \geq 0$. There is one thing here which I do not understand. If you compare this with the application of the explicit formulas in Poitou's talk then you will see that he puts $s = 1 + n^{-1}$, where $n = [K : \mathbb{Q}]$, but not $s \to \infty$. In his case (**) looks like

$$"g" = -\frac{1}{s} - \frac{1}{s-1} + (>0)$$

and then he divides by n, so

$$\frac{"g"}{n} = -\frac{1}{ns} - \frac{1}{n(s-1)} + (>0)$$

and he takes s as above. Does this mean that the same type of argument can be used in the function field case for (**) ? In Serre's article (*C.R. Acad. Sci. Paris* **296** (1983),397–402) there is some analogy between the degree $n = [K : \mathbb{Q}]$ in the number field case and the number of rational points ($= |B(\mathbb{F}_q)|$) in the geometrical case. So it might be that his use of explicit formulas in the geometric case is closer to the arguments in Poitou's talk.

3. Surfaces over $\mathbb{F}_q$ and the Van de Ven-Bogomolov-Miyaoka-Yau-inequality.

There exist surfaces over $\mathbb{F}_q$ such that the inequality $c_1^2 \leq ac_2$ with any positive a is not valid for them, and moreover there is a family of surfaces such that $c_2 =$ const. and $c_1^2 \to +\infty$. It means that that it is impossible to get an inequality $c_1^2 \leq f(c_2)$ for them. The examples are in my paper (*Izvestia* 1972) and they have a very non-reduced Picard variety Pic(V). Namely, $\dim Pic(V) = b_1(V)$ is fixed for the family, but $\dim T_{Pic(V),0} = \dim H^1(V, O_V)$ tends to infinity. Well, this suggests that some inequality is still valid in char $p > 0$, but with an additional term which measures the non-smoothness of $Pic(V)$. It seems that the same situation occurs in the closely related problem of Kodaira's Vanishing Theorem. Mumford has shown that "*Pic* is reduced" implies Kodaira Vanishing and also the known examples where Kodaira's Theorem is false are related to the "pathological" surfaces mentioned above (cf. Raynaud's example, see Szpiro's Seminaire in Asterisque no. 86). It is reasonable to expect that the VdV-B-M-Y-inequality (or some version of it) is valid for all the surfaces with reduced *Pic*. It suffices to compare with the number field case because we know that for these "pathological" examples Mordell's Conjecture is also false (in a rather strong way : the heights are unbounded).

After this introduction, let me prove the following inequality for all surfaces over $\mathbb{F}_q$ (more precisely, for the fibrations $f: V \to B$ as above)

$$\omega_{V/B}^2 \leq 5\delta + 3\chi_0 + 12g + 6(2g(B) - 2) + 12 + 6\tau, \qquad (*)$$

where $\tau = 2h^{1,0} - b_1(V)$ shows the deviation of V from the classical Hodge Theorem over $\mathbb{C}$.

Proof of (*). Actually we prove the inequality

$$c_1^2 \leq 5c_2 + 6b_1(V) + 6\tau \qquad (**)$$

which is valid for all surfaces and is equivalent to (*) for fibrations. We know that

$$c_2(V) = \chi(V, O_V) - \chi(V, \Omega_V^1) + \chi(V, \Omega_V^2).$$

By the Duality Theorem $\chi(V, O_V) = \chi(V, \Omega_V^2)$, so

$$c_2 = 2\chi(O) - \chi(\Omega^1). \qquad (1)$$

Now

$$\chi(V, \Omega_V^1) = h^0(\Omega^1) - h^1(\Omega^1) + h^2(\Omega^1)$$

and by duality $h^0(\Omega^1) = h^2(\Omega^1)$, so

$$\chi(\Omega^1) = 2h^{1,0} - h^1(\Omega^1). \tag{2}$$

Because

$$h^1(\Omega^1) \geq 0 \tag{***}$$

we have

$$\chi(\Omega^1) \leq 2h^{1,0}. \tag{3}$$

Taking (1), (2) and (3) together and using Noether's formula $12\chi(O) = c_1^2 + c_2$ we see that

$$c_1^2 \leq 5c_2 + 12h^{1,0},$$

which is our (**).

Remark 1. This inequality is not the VdV-B-M-Y-inequality, but in char $= 0$ it can be used in completely the same way to deduce all the interesting results. It might be possible to prove in that way that for surfaces in char $p > 0$ with reduced Pic an inequality like (*) is valid but without τ.

Remark. 2. Having in mind the number field case we see that our proof contains the cotangent bundle Ω^1, the analogue of which is unknown in the arithmetical situation. But we really do not need the bundle itself, but the numbers which give the dimensions of its cohomologies.

Then all the steps of the proof are general facts like duality and which give *equalities.* So essentially the inequality (**) is a corollary of the inequality (***) which has the same appearance and meaning as the inequality $g \geq 0$ for curves (i.e. the dimension of some space is non-negative).

Is it possible to find the same approach to (***) as made for the curves in 2) ?

4. A discussion over $\mathbb{F}_q$ and over $\mathbb{Z}$.

We would like to relate the numbers $h^{p,q} = \dim H^q(V, \Omega^p)$ not with the cotangent bundle Ω^1 (note that O_V and Ω^2 are defined for arithmetic surfaces) but with the zeta function $\zeta_V(s)$. If V is a surface over $\mathbb{C}$ then we have the Hodge decomposition

$$H^m(V, \mathbb{C}) = \bigoplus_{p,q} H^q(V, \Omega^p) \tag{*}$$

Some substitute for it was studied for surfaces over $\mathbb{F}_q$ using cristalline cohomology $H^m_{cris}(V/W)$. These are modules over the Witt ring $W =$

$W(\mathbb{F}_q)$ and it is possible to define complexes C^p, $p = 0,1,2$ which correspond to Ω^p (Mazur, Bloch, Raynaud, Illusie). There is a spectral sequence $H^q(V, C^p) \Rightarrow H^{p+q}_{cris}(V/W)$ which defines the slope filtration in H^*_{cris} and consequently some decomposition of $\zeta_V(s)$ (at least modulo some conditions on V). The point is that the slope filtration is invariant under the Frobenius automorphism. So we have for our zeta function an expression like

$$\zeta_V(s) = \prod_{p,q} \zeta_V^{(p,q)}(s), \qquad (*')$$

where $\zeta_V^{(p,q)}(s)$ is the characteristic polynomial on the corresponding subspace of the slope filtration. If we restrict ourselves to $\zeta_V^{(1,1)}$ then I believe it satisfies a functional equation of the following type

$$\zeta_V^{(1,1)}(s) = q^{h^{(1,1)}(V)(1-s)} \zeta_V^{(1,1)}(2-s). \qquad (**)$$

Here it is more correct to write $H^1(V, C^1)$ than $h^{1,1}$, but for "general" surfaces they are the same by Mazur's Theorem.

Now we can of course deduce the inequality (***) from 3) using (**) exactly as in 2). This is really useless in the geometrical case. But in the number field case the problem is meaningful. Let X be a curve over a number field K of genus $g \geq 1$ (even $g = 1$ is interesting, so I was wrong claiming the opposite during our discussion). Is it possible to decompose $\zeta_X(s)$ (the usual Hasse-Weil zeta function) into pieces like the above such that the full (hypothetical) functional equation will be the product of the new ones ? It might be too optimistic to look for $\zeta_X^{(p,q)}(s)$ for all p, q. Maybe only something corresponding to a given p exists. On the other hand a decomposition as above exists for surfaces if we consider cohomology with p-adic coefficients (actually $W(\mathbb{F}_q)$). It seems unlikely that the "pieces" $\zeta_V^{(p,q)}$ are defined over $\mathbb{Z}$ (so $\zeta_V^{(p,q)} \notin \mathbb{Z}[q^{-s}]$) even in the "general" case. But I do not know. [You asked about the Euler product for these zeta functions like $\zeta_V^{(p,q)}(s)$. It seems that something like that should exist. In usual Hodge Theory we have the Atiyah-Bott fixed point Theorem which gives the Euler product for what you may call the zeta functions of automorphisms over $\mathbb{C}$. But I am completely ignorant about the situation with cristalline cohomology.]

The decompositiion $(*')$ can be called the multiplicative one. The other idea is to look for some pieces inside $\zeta(s)$ and we call it the additive decomposition. Returning to 2) we see that the function $P(s) = (1 - q^{-s})(1 - q^{1-s})\zeta_B(s)$ was important. But we could also use the function $F(s)$ defined in the equality

$$\zeta_V(s) = \frac{a}{1 - q^{-s}} + \frac{b}{1 - q^{1-s}} + F(s)$$

which is the standard decomposition of this rational function in q^{-s}.

If we have a surface V over $\mathbb{F}_q$ then

$$\zeta_V(s) = \frac{a}{1-q^{-s}} + \frac{B(s)}{(1-q^{1-s})^\rho} + \frac{c}{1-q^{2-s}} + \frac{P(s)}{Q(s)} + F(s),$$

where a, c are constants , $\rho = \mathrm{rank} NS(V)$, $\deg(P(s)) < \rho$, $Q(s)$ is the other part of $\det(1 - Frob|H^2_{et}(V))$ after dividing by $(1-q^{1-s})^\rho$, $\deg(P(s)) < \deg(Q(s))$ and $F(s)$ is a polynomial. So all of these B, P, Q and F are entire functions. This corresponds to the following picture of the sigularities of $\zeta_V(s)$

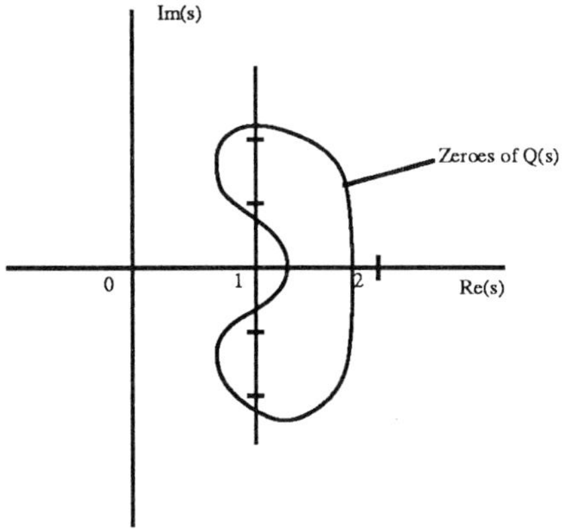

If you like you can decompose $Q(s)$ into irreducible parts further. The pieces $P, Q, ..$ have the advantage over $\zeta^{(p,q)}$ that they are defined over $\mathbb{Q}$. And I believe that some of them are related to $\chi(\Omega^1)$. A weak indication for this is the Artin-Tate conjecture for $\zeta_V(s)$ which states that

$$B(1) = q^{\chi(O_V)} . \frac{|\mathrm{Pic}(V)_{tor}|^2}{|H^0(V, O_V^*)|^2 |B_2(V)| \det NS(V)},$$

The numbers here have different origins. The quotient on the right hand side is actually a measure for some quotient space (like $\frac{h}{q-1} = \mu(A^{*(1)}/K^*)$ in dimension 1 over $\mathbb{F}_q$, $h = |Pic^0(V)|$ is a class number, $A^{*(1)}$ are the ideles of volume 1, and the whole expression is the residue of $\zeta_B(s)$ in $s = 0$), but $q^{\chi(O_V)}$ seems to correspond to some functional equation. Let me name this expectation the principal parts conjecture and finish at this point. Now here is my train to Moscow.

Sincerely yours,

A.N. Parshin.

Additional References

[1] A.N. Parshin. The Bogomolov-Miyaoka-Yau inequality for the arithmetic surfaces and its applications. Sém. de Th. des Nombres, Paris, 1986-87, pp. 299-312. Progress in Math. **75**,1989, Birkhäuser Boston.

[2] A.N. Parshin. On an application of the ramified coverings in the theory of diophantine equations. Math. Sb. **t. 180**, (1989), pp.244-259. (in Russian).

The old subvariety of $J_o(pq)$

KENNETH A. RIBET*

Introduction

Let p and q be distinct primes. The old part of $J_o(pq)$ is the abelian subvariety $A + B$ of $J_o(pq)$ generated by the images

$$A = \text{Image}(J_o(p)^2 \xrightarrow{\alpha} J_o(pq)), \qquad B = \text{Image}(J_o(q)^2 \xrightarrow{\beta} J_o(pq))$$

of the two indicated degeneracy maps. Here, $J_o(N)$ denotes the Jacobian $\text{Pic}^o(X_o(N))$ of the standard modular curve $X_o(N)$, for each integer $N \geq 1$. Also, we have written $J_o(p)^2$ for the product $J_o(p) \times J_o(p)$, and have used analogous notation for $J_o(q)^2$. The definitions of α and β will be given below; see also [6], §2a.

The structure of A was determined in [14]. Namely, the kernel of α is the Shimura subgroup Σ_p of $J_o(p)$, viewed as a subgroup of $J_o(p)^2$ via the antidiagonal embedding $x \mapsto (x, -x)$. Thus we have $A = J_o(p)^2/\Sigma_p$ and, analogously, $B = J_o(q)^2/\Sigma_q$. Since A and B are known, we consider that to understand $A + B$ is to understand $A \cap B$, which is a finite abelian group. The main purpose of this note is to identify $A \cap B$, up to groups of 2-power order. In other words, we identify the ℓ-primary part of $A \cap B$ for each *odd* prime ℓ.

Let C_p be the cuspidal subgroup of $J_o(p)$. This group is cyclic of order $\text{num}(\frac{p-1}{12})$, and appears frequently in [5]. (The symbol "num" denotes the *numerator* of a rational number. Thus, for $r \in \mathbf{Q}^*$, $\text{num}(r)$ is the order of $\frac{1}{r}$ in $\mathbf{Q}/\mathbf{Z}$.) For the moment, view C_p in $J_o(p)^2$ by the antidiagonal embedding, and let $\bar{C}_p$ be the image of the antidiagonal C_p in A. Since

*Partially supported by the National Science Foundation

$C_p \cap \Sigma_p$ is known to be the group $C_p[2]$ of elements of order dividing 2 in C_p ([5], p. 102), the group $\bar{C}_p$ is cyclic of order num $\frac{p-1}{24}$. After consulting Ogg [9], and performing some computations, one checks that $\bar{C}_p$ is a subgroup of the cyclic subgroup $\mathcal{C}$ of $J_o(pq)$ generated by the class of the divisor

$$P_1 - P_p - P_q + P_{pq}$$

on $X_o(pq)$. (See §5 below.) According to [9], p. 459, the order of $\mathcal{C}$ is $\mathrm{num}\left(\frac{(p-1)(q-1)}{24}\right)$.

Let $\bar{C}_q \subset B$ be the analogue of $\bar{C}_p$ with p replaced by q. Then $\bar{C}_q$ has order $\mathrm{num}(\frac{q-1}{24})$, and again lies in the cyclic group $\mathcal{C}$. (The primes p and q play symmetric roles in the formation of $\mathcal{C}$.) It follows that the group $\bar{C}_p \cap \bar{C}_q$ has order

$$n := \gcd\left(\mathrm{num}\left(\tfrac{p-1}{24}\right), \mathrm{num}\left(\tfrac{q-1}{24}\right)\right).$$

Our main result is

THEOREM 1 *The finite abelian group $A \cap B$ and its subgroup $\bar{C}_p \cap \bar{C}_q$ are equal up to 2-groups. In other words, the quotient $Q = (A \cap B)/(\bar{C}_p \cap \bar{C}_q)$ has 2-power order.*

COROLLARY 1 *The order of the odd part of $A \cap B$ is the odd part of the integer n.*

As explained above this Corollary follows from the Theorem, together with the computation of §5.

A simple application of Theorem 1 concerns the kernel κ of the natural map

$$\gamma : J_o(p)^2 \times J_o(q)^2 \longrightarrow J_o(pq)$$

which is obtained from α and β. The image of γ is the abelian variety $A + B$ mentioned above; it is the old subvariety of $J_o(pq)$. View γ as the composite of the surjection $\alpha \times \beta : J_o(p)^2 \times J_o(q)^2 \to A \times B$ and the map $A \times B \to J_o(pq)$, $(a, b) \mapsto a + b$, whose kernel is identified with $A \cap B$ by the map $x \in A \cap B \mapsto (x, -x) \in A \times B$. We find an exact sequence

$$0 \to \Sigma_p \times \Sigma_q \to \kappa \to A \cap B \to 0.$$

Let κ_o be the inverse image of $\bar{C}_p \cap \bar{C}_q$ in κ; then we have an exact sequence

$$0 \to \Sigma_p \times \Sigma_q \to \kappa_o \to \bar{C}_p \cap \bar{C}_q \to 0.$$

This sequence is "nearly" split in the sense that there is a cyclic subgroup of κ_o which maps onto $\bar{C}_p \cap \bar{C}_q$ and whose intersection with $\Sigma_p \times \Sigma_q$ has order dividing 2. Indeed, to find such a subgroup, we can choose a generator t of $\bar{C}_p \cap \bar{C}_q$ and lifts x and y of t in C_p and C_q, respectively. The element $(x, -x, -y, y)$ of $J_o(p)^2 \times J_o(q)^2$ maps to the element $(t, -t)$ of $A \times B$, which we have identified with $t \in A \cap B$; it therefore is a lift of t to κ_o. The cyclic subgroup of $J_o(p)^2 \times J_o(q)^2$ which is generated by $(x, -x, -y, y)$ has an intersection with $\Sigma_p \times \Sigma_q$ which is of order 1 or 2, since $\Sigma_p \cap C_p = C_p[2]$ in $J_o(p)$, and $\Sigma_q \cap C_q = C_q[2]$ in $J_o(q)$.

COROLLARY **2** *The groups κ_o and κ are equal up to groups of 2-power order. More precisely, κ/κ_o is a 2-abelian group.*

Proof. Indeed, the indicated quotient is isomorphic to the quotient Q which appears in Theorem 1. ■

Another application of Theorem 1 concerns a question which was raised by Mazur ([6], §2b, Remark). For brevity, let us set $J = J_o(pq)$ and let J_{old} be the old subvariety $A + B$ of J. Let $J^{\text{new}} = J/J_{\text{old}}$, so that we have a tautological exact sequence

$$0 \to J_{\text{old}} \to J \to J^{\text{new}} \to 0.$$

Dualizing, we obtain a second sequence

$$0 \to (J^{\text{new}})^\vee \to J^\vee \to (J_{\text{old}})^\vee \to 0.$$

Since there is a canonical polarization $J \approx J^\vee$ (the *theta polarization*, coming from the fact that J is a Jacobian), we may regard $(J^{\text{new}})^\vee$ as an abelian subvariety of J. This subvariety of J is the *new subvariety* J_{new} of J, and the quotient J/J_{new} is the old quotient J^{old} of J. The composite of the inclusion $J_{\text{old}} \hookrightarrow J$ and the projection $J \to J^{\text{old}}$ is an isogeny

$$\lambda : J_{\text{old}} \to J^{\text{old}}.$$

Mazur asks for information about the *degree* of λ.

By the reasoning employed in §3 of [14], we obtain a direct relation between the kernel of λ and the group κ which appears above. Namely, let Θ be a line bundle on J corresponding to the "theta divisor" of J, and let M be the pullback of Θ to $J_{\text{old}} \subseteq J$. The isogeny λ is then the polarization ϕ_M which is attached to M (as defined in [8], Chapter II, §6). Let L be the pullback of Θ to $J_o(p)^2 \times J_o(q)^2$, and let $\Omega = K(L)$ be the kernel of the polarization

$$\phi_L : J_o(p)^2 \times J_o(q)^2 \to (J_o(p)^2 \times J_o(q)^2)^\vee$$

arising from L. The group Ω contains κ, and Ω is endowed with a canonical skew-symmetric $\mathbf{G_m}$-valued pairing. Let $\kappa^\perp$ be the annihilator of κ in this pairing. As explained in [8], §23, we have $\kappa^\perp \supseteq \kappa$, and a canonical isomorphism

$$\ker(\lambda) \approx \kappa^\perp/\kappa.$$

In particular, we have

$$\text{degree}(\lambda) = \text{card}(\Omega)/\,\text{card}(\kappa)^2.$$

To identify Ω, we rewrite $(J_o(p)^2 \times J_o(q)^2)^\vee$ as $J_o(p)^2 \times J_o(q)^2$, again using the autoduality of the Jacobian, and view ϕ_L as an endomorphism of $J_o(p)^2 \times J_o(q)^2$. Note, for the purposes of orientation, that any such endomorphism decomposes *a priori* as the "external product" of an endomorphism of $J_o(p)^2$ and an endomorphism of $J_o(q)^2$, since there are no homomorphisms in either direction between $J_o(p)$ and $J_o(q)$. (One can see this, for example, from the fact that $J_o(p)$ has good reduction at q, while $J_o(q)$ has purely toric reduction at q.) Hence Ω is automatically the product of a subgroup Ω_p of $J_o(p)^2$ and a subgroup Ω_q of $J_o(q)^2$. By the method of [14], §3, we find that ϕ_L may be decomposed as the the product of the endomorphism $\begin{pmatrix} 1+q & T_q \\ T_q & 1+q \end{pmatrix}$ of $J_o(p)^2$ and the endomorphism $\begin{pmatrix} 1+p & T_p \\ T_p & 1+p \end{pmatrix}$ of $J_o(q)^2$. These endomorphisms are both isogenies, and their degrees are respectively

$$\prod_f \left((1+q)^2 - a_q(f)^2\right)^2, \qquad \prod_g \left((1+p)^2 - a_p(g)^2\right)^2,$$

where f and g run over the sets of weight-2 newforms on $\Gamma_o(p)$ and $\Gamma_o(q)$, respectively. The notation $a_q(f)$, for instance, indicates the qth coefficient of the Fourier expansion of f. Hence we have

$$\text{card}(\Omega) = \prod_f \left((1+q)^2 - a_q(f)^2\right)^2 \cdot \prod_g \left((1+p)^2 - a_p(g)^2\right)^2.$$

Meanwhile, we have determined that card(κ) is the product of an integer of the form 2^t ($t \geq 0$) with the quantity

$$\text{card}(\kappa_o) = \text{num}\left(\tfrac{p-1}{12}\right) \cdot \text{num}\left(\tfrac{q-1}{12}\right) \cdot \text{gcd}\left(\text{num}\left(\tfrac{p-1}{24}\right), \text{num}\left(\tfrac{q-1}{24}\right)\right).$$

Refer to card(κ_o) as P. Summing up the discussion, we have

THEOREM **2** *Let D = degree(λ) be the order of the kernel of the natural map $J_{\text{old}} \to J^{\text{old}}$. Then D,* a priori *a perfect square, divides the integer*

$$\frac{\prod_f \left((1+q)^2 - a_q(f)^2\right)^2 \cdot \prod_g \left((1+p)^2 - a_p(g)^2\right)^2}{P^2}.$$

The ratio of this integer to D is a power of 2.

We prove Theorem 1 by arithmetic methods, combining the main theorem of [15] with an assortment of results from [5]. In particular, we rely on the results of [5] concerning: pure admissible groups, Ogg's Conjecture (Conjecture 2 of [10]), and a "twisted version" of Ogg's Conjecture *(loc. cit.)*. Since the statement of the theorem is purely transcendental, one imagines that the theorem may be proved by transcendental methods. It would be of considerable interest to find such a proof, which would presumably identify all of $A \cap B$, as opposed to its odd part.

It is a pleasure to thank the organizers of the TEXel conference for their kind invitation to an especially enjoyable and productive mathematical encounter. The results contained in this article were discovered in large part at the Weizmann Institute, during a workshop on Iwasawa Theory in May, 1989. The author wishes to thank this institution, and especially Shai Haran, for the hospitality. Finally, thanks go to Ling San for a careful reading of preliminary versions of this article, and to the referee, for helpful comments.

1 Hecke operators on $A \cap B$

For each integer $N \geq 1$, the modular curve $X_o(N)$ carries a family of Hecke correspondences T_n ($n \geq 1$). Further, for each positive divisor D of N such that D and N/D are relatively prime, one has an Atkin-Lehner involution w_D on $X_o(N)$. (See, for example, [7] for a discussion of these operators in various guises.) These operators induce endomorphisms of $J_o(N) = \mathrm{Pic}^o(X_o(N))$ which are again denoted by the symbols T_n and w_D. The subring of $\mathrm{End}(J_o(N))$ generated by the T_n is denoted $\mathbf{T}_N$. This ring is already generated by the operators T_ℓ for ℓ *prime*. If ℓ is a divisor of N, the operator T_ℓ is often denoted U_ℓ and referred to as an Atkin-Lehner operator.

The modular curves $X_o(N)$ for varying N are connected by *degeneracy operators*, which are discussed, for instance, in [6]. Recall that if N is a product DM, then there is a degeneracy operator $\pi_d : X_o(N) \to X_o(M)$ for each positive divisor d of D. By pullback, we obtain homomorphisms

$$\pi_d^* : J_o(M) \to J_o(N)$$

for each d. Assembling together $\pi_1^*, \pi_q^* : J_o(p) \rightrightarrows J_o(pq)$, we define the map

$$\alpha = \pi_1^* \times \pi_q^* : J_o(p)^2 \to J_o(pq).$$

The map $\beta : J_o(q)^2 \to J_o(pq)$ is defined similarly.

"Formulaire"

The compatibility of α and β with the various operators T_n and w_D is well known. Here is a summary of the behavior of these operators under α (for β, permute the roles of p and q):

1. We have $T_n(\alpha(x,y)) = \alpha(T_n x, T_n y)$ for all n prime to q, and $x, y \in J_o(p)$. In other words, for n prime to q we have $T_n \circ \alpha = \alpha \circ T_n$, where the latter T_n is the Hecke operator labeled T_n in $\mathbf{T}_p$, which is understood to be acting diagonally on $J_o(p)^2$.

2. We have $\alpha \circ w_p = w_p \circ \alpha$.

3. The qth Atkin-Lehner involution w_q on $J_o(pq)$ satisfies $w_q(\alpha(x,y)) = \alpha(y,x)$ for $x, y \in J_o(p)$. Equivalently, we have $w_q \circ \pi_q^* = \pi_1^*$ and $w_q \circ \pi_1^* = \pi_q^*$.

4. The qth Hecke operators T_q on $J_o(p)$ and $J_o(pq)$ satisfy

$$T_q(\alpha(x,y)) = \alpha(T_q x + qy, -x).$$

The last formula is probably clearer if we use the alternative notation U_q for the qth Hecke operator on $J_o(pq)$:

$$U_q(\alpha(x,y)) = \alpha(T_q x + qy, -x).$$

It is frequently advantageous for calculations to use the symbols U_p and U_q for the pth and q^{th} Hecke operators on $J_o(pq)$, reserving T_p and T_q for the pth Hecke operator on $J_o(q)$ and the qth Hecke operator on $J_o(p)$, respectively. In a similar vein, it is probably best to refer to the pth Hecke operator of $J_o(p)$ as U_p, and to the qth Hecke operator of $J_o(q)$ as U_q.

The formulas above show clearly that the subvariety A of $J_o(pq)$ is stable under the ring $\mathbf{T}_{pq}$ and under the involutions w_p and w_q. By symmetry, the intersection $A \cap B$ is $\mathbf{T}_{pq}$-stable, so that it is naturally a module for the algebra $\mathbf{T}_{pq}$.

It is important to note that $A \cap B$ carries, as well, natural actions of the two rings $\mathbf{T}_p$ and $\mathbf{T}_q$. To see this, it is enough, by symmetry, to exhibit a natural action of $\mathbf{T}_p$ on $A \cap B$. The ring $\mathbf{T}_p$ acts diagonally on $J_o(p)^2$, and Σ_p is $\mathbf{T}_p$-stable in $J_o(p)^2$. Therefore, there is a natural action of $\mathbf{T}_p$ on A, and the claim is that $A \cap B$ is stable under this action. The only subtle point is the stability of $A \cap B$ under the operator labeled T_q in $\mathbf{T}_p$, which does *not* coincide in general on A with the operator U_q coming from $\mathbf{T}_{pq}$.

To treat this point, we use the last of the above formulas, plus the Cayley-Hamilton Theorem, to establish the identity $U_q^2 - U_q T_q + q = 0$ on A. On B, U_q is an involution: the negative of the involution w_q. (This follows, for instance, from the proof of Proposition 3.7 of [15]. The endomorphism $w_q + U_q$ of $J_o(q)$ factors through the degeneracy map $\pi^* : J_o(1) \to J_o(q)$, whose source is 0.) We therefore have

$$T_q = U_q(q+1) = -w_q(q+1)$$

on $A \cap B$.

2 Galois action on $A \cap B$

In the above discussion, we have considered $A \cap B$ as a $\mathbf{T}_p$-stable submodule of A. A closely related module is the inverse image $(A \cap B)^{\sim}$ of $A \cap B$ in $J_o(p)^2$. Thus $(A \cap B)^{\sim}$ is an extension of $A \cap B$ by the Shimura subgroup Σ_p of $J_o(p)$, which we identify with its antidiagonal image in $J_o(p)^2$. The group $(A \cap B)^{\sim}$ is a finite $\mathbf{T}_p$-stable submodule of $J_o(p)^2$. *Until further notice, we shall write simply* $\mathbf{T}$ *for the Hecke algebra* $\mathbf{T}_p$.

Up to now, we have tacitly regarded the curves $X_o(p)$, $X_o(q)$, and $X_o(pq)$, and their Jacobians, as being defined over $\mathbf{C}$. However, one knows from work of Shimura (see, e.g., [18]) that these curves exist over $\mathbf{Q}$. (In fact, by [1] there are even good models for these curves over $\mathbf{Z}$. See also [4].) One sees from their modular definitions that the various Hecke operators, Atkin-Lehner involutions, and degeneracy operators we have considered are all defined over $\mathbf{Q}$. It follows from this that the abelian subvarieties A and B of $J_o(pq)$ are defined over $\mathbf{Q}$, so that the intersection $A \cap B$ is defined over $\mathbf{Q}$. We view it as a finite $\mathrm{Gal}(\overline{\mathbf{Q}}/\mathbf{Q})$-module with an equivariant action of the ring $\mathbf{T}$, or, equivalently, as a $\mathbf{T}[\mathrm{Gal}(\overline{\mathbf{Q}}/\mathbf{Q})]$-module. From the definition of $(A \cap B)^{\sim}$ as an inverse image, we see that this subgroup of $J_o(p)^2$, with its $\mathbf{T}$-action, is defined over $\mathbf{Q}$.

THEOREM 3 *The* $\mathrm{Gal}(\overline{\mathbf{Q}}/\mathbf{Q})$*-modules* $A \cap B$ *and* $(A \cap B)^{\sim}$ *extend to finite flat commutative group schemes over* $\mathrm{Spec}(\mathbf{Z})$.

Proof. The theorem means that there are groups schemes $\mathcal{G}_1$ and $\mathcal{G}_2$ over $\mathrm{Spec}(\mathbf{Z})$ whose associated Galois modules $\mathcal{G}_i(\overline{\mathbf{Q}})$ are isomorphic to $A \cap B$ and $(A \cap B)^{\sim}$, respectively.

The group $A \cap B$ extends to a finite flat group scheme over $\mathrm{Spec}(\mathbf{Z}[\frac{1}{p}])$ because it is a rational subgroup of the abelian variety A, which has good reduction outside p. Symmetrically, $A \cap B$ extends to a finite flat group scheme over $\mathrm{Spec}(\mathbf{Z}[\frac{1}{q}])$. ¿From this, we may deduce that it extends to a finite flat group scheme over $\mathrm{Spec}(\mathbf{Z})$. (For example, we can apply the discussion of [5], Chapter I, §1 to the ℓ-primary part of $A \cap B$, for each prime number ℓ.)

We have an exact sequence

$$0 \to \Sigma_p \to (A \cap B)^{\sim} \to A \cap B \to 0.$$

To show that $(A \cap B)^{\sim}$ extends to Spec($\mathbf{Z}$), we may treat separately the ℓ-primary components of $(A \cap B)^{\sim}$. The assertion to be proved is obvious for those ℓ which are prime to the order $n_p = \text{num}\left(\frac{p-1}{12}\right)$ of Σ_p, since the ℓ-primary components of $A \cap B$ and $(A \cap B)^{\sim}$ are isomorphic in that case. It thus suffices to treat the prime-to-p part of $(A \cap B)^{\sim}$, which is a finite Gal($\overline{\mathbf{Q}}/\mathbf{Q}$)-stable submodule $R \supseteq \Sigma_p$ of $J_o(p)^2(\overline{\mathbf{Q}})$.

We are required to show that R is *unramified at* p. Fix a decomposition group $D = \text{Gal}(\overline{\mathbf{Q}}_p/\mathbf{Q}_p) \subset \text{Gal}(\overline{\mathbf{Q}}/\mathbf{Q})$ for p in Gal($\overline{\mathbf{Q}}/\mathbf{Q}$), and let I be the inertia subgroup of D. We wish to show that I acts trivially on R. More generally:

LEMMA 1 *Let $G \supseteq \Sigma_p$ be a finite I-stable subgroup of $J_o(p)^2(\overline{\mathbf{Q}}_p)$, whose order is prime to p. Assume that I acts trivially on G/Σ_p. Then I acts trivially on G.*

To prove the lemma, we first note the following facts, which are variants for $J_o(p)^2$ of results proved by Mazur [5] for $J_o(p)$:

1. The group Σ_p extends to a finite flat subgroup of the Néron model $\mathcal{J}$ of $J_o(p)^2$ over Spec($\mathbf{Z}$). (Compare [5], p. 100.)

2. In characteristic p, Σ_p has trivial intersection with the connected component $\mathcal{J}^o_{/\mathbf{F}_p}$ of $\mathcal{J}$. (Cf. [5], p. 101.)

In the latter statement, the group $T = \mathcal{J}^o_{/\mathbf{F}_p}$ is a torus over $\mathbf{F}_p$. The group $T(\overline{\mathbf{F}}_p)$, which is a torsion abelian group with trivial p-primary component, may be canonically identified with a subgroup of $J_o(p)^2(\overline{\mathbf{Q}}_p)^I$ (for example, by [17], Lemma 2). The second assertion gives the equality $\Sigma_p \cap T(\overline{\mathbf{F}}_p) = 0$ inside $J_o(p)^2(\overline{\mathbf{Q}}_p)$.

With these preliminary facts recorded, we may now prove the lemma by a variant of the argument given for Lemma 16.5 of [5], Chapter II. Take $g \in G$ and $\gamma \in I$. Since I acts trivially on G/Σ_p, we have $(i-1)g \in \Sigma_p$. On the other hand, $(i-1)g$ lies in the group $T(\overline{\mathbf{F}}_p)$. This follows from the fact that $J_o(p)^2$ has purely toric reduction at p, as can be seen from the discussion in Exposé IX, §7 of [3] or the exact sequence which is given as Lemma 3.3.1 of [13]. Hence $(i-1)g = 0$, which gives the desired statement that i acts trivially on g. ■

3 Maximal ideals of the Hecke algebra $\mathbf{T}_p$

The *Eisenstein ideal* of $\mathbf{T} = \mathbf{T}_p$ is the ideal I generated by the elements $T_\ell - \ell - 1$ for prime numbers $\ell \neq p$, together with the difference $U_p - 1$ ([5], p. 95). The Eisenstein primes of $\mathbf{T}$ are the maximal ideals m of $\mathbf{T}$ which contain I. These ideals are in 1-1 correspondence with the prime numbers dividing $n_p = \text{num}\left(\frac{p-1}{12}\right)$, a prime number $\ell \mid n_p$ corresponding to the maximal ideal $\mathsf{m} = (I, \ell)$.

For each maximal ideal m of $\mathbf{T}$, let ρ_m be the usual semisimple two-dimensional representation of $\text{Gal}(\overline{\mathbf{Q}}/\mathbf{Q})$ over $k_\mathsf{m} = \mathbf{T}/\mathsf{m}$ which is associated to m by the constructions of [2]. Thus, ρ_m is unramified outside the primes p and ℓ, where ℓ is the characteristic of the finite field k_m. For r a prime other than ℓ or p, the characteristic polynomial of $\rho_\mathsf{m}(\mathsf{Frob}_r)$, where Frob_r is a Frobenius element for r in the Galois group $\text{Gal}(\overline{\mathbf{Q}}/\mathbf{Q})$, is the polynomial $X^2 - T_r X + r \in k_\mathsf{m}[X]$. One knows ([5], Chap. II, §14) that ρ_m is reducible over k_m if and only m is Eisenstein. In this case, k_m is the prime field $\mathbf{F}_\ell$, and ρ_m is isomorphic to the direct sum of the trivial 1-dimensional representation and the 1-dimensional representation μ_ℓ of $\text{Gal}(\overline{\mathbf{Q}}/\mathbf{Q})$.

Recall that ρ_m is *finite* at p (cf. [16]) if and only if: the restriction of ρ_m to a decomposition group $\text{Gal}(\overline{\mathbf{Q}}_p/\mathbf{Q}_p)$ for p in $\text{Gal}(\overline{\mathbf{Q}}/\mathbf{Q})$ is isomorphic to the representation arising from a k_m-vector space scheme of rank 2 (in the sense of [12]) over $\mathbf{Z}_p$. For $\ell \neq p$, ρ_m is finite at p if and only if it is unramified at p.

THEOREM 4 *Assume that $\ell \neq 2$. Suppose that ρ_m is finite at p. Then m is Eisenstein.*

Proof. Assume that ρ_m is finite at p, but not Eisenstein. Then, by the main theorem (Theorem 1.1) of [15], the representation ρ_m is "modular of level 1." In particular, ρ_m may be realized by a group of ℓ-torsion points of the abelian variety $J_o(1)$. This is absurd, since $J_o(1)$ is 0. ■

4 Proof of the main theorem

Let M be the "odd part" of $(A \cap B)^{\sim}$, i.e., the direct sum of the ℓ-primary subgroups of $(A \cap B)^{\sim}$, for ℓ odd. Our aim is to show that M is "small."

To do this, we first control the set of prime ideals of $\mathbf{T}$ which are in the support of M:

PROPOSITION 1 *If* $\mathfrak{m}$ *is a maximal ideal of* $\mathbf{T}$ *in the support of* M*, then* $\mathfrak{m}$ *is an Eisenstein prime.*

Proof. Let $\mathfrak{m}$ be in the support of M. Then, by the definition of M, $\mathfrak{m}$ is prime to 2. Let $M[\mathfrak{m}]$ be the kernel of $\mathfrak{m}$ on M, i.e., the set of $m \in M$ which are killed by all elements of $\mathfrak{m}$. Since M is finite, and $\mathfrak{m}$ lies in the support of M, $M[\mathfrak{m}]$ is non-zero. Assume that $\mathfrak{m}$ is in the support of M and that $\mathfrak{m}$ is non-Eisenstein. Then a well known argument of Mazur ([5], proof of Proposition 14.2 of Chapter II) shows that the $k_{\mathfrak{m}}[\mathrm{Gal}(\overline{\mathbf{Q}}/\mathbf{Q})]$-module $M[\mathfrak{m}]$ is a successive extension of copies of the representation $\rho_{\mathfrak{m}}$. In other words, let V be a $k_{\mathfrak{m}}[\mathrm{Gal}(\overline{\mathbf{Q}}/\mathbf{Q})]$-module which affords $\rho_{\mathfrak{m}}$. Then the semisimplification of $M[\mathfrak{m}]$ is some (non-zero) power of V.

In particular, we can find an embedding $V \hookrightarrow M$. By Theorem 3, M extends to a finite flat group scheme $\mathcal{M}$ over $\mathrm{Spec}(\mathbf{Z})$. The Zariski closure of V in $\mathcal{M}$ is then a finite flat group scheme $\mathcal{V}$ over $\mathrm{Spec}(\mathbf{Z})$ which prolongs V. Thus $\rho_{\mathfrak{m}}$ is finite at p, which contradicts Theorem 4. ■

COROLLARY *For each prime* ℓ*, let* M_ℓ *be the* ℓ*-primary part of the abelian group* M*. Then* M_ℓ *is trivial unless* ℓ *is an odd prime dividing* n_p*, in which case the semisimplification of* M_ℓ *is a direct sum of modules of the form* μ_ℓ *and* $\mathbf{Z}/\ell\mathbf{Z}$*.*

Proof. By construction, the order of M is odd. By the Proposition, only primes ℓ dividing n_p can divide the order of M. Moreover, for $\ell \mid n_p$, only the Eisenstein prime $\mathfrak{m} = (I, \ell)$ can intervene in the support of M_ℓ. Hence M_ℓ is annihilated by some power of $\mathfrak{m}$, which means that $M_\ell \subseteq J_o(p)^2[\mathfrak{m}^\nu]$ for some integer $\nu \geq 0$. All Jordan-Hölder constituents of the latter module are of the form μ_ℓ and $\mathbf{Z}/\ell\mathbf{Z}$ ([5], Chapter II, Proposition 14.1). ■

THEOREM 5 *The module* $M \subset J_o(p) \times J_o(p)$ *is contained in the direct sum* $N \oplus N$*, where* N *is the submodule* $\Sigma_p + C_p$ *of* $J_o(p)$*.*

Proof. The Gal($\overline{\mathbf{Q}}/\mathbf{Q}$)-module M extends to a finite flat group scheme $\mathcal{M}$ over Spec($\mathbf{Z}$) (Theorem 3). In the language of [5], Chapter I, §1(f), the above Corollary states that the ℓ-primary parts of $\mathcal{M}$ are "admissible." Proposition 4.5 of [5], Chapter I then tells us that $\mathcal{M}$ is *pure* in the sense that it is the direct product of a constant group by a group whose dual is constant. (A group whose dual is constant is called a "μ-type group" in [5].)

The largest constant subgroup of $J_o(p)$ is C_p ([5], Chapter III, Theorem 1.2), while the largest μ-type subgroup of $J_o(p)$ is Σ_p ([5], Chapter III, Theorem 1.3.) ■

Note that the sum $\Sigma_p + C_p$ inside $J_o(p)$ is very nearly a direct sum. The intersection $\Sigma_p \cap C_p$ is the group of elements of order dividing 2 in C_p ([5], Chapter II, Proposition 11.11). This group has order 2 if $p \equiv 1 \pmod 8$ and is trivial otherwise.

The Theorem implies that M is contained in the direct sum $J_o(p)[I] \oplus J_o(p)[I]$, where I is again the Eisenstein ideal.

We now prove the main result (Theorem 1), whose statement we reformulate as follows:

> *The odd part of $A \cap B$ is contained in the intersection of the two groups $\bar{C}_p = \alpha(C_p^-)$ and $\bar{C}_q = \beta(C_q^-)$, the exponent $^-$ indicating that C_p and C_q have been embedded antidiagonally in $J_o(p)^2$ and $J_o(q)^2$, respectively.*

Proof. By symmetry, it suffices to show that the odd part of $A \cap B$ is contained in $\bar{C}_p$. We know by Theorem 5 that the odd part of $A \cap B$ is contained in $\alpha(N \oplus N)$. Since α kills the antidiagonal Σ_p^-, the group $\alpha(N \oplus N)$ is, neglecting 2-abelian groups, the sum

$$\alpha(C_p^+) + \alpha(C_p^-) + \alpha(\Sigma_p^+),$$

where the exponent $^+$ is now used for the diagonal embedding. The prime-to-2 part of this sum is direct. By the *formulaire* presented above, the Atkin-Lehner involution w_q operates as $+1$ on the groups with exponent $^+$ and as -1 on the group with exponent $^-$. However, w_p acts on $J_o(p)[I]$ as -1. Therefore, w_p acts on the displayed sum as -1, so that w_p is -1 on the odd part of $A \cap B$.

By symmetry, w_q must act as -1 on the odd part of $A \cap B$. Therefore, this odd part is contained in $\alpha(C_p^-)$, as was claimed. ■

5 Computations with cusps

The aim of this § is to justify the claim, made in the introduction, that the subgroup $\bar{C}_p = \alpha(C_p^-)$ of $J_o(pq)$ lies in the cyclic subgroup of $J_o(pq)$ generated by the class of the divisor $P_1 - P_p - P_q + P_{pq}$. This divisor is formed from the four cusps of the curve $X_o(pq)$, which are in natural 1-1 correspondence with the positive divisors of pq. We have used the notation of Ogg [9], who writes P_d for the cusp corresponding to the divisor d. This notation will apply also for the modular curve $X_o(p)$; thus we will consider that C_p is the cyclic subgroup of $J_o(p)$ generated by the class of the divisor $P_1 - P_p$ on $X_o(p)$. We recall also that the the map α is constructed from the two degeneracy coverings

$$\pi_1, \pi_q : X_o(pq) \rightrightarrows X_o(p)$$

and that the $^-$ in C_p^- indicates the antidiagonal embedding. Therefore, $\bar{C}_p$ is the cyclic group generated by $(\pi_1^* - \pi_q^*)(\overline{P_1 - P_p})$; the "bar" over $P_1 - P_p$ is used here in denote the class of the indicated divisor.

To study this divisor, we will consider the maps π_1^* and π_q^* which are induced by the degeneracy maps *on the level of divisors*. The only points of $X_o(pq)$ lying over the cusp P_1 of $X_o(p)$ are the cusps P_1 and P_q of $X_o(pq)$. Hence we have $\pi_1^*(P_1) = aP_1 + bP_q$ for some integers $a, b \geq 0$; these integers sum to $q+1$, the degree of the covering π_1. [The actual values of a and b, which are not needed here, are the ramification indices of P_1 and P_q in the covering $\pi_1 : X_o(pq) \to X_o(p)$. They are 1 and q, up to permutation. The author computed them by calculating the divisors of the function $\Delta(z)/\Delta(pz)$ on the two curves $X_o(p)$ and $X_o(pq)$, employing the techniques presented in [11]. An alternative approach, suggested by the referee, is to identify a and b with the ramification indices of P_1 and P_q in the covering $\pi_1 : X_o(q) \to X_o(1)$, and to compute these latter indices by techniques involving fundamental domains.]

The covering $\pi_1 : X_o(pq) \to X_o(p)$ is equivariant with respect to the Atkin-Lehner involutions w_p on $X_o(p)$ and $X_o(pq)$. Further, on both of these curves, w_p permutes the cusp labeled P_1 with the cusp labeled P_p. Finally, the involution w_p on $X_o(pq)$ permutes the cusps P_q and P_{pq}. Therefore, $\pi_1^*(P_p) = aP_p + bP_{pq}$. On the other hand, we have $\pi_1 w_q = \pi_q$, and the involution w_q of $X_o(pq)$ permutes P_1 with P_q and P_p with P_{pq}. Therefore, we have:

$$\pi_q^*(P_1) = aP_q + bP_1, \qquad \pi_q^*(P_p) = aP_{pq} + bP_p.$$

Combining everything together gives

$$(\pi_1^* - \pi_p^*)(P_1 - P_p) = (a - b)(P_1 - P_p - P_q + P_{pq}).$$

By passing to the level of divisor classes, we obtain the desired result.

References

[1] Deligne, P., Rapoport, M.: Schémas de modules de courbes elliptiques. Lecture Notes in Mathematics **349**, 143–316 (1973)

[2] Deligne, P., Serre, J-P.: Formes modulaires de poids 1. Ann. Sci. Ecole Norm. Sup. **7**, 507–530 (1974)

[3] (SGA 7 I) Grothendieck, A.: Groupes de monodromie en géométrie algébrique. Lecture Notes in Mathematics **288**. Berlin-Heidelberg-New York: Springer 1972

[4] Katz, N. M., Mazur, B.: Arithmetic Moduli of Elliptic Curves. Annals of Math. Studies **108**. Princeton: Princeton University Press 1985

[5] Mazur, B.: Modular curves and the Eisenstein ideal. Publ. Math. IHES **47**, 33–186 (1977)

[6] Mazur, B.: Rational isogenies of prime degree. Invent. Math. **44**, 129–162 (1978)

[7] Mazur, B., Wiles, A.: Class fields of abelian extensions of **Q**. Invent. Math. **76**, 179–330 (1984)

[8] Mumford, D.: Abelian Varieties. London: Oxford University Press 1970

[9] Ogg, A.: Hyperelliptic modular curves. Bull. Soc. Math. France **102**, 449–462 (1974)

[10] Ogg, A.: Diophantine equations and modular forms. Bull. AMS **81**, 14–27 (1975)

[11] Ogg, A.: Rational points on certain elliptic modular curves. Proc. Symp. Pure Math. **24**, 221–231 (1973)

[12] Raynaud, M.: Schémas en groupes de type $(p, \ldots, p)$. Bull. Soc. Math. France **102**, 241–280 (1974)

[13] Ribet, K.: Galois action on division points of abelian varieties with real multiplications. Am. J. Math. **98**, 751–804 (1976)

[14] Ribet, K.: Congruence relations between modular forms. Proc. International Congress of Mathematicians 1983, 503–514

[15] Ribet, K.: On modular representations of $\mathrm{Gal}(\overline{\mathbf{Q}}/\mathbf{Q})$ arising from modular forms. Invent. Math. To appear

[16] Serre, J-P.: Sur les représentations modulaires de degré 2 de $\mathrm{Gal}(\overline{\mathbf{Q}}/\mathbf{Q})$. Duke Math. J. **54**, 179–230 (1987)

[17] Serre, J-P., Tate, J.: Good reduction of abelian varieties. Annals of Math. **88**, 492–517 (1968)

[18] Shimura, G.: Introduction to the Arithmetic Theory of Automorphic Functions. Princeton: Princeton University Press 1971

Kolyvagin's System of Gauss Sums

KARL RUBIN*

Introduction.

Recently in [1] Kolyvagin introduced a remarkable inductive procedure which improves upon Stickelberger's theorem and results of Thaine [6] on ideal class groups of cyclotomic fields. For every Dirichlet character χ modulo a prime p Kolyvagin was able to determine the order of the χ-component of the p-part of the ideal class group of $\mathbf{Q}(\boldsymbol{\mu}_p)$. These orders were already known from the work of Mazur and Wiles [3], but Kolyvagin's proof is very much simpler. Kolyvagin's method also determines the abelian group structure of these ideal class groups in terms of Stickelberger ideals.

The proof divides naturally into two cases, according as to whether the character χ is even or odd. When χ is even the proof uses cyclotomic units in an infinite family of abelian fields containing $\mathbf{Q}(\boldsymbol{\mu}_p)$; an account of this method and a generalization to $\mathbf{Q}(\boldsymbol{\mu}_{p^\infty})$ (the Mazur-Wiles theorem, Iwawawa's "main conjecture") are given in [4]. When χ is odd, the proof relies on Gauss sums in an infinite family of abelian fields containing $\mathbf{Q}(\boldsymbol{\mu}_p)$, as does the proof of Stickelberger's theorem. In this paper we give an exposition of Kolyvagin's results in the case of odd characters, dealing for simplicity only with the field $\mathbf{Q}(\boldsymbol{\mu}_p)$. For the main results see Theorems 4.3 and 4.4.

I would like to thank René Schoof and Alice Silverberg for helpful conversations.

§1. Setting and notation.

Fix a rational prime $p > 2$ and let $F = \mathbf{Q}(\boldsymbol{\mu}_p)$. Fix a large power M of p (to be specified later) and let $\mathcal{S}$ denote the set of positive squarefree integers divisible only by primes $l \equiv 1 \pmod{M}$. In particular these l split

*partially supported by NSF grants

completely in $F/\mathbf{Q}$. For every $n \in \mathcal{S}$ write

$$G_n = \mathrm{Gal}(F(\boldsymbol{\mu}_n)/F) \cong \mathrm{Gal}(\mathbf{Q}(\boldsymbol{\mu}_n)/\mathbf{Q})$$

and write $\mathbf{N}_n$ for the norm operator

$$\mathbf{N}_n = \sum_{\tau \in G_n} \tau \in \mathbf{Z}[G_n].$$

There is a natural isomorphism $G_n = \prod_{l|n} G_l$ (product over primes l dividing n), and $\mathbf{N}_n = \prod_{l|n} \mathbf{N}_l \in \mathbf{Z}[G_n]$. If $mn \in \mathcal{S}$ we will identify G_n with $\mathrm{Gal}(F(\boldsymbol{\mu}_{mn})/F(\boldsymbol{\mu}_m)) \subset G_{mn}$, and if $l \in \mathcal{S}$ is prime, $l \nmid n$, then we will write $\mathrm{Fr}(l)$ for the Frobenius of l in G_n, the automorphism which sends each n-th root of unity to its l-th power. For every prime $l \in \mathcal{S}$ fix a generator σ_l of G_l (which is cyclic of order $l-1$) and define

$$\mathbf{D}_l = \sum_{i=1}^{l-2} i\sigma_l^i \in \mathbf{Z}[G_l].$$

This "operator" is constructed to satisfy the identity

$$(\sigma_l - 1)\mathbf{D}_l = (l-1) - \mathbf{N}_l \tag{1}$$

in $\mathbf{Z}[G_l]$. For $n \in \mathcal{S}$ define $\mathbf{D}_n = \prod_{l|n} \mathbf{D}_l \in \mathbf{Z}[G_n]$.

Suppose $n \in \mathcal{S}$, $r \equiv 1 \pmod{pn}$ is prime, ζ_r is a primitive r-th root of unity, and $\mathfrak{r}$ is a prime of $F(\boldsymbol{\mu}_n)$ above r. Define the Gauss sum

$$g(n, \mathfrak{r}, \zeta_r) = \sum_{a=1}^{r-1} \varepsilon_{n,\mathfrak{r}}(a)\zeta_r^a \in F(\boldsymbol{\mu}_{rn})^\times$$

where $\varepsilon_{n,\mathfrak{r}} : (\mathbf{Z}/r\mathbf{Z})^\times \to \boldsymbol{\mu}_{pn}$ is the character satisfying

$$\varepsilon_{n,\mathfrak{r}}(a) \equiv a^{-(r-1)/pn} \pmod{\mathfrak{r}}.$$

These numbers transform in the following way under Galois automorphisms:

$$g(n, \mathfrak{r}, \zeta_r)^\sigma = g(n, \mathfrak{r}, \zeta_r^\sigma) = \varepsilon_{n,\mathfrak{r}}(\sigma)^{-1} g(n, \mathfrak{r}, \zeta_r) \quad \text{for } \sigma \in G_r, \tag{2}$$

$$g(n, \mathfrak{r}, \zeta_r)^\sigma = \sum_{a=1}^{r-1} \varepsilon_{n,\mathfrak{r}}^\sigma(a)\zeta_r^a = g(n, \mathfrak{r}^\sigma, \zeta_r) \quad \text{for } \sigma \in G_n. \tag{3}$$

Write $\Delta = \mathrm{Gal}(F/\mathbf{Q}) \cong (\mathbf{Z}/p\mathbf{Z})^\times$. We can view Δ as a subgroup of $\mathrm{Gal}(\mathbf{Q}^{ab}/\mathbf{Q})$ by identifying it with the tame inertia group of p. This allows us to identify Δ with a subgroup of $\mathrm{Gal}(F(\boldsymbol{\mu}_n)/\mathbf{Q})$ for every integer n. For every character $\chi : \Delta \to \mathbf{Z}_p^\times$ define

$$e_\chi = \frac{1}{p-1} \sum_{\sigma \in \Delta} \chi^{-1}(\sigma)\sigma \in \mathbf{Z}_p[\Delta] \subset \mathbf{Z}_p[\mathrm{Gal}(\mathbf{Q}^{ab}/\mathbf{Q})].$$

If Y is a $\mathbf{Z}_p[\mathrm{Gal}(\mathbf{Q}^{ab}/\mathbf{Q})]$-module, we write $Y^\chi = e_\chi Y$, the maximal submodule of Y on which Δ acts via χ.

Fix once and for all a generator δ of Δ, and for every $n \in \mathcal{S}$ fix an integer b_n such that

$$\zeta^\delta = \zeta^{b_n} \quad \text{for all} \quad \zeta \in \boldsymbol{\mu}_{Mpn}.$$

For every $n \in \mathcal{S}$ and prime $\mathfrak{r}$ of $F(\boldsymbol{\mu}_n)$, $\mathfrak{r} \mid r \equiv 1 \pmod{pn}$, define

$$\alpha(n, \mathfrak{r}) = g(n, \mathfrak{r}, \zeta_r)^{\delta - b_n}$$

with some choice of ζ_r. Since $\delta - b_n$ annihilates $\boldsymbol{\mu}_{pn}$, (2) shows that $\alpha(n, \mathfrak{r}) \in F(\boldsymbol{\mu}_n)^\times$ and $\alpha(n, \mathfrak{r})$ is independent of the choice of ζ_r.

For $n \in \mathcal{S}$ define the Stickelberger element

$$s(n) = \Big(\sum_{\substack{a=1 \\ (a,pn)=1}}^{pn} a\tau_a^{-1} \Big) \in \mathbf{Z}[\mathrm{Gal}(F(\boldsymbol{\mu}_n)/\mathbf{Q})]$$

where $\tau_a \in \mathrm{Gal}(F(\boldsymbol{\mu}_n)/\mathbf{Q})$ is the automorphism which sends each $\zeta \in \boldsymbol{\mu}_{pn}$ to ζ^a, and define

$$\theta(n) = \frac{1}{pn}(\delta - b_n)s(n), \quad \theta(n, \chi) = e_\chi \theta(n)$$

for every character χ of Δ. It is well-known (or an easy exercise) that $\theta(n) \in \mathbf{Z}[\mathrm{Gal}(F(\boldsymbol{\mu}_n)/\mathbf{Q})]$, so $\theta(n, \chi) \in \mathbf{Z}_p[\mathrm{Gal}(F(\boldsymbol{\mu}_n)/\mathbf{Q})]^\chi$.

Proposition 1.1. *Suppose $n \in \mathcal{S}$ and $\mathfrak{r}$ is a prime of $F(\boldsymbol{\mu}_n)$ lying above a rational prime congruent to 1 modulo pn.*

(i) *If l is a prime dividing n, then $\alpha(n, \mathfrak{r})^{\mathbf{N}_l} = \alpha(n/l, \mathbf{N}_l\mathfrak{r})^{1 - \mathrm{Fr}(l)^{-1}} \beta^M$ for some $\beta \in F(\boldsymbol{\mu}_{n/l})^\times$.*

(ii) *The principal ideal $(\alpha(n, \mathfrak{r})) = \theta(n)\mathfrak{r}$.*

Proof. Using the Davenport-Hasse distribution relation ([2] Chapter 2, Theorem 10.1) and (2) we have

$$\begin{aligned}
\alpha(n,\mathfrak{r})^{\mathbf{N}_l} &= \prod_{\sigma\in G_l}(\sum_{a=1}^{r-1}\varepsilon_{n,\mathfrak{r}}^{\sigma}(a)\zeta_r^{a})^{\delta-b_n} \\
&= (g(n/l,\mathbf{N}_l\mathfrak{r},\zeta_r^{l})^{-\mathrm{Fr}(l)^{-1}}\prod_{\psi:G_r\to\boldsymbol{\mu}_l}(\sum_{a=1}^{r-1}\psi\varepsilon_{n,\mathfrak{r}}(a)\zeta_r^{a}))^{\delta-b_n} \\
&= (g(n/l,\mathbf{N}_l\mathfrak{r},\zeta_r)^{1-\mathrm{Fr}(l)^{-1}}\varepsilon_{n,\mathfrak{r}}(l)^{(1-l)}r^{(l-1)/2})^{\delta-b_n}.
\end{aligned}$$

Since $M|(l-1)/2$ and $(Mn/l)|(b_n-b_{n/l})$, this proves (i). The second assertion is the classical ideal factorization of the Gauss sums; see for example [2] Chapter 2, Theorem 2.2. ■

§2. Properties of Kolyvagin's system of Gauss sums.

Lemma 2.1. *If $n\in\mathcal{S}$ and $\mathfrak{r}$ is a prime of $F(\boldsymbol{\mu}_n)$ lying above a rational prime congruent to 1 modulo pn, then*
(i) $\alpha(n,\mathfrak{r})^{\mathbf{D}_n}\in[F(\boldsymbol{\mu}_n)^\times/(F(\boldsymbol{\mu}_n)^\times)^M]^{G_n}$,
(ii) $\mathbf{D}_n\theta(n)\in(\mathbf{Z}/M\mathbf{Z})[Gal(F(\boldsymbol{\mu}_n)/\mathbf{Q})]^{G_n}$.

Proof. We will show (i) by induction on the number of primes dividing n. If $n=1$ there is nothing to prove. If $l|n$, then by (1) and Proposition 1.1(i)

$$\begin{aligned}
\alpha(n,\mathfrak{r})^{\mathbf{D}_n(\sigma_l-1)} &= \alpha(n,\mathfrak{r})^{\mathbf{D}_{n/l}((l-1)-\mathbf{N}_l)} \\
&\equiv \alpha(n/l,\mathbf{N}_l\mathfrak{r})^{\mathbf{D}_{n/l}(\mathrm{Fr}(l)^{-1}-1)} \pmod{(F(\boldsymbol{\mu}_n)^\times)^M}.
\end{aligned}$$

Since $\mathrm{Fr}(l)$ fixes F, our induction hypothesis shows

$$\alpha(n/l,\mathbf{N}_l\mathfrak{r})^{\mathbf{D}_{n/l}(\mathrm{Fr}(l)^{-1}-1)}\in(F(\boldsymbol{\mu}_{n/l})^\times)^M.$$

These σ_l generate G_n, so this proves (i). The second assertion can be shown similarly; alternatively it follows from (i) by Proposition 1.1(ii). ■

From now on we fix an odd character χ of Δ and suppose that $\chi\neq\omega$ where ω is the Teichmuller character giving the action of Δ on $\boldsymbol{\mu}_p$.

Lemma 2.2. *If F' is an extension of F, abelian over $\mathbf{Q}$, then the canonical map*

$$(F^\times/(F^\times)^M)^\chi\to([F'^\times/(F'^\times)^M]^{\mathrm{Gal}(F'/F)})^\chi$$

is an isomorphism.

Proof. Using the canonical isomorphisms

$$F^\times/(F^\times)^M \cong H^1(\overline{F}/F, \boldsymbol{\mu}_M) \text{ and } F'^\times/(F'^\times)^M \cong H^1(\overline{F}/F', \boldsymbol{\mu}_M),$$

we get an inflation-restriction exact sequence

$$H^1(F'/F, \boldsymbol{\mu}_M(F')) \to F^\times/(F^\times)^M \to \\ \to [F'^\times/(F'^\times)^M]^{\mathrm{Gal}(F'/F)} \to H^2(F'/F, \boldsymbol{\mu}_M(F')).$$

Since $\chi \neq \omega$ the outer terms have trivial χ-component. ∎

If $n \in \mathcal{S}$, $\mathfrak{r}$ is a prime of $F(\boldsymbol{\mu}_n)$ lying above a rational prime $r \equiv 1 \pmod{pn}$, and ρ denotes the prime of F below $\mathfrak{r}$, we will write $\kappa(n,\rho) \in (F^\times/(F^\times)^M)^\chi$ for the image of $\alpha(n,\mathfrak{r})^{e_\chi \mathbf{D}_n}$ under the isomorphism of Lemma 2.2 with $F' = F(\boldsymbol{\mu}_n)$, using Lemma 2.1(i). By Lemma 2.1(i) and (3), this image depends on ρ, not on $\mathfrak{r}$, so this notation is well-defined. We also define $\delta(n) \in (\mathbf{Z}/M\mathbf{Z})[\Delta]^\chi$ to be the element satisfying

$$\delta(n)\mathbf{N}_n = \mathbf{D}_n \theta(n,\chi).$$

The existence of $\delta(n)$ is guaranteed by Lemma 2.1(ii) since

$$(\mathbf{Z}/M\mathbf{Z})[\mathrm{Gal}(F(\boldsymbol{\mu}_n)/\mathbf{Q})]^{G_n} = \mathbf{N}_n(\mathbf{Z}/M\mathbf{Z})[\Delta].$$

Each $\kappa(n,\rho)$ gives a principal ideal of F (modulo M-th powers of ideals) which can be viewed as a relation in the ideal class group of F. These relations will be used to bound the size of the ideal class group. To do this, we must understand the prime factorizations of these ideals and also how to choose n and ρ so as to get useful relations.

Let $\mathcal{O}_F$ denote the ring of integers of F, and write $\mathcal{I} = \bigoplus_\lambda \mathbf{Z}\lambda$ for the group of fractional ideals of F, written additively. For every rational prime l write $\mathcal{I}_l = \bigoplus_{\lambda | l} \mathbf{Z}\lambda$, so $\mathcal{I} = \bigoplus_l \mathcal{I}_l$, and if $y \in F^\times$ let $(y) \in \mathcal{I}$ denote the principal ideal generated by y, and $[y] \in \mathcal{I}/M\mathcal{I}$ and $[y]_l \in \mathcal{I}_l/M\mathcal{I}_l$ the projections of (y). Note that $[y]$ and $[y]_l$ are also well-defined for $y \in F^\times/(F^\times)^M$.

Proposition 2.3. *For every $n \in \mathcal{S}$ and prime ρ of F lying above a rational prime $r \equiv 1 \pmod{pn}$, there is a unique $\kappa(n,\rho) \in (F^\times/(F^\times)^M)^\chi$ such that*

$$\kappa(n,\rho) \equiv \alpha(n,\mathfrak{r})^{e_\chi \mathbf{D}_n} \pmod{(F(\boldsymbol{\mu}_n)^\times)^M} \text{ for every } \mathfrak{r} \text{ of } F(\boldsymbol{\mu}_n) \text{ above } \rho.$$

Further,
(i) *If* $l \nmid nr$ *then* $[\kappa(n,\rho)]_l = 0$,
(ii) $[\kappa(n,\rho)]_r = \delta(n)\rho$.

Proof. The first statement is the definition of $\kappa(n,\rho)$. Both (i) and (ii) follow from Proposition 1.1(ii), since all primes not dividing n are unramified in $F(\boldsymbol{\mu}_n)/F$. ■

For each prime $l \in \mathcal{S}$ let $g_l \in (\mathbf{Z}/l\mathbf{Z})^\times$ be the generator corresponding to our chosen generator σ_l of G_l; that is, $\zeta^{\sigma_l} = \zeta^{g_l}$ for $\zeta \in \boldsymbol{\mu}_l$. If K is a number field and λ is a prime of K of degree 1 lying above l, define a "logarithm" map to the base g_l

$$\nu_\lambda : \{y \in F^\times : \mathrm{ord}_\lambda(y) = 0\} \to \mathbf{Z}/(l-1)\mathbf{Z}$$

by $g_l^{\nu_\lambda(u)} \equiv u \pmod{\lambda}$. If $l \in \mathcal{S}$ is prime and λ is a prime of F above l, we extend this logarithm map to

$$\varphi_\lambda : \{y \in F^\times/(F^\times)^M : [y]_l = 0\} \to (\mathbf{Z}/M\mathbf{Z})[\Delta]$$

by

$$\varphi_\lambda(y) \equiv \sum_{\sigma\in\Delta} \nu_{\lambda^\sigma}(\overline{y})\sigma \pmod{M},$$

where $\overline{y}$ is a lift of y to $F^\times$ which is prime to l.

Theorem 2.4. *Suppose $n \in \mathcal{S}$, and ρ and ρ' are primes of F lying above distinct rational primes $r \equiv r' \equiv 1 \pmod{np^2M}$. Write $A(n)$ for the p-part of the ideal class group of $F(\boldsymbol{\mu}_n)$, and suppose further that there are primes $\mathfrak{r}$ and $\mathfrak{r}'$ of $F(\boldsymbol{\mu}_n)$ above ρ and ρ', respectively, whose projections into $A(n)^\chi$ are the same. Then*

$$\varphi_\rho(\kappa(n,\rho')) \equiv \delta(nr) \pmod{\delta(n)(\mathbf{Z}/M\mathbf{Z})[\Delta]^\chi}.$$

The remainder of this section will be devoted to proving this theorem. Proposition 2.3 and Theorem 2.4 (the analogue for Gauss sums of Proposition 2.4 of [4]) are crucial to Kolyvagin's induction argument given in §4.

Lemma 2.5. (Kolyvagin) *Suppose $m \in \mathbf{Z}$ and $r \in \mathcal{S}$ is prime, $r \equiv 1 \pmod{2m}$. Write $t = (r-1)/m$. Then for every a, $1 \le a \le m-1$, such that $(a,m) = 1$,*

$$\sum_{\substack{b=1 \\ b\equiv a \,(\mathrm{mod}\, m),\, r\nmid b}}^{mr} b\nu_r(b) \equiv x + m\nu_r(-1/(at)!) \pmod{t/2}$$

where $x \in \mathbf{Z}/(t/2)\mathbf{Z}$ is independent of a. Here ν_r is the logarithm map defined above.

Proof. This computation is due to Kolyvagin ([1], §2.1); we reproduce it here for completeness. For $b \in \mathbf{Z}$ write $\langle b \rangle$ for the least non-negative residue of b modulo m. Then

$$\sum_{\substack{b=1 \\ b \equiv a \pmod m), r \nmid b}}^{mr} b\nu_r(b) = \nu_r\Big(\prod_{\substack{b=1 \\ b \equiv a \pmod m), r \nmid b}}^{mr} b^b\Big) = \nu_r\Big(\prod_{c=1}^{r-1} c^{c+r\langle a-c \rangle}\Big)$$

since $r \equiv 1 \pmod m$. Also,

$$\prod_{c=1}^{r-1} c^{c+r\langle a-c \rangle} \equiv \prod_{c=1}^{r-1} c^{c+\langle a-c \rangle} = \prod_{c=1}^{r-1} c^{c-\langle c \rangle} \prod_{c=1}^{r-1} c^a \prod_{\substack{c=1 \\ \langle c \rangle > a}}^{r-1} c^m \pmod r.$$

The first product on the right-hand side is independent of a. The second is $((r-1)!)^a \equiv (-1)^a \pmod r$, and $\nu_r(-1) = (r-1)/2$ so this vanishes modulo $t/2$. For the last product, consider the two subsets of $\{1, 2, \ldots, r-1\}$

$$C = \{c : 1 \le c \le r-1 \text{ and } \langle c \rangle > a\} = \{d + me : a < d < m,\ 0 \le e < t\},$$

$$C' = \{c : at < c' \le r-1-t\} = \{dt - e : a < d < m,\ 0 \le e < t\}.$$

Multiplication by t gives a bijection from C to C' modulo r. Therefore

$$\prod_{\langle c \rangle > a} c \equiv t^{-t(m-1-a)}(r-1-t)!/(at)! \pmod r.$$

Up to a constant independent of a this is $t^{at}(-1/(at)!)$, and $\nu_r(t^{at}) \equiv 0 \pmod t$. This proves the desired congruence. ■

For $n \in \mathbf{Z}$ and primes $r \equiv 1 \pmod{nM}$ define elements of $(\mathbf{Z}/M\mathbf{Z})[\mathrm{Gal}(F(\boldsymbol{\mu}_n)/\mathbf{Q})]$ by

$$s'(n,r) \equiv \sum_{a \in (\mathbf{Z}/pn\mathbf{Z})^\times} \nu_r(-1/(at)!)\tau_a^{-1} \pmod M,$$

$$\theta'(n,r,\chi) = e_\chi(\delta - b_n)s'(n,r)$$

where $\tau_a \in \mathrm{Gal}(F(\boldsymbol{\mu}_n)/\mathbf{Q})$ acts on $\boldsymbol{\mu}_{pn}$ by $\zeta \mapsto \zeta^a$, and $t = (r-1)/pn$.

Corollary 2.6. *Suppose $n \in \mathcal{S}$ and $r \equiv 1 \pmod{np^2M}$. Then*

$$\mathbf{D}_r\theta(nr,\chi) = \theta'(n,r,\chi)\mathbf{N}_r \quad in \quad (\mathbf{Z}/M\mathbf{Z})[Gal(F(\boldsymbol{\mu}_{nr})/\mathbf{Q})].$$

Proof. We can write $\mathbf{D}_r = \sum_{b\in(\mathbf{Z}/r\mathbf{Z})^\times} \nu_r(b)\tau_b$, and one checks easily that

$$\mathbf{D}_r s(nr) = \sum_{a\in(\mathbf{Z}/pnr\mathbf{Z})^\times} f(a)\tau_a^{-1}$$

where

$$f(a) = \sum_{\substack{b=1 \\ b\equiv a \pmod{pn},\, r\nmid b}}^{pnr} b\nu_r(a^{-1}b) = \sum b\nu_r(b) - \nu_r(a)\sum b$$
$$\equiv \sum b\nu_r(b) \pmod{(r-1)/2}.$$

Thus by Lemma 2.5 with $m = pn$,

$$\mathbf{D}_r s(nr) \equiv (x\mathbf{N}_{nr}\sum_{\sigma\in\Delta}\sigma) + pns'(n,r))\mathbf{N}_r \pmod{pM}.$$

Since $\chi \neq 1$, $r \equiv 1 \pmod M$ and $b_{nr} \equiv b_n \pmod{pM}$,

$$\mathbf{D}_r\theta(nr,\chi) = (pnr)^{-1}e_\chi(\delta - b_{nr})\mathbf{D}_r s(nr) \equiv \theta'(n,r,\chi)\mathbf{N}_r \pmod M. \quad \blacksquare$$

The next proposition gives Kolyvagin's remarkable connection between two classical properties of Gauss sums: their ideal factorizations and their residue classes modulo the next higher power of each prime divisor. This is the key to proving Theorem 2.4. Recall φ_ρ is the map defined before the statement of Theorem 2.4.

Proposition 2.7. (Kolyvagin) *Suppose $n \in \mathcal{S}$, and ρ is a prime of F lying above a rational prime $r \equiv 1 \pmod{np^2M}$. Then there is a $\pi \in F^\times$ such that $[\pi]_r = \rho$, $[\kappa(n,\rho)/\pi^{\delta(n)}]_r = 0$, and*

$$\varphi_\rho(\kappa(n,\rho)/\pi^{\delta(n)}) = \delta(nr).$$

Proof. Fix a primitive r-th root of unity ζ_r and a prime $\mathfrak{r}$ of $F(\boldsymbol{\mu}_n)$ above ρ, and write $\mathfrak{R}$ for the prime of $F(\boldsymbol{\mu}_{nr})$ above $\mathfrak{r}$. Fix an element Π of $F(\boldsymbol{\mu}_{nr})^\times$, integral at all primes above r, such that

$$\Pi \equiv \zeta_r - 1 \pmod{\mathfrak{R}^2} \quad \text{and} \quad \Pi \equiv 1 \pmod{\mathfrak{R}^\tau}$$

for all $\tau \neq 1$ in $\mathrm{Gal}(F(\boldsymbol{\mu}_n)/\mathbf{Q})$. By a classical property of Gauss sums ([2] Chapter 1, Theorem 2.1), if $1 \leq a < pn$, $\tau = \tau_a \in \mathrm{Gal}(F(\boldsymbol{\mu}_n)/\mathbf{Q})$ acts on $\boldsymbol{\mu}_{pn}$ by $\zeta \mapsto \zeta^a$, and $t = (r-1)/pn$, then

$$g(n, \mathfrak{r}, \zeta_r)/(\Pi^{\tau^{-1}})^{at} \equiv g(n, \mathfrak{r}, \zeta_r)/\Pi^{s(n)t} \equiv -1/(at)! \pmod{\mathfrak{R}^{\tau^{-1}}}. \quad (4)$$

Define a Galois-equivariant map

$$\psi : \{y \in F(\boldsymbol{\mu}_{nr})^{\times} : y \text{ is prime to } r\} \to (\mathbf{Z}/M\mathbf{Z})[\mathrm{Gal}(F(\boldsymbol{\mu}_n)/\mathbf{Q})],$$

using the logarithm maps $\nu_{\mathfrak{R}^\tau}$ defined above, by

$$\psi(y) \equiv \sum_{\tau \in \mathrm{Gal}(F(\boldsymbol{\mu}_n)/\mathbf{Q})} \nu_{\mathfrak{R}^\tau}(y)\tau \pmod{M}.$$

Then for $y \in F^{\times}$ prime to r, $\psi(y) = \varphi_\rho(y)\mathbf{N}_n$. By (4),

$$\psi(g(n, \mathfrak{r}, \zeta_r)/\Pi^{s(n)t}) = s'(n, r)$$

so by Corollary 2.6, applying $e_\chi(\delta - b_n)\mathbf{N}_r$ to both sides,

$$e_\chi \psi(\alpha(n, \mathfrak{r})/\Pi^{(r-1)\theta(n)})\mathbf{N}_r = \theta'(n, r, \chi)\mathbf{N}_r = \mathbf{D}_r\theta(nr, \chi)$$

in $(\mathbf{Z}/M\mathbf{Z})[\mathrm{Gal}(F(\boldsymbol{\mu}_{nr})/\mathbf{Q})]$. Now applying $\mathbf{D}_n$ to both sides shows

$$\psi(\kappa(n, \rho)/\Pi_0^{\delta(n)\mathbf{N}_n}) = \delta(nr)\mathbf{N}_n,$$

where $\Pi_0 \in F(\boldsymbol{\mu}_n)^{\times}$ and $\Pi_0/\Pi^{r-1} \equiv 1 \pmod{^{\times} r}$. Thus

$$\varphi_\rho(\kappa(n, \rho)/(\Pi_0^{\mathbf{N}_n})^{\delta(n)}) = \delta(nr). \qquad \blacksquare$$

Proposition 2.8. *Suppose $n \in \mathcal{S}$, and ρ and ρ' are primes of F lying above distinct rational primes $r \equiv r' \equiv 1 \pmod{np^2M}$. Write $A(n)$ for the p-part of the ideal class group of $F(\boldsymbol{\mu}_n)$, and suppose further that there are primes $\mathfrak{r}$ and $\mathfrak{r}'$ of $F(\boldsymbol{\mu}_n)$ above ρ and ρ', respectively, whose projections into $A(n)^\chi$ are the same. Then*

$$\kappa(n, \rho')/\kappa(n, \rho) \in (F^{\times})^{\delta(n)}(F^{\times})^M/(F^{\times})^M.$$

Proof. Write $\mathcal{I}(F(\boldsymbol{\mu}_n))$ for the group of fractional ideals of $F(\boldsymbol{\mu}_n)$ (written additively) and $h_n = \#(A(n))$. Since $\mathfrak{r}' - \mathfrak{r}$ projects to 0 in $A(n)^\chi$, we can choose $\beta \in F(\boldsymbol{\mu}_n)^{\times}$ so that its principal ideal satisfies

$$(\beta) \equiv e_\chi(\mathfrak{r}' - \mathfrak{r}) \pmod{Mh_n\mathcal{I}(F(\boldsymbol{\mu}_n))}.$$

Choose any $d_\chi \in \mathbf{Z}[\mathrm{Gal}(F(\boldsymbol{\mu}_n)/\mathbf{Q})]$ such that $d_\chi \equiv e_\chi \pmod{Mh_n}$. By Proposition 1.1(ii), the principal ideal $((\beta^{\theta(n)}\alpha(n,\mathfrak{r})/\alpha(n,\mathfrak{r}'))^{d_\chi})$ belongs to $Mh_n\mathcal{I}(F(\boldsymbol{\mu}_n))$. Every principal ideal in $Mh_n\mathcal{I}(F(\boldsymbol{\mu}_n))$ is the M-th power of a principal ideal, so

$$(\beta^{\theta(n)}\alpha(n,\mathfrak{r})/\alpha(n,\mathfrak{r}'))^{d_\chi} \in (F(\boldsymbol{\mu}_n)^\times)^M \mathcal{O}^\times_{F(\boldsymbol{\mu}_n)}.$$

By construction

$$(\beta^{\theta(n)}\alpha(n,\mathfrak{r})/\alpha(n,\mathfrak{r}'))^{d_\chi} \in [F(\boldsymbol{\mu}_n)^\times/(F(\boldsymbol{\mu}_n)^\times)^M]^\chi,$$

and since χ is odd and different from ω, $[\mathcal{O}^\times_{F(\boldsymbol{\mu}_n)}/(\mathcal{O}^\times_{F(\boldsymbol{\mu}_n)})^M]^\chi$ is trivial. Therefore (using Lemma 2.2)

$$\kappa(n,\rho')/\kappa(n,\rho) = \beta^{\mathbf{D}_n\theta(n,\chi)} = (\beta^{\mathbf{N}_n})^{\delta(n)} \text{ in } F^\times/(F^\times)^M. \qquad ■$$

Proof of Theorem 2.4. With π as in Proposition 2.7,

$$\kappa(n,\rho') = (\kappa(n,\rho)/\pi^{\delta(n)})(\pi^{\delta(n)}\kappa(n,\rho')/\kappa(n,\rho))$$

so the theorem is immediate from Propositions 2.7 and 2.8. ■

§3. An application of the Chebotarev theorem.

The next theorem, together with Proposition 2.3 and Theorem 2.4, will enable us to construct all the relations we need in the ideal class group of F. A variant of it appears in work of Thaine ([6] Proposition 4). Recall that χ is an odd character of Δ, $\chi \neq \omega$. For every positive integer n let $A(n)$ denote the p-part of the ideal class group of $F(\boldsymbol{\mu}_n)$.

Theorem 3.1. *Suppose one is given positive integers k and n, $\mathfrak{c} \in A(n)^\chi$, a power M of p, and $\beta \in (F^\times/(F^\times)^M)^\chi$. Let t denote the largest divisor of M such that $\beta \in (F^\times)^t/(F^\times)^M$. Then there are infinitely many primes λ of F such that (writing l for the rational prime below λ)*

(i) *there is a prime of $F(\boldsymbol{\mu}_n)$ above λ whose projection into $A(n)^\chi$ is $\mathfrak{c}$,*
(ii) $l \equiv 1 \pmod{Mkn}$,
(iii) $[\beta]_l = 0$ *and there is a $u \in (\mathbf{Z}/M\mathbf{Z})^\times$ such that $\varphi_\lambda(\beta) = ute_\chi$.*

Proof. Let H be the subfield of the Hilbert class field of $F(\boldsymbol{\mu}_n)$ such that $\mathrm{Gal}(H/F(\boldsymbol{\mu}_n))$ is identified with $A(n)^\chi$ by class field theory, and let $F' = F(\boldsymbol{\mu}_{Mkn})$. Let W be the subgroup of $(F^\times/(F^\times)^M)^\chi$ generated by β.

Since $\chi \neq \omega$, W has order M/t. Using Lemma 2.2, Kummer theory gives a $\mathrm{Gal}(F'/\mathbf{Q})$-equivariant isomorphism

$$\mathrm{Gal}(F'(\beta^{1/M})/F') \cong \mathrm{Hom}(W, \boldsymbol{\mu}_M)$$

which shows that $\mathrm{Gal}(F'(\beta^{1/M})/F')$ is cyclic of order M/t and that Δ acts on $\mathrm{Gal}(F'(\beta^{1/M})/F')$ via $\omega\chi^{-1}$. Since Δ acts on $\mathrm{Gal}(HF'/F') \subset A(n)^{\chi}$ via χ and $\chi \neq \omega\chi^{-1}$, $F'(\beta^{1/M}) \cap HF' = F'$. Since $\mathrm{Gal}(F'/F(\boldsymbol{\mu}_n))$ is generated by inertia groups, there is no nontrivial unramified extension of $F(\boldsymbol{\mu}_n)$ in F', so $F'(\beta^{1/M}) \cap H = F(\boldsymbol{\mu}_n)$.

Thus we can choose

$$\delta \in \mathrm{Gal}(HF'(\beta^{1/M})/F(\boldsymbol{\mu}_n))$$

such that δ restricts to a generator of $\mathrm{Gal}(F'(\beta^{1/M})/F')$ on $F'(\beta^{1/M})$ and to $\mathfrak{c}$ on H. Let $\mathfrak{q}$ be any prime of $F(\boldsymbol{\mu}_n)$ of degree 1 whose Frobenius in $\mathrm{Gal}(HF'(\beta^{1/M})/F(\boldsymbol{\mu}_n))$ is the conjugacy class of δ, and such that all conjugates of $\mathfrak{q}$ are unramified in $F'(\beta^{1/M})/\mathbf{Q}$. The Chebotarev theorem guarantees the existence of infinitely many such $\mathfrak{q}$. Let l and λ denote the primes of $\mathbf{Q}$ and F, respectively, below $\mathfrak{q}$. We must verify that (i), (ii), and (iii) are satisfied.

The identification of $A(n)^{\chi}$ with $\mathrm{Gal}(H/F(\boldsymbol{\mu}_n))$ sends the class of $\mathfrak{q}$ to the Frobenius of $\mathfrak{q}$, so (i) is immediate. Also, since δ is trivial on F', l splits completely in $\mathbf{Q}(\boldsymbol{\mu}_{Mkn})/\mathbf{Q}$ which proves (ii). The first assertion of (iii), that $[\beta]_l = 0$, holds because l is unramified in $F'(\beta^{1/M})/\mathbf{Q}$.

The Frobenius of λ has order M/t on $F'(\beta^{1/M})$, so β has order M/t in $(\mathcal{O}_F/\lambda)^{\times}/((\mathcal{O}_F/\lambda)^{\times})^M$. Thus the coefficient of 1 in $\varphi_{\lambda}(\beta)$, namely $\nu_{\lambda}(\beta)$, is a unit times t. But $\varphi_{\lambda}(\beta) \in (\mathbf{Z}/M\mathbf{Z})[\Delta]^{\chi} = (\mathbf{Z}/M\mathbf{Z})e_{\chi}$, so this proves (iii). ∎

§4. The ideal class group of $\mathbf{Q}(\boldsymbol{\mu}_p)$.

We now use the machinery of the previous sections to study the ideal class group of $F = \mathbf{Q}(\boldsymbol{\mu}_p)$. Our method is essentially the same as that of Kolyvagin in [1]. I would like to thank René Schoof for suggestions which improved the exposition of this section, and for his preprint [5] which inspired the statement of Theorem 4.4.

Let A denote the p-part of the ideal class group of F, and for every $n \in \mathcal{S}$ let $d(n)$ denote the largest divisor of M which divides $\delta(n) \in (\mathbf{Z}/M\mathbf{Z})[\Delta]^{\chi} \cong (\mathbf{Z}/M\mathbf{Z})e_{\chi}$. For every odd character ψ of Δ define the generalized Bernoulli number

$$\mathbf{B}_{1,\psi} = \frac{1}{p}\sum_{a=1}^{p-1} a\psi(a).$$

Then $\mathbf{B}_{1,\psi} \in \mathbf{Z}_p$ if $\psi \neq \omega$. By definition,

$$\delta(1) = e_\chi(\delta - b_1) \sum_{a=1}^{p-1} \frac{a}{p} \tau_a^{-1} = (\chi(\delta) - b_1)\mathbf{B}_{1,\chi^{-1}} e_\chi.$$

Since δ generates Δ, $b_1 \equiv \omega(\delta) \pmod{p}$, and $\chi \neq \omega$, we have

$$(\chi(\delta) - b_1) \in \mathbf{Z}_p^\times.$$

Thus

$$d(1)(\mathbf{Z}/M\mathbf{Z}) = \mathbf{B}_{1,\chi^{-1}}(\mathbf{Z}/M\mathbf{Z}).$$

Theorem 4.1. *Suppose $d(1)\#(A^\chi)p \mid M$, $n \in \mathcal{S}$, and $\mathfrak{c} \in A^\chi$. Let B denote the subgroup of A generated by the classes of primes of F dividing n. If $d(n) \mid d(1)$ and $\mathfrak{c} \notin B^\chi$, then there is a prime λ of F lying above $l \in \mathcal{S}$ such that*

(i) *the projection of the class of λ into A^χ is $\mathfrak{c}$,*
(ii) *$d(nl) \mid d(n)$ and $d(n)/d(nl)$ annihilates $\mathfrak{c}$ in A^χ/B^χ.*

Proof. Choose a prime ρ of F in the class $\mathfrak{c}$ lying above a rational prime $r \equiv 1 \pmod{np^2M}$. That this is possible is a (simple) special case of Theorem 3.1. Let t denote the largest divisor of M such that $\kappa(n, \rho) \in (F^\times)^t/(F^\times)^M$. By Proposition 2.3,

$$[\kappa(n, \rho)] = \delta(n)\rho + \sum_{l|n} [\kappa(n, \rho)]_l \quad \text{in} \quad \mathcal{I}/M\mathcal{I},$$

so we clearly have $t \mid d(n)$. Since $d(n) \mid d(1)$, M/t annihilates A^χ and we can divide this relation by t and project into A^χ/B^χ to conclude

$$(d(n)/t)\mathfrak{c} = 0 \quad \text{in} \quad A^\chi/B^\chi. \tag{5}$$

Since $\mathfrak{c} \neq 0$ in A^χ/B^χ we have $pt \mid d(n)$.

Fix a prime $\mathfrak{r}$ of $F(\boldsymbol{\mu}_n)$ above ρ. Applying Theorem 3.1 with $\beta = \kappa(n, \rho)$ yields a prime λ of F lying above a rational prime $l \equiv 1 \pmod{np^2M}$ such that some prime of $F(\boldsymbol{\mu}_n)$ above λ lies in the same class (in the χ-component of the class group of $F(\boldsymbol{\mu}_n)$) as $\mathfrak{r}$, and such that

$$\varphi_\lambda(\kappa(n, \rho)) = ute_\chi$$

for some $u \in (\mathbf{Z}/M\mathbf{Z})^\times$. Thus by Theorem 2.4,

$$ute_\chi = \varphi_\lambda(\kappa(n, \rho)) \equiv \delta(nl) \pmod{\delta(n)(\mathbf{Z}/M\mathbf{Z})[\Delta]^\chi}.$$

Since $pt \mid d(n)$, we have $t = d(nl)$ and the theorem follows from (5). ■

Corollary 4.2. *Suppose $d(1)\#(A^\chi)p \mid M$ and $n \in \mathcal{S}$. Let B denote the subgroup of A generated by the classes of primes of F dividing n. Then*

$$\#(A^\chi/B^\chi) \mid d(n).$$

Proof. First suppose $d(n) \mid d(1)$. Let $\mathfrak{c}_1, \ldots, \mathfrak{c}_k \in A^\chi$ be a collection of ideal classes which generate A^χ/B^χ. Write c_i for the order of $\mathfrak{c}_i$ in $A^\chi/\langle B^\chi, \mathfrak{c}_1, \ldots, \mathfrak{c}_{i-1}\rangle$, so $\#(A^\chi/B^\chi) = \prod_i c_i$. We assume that our set $\{\mathfrak{c}_1, \ldots, \mathfrak{c}_k\}$ is minimal in the sense that $c_i \neq 1$ for all i.

We can apply Theorem 4.1 inductively to choose primes $\lambda_1, \ldots, \lambda_k$ of F above rational primes $l_1, \ldots, l_k \in \mathcal{S}$ such that

$$\begin{aligned} &\lambda_i \text{ projects to } \mathfrak{c}_i, \quad d(n_i) \mid d(n_{i-1}), \quad \text{and} \\ &(d(n_{i-1})/d(n_i))\mathfrak{c}_i = 0 \text{ in } A^\chi/\langle B^\chi, \mathfrak{c}_1, \ldots, \mathfrak{c}_{i-1}\rangle \end{aligned}$$

for $i = 1, 2, \ldots, k$, where $n_i = n \prod_{j \le i} l_j$, $n_0 = n$. Then $c_i \mid (d(n_{i-1})/d(n_i))$ for every i, so

$$\#(A^\chi/B^\chi) \mid \prod_{i=1}^{k}(d(n_{i-1})/d(n_i)) \mid d(n).$$

This proves the theorem if $d(n) \mid d(1)$. In particular it holds for $n = 1$, so $\#(A^\chi) \mid d(1)$. Now if $d(n) \nmid d(1)$ then

$$\#(A^\chi/B^\chi) \mid \#(A^\chi) \mid d(1) \mid d(n). \qquad ■$$

Theorem 4.3. (Mazur-Wiles, Kolyvagin) *For every odd character $\chi \neq \omega$ of Δ,*

$$\#(A^\chi) = p^{b(\chi)}$$

where $b(\chi) = \mathrm{ord}_p(\mathbf{B}_{1,\chi^{-1}})$.

Proof. For $a, b \in \mathbf{Z}_p$ write $a \sim b$ to mean that $a/b \in \mathbf{Z}_p^\times$. Thus by Corollary 4.2 with $n = 1$ and the remarks before Theorem 4.1, for every odd character $\chi \neq \omega$ of Δ,

$$\#(A^\chi) \mid \mathbf{B}_{1,\chi^{-1}} \text{ in } \mathbf{Z}_p. \tag{6}$$

As is well known (for example [2] Corollary 2, §3, Chapter 1),

$$\#(A^\omega) \sim p\mathbf{B}_{1,\omega^{-1}} \sim 1.$$

Combining this with the analytic class number formula ([2] Chapter 3 §3, **CNF$^-$**),

$$\prod_{\chi\,\mathrm{odd},\chi\neq\omega} \#(A^\chi) = \prod_{\mathrm{odd}\,\chi} \#(A^\chi) \sim p \prod_{\mathrm{odd}\,\chi} \mathbf{B}_{1,\chi} \sim \prod_{\chi\,\mathrm{odd},\chi\neq\omega} \mathbf{B}_{1,\chi^{-1}},$$

so theorem follows from the divisibilities (6). ∎

Remark. It follows from Theorem 4.3 that in the proof of Corollary 4.2 with $n = 1$ we must have $c_i = (d(n_{i-1})/d(n_i))$ for every i, and $d(n_k) = 1$.

Fix a power M of p larger than $p^{\mathrm{ord}_p(\mathbf{B}_{1,\chi^{-1}})}\#(A^\chi)$. For each nonnegative integer k define the "higher Stickelberger ideal" $S_k^\chi \subset (\mathbf{Z}/M\mathbf{Z})[\Delta]^\chi$ to be the ideal generated by

$$\{\delta(n) : n \in \mathcal{S} \text{ and } n \text{ is divisible by exactly } k \text{ primes}\}.$$

Then $S_0^\chi = \delta(1)(\mathbf{Z}/M\mathbf{Z})[\Delta]^\chi$ is the χ-part of the usual Stickelberger ideal. The following theorem on the group structure of A^χ is implicit in [1] (§2, especially Theorem 7, and §2.1). Compare this result with the conjecture of Schoof in §3 of [5].

Theorem 4.4. *We have*

$$S_0^\chi \subset S_1^\chi \subset S_2^\chi \subset \ldots, \quad \#(S_1^\chi/S_0^\chi) \geq \#(S_2^\chi/S_1^\chi) \geq \ldots,$$

and

$$A^\chi \cong \bigoplus_{i=0}^{\infty} (S_{i+1}^\chi/S_i^\chi).$$

Proof. That $S_i^\chi \subset S_{i+1}^\chi$ is clear from Theorem 4.1. Write

$$A^\chi \cong \bigoplus_{i=1}^{k} (\mathbf{Z}/c_i\mathbf{Z}) \quad \text{with} \quad c_1 \geq c_2 \geq \ldots \geq c_k,$$

and let $\mathfrak{c}_i$ be a generator of the i-th summand. Let $n_1, \ldots, n_k$ be the integers constructed in the proof of Corollary 4.2 with $n = n_0 = 1$ and these $\mathfrak{c}_i$. Then n_i is a product of exactly i primes, and by the remark following the proof of Theorem 4.3,

$$d(n_i) = \prod_{j>i} c_j$$

for $0 \le i \le k$. Conversely, suppose $n \in \mathcal{S}$ is a product of exactly i primes. If B is the subgroup of A generated by the classes of primes of F dividing n, then $\#(B^\chi) \mid \prod_{j \le i} c_j$ so by Corollary 4.2,

$$\prod_{j>i} c_j \mid d(n).$$

Thus $S_i^\chi = (\prod_{j>i} c_j)(\mathbf{Z}/M\mathbf{Z})[\Delta]^\chi$ for $i \le k$, which proves the theorem. ■

Remarks. 1) A priori, S_k^χ depends on M because $\mathcal{S}$ does. However, Theorem 4.4 shows that for $M > p^{\operatorname{ord}_p(\mathbf{B}_{1\chi^{-1}})}\#(A^\chi)$, S_k^χ is independent of M. This gives a natural definition of higher Stickelberger ideals in $\mathbf{Z}_p[\Delta]^\chi$.

2) If we define $S_k^\omega = \mathbf{Z}_p[\Delta]^\omega$ for every k, and $S_k = \bigoplus_{\text{odd } \chi} S_k^\chi \subset \mathbf{Z}_p[\Delta]$, then Theorem 4.4 gives a $\mathbf{Z}_p[\Delta]$-isomorphism

$$A^- \cong \bigoplus_{i=0}^{\infty} (S_{i+1}/S_i)$$

where A^- denotes the (-1)-component of A for the action of complex conjugation.

3) All the results of this paper would work equally well with $\mathbf{Q}(\boldsymbol{\mu}_p)$ replaced by $\mathbf{Q}(\boldsymbol{\mu}_m)$, as long as $p \nmid [\mathbf{Q}(\boldsymbol{\mu}_m) : \mathbf{Q}]$.

REFERENCES

[1] V. A. Kolyvagin: Euler systems. *To appear.*

[2] S. Lang: *Cyclotomic Fields.* Grad. Texts in Math. **59**, Springer-Verlag: New York (1978) (also 2nd edition, Grad. Texts in Math. **121**, Springer-Verlag: New York (1990)).

[3] B. Mazur, A. Wiles: Class fields of abelian extensions of **Q**. *Invent. Math.* **76** (1984) 179-330.

[4] K. Rubin: The Main Conjecture. Appendix to: *Cyclotomic Fields I and II*, combined 2nd edition. Grad. Texts in Math. **121**, Springer-Verlag: New York (1990) 397-419.
[5] R. Schoof: The structure of the minus class groups of abelian number fields. University Utrecht preprint #588, October 1989.
[6] F. Thaine: On the ideal class groups of real abelian number fields. *Annals of Math.* **128** (1988) 1-18.

The exponents of the groups of points on the reductions of an elliptic curve

RENÉ SCHOOF*

Abstract.

Let E denote an elliptic curve over $\mathbf{Q}$ without complex multiplication. It is shown that the exponents of the groups $E(\mathbf{F}_p)$ grow at least as fast as $\frac{\sqrt{p}\log p}{(\log\log p)^2}$.

1. Introduction.

The exponent $\exp(A)$ of a finite abelian group A is the smallest positive integer m for which $ma = 0$ for all $a \in A$. In this note we will study the exponents of the groups $E(\mathbf{F}_p)$ of points modulo a prime of good reduction p of an elliptic curve E which is defined over $\mathbf{Q}$.

By the Riemann hypothesis for elliptic curves over finite fields [6, Ch.5.Thm.1.1], proved by H. Hasse in 1933, we have that

$$|\#E(\mathbf{F}_p) - (p+1)| < 2\sqrt{p}.$$

Since the group $E(\mathbf{F}_p)$ can be generated by two points [6, Ch.3.Cor.6.4] we see that

$$\sqrt{p} - 1 < \exp(E(\mathbf{F}_p)) < (\sqrt{p}+1)^2.$$

We will show that for elliptic curves without complex multiplication the lower bound can be somewhat improved:

Theorem 1.1. *Let E be an elliptic curve over $\mathbf{Q}$ without complex multiplication. There exists a constant $C_E > 0$ such that*

$$\frac{\exp(E(\mathbf{F}_p))}{\sqrt{p}} > C_E \frac{\log p}{(\log\log p)^2}$$

* supported by the Netherlands Organization of Scientific Research.

for every prime p of good reduction.

Note that "complex multiplication" means complex multiplication over $\mathbf{C}$. Theorem 1.1 is proved in section 2. It is a consequence of the fact that the absolute values of the discriminants Δ_p of the endomorphism rings of the curve E (mod p) grow as p goes to infinity. We will show that

$$|\Delta_p| \geq C'_E \frac{(\log p)^2}{(\log\log p)^4}$$

for some constant $C'_E > 0$.

In section 3 we will investigate how small the exponents $\exp(E(\mathbf{F}_p))$ and the discriminants Δ_p can be. Assuming certain generalized Riemann Hypotheses we can show that there exists a constant c_E such that for infinitely many primes p the quantity $\exp(E(\mathbf{F}_p))/\sqrt{p}$ is less than $c_E p^{3/8} \log p$. Similarly we can show that there exists a constant c'_E such that $|\Delta_p| \leq c'_E p^{2/3} \log p$ for infinitely many primes p.

I would like to thank Hendrik Lenstra for the elegant proof of Lemma 2.1 and the referee for many helpful suggestions.

2. Large exponents.

In this section we will prove Theorem 1. We need two lemmas.

Lemma 2.1. *Let R be a quadratic order of discriminant Δ and let $\chi(x)$ denote the quadratic residue symbol $(\frac{\Delta}{x})$. Let furthermore ℓ be a prime and let O denote $R \otimes \mathbf{Z}_\ell$. We have*

$$\sum_{\mathfrak{a}} \frac{1}{N\mathfrak{a}} = \frac{\ell+1}{\ell - \chi(\ell)}$$

where the sum runs over all invertible ideals $\mathfrak{a}$ in O which are not contained in ℓO and the norm $N\mathfrak{a}$ is just $\#(R/\mathfrak{a})$.

Proof. When $\chi(\ell) = 1$ he ring O is isomorphic to $\mathbf{Z}_\ell \times \mathbf{Z}_\ell$, when $\chi(\ell) = -1$ it is a local ring with maximal ideal generated by ℓ and when $\chi(\ell) = 0$ it is a local ring with maximal ideal $\mathfrak{l}$ of index ℓ.

Let μ denote a Haar measure on O. We have

$$\begin{aligned}
\mu(O^*) = \mu(\mathbf{Z}_\ell^*)^2 &= (1 - \frac{1}{\ell})^2 \mu(O) \qquad \text{if } \chi(\ell) = 1,\\
&= \mu(O) - \mu(\ell O) = (1 - \frac{1}{\ell^2})\mu(O) \qquad \text{if } \chi(\ell) = -1,\\
&= \mu(O) - \mu(\mathfrak{l}) = (1 - \frac{1}{\ell})\mu(O) \qquad \text{if } \chi(\ell) = 0.
\end{aligned}$$

In other words

$$\mu(O^*) = (1 - \frac{1}{\ell})(1 - \frac{\chi(\ell)}{\ell})\mu(O). \tag{1}$$

The ring O is a disjoint union of ℓO, the zero divisors not in ℓO and sets of the form αO^* where α is not a zero divisor and α is not contained in ℓO. Since the zero divisors have measure zero we find that

$$\mu(O) = \mu(\ell O) + \sum_{\alpha} \mu(\alpha O^*).$$

Therefore

$$(1 - \frac{1}{\ell^2})\mu(O) = \mu(O^*) \sum_{\alpha} \frac{1}{N\alpha}$$

and hence by (1) that

$$\sum_{\alpha} \frac{1}{N\alpha} = \frac{\ell + 1}{\ell - \chi(\ell)}$$

where the summation runs over $\alpha \in O - \ell O$ which are not zero divisors; they are counted modulo units of O. Since O is a semi-local ring the invertible O-ideals are precisely the ideals of the form αO where α is not a zero divisor. The result now follows at once.

It was pointed out by the referee that the following lemma on binary quadratic forms can also be proved in an elementary way i.e. without invoking Lemma 2.1.

Lemma 2.2. *There exists an absolute constant C such that for every $\Delta \in \mathbf{Z}_{<0}$ congruent to 0 or 1 (mod 4) we have that*

$$\sum \frac{1}{a} \leq C \log^2 |\Delta|$$

where the sum runs over all reduced positive definite primitive binary quadratic forms $aX^2 + bXY + cY^2$ of discriminant Δ.

Proof. It is easy to see that for a reduced positive definite quadratic form $aX^2 + bXY + cY^2$ one has that $a \leq \sqrt{|\Delta|/3}$. Using the dictionary between quadratic forms of discriminant Δ and ideals in the order of discriminant Δ one sees that

$$\sum \frac{1}{a} = \sum_{\mathfrak{a}} \frac{1}{N\mathfrak{a}}$$

where the second summation runs over all invertible ideals $\mathfrak{a}$ in the order of discriminant Δ which are *primitive* and have their norms not exceeding

$\sqrt{|\Delta|/3}$. Here an ideal in an order R is called primitive if it is not contained in mR for any integer $m > 1$.

Decomposing the ideals into products of prime ideals one finds that this sum does not exceed

$$\prod_{\ell} \sum_{\mathfrak{a}} \frac{1}{N\mathfrak{a}}$$

where the product runs over the primes ℓ less than $\sqrt{|\Delta|/3}$ and the sums run over all primitive ideals $\mathfrak{a}$ having norm a power of ℓ. These sums can be computed in the semi-local rings $R \otimes \mathbf{Z}_{\ell}$. Therefore we obtain as an application of lemma 2.1 that

$$\sum \frac{1}{a} \leq \prod_{\ell} \frac{\ell+1}{\ell-\chi(\ell)} \leq \prod_{\ell} (1-\frac{1}{\ell})^{-2}.$$

By Mertens' Theorem [3,Thm 429] one has that the product $\prod(1-\frac{1}{\ell})^{-1}$ is $O(\log\sqrt{|\Delta|/3})$ and therefore we find that $\sum \frac{1}{a} \leq C\log^2|\Delta|$ for some universal constant C. This proves the lemma.

For $z \in \mathbf{C}$ with $\mathrm{Re} z > 0$ let $j(z)$ denote the modular function

$$j(z) = \frac{(1+240\sum_{n=1}^{\infty} \frac{n^3 q^n}{1-q^n})^3}{q\prod_{n=1}^{\infty}(1-q^n)^{24}}$$

where q denotes $e^{2\pi i z}$. The Fourier expansion of the j-functions begins as $q^{-1} + 744 + 196884q + \ldots$. For z in the standard fundamental domain for the action of $\mathrm{SL}_2(\mathbf{Z})$ on the upper half plane the quantity $|qj(z)|$ is bounded. In fact, it is easy to see that there exists a constant $C \in \mathbf{R}$ such that $|qj(z)| < C$ for all $z \in \mathbf{C}$ with imaginary part at least $\frac{1}{2}\sqrt{3}$.

Proposition 2.3. *Let $\Delta \in \mathbf{Z}_{<0}$ be congruent to 0 or 1 (mod 4) and let $n, m \in \mathbf{Z}$. Then*

$$\prod_{aX^2+bXY+cY^2} |m - n\cdot j(\frac{-b+i\sqrt{|\Delta|}}{2a})| \leq e^{C\sqrt{|\Delta|}\log^2|\Delta|}$$

where C is a constant depending only on n and m. The product runs over the $SL_2(\mathbf{Z})$-equivalence classes of positive definite primitive binary quadratic forms of discriminant Δ .

Proof. A quadratic form $aX^2+bXY+cY^2$ of discriminant Δ is reduced if and only if the number $\frac{-b+i\sqrt{(|\Delta|}}{2a}$ is in the standard fundamental domain for the action of $SL_2(\mathbf{Z})$ on the upper half plane. Therefore we have, for some absolute constants C and C_1, that

$$|m - n \cdot j(\frac{-b + i\sqrt{|\Delta|}}{2a})| \leq |m| + |n|C|q|$$
$$\leq C_1 \max(|n|, |m|) e^{\frac{\pi\sqrt{|\Delta|}}{a}}.$$

Every equivalence class contains exactly one reduced form and we conclude that

$$\prod |m - n \cdot j(\frac{-b + i\sqrt{|\Delta}}{2a})| \leq (C_1 \max(|n|, |m|))^{h(\Delta)} e^{\pi\sqrt{|\Delta|}\sum \frac{1}{a}}.$$

For a reduced form $aX^2 + bXY + cY^2$ one has that $a \leq \sqrt{|\Delta/3|}$. Therefore it follows at once from Lemma 2.2 that there is a constant C_2 such that the class number $h(\Delta)$ is bounded by $C_2\sqrt{|\Delta|}\log^2|\Delta|$ for all Δ. By Lemma 2.2 there exists an absolute constant C_3 such that $\sum \frac{1}{a} \leq C_3 \log^2|\Delta|$. This is easily seen to imply that

$$\prod |m - n \cdot j(\frac{-b + i\sqrt{|\Delta|}}{2a})| \leq e^{C_4\sqrt{|\Delta|}\log^2|\Delta|}$$

for some absolute constant C_4. This proves the proposition.

Corollary 2.4. *Let E be an elliptic curve over $\mathbf{Q}$ without complex multiplication. For every prime p of good reduction let Δ_p denote the discriminant of the endomorphism ring of the reduced curve $E_{/\mathbf{F}_p}$. There exists a constant $C'_E > 0$ such that*

$$|\Delta_p| > C'_E \frac{(\log p)^2}{(\log\log p)^4} \qquad \text{for all } p.$$

Proof. We may assume that $p \geq 5$. If p is a prime of supersingular reduction we have that $\Delta_p = -p$ or $-4p$. If p is a prime of ordinary reduction we observe that the j-invariant $j(E)$ of the curve E is congruent to the j-invariant $j_p \in \bar{\mathbf{Q}}$ of an elliptic curve with complex multiplication by the quadratic order of discriminant Δ_p. We conclude, since $j(E) - j_p$ is not zero, that $p \leq \prod |n - mj'|$ where n and m are integers such that $j(E) = n/m$ and where the product runs over the conjugates j' of j_p. These conjugates are precisely the $j(\frac{-b+i\sqrt{|\Delta|}}{2a})$ with $aX^2 + bXY + cY^2$ a primitive binary quadratic form of discriminant Δ_p. It follows from Proposition 2.3 that

$$p \leq e^{C\sqrt{|\Delta_p|}\log^2|\Delta_p|}$$

which easily implies the required result.

Proof of Theorem 1.1. We may assume that p is a prime of good reduction. Let Δ_p denote the discriminant of the ring of endomorphisms O of E over $\mathbf{F}_p$. We write

$$E(\mathbf{F}_p) \cong \mathbf{Z}/nd\mathbf{Z} \oplus \mathbf{Z}/d\mathbf{Z}$$

so that $\exp(E(\mathbf{F}_p)) = nd$. We have

$$\exp(E(\mathbf{F}_p))^2 = n^2 d^2 = n\#E(\mathbf{F}_p) \geq n(\sqrt{p}-1)^2$$

by Hasse's inequality. Since the d-torsion points of E are defined over $\mathbf{F}_p$ we must have that

$$\phi_p = 1 + d\psi \qquad \text{for some } \psi \in O.$$

Since $\phi_p \notin \mathbf{Z}$ the endomorphism ψ is not in $\mathbf{Z}$ either and we find that

$$nd^2 = \#E(\mathbf{F}_p) = (\phi - 1)(\bar{\phi} - 1) = d^2\psi\bar{\psi} \geq d^2\frac{|\Delta_p|}{4}.$$

Therefore $n \geq \frac{|\Delta_p|}{4}$ and we get

$$\exp(E(\mathbf{F}_p)) \geq \frac{1}{2}\sqrt{|\Delta_p|}(\sqrt{p}-1).$$

The result now follows at once from Corollary 2.4.

Remark 2.5. Theorem 1.1 is very probably false when the elliptic curve E has complex multiplication: consider the elliptic curve given by $Y^2 = X^3 - X$. It has complex multiplication by the ring $\mathbf{Z}[i]$. This curve has its 2-torsion points rational over $\mathbf{Q}$ while the field of definition of the 4-torsion points is $\mathbf{Q}(\zeta_8)$, the field of 8th roots of unity. We conclude that for primes p that are congruent to 1 (mod 8), the Frobenius endomorphism ϕ_p is congruent to 1 (mod 4) in $\mathbf{Z}[i]$.

Suppose p is an odd prime of the form n^2+1. In this case n is even and p splits in $\mathbf{Z}[i]$ as $(1-ni)(1+ni)$. Therefore the Frobenius endomorphism ϕ_p of a prime like this must be one of $\pm 1 \pm ni$ or $\pm i \pm n$. One concludes that for primes p of the form n^2+1 with $n \equiv 0 \pmod 4$ one has that $\phi_p = 1 \pm ni$ and

$$E(\mathbf{F}_p) \cong \mathbf{Z}/n\mathbf{Z} \oplus \mathbf{Z}/n\mathbf{Z}$$

showing that $\exp(E(\mathbf{F}_p)) \approx \sqrt{p}$ Since it is very likely that there are infinitely many primes of the form n^2+1 with $n \equiv 0 \pmod 4$, it is also likely that Theorem 1.1 is false for elliptic curves with complex multiplication. It is easy to show that certain standard conjectures on the distribution of prime numbers in quadratic progressions imply that for every elliptic curve E over $\mathbf{Q}$ with complex multiplication the exponent of $E(\mathbf{F}_p)$ is $\sqrt{p}+o(1)$ for infinitely many primes p. It would be interesting to try and show without any unproved hypotheses that for elliptic curves over $\mathbf{Q}$ with complex multiplication there exist infinitely many primes p for which $\exp(E(\mathbf{F}_p))/\sqrt{p}$ is bounded.

3. Small exponents.

In this section we will show that for an elliptic curve E over $\mathbf{Q}$ there exist infinitely many primes p for which $\exp(E(\mathbf{F}_p))$ is relatively small. We will derive the results from an effective Čebotarev Density Theorem [4]. A strong and realistic version of this theorem can at present only be proved assuming the truth of certain Generalized Riemann Hypotheses (GRH). Propositions 3.3 and 3.4 are therefore only valid under these hypotheses.

Let K be a finite Galois extension of $\mathbf{Q}$ with $G = \mathrm{Gal}(K/\mathbf{Q})$. The discriminant of K is denoted by Δ_K and we let $\delta_K = |\Delta_K|^{1/n}$ where $n = [K:\mathbf{Q}]$ denote the root discriminant of K. For a prime p let e_p denote the ramification index of p in K over $\mathbf{Q}$ and let r_p denote $v_{\mathfrak{p}}(\mathfrak{D}_{K/\mathbf{Q}})$; here $\mathfrak{D}_{K/\mathbf{Q}}$ denotes the different of K over $\mathbf{Q}$ and $v_{\mathfrak{p}}$ denotes the normalized valuation associated to a prime $\mathfrak{p}$ over p in K. It is elementary to check that

$$\delta_K = \prod_p p^{r_p/e_p}$$

where the product runs over all primes p.

Prposition 3.1. *Let E be an elliptic curve over $\mathbf{Q}$ and let ℓ be a prime. By K we denote the field generated by the group of ℓ-torsion points $E[\ell]$. We have*

$$\delta_K < |\Delta_E|\ell^2;$$

here Δ_E denotes the discriminant of the curve E.

Proof. The field K is a Galois extension of $\mathbf{Q}$. We put $G = \mathrm{Gal}(K/\mathbf{Q})$. Using the notation introduced above we have that $r_p/e_p = 0$ for every unramified prime p. We will estimate the ratios r_p/e_p for all primes p that ramify in K over $\mathbf{Q}$.

case 1. $p \neq \ell$.

Since p is ramified in K we must have that p is a prime of bad reduction. We use Ogg's formula [5]:

$$\mathrm{ord}_p \Delta_E = \#C + \epsilon + \delta - 1$$

where C denotes the set of components of the Néron minimal model of E over $\mathbf{Z}$ and $\epsilon = 1$ or 2 according as the reduced curve $E \pmod p$ has a double point or a cusp singularity. In either case we have

$$\mathrm{ord}_p \Delta_E \geq \delta + 1$$

where δ is Serre's measure of wild ramification [5]:

$$\delta = \frac{1}{e_p} \sum_{j=1}^{\infty} \#G_j \ \mathrm{codim}(E[\ell]^{G_j}).$$

Here $G_j = \{\sigma \in G : \sigma(x) \equiv x \pmod{\mathfrak{p}^j}$ for all integral $x\}$ denotes the j-th higher ramification group. Let i denote the largest integer for which $G_i \neq \{1\}$. We clearly have

$$\delta \geq \frac{1}{e_p} \sum_{j=1}^{i} \#G_j.$$

Therefore

$$r_p = v_{\mathfrak{p}}(\mathfrak{D}_{K/\mathbf{Q}}) = \sum_{j=0}^{\infty} (\#G_j - 1) \leq e_p - 1 + e_p\delta < e_p \mathrm{ord}_p \Delta_E$$

and hence

$$\frac{r_p}{e_p} < \mathrm{ord}_p \Delta_E.$$

case 2. $p = \ell$.
If ℓ is tamely ramified in K we clearly have that

$$\frac{r_\ell}{e_\ell} = 1 - \frac{1}{e_\ell} < 2.$$

If ℓ is wildly ramified we let $\mathfrak{l}$ denote a prime over ℓ in K and we consider the local field extension $K_{\mathfrak{l}}$ over $\mathbf{Q}_\ell$ with Galois group G_ℓ. The group G_ℓ considered as a subgroup of $GL_2(\mathbf{F}_\ell)$ contains the subgroup

$N = \{\begin{pmatrix} 1 & a \\ 0 & 1 \end{pmatrix} : a \in \mathbf{F}_\ell\}$ as a normal subgroup. Therefore G_ℓ is contained in the normalizer of $\begin{pmatrix} 1 & 1 \\ 0 & 1 \end{pmatrix}$ which is the group of upper triangular matrices. By the non-degeneracy of the Weil-pairing the determinant maps G/N onto $(\mathbf{Z}/\ell\mathbf{Z})^*$ and we find that $K_\mathfrak{l}$ is a cyclic extension of degree ℓ of the field $F(\zeta_\ell)$ where F is an unramified extension of $\mathbf{Q}_\ell$. By local class field theory the conductor of $K_\mathfrak{l}$ over $F(\zeta_\ell)$ is at most $(\zeta_\ell - 1)^{\ell+1}$. One easily checks that $r_\ell \leq 2\ell^2 - 2\ell - 1$ and, since $e_\ell \leq \ell(\ell-1)$, that

$$\frac{r_\ell}{e_\ell} < 2.$$

Combining everything we obtain

$$\delta_K < |\Delta_E|\ell^2$$

as required.

Remark 3.2. When E has good reduction at ℓ one has in fact that

$$\delta_K < |\Delta_E|\ell^{1+\frac{1}{\ell-1}}$$

(see [2].) The result in Prop.3.1 is sufficiently strong for our purposes.

Assuming the Generalized Riemann Hypotheses (*GRH*) we can now show the following:

Proposition 3.3. *(Assuming GRH) Let E be an elliptic curve over $\mathbf{Q}$. There exists a constant c_E such that*

$$\frac{\exp(E(\mathbf{F}_p))}{\sqrt{p}} < c_E p^{3/8} \log p$$

for infinitely many primes p.

Proof. Let ℓ be a prime and let K denote the field generated by the ℓ-torsion points of E. Since the cardinality of $G = \mathrm{Gal}(K/\mathbf{Q})$ is at most $\#GL_2(\mathbf{F}_\ell) < \ell^4$ we have by Prop.3.1 that the discriminant Δ_K of K satisfies

$$|\Delta_K| \leq |\Delta_E \ell^2|^{\ell^4}.$$

By the effective Čebotarev Density Theorem [4] we find that there exists a prime $p < c\ell^8 \log^2 |\Delta_E \ell^2|$ that splits completely in K. Here c denotes some absolute constant. The ℓ-torsion points are rational over $\mathbf{F}_p$. Therefore

$$\exp(E(\mathbf{F}_p)) \leq \frac{(\sqrt{p}+1)^2}{\ell}$$

which is easily seen to imply that

$$\frac{\exp(E(\mathbf{F}_p))}{\sqrt{p}} \leq c_E p^{3/8} \log p$$

for some constant c_E which only depends on E. Since one finds infinitely many primes this way, the result follows.

Proposition 3.4. *(Assuming GRH) Let E be an elliptic curve over $\mathbf{Q}$. There exists a constant c'_E such that*

$$|\Delta_p| < c'_E p^{2/3} \log p$$

for infinitely many primes p.

Proof. Let ℓ be a prime and let K' denote the subfield of the field K in the previous Proposition which is invariant under the centre of $GL_2(\mathbf{F}_\ell)$. We have that $[K' : \mathbf{Q}] \leq \#PGL_2(\mathbf{F}_\ell) < \ell^3$ and therefore

$$|\Delta_{K'}| \leq |\Delta_E \ell^2|^{\ell^3}.$$

There exists a prime $p < c\ell^6 \log^2 |\Delta_E \ell^2|$ that splits completely in K'. The Frobenius endomorphism ϕ_p of p is congruent to an integer $m \pmod{\ell}$ in $\mathrm{End}(E_{/\mathbf{F}_p})$. Since $\phi_p \notin \mathbf{Z}$ we have in $\mathrm{End}(E_{/\mathbf{F}_p})$ the relation $\phi_p = m + \ell\psi$ with $\psi \notin \mathbf{Z}$. As in the proof of Theorem 1.1 it follows that $\frac{|\Delta|}{4}\ell^2 \leq p$ and hence that

$$|\Delta_p| \leq c'_E p^{2/3} \log p$$

for some constant c'_E which only depends on E. This proves Prop.3.4.

Conjecture 3.5 One would conjecture that for infinitely many primes ℓ as in Propositions 3.3 and 3.4 one can find primes p completely splitting in K or K' which do not exceed $\log|\Delta_K|$ or $\log|\Delta_{K'}|$ respectively. This would imply that

$$\frac{\exp(E(\mathbf{F}_p))}{\sqrt{p}} \leq c_E p^{1/4} \log p$$

and

$$|\Delta_p| \leq c'_E p^{1/3} \log p$$

for infinitely many primes p.

This conjecture seems to be related to an elliptic analogue of the problem of estimating the largest prime divisor of $p-1$, see [1]. I have no idea how the exponents and the $|\Delta_p|$ are distributed as p varies.

BIBLIOGRAPHY

[1] Deshouillers, J.-M.: Théorème de Fermat: La contribution de Fouvry. *Sém. Bourbaki, exp. 648 (juin 1985).* Astérisque **133/134**, (1986), 309–318.

[2] Fontaine, J.-M.: Il n'y a pas de variété abélienne sur **Z**, *Invent. Math.* **81**, (1985), 515–538.

[3] Hardy, G. and Wright, E.: *An Introduction to the Theory of Numbers*, Oxford Univ. Press, Oxford 1960.

[4] Lagarias, J.C., Montgomery, H.L. and Odlyzko, A.M.: A Bound for the Least Prime Ideal in the Chebotarev Density Theorem, *Invent. Math.* **54**, (1979), 271–296.

[5] Ogg, A.: Elliptic curves and wild ramification, *Amer J. of Math.* **89**, (1967), 1–21.

[6] Silverman, J.: *The Arithmetic of Elliptic curves*, Graduate Texts in Math. **106**, Springer-Verlag, New York 1986.

The Generalized De Rham-Witt Complex and Congruence Differential Equations

JAN STIENSTRA

1 Introduction.

The De Rham-Witt complex is a powerful instrument for studying the crystalline cohomology of a smooth projective variety over a perfect field of positive characteristic. In [9] the De Rham-Witt complex is constructed for schemes on which some prime number p is zero. Here in section 2 we construct on every scheme X on which 2 is invertible the *generalized De Rham-Witt complex* $\underline{\mathcal{W}\Omega}^{\cdot}_{X}$; this is a Zariski sheaf of anti-commutative differential graded algebras with the additional structures and properties described in (2.1)–(2.6). Section 3 gives the (obvious) definition of the *relative generalized De Rham-Witt complex* $\underline{\mathcal{W}\Omega}^{\cdot}_{X/S}$ for $f : X \to S$ a morphism of schemes over $\mathbf{Z}[\frac{1}{2}]$.

Now consider a proper smooth morphism $f : X \to S$ of smooth schemes over some open part of Spec $\mathbf{Z}[\frac{1}{2}]$. For simplicity we assume that S is affine. Let s be a closed point of S with residue field $k(s)$ of characteristic $p > 2$. Let X_s be the fiber of f over s. Using the functoriality of the constructions and the projection onto the p-typical part (see (2.4)) one obtains for all $m \geq 0$ a *specialization map*

$$\mathbf{H}^m(X, \underline{\mathcal{W}\Omega}^{\cdot}_{X/S}) \to \mathbf{H}^m(X_s, \mathcal{W}\Omega^{\cdot}_{X_s})$$

compatible with the Frobenius endomorphisms $\underline{F}_p$ on its source and target. Here $\mathcal{W}\Omega^{\cdot}_{X_s}$ is the classical De Rham-Witt complex on X_s. Since X_s is a smooth proper scheme over the perfect field $k(s)$ one knows from ([9] II(1.4),(2.8)) that $\mathbf{H}^m(X_s, \mathcal{W}\Omega^{\cdot}_{X_s})$ is isomorphic to the crystalline cohomology $\mathrm{H}^m_{\mathrm{crys}}(X_s)$ of X_s.

On the other hand from (2.2) one gets a homomorphism

$$\mathbf{H}^m(X, \underline{\mathcal{W}\Omega}^{\cdot}_{X/S}) \to \mathbf{H}^m(X, \Omega^{\cdot}_{X/S})$$

for every $m \geq 0$. In order to turn this effectively into a result on the interaction of Frobenius on the crystaline cohomology of the fibers X_s and the Gauss-Manin connection on the De Rham cohomology of X/S we must however replace our complexes by the projective systems of complexes

$$\begin{array}{ll} \{\underline{W\Omega}^{\cdot}_X \bmod N\}_{N\in\mathbf{N}} , & \{\underline{W\Omega}^{\cdot}_{X/S} \bmod N\}_{N\in\mathbf{N}} , \\ \{\Omega^{\cdot}_{X/S} \bmod N\}_{N\in\mathbf{N}} , & \{W\Omega^{\cdot}_{X_s} \bmod N\}_{N\in\mathbf{N}} , \end{array}$$

indexed by the multiplicative monoid of the positive integers. This respects the relation with crystalline cohomology: see (5.9). The reason for looking at the complexes modulo N is that now the Frobenius homomorphism F_N induces a homomorphism

$$F_N^* : \mathrm{H}^m(X, \underline{W\mathcal{O}}_X) \to \mathbf{H}^m(X, \underline{W\Omega}^{\cdot}_X \bmod N)$$

for every $m \geq 0$ (see (4.5)). This simple observation is the key to theorem (4.6) and its corollary (4.7) which states that matrices which come from the action of the Frobenius operator F_N on the generalized Witt vector cohomology group $\mathrm{H}^m(X, \underline{W\mathcal{O}}_X)$ provide solutions to the differential equations taken modulo N associated with the Gauss-Manin connection acting on $\mathrm{H}^{r-m}(X, \Omega^r_{X/S})$. In a number of interesting examples these matrices can explicitly be calculated via Čech cocycles for generalized Witt vector cohomology (see [18] (5.6)).

A result of this type was first observed by Igusa for the Legendre family of elliptic curves $y^2 = x(x-1)(x-\lambda)$ over $S = \operatorname{Spec} \mathbf{Z}[\lambda, (2\lambda(1-\lambda))^{-1}]$ with $N = p$ an odd prime number. Manin [15] proved it for general smooth families of curves in characteristic $p > 0$. Katz ([11] prop. (2.3.6.3)) generalized it to higher dimensions, but still in characteristic $p > 0$. In [12] Katz reinterpreted Igusa's observation and greatly generalized it to congruence differential equations modulo arbitrary N for the coefficients of formal expansions of differential forms.

In [13] Katz used the expansion coefficients of differential forms to describe the *top slope quotient crystal* of $\mathbf{H}^m_{\mathrm{DR}}(X/S)$ for certain families of varieties X/S. In section 5 we prove a similar result for the *unit root subcrystal*. One can imagine a kind of Hodge symmetry relating (5.6) and the main theorem of [13].

Our work on $\underline{W\Omega}^{\cdot}_{X/S}$ was in part motivated by the remark in [13] p.246 that the result of op. cit. might perhaps be considered as evidence for the existence of a theory of De Rham-Witt with parameters. A second motivation comes from the comparison isomorphism between crystalline cohomology in characteristic p and De Rham cohomology in characteristic 0 (see [1] (7.26.3)). There the De Rham side has no natural Frobenius and the crystalline side has no Hodge filtration. $\underline{W\Omega}^{\cdot}_{X/S}$ is an object with Frobenius and good filtrations of its own, which sees both $\mathrm{H}^*_{\mathrm{crys}}$ and $\mathbf{H}^*_{\mathrm{DR}}$; see (5.9) for a more explicit example of how this works out.

2 Construction of the generalized De Rham-Witt complex.

In this section we give the construction of the generalized De Rham-Witt complex $\underline{W\Omega}^{\cdot}_X$ on a scheme X on which 2 is invertible. This is a Zariski sheaf of anti-commutative differential graded algebras with the additional structures and properties described in (2.1)–(2.6).

2.1 Let $\underline{W\Omega}^i_X$ be the homogeneous component of degree i of $\underline{W\Omega}^{\cdot}_X$. Then $\underline{W\Omega}^i_X = 0$ for $i < 0$ and $\underline{W\Omega}^0_X$ is the sheaf of generalized Witt vectors on X (cf. [3,8,2]). We shall usually write $\underline{W\mathcal{O}_X}$ instead of $\underline{W\Omega}^0_X$.

2.2 Let $\Omega^{\cdot}_X$ be the absolute De Rham complex on X i.e. the complex of differential forms relative to $\mathbf{Z}$. Put $\widetilde{\Omega}^i_X := \Omega^i_X/(i!\text{-torsion in } \Omega^i_X)$ and $\widetilde{\Omega}^{\cdot}_X := \bigoplus_{i\geq 0} \widetilde{\Omega}^i_X$. Then there is a homomorphism of sheaves of differential graded algebras

$$\pi : \underline{W\Omega}^{\cdot}_X \to \widetilde{\Omega}^{\cdot}_X;$$

in degree 0 this is the projection onto the first Witt vector coordinate $\underline{W\mathcal{O}_X} \to \mathcal{O}_X$.

2.3 For every integer $m \geq 1$ and every $i \geq 0$ there are homomorphisms of additive groups

$$F_m, V_m : \underline{W\Omega}^i_X \to \underline{W\Omega}^i_X$$

satisfying the following relations

$$\begin{array}{lll} F_m V_m = m, & F_m F_n = F_{mn}, & V_m V_n = V_{mn}, \\ V_m d = m d V_m, & d F_m = m F_m d, & F_m d V_m = d, \\ F_m(ab) = (F_m a)(F_m b), & V_m(a(F_m b)) = (V_m a)b, & \end{array}$$

for all m, n and for all sections a, b of $\underline{W\Omega}^{\cdot}_X$, and

$$V_n F_m = F_m V_n \quad \text{if} \quad (n, m) = 1;$$

here $d : \underline{\mathcal{W}\Omega}^i_X \to \underline{\mathcal{W}\Omega}^{i+1}_X$ is the differential of the differential graded algebra $\underline{\mathcal{W}\Omega}^{\cdot}_X$. On the sheaf of generalized Witt vectors $\underline{\mathcal{W}\mathcal{O}}_X$ the operators F_m and V_m coincide with the usual Frobenius and Verschiebung operators (cf.[3,8,2]).

Obviously F_m does not commute with d. However one obtains an endomorphism $\underline{F}_m$ of the sheaf of differential graded algebras $\underline{\mathcal{W}\Omega}^{\cdot}_X$ by taking $\underline{F}_m = m^i F_m$ on $\underline{\mathcal{W}\Omega}^i_X$.

2.4 Let p be an odd prime number and let X be a scheme of characteristic p. Then there is an idempotent endomorphism E_p of the differential graded algebra $\underline{\mathcal{W}\Omega}^{\cdot}_X$ which projects $\underline{\mathcal{W}\Omega}^{\cdot}_X$ onto its p-typical part: $E_p\underline{\mathcal{W}\Omega}^{\cdot}_X = \cap\{\ker F_m \mid m \text{ prime} \neq p\}$. So $E_p\underline{\mathcal{W}\Omega}^{\cdot}_X$ is a sheaf of anti-commutative differential graded algebras. Its component in degree 0 is the sheaf $\mathcal{W}\mathcal{O}_X$ of p-typical Witt vectors on X. Since E_p commutes with F_p and V_p, the operators F_p and V_p act on $E_p\underline{\mathcal{W}\Omega}^{\cdot}_X$ and here in characteristic p they commute: $V_pF_p = F_pV_p = p$. There is an isomorphism of sheaves of differential graded algebras

$$\underline{\mathcal{W}\Omega}^{\cdot}_X \simeq (E_p\underline{\mathcal{W}\Omega}^{\cdot}_X)^{\mathbf{N}\setminus p\mathbf{N}},$$

where the right hand side is the product of copies of $E_p\underline{\mathcal{W}\Omega}^{\cdot}_X$ indexed by the set of positive integers prime to p. Let $\mathcal{W}\Omega^{\cdot}_X$ be the De Rham-Witt complex on X constructed by Deligne and Illusie [9]. There is a surjective homomorphism of sheaves of differential graded algebras from $\mathcal{W}\Omega^{\cdot}_X$ onto $E_p\underline{\mathcal{W}\Omega}^{\cdot}_X$ compatible with the operators F_p and V_p on both sides. If X is a smooth scheme over a perfect field of characteristic p this is an isomorphism:

$$\mathcal{W}\Omega^{\cdot}_X \simeq E_p\underline{\mathcal{W}\Omega}^{\cdot}_X.$$

2.5 Let X be a scheme over $\mathbf{Q}$. Then there is an idempotent endomorphism E_0 of the differential graded algebra $\underline{\mathcal{W}\Omega}^{\cdot}_X$ with image $E_0\underline{\mathcal{W}\Omega}^{\cdot}_X = \bigcap_{m>1}\ker F_m$. There are isomorphisms of sheaves of differential graded algebras

$$E_0\underline{\mathcal{W}\Omega}^{\cdot}_X \simeq \Omega^{\cdot}_X, \qquad \underline{\mathcal{W}\Omega}^{\cdot}_X \simeq (\Omega^{\cdot}_X)^{\mathbf{N}}.$$

2.6 The constructions are functorial: let $f : Y \to X$ be a morphism of schemes over $\mathbf{Z}[\frac{1}{2}]$. Then there is a homomorphism $\underline{\mathcal{W}\Omega}^{\cdot}_X \to f_*\underline{\mathcal{W}\Omega}^{\cdot}_Y$ of sheaves of differential graded algebras on X compatible with the operators F_m and V_m on both sides. On (hyper-) cohomology groups this induces homomorphisms like

$$\mathrm{H}^n(X, \underline{\mathcal{W}\Omega}^i_X) \to \mathrm{H}^n(Y, \underline{\mathcal{W}\Omega}^i_Y), \qquad \mathbf{H}^n(X, \underline{\mathcal{W}\Omega}^{\cdot}_X) \to \mathbf{H}^n(Y, \underline{\mathcal{W}\Omega}^{\cdot}_Y).$$

The construction of the generalized De Rham-Witt complex with its additional structures is essentially given in [19]. However in op. cit. it is specialized to characteristic p situations too early for our present purpose. Therefore we shall briefly recall the constructions in in such a way that the general statements in (2.1) –(2.6) become completely justified.

2.7 Let A be a commutative ring with 1. In [19] K-theory is used to construct an anti-commutative graded ring with 1

$$\widetilde{K}_*(\mathrm{End}(A)) = \bigoplus_{i\geq 0} \widetilde{K}_i(\mathrm{End}(A))$$

equipped with homomorphisms for the additive structure

$$F_m, V_m : \widetilde{K}_i(\mathrm{End}(A)) \to \widetilde{K}_i(\mathrm{End}(A)), \qquad i \geq 0,$$

for every positive integer m, and with a derivation

$$d : \widetilde{K}_i(\mathrm{End}(A)) \to \widetilde{K}_{i+1}(\mathrm{End}(A)), \qquad i \geq 0.$$

The relations listed in (2.3) hold also for F_m, V_m and d on $\widetilde{K}_*(\mathrm{End}(A))$ except for $F_m dV_m = d$ which here only holds for odd m, and for $d^2 = 0$ which here is weakened to $2d^2 = 0$ (see [19] theorem (1.8)).

2.8 In [20] the graded ring $\widetilde{K}_*(\mathrm{End}(A))$ is equipped with a decreasing filtration by homogeneous ideals $\{\mathrm{Fil}_n \widetilde{K}_*(\mathrm{End}(A))\}_{n\geq 1}$ with $\mathrm{Fil}_1 \widetilde{K}_*(\mathrm{End}(A)) = \widetilde{K}_*(\mathrm{End}(A))$. Define

$$\widetilde{K}_*(\mathrm{End}(A))^c := \lim_{\leftarrow n} [\widetilde{K}_*(\mathrm{End}(A))/\mathrm{Fil}_n \widetilde{K}_*(\mathrm{End}(A))]$$

with the topology of a projective limit of discrete spaces. From the proof of the proposition in [20] one gets

$$F_m(\mathrm{Fil}_{mn}) \subset \mathrm{Fil}_n, \qquad V_m(\mathrm{Fil}_n) \subset \mathrm{Fil}_{mn}, \qquad d(\mathrm{Fil}_{qn}) \subset \mathrm{Fil}_n$$

for all m, $n \geq 1$ and for some positive integer q independent of n and A. Thus F_m, V_m and d extend to continuous operators on $\widetilde{K}_*(\mathrm{End}(A))^c$ satisfying the same relations as in (2.7). Moreover, $\widetilde{K}_0(\mathrm{End}(A))^c$ is canonically isomorphic with the ring of generalized Witt vectors over A ([20] p.220).

2.9 If 2 is invertible in A it is also invertible in $\widetilde{K}_0(\mathrm{End}(A))^c$. Therefore all relations in (2.3)(see also (2.7)) are valid for the operators acting on $\widetilde{K}_*(\mathrm{End}(A))^c$.

2.10 From now on we assume that 2 is invertible in A.

Definition $\underline{W\Omega}^{\cdot}_A$:= closure of the graded subring of $\widetilde{K}_*(\mathrm{End}(A))^c$ generated by $\widetilde{K}_0(\mathrm{End}(A))^c$ and $d\widetilde{K}_0(\mathrm{End}(A))^c$.

2.11 In particular, $\underline{W\Omega}^0_A$ is the ring of generalized Witt vectors over A. Its additive group is isomorphic with the multiplicative group $(1+tA[[t]])^*$. Every element of the latter group can be written uniquely in the form $\prod_{n\geq 1}(1-a_nt^n)^{-1}$ with all $a_n \in A$. The elements of $\underline{W\Omega}^0_A$ can therefore be written uniquely as $\sum_{n\geq 1} V_n \underline{\underline{a_n}}$, where $\underline{\underline{a}}$ is the Witt vector which corresponds to the power series $(1-at)^{-1}$.

2.12 Proposition *For all* $i \geq 0$ *and* $m \geq 1$ *one has*

$$d\underline{W\Omega}^i_A \subset \underline{W\Omega}^{i+1}_A, \qquad V_m \underline{W\Omega}^i_A \subset \underline{W\Omega}^i_A, \qquad F_m \underline{W\Omega}^i_A \subset \underline{W\Omega}^i_A.$$

Proof The results for d and V_m follow immediately from the relations in (2.3). The problem for F_m is easily reduced to showing $F_m dV_n \underline{\underline{a}} \in \underline{W\Omega}^1_A$ for all $m,n \geq 1$ and $a \in A$; the Witt vector $\underline{\underline{a}}$ is defined in (2.11). In view of the relations in (2.3) we may even assume $(m,n)=1$. Choose integers q and r such that $qn+rm=1$. Then

$$F_m dV_n \underline{\underline{a}} = qV_n F_m d\underline{\underline{a}} + rdF_m V_n \underline{\underline{a}}.$$

Formula (8.3.3) in [19] shows

$$F_m d\underline{\underline{a}} = \underline{\underline{a}}^{m-1} d\underline{\underline{a}}$$

Thus we find

$$F_m dV_n \underline{\underline{a}} = q(V_n \underline{\underline{a}}^{m-1}) d(V_n \underline{\underline{a}}) + rdF_m V_n \underline{\underline{a}} \in \underline{W\Omega}^1_A \tag{1}$$

□

2.13 The preceding construction depends functorially on A: if $g : A \to B$ is a homomorphism of $\mathbf{Z}[\frac{1}{2}]$-algebras, then there is a continuous homomorphism of graded topological rings

$$g_* : \underline{W\Omega}^{\cdot}_A \to \underline{W\Omega}^{\cdot}_B$$

which commutes with the operators d, F_m and V_m $(m \geq 1)$. It sends the Witt vector $\underline{\underline{a}} \in \underline{W\Omega}^0_A$ to $\underline{\underline{g(a)}} \in \underline{W\Omega}^0_B$.

2.14 We get on every scheme X over $\mathbf{Z}[\frac{1}{2}]$ a pre-sheaf for the Zariski topology

$$(\text{Zariski open } U \subset X) \mapsto \underline{\mathcal{W}\Omega}^{\cdot}_{\Gamma(U,\mathcal{O}_X)}.$$

We define

$$\underline{\mathcal{W}\Omega}^{\cdot}_X := \text{ the sheaf associated with the above pre-sheaf.}$$

and call this the *generalized De Rham-Witt complex* of X. We shall usually write $\underline{\mathcal{W}\mathcal{O}}_X$ instead of $\underline{\mathcal{W}\Omega}^0_X$.

This completes the construction of $\underline{\mathcal{W}\Omega}^{\cdot}_X$ and of the operators d, F_m and V_m ($m \geq 1$) acting on it. It follows from (2.7) and (2.9) that $\underline{\mathcal{W}\Omega}^{\cdot}_X$ with d is a sheaf of anti-commutative differential graded algebras and that the relations in (2.3) hold. Moreover (2.8) proves (2.1); in particular, $\underline{\mathcal{W}\mathcal{O}}_X$ is the sheaf of generalized Witt vectors on X. The functoriality property in (2.6) is a consequence of (2.13).

We now turn to the construction of the homomorphism π in (2.2).

2.15 Let A be a commutative ring with 1 and 2 invertible in A. By [19](3.4) (see also [20]) we have a bilinear pairing

$$\langle , \rangle : \widetilde{K}_i(\mathrm{End}(A))^c \times \widetilde{K}_0(\mathrm{Nil}(\mathbf{Z}[t]/(t^2))) \to K_{i+1}(A[t]/(t^2)).$$

From this we get in particular a homomorphism

$$\langle , \underline{\underline{t}} \rangle : \widetilde{K}_i(\mathrm{End}(A))^c \to K_{i+1}(A[t]/(t^2))$$

where $\underline{\underline{t}}$ is the element of $\widetilde{K}_0(\mathrm{Nil}(\mathbf{Z}[t]/(t^2)))$ defined in [19](5.2). On the other hand one has Gersten's map (see [2] p.206 (3.2),(3.3))

$$\mathrm{dlog} \ : K_{i+1}(A[t]/(t^2)) \to \Omega^{i+1}_{A[t]/(t^2)};$$

here we work with differential forms relative to $\mathbf{Z}$. Let

$$\rho : \underline{\mathcal{W}\Omega}^i_A \to \Omega^{i+1}_{A[t]/(t^2)}$$

be the composite $\mathrm{dlog}\langle , \underline{\underline{t}} \rangle$ restricted to $\underline{\mathcal{W}\Omega}^i_A$.

2.16 The group $\underline{\mathcal{W}\Omega}^i_A$ is topologically generated by the elements

$$(V_{n_0}\underline{\underline{a_0}})d(V_{n_1}\underline{\underline{a_1}}) \cdots d(V_{n_i}\underline{\underline{a_i}})$$

with $n_0, \dots, n_i \in \mathrm{N}, a_0, \dots, a_i \in A$ (cf.(2.11)).

Lemma *In the above situation let* $\alpha = (V_{n_0}\underline{\underline{a_0}})d(V_{n_1}\underline{\underline{a_1}})\cdots d(V_{n_i}\underline{\underline{a_i}})$. *If all* $n_j = 1$ *then*

$$\rho(\alpha) = (-1)^i i! d(-ta_0 da_1 \wedge \cdots \wedge da_i).$$

Otherwise $\rho(\alpha) = 0$.

Proof Let $n = \max(n_0, \ldots, n_i)$. Assume first $n \geq 2$. Using (2.3) one easily rewrites α in the form $\alpha = V_n\beta + dV_n\gamma$ with $\beta \in \underline{W\Omega}^i_A, \gamma \in \underline{W\Omega}^{i-1}_A$. From [19](3.2) one gets $\langle\alpha, \underline{\underline{t}}\rangle = \langle\beta, F_n\underline{\underline{t}}\rangle + (-1)^i\langle\gamma, F_n d\underline{\underline{t}}\rangle$. Loc.cit. (1.6) and (5.2) show $F_n\underline{\underline{t}} = \underline{\underline{t}}^n = 0$. Loc.cit. (8.3.3) yields $F_n d\underline{\underline{t}} = \underline{\underline{t}}^{n-1}d\underline{\underline{t}} = 0$ for $n \geq 3$. For $n = 2$ we compute $F_2 d\underline{\underline{t}} = 2^{-1}dF_2\underline{\underline{t}} = 0$. This proves $\rho(\alpha) = 0$ if $n \geq 2$. The formula for $\rho(\alpha)$ in case $n = 1$ follows from [19](7.6) and [2] p.206 (3.3); more precisely the argument is as follows. By functoriality it suffices to prove the formula for the case that $a_0, \ldots, a_i$ are the indeterminates in the polynomial ring $P := \mathbf{Z}[\frac{1}{2}][a_0, \ldots, a_i]$. Set $Q := P[a_0^{-1}, \ldots, a_i^{-1}]$. Using the injectivity of the natural homomorphism

$$\Omega^{i+1}_{P[t]/(t^2)} \to \Omega^{i+1}_{Q[t]/(t^2)}$$

and functoriality we see that it suffices to prove the formula with $a_0, \ldots, a_i$ in Q. Then [19](3.1) and the proof of [19](7.6) give

$$\langle\alpha, \underline{\underline{t}}\rangle = \{1 - ta_0a_1\cdots a_i, a_1, \ldots, a_i\}.$$

The right hand side is a Steinberg symbol in $K_{i+1}(Q[t]/(t^2))$. Applying [2] p.206 (3.3) to compute the dlog of this Steinberg symbol we find

$$\rho(\alpha) = (-1)^i i!(1 - ta_0a_1\cdots a_i)^{-1}d(-ta_0) \wedge da_1 \wedge \cdots \wedge da_i$$

This is equal to $(-1)^i i! d(-ta_0 da_1 \wedge \cdots \wedge da_i)$ because $t^2 = 0$ and $\frac{1}{2} \in Q$. □

2.17 Lemma *Let* A *be as in* (2.15). *Define*

$$\psi : \Omega^i_A \to \Omega^{i+1}_{A[t]/(t^2)}, \qquad \psi(\eta) = (-1)^i i! d(-t\eta).$$

Then

$$\ker\psi = (i! - \text{torsion in } \Omega^i_A) := \ker(i! : \Omega^i_A \to \Omega^i_A).$$

Proof Consider the map $\Omega^1_{A[t]/(t^2)} \to \Omega^1_A \oplus \Omega^1_{A[t]/(t^2)/A}$ which is the direct sum of the map induced by $t \mapsto 0$ and the map taking differentials relative to A. Its $(i+1)$-fold exterior power over $A[t]/(t^2)$ is a map $\Omega^{i+1}_{A[t]/(t^2)} \to \Omega^{i+1}_A \oplus \Omega^i_A dt$ which sends $d(t\eta)$ to $(-1)^i\eta dt$. The lemma is now clear. □

2.18 Define $\widetilde{\Omega}^i_A := \Omega^i_A/(i!\text{-torsion in } \Omega^i_A)$. Then the map ψ from (2.17) induces an isomorphism $\overline{\psi} : \widetilde{\Omega}^i_A \xrightarrow{\sim} \text{image } \psi$. From (2.16) one sees that the image of ρ is contained in the image of ψ. So we can compose ρ with $\overline{\psi}^{-1}$. Define

$$\pi_i := \overline{\psi}^{-1}\rho : \underline{W\Omega}^i_A \to \widetilde{\Omega}^i_A.$$

2.19 From (2.16) one obtains explicit formulas for π_i:

$$\pi_i((V_{n_0}\underline{\underline{a_0}})d(V_{n_1}\underline{\underline{a_1}})\cdots d(V_{n_i}\underline{\underline{a_i}})) = 0 \text{ if some } n_j \neq 1$$

$$\pi_i(\underline{\underline{a_0}}d\underline{\underline{a_1}}\cdots d\underline{\underline{a_i}}) = a_0 da_1 \wedge \cdots \wedge da_i.$$

These formulas show that π_i is surjective. They also show that the direct sum π of the maps π_i is a homomorphism of differential graded algebras $\pi : \underline{W\Omega}^{\cdot}_A \to \widetilde{\Omega}^{\cdot}_A$.

2.20 Let X be a scheme over $\mathbf{Z}[\frac{1}{2}]$. Then sheafification of the above construction provides the homomorphism of sheaves of differential graded algebras on X

$$\pi : \underline{W\Omega}^{\cdot}_X \to \widetilde{\Omega}^{\cdot}_X$$

for (2.2).

2.21 Let P be a set of prime numbers and let X be a scheme such that every prime number in P is invertible in $\mathcal{O}_X$. Then every $l \in P$ is also invertible in $\underline{W\mathcal{O}}_X$. Moreover with notations as in (2.8) we have $F_l\text{Fil}_1 \subset \text{Fil}_1$ and $V_l\text{Fil}_1 \subset \text{Fil}_l$ for every l. Therefore the expression

$$E^P := \prod_{l \in P}(1 - l^{-1}V_lF_l)$$

defines an operator on $\underline{W\Omega}^{\cdot}_X$. One easily checks that it is an idempotent operator, that it commutes with d, V_p and F_p for all primes $p \notin P$ and that $E^P(ab) = (E^Pa)(E^Pb)$ for all sections a,b of $\underline{W\Omega}^{\cdot}_X$. Furthermore it is clear that for every $l \in P$ the image of E^P is contained in $\ker F_l$ and that E^P is the identity on $\ker F_l$. Consequently

$$E^P\underline{W\Omega}^{\cdot}_X = \bigcap_{l \in P} \ker F_l.$$

Let $\overline{P} \subset \mathbf{N}$ be the multiplicatively closed subset with 1 generated by P. Then there is an isomorphism of sheaves of differential graded algebras

$$\underline{W\Omega}^{\cdot}_X \simeq (E^P\underline{W\Omega}^{\cdot}_X)^{\overline{P}};$$

on homogeneous sections of degree i the map $\rightarrow$ sends a to $(m^i E^P F_m a)_{m\in\overline{P}}$ and the map $\leftarrow$ sends $(b_m)_{m\in\overline{P}}$ to $\sum_{m\in\overline{P}} m^{-i-1} V_m b_m$. All this is an easy consequence of the relations in (2.3). We apply it in the situation of (2.5) (resp. (2.4)) with P the set of all primes (resp. all primes $\neq p$) and write E_0 (resp. E_p) for E^P. The results relating in (2.4) $E_p\underline{W\Omega}^{\cdot}_X$ to the De Rham-Witt complex of Deligne and Illusie are proved in [19] section 8. The isomorphism

$$E_0\underline{W\Omega}^{\cdot}_X \simeq \Omega^{\cdot}_X$$

in (2.5) is proved as follows. Note that $E_0 V_m = 0$ for all $m \geq 2$. Combining this with the definition of the ring structure on generalized Witt vectors one sees that there is a ring homomorphism

$$\lambda : \mathcal{O}_X \rightarrow E_0\underline{W\mathcal{O}}_X$$

which on sections is defined by $\lambda(a) = E_0\underline{\underline{a}}$ (see (2.11) for $\underline{\underline{a}}$). Because of the universal property of $\Omega^{\cdot}_X$ this homomorphism from $\mathcal{O}_X$ into the degree 0 component of the differential graded algebra $E_0\underline{W\Omega}^{\cdot}_X$ extends uniquely to a homomorphism of differential graded algebras

$$\lambda : \Omega^{\cdot}_X \rightarrow E_0\underline{W\Omega}^{\cdot}_X.$$

This homomorphism is surjective because $\underline{W\Omega}^{\cdot}_X$ is topologically generated by the sections described in (2.16) and because $E_0 V_m = 0$ for $m \geq 2$. A simple computation shows that $\pi\lambda$ is the identity map on $\Omega^{\cdot}_X$, where π is the homomorphism π from (2.2) restricted to the image of E_0. This proves that π induces an isomorphism $E_0\underline{W\Omega}^{\cdot}_X \simeq \Omega^{\cdot}_X$.

3 The relative generalized De Rham-Witt complex.

3.1 Let $f : X \rightarrow S$ be a morphism of schemes over $\mathbf{Z}[\frac{1}{2}]$. We define the relative generalized De Rham-Witt complex $\underline{W\Omega}^{\cdot}_{X/S}$ on X to be the quotient of $\underline{W\Omega}^{\cdot}_X$ by the closure of the ideal generated by $d(f^{-1}\underline{W\mathcal{O}}_S)$. It is clear that $\underline{W\Omega}^{\cdot}_{X/S}$ is a sheaf of anti-commutative differential graded algebras with $\underline{W\Omega}^0_{X/S} = \underline{W\mathcal{O}}_X$. The homomorphism $\pi : \underline{W\Omega}^{\cdot}_X \rightarrow \widetilde{\Omega}^{\cdot}_X$ induces a homomorphism

$$\pi : \underline{W\Omega}^{\cdot}_{X/S} \rightarrow \widetilde{\Omega}^{\cdot}_{X/S},$$

where $\Omega^{\cdot}_{X/S}$ is the usual relative De Rham complex of X/S and $\widetilde{\Omega}^i_{X/S} := \Omega^i_{X/S}/(i! - \text{torsion})$. Using the relations in (2.3) and formula (1) in (2.12) one easily checks that the operators F_m and V_m on $\underline{W\Omega}^{\cdot}_X$ map the ideal $(d(f^{-1}\underline{W\mathcal{O}}_S)) \cdot \underline{W\Omega}^{\cdot}_X$ into itself and thus induce operators F_m and V_m on $\underline{W\Omega}^{\cdot}_{X/S}$. The relations in (2.3) pass without change to $\underline{W\Omega}^{\cdot}_{X/S}$. Notice also the analogue of the functoriality property (2.6): a commutative square

$$\begin{array}{ccc} Y & \xrightarrow{g} & X \\ \downarrow & & \downarrow \\ T & \longrightarrow & S \end{array}$$

leads naturally to a homomorphism $\underline{W\Omega}^{\cdot}_{X/S} \to g_*\underline{W\Omega}^{\cdot}_{Y/T}$.

3.2 Suppose S is the spectrum of a perfect field of characteristic $p > 2$. Then F_p is surjective on $\underline{W\mathcal{O}}_S$. Because of $d\,F_p^r = p^r\,F_p^r\,d = V_p^r\,F_p^{2r}\,d$ in characteristic p and $V_p^r\,\mathrm{Fil}_1 \subset \mathrm{Fil}_{p^r}$ (see (2.8)), the subsheaf $d(f^{-1}\underline{W\mathcal{O}}_S)$ of $\underline{W\Omega}^{\cdot}_X$ is zero. So

$\underline{W\Omega}^{\cdot}_{X/S} = \underline{W\Omega}^{\cdot}_X$ if S is the spectrum of a perfect field of characteristic $p > 2$.

4 Congruence differential equations.

4.1 Let $S = \operatorname{Spec} A$ be an affine scheme which is smooth over an open part of $\operatorname{Spec} \mathbf{Z}[\frac{1}{2}]$. Let $f : X \to S$ be a projective smooth morphism of relative dimension r. We assume that all Hodge cohomology groups $\mathrm{H}^j(X, \Omega^i_{X/S})$ are free A-modules and $\mathrm{H}^r(X, \Omega^r_{X/S}) \simeq A$.

4.2 These hypotheses imply that the Hodge-De Rham spectral sequence $\mathrm{E}_1^{ij} := \mathrm{H}^j(X, \Omega^i_{X/S}) \Rightarrow \mathbf{H}^{i+j}(X, \Omega^{\cdot}_{X/S})$ degenerates at E_1 (note that A is flat over $\mathbf{Z}$ and use [4] th.(5.1)). So in particular all De Rham cohomology groups $\mathbf{H}^m(X, \Omega^{\cdot}_{X/S})$ are also free A-modules and $\mathbf{H}^{2r}(X, \Omega^{\cdot}_{X/S}) \simeq A$. Moreover the homomorphism

$$\beta : \mathbf{H}^m(X, \Omega^{\cdot}_{X/S}) \to \mathrm{H}^m(X, \mathcal{O}_X),$$

induced by the projection of the complex $\Omega^{\cdot}_{X/S}$ onto its degree 0 component $\mathcal{O}_X$, is surjective and the homomorphism

$$\mathrm{H}^{r-m}(X, \Omega^r_{X/S}) \to \mathbf{H}^{2r-m}(X, \Omega^{\cdot}_{X/S}),$$

induced by the inclusion of $\Omega^r_{X/S}$ as degree r component into $\Omega^{\cdot}_{X/S}$, is injective for every $m \geq 0$. One has a perfect pairing

$$\langle , \rangle : \mathbf{H}^m(X, \Omega^{\cdot}_{X/S}) \times \mathbf{H}^{2r-m}(X, \Omega^{\cdot}_{X/S}) \to \mathbf{H}^{2r}(X, \Omega^{\cdot}_{X/S}) \simeq A$$

which induces the duality

$$\mathrm{H}^{r-m}(X, \Omega^r_{X/S}) = \mathrm{H}^m(X, \mathcal{O}_X)^\vee .$$

4.3 Recall the Katz-Oda construction of the Gauss-Manin connection [14, 11]. The Koszul filtration $\{K^{i\cdot}\}_{i\geq 0}$ on the absolute De Rham complex $\Omega^\cdot_X$ is defined by

$$K^{i\cdot} := \mathrm{image}(f^*\Omega^i_S \otimes \Omega^{\cdot - i}_X \to \Omega^\cdot_X).$$

It satisfies

$$K^{0\cdot}/K^{1\cdot} \simeq \Omega^\cdot_{X/S}, \qquad K^{1\cdot}/K^{2\cdot} \simeq f^*\Omega^1_S \otimes \Omega^{\cdot-1}_{X/S}.$$

The boundary map in the long exact hypercohomology sequence associated with the exact sequence of complexes $0 \to K^{1\cdot}/K^{2\cdot} \to K^{0\cdot}/K^{2\cdot} \to K^{0\cdot}/K^{1\cdot} \to 0$ yields the Gauss-Manin connection

$$\nabla : \mathbf{H}^m(X, \Omega^\cdot_{X/S}) \to \Omega^1_S \otimes \mathbf{H}^m(X, \Omega^\cdot_{X/S})$$

for $m \geq 0$.

These constructions work equally well if we take the complexes modulo a positive integer N. They then provide the Gauss-Manin connection ∇ on $\mathbf{H}^m(X, \Omega^\cdot_{X/S} \bmod N)$ and show that the image of $\mathbf{H}^m(X, \Omega^\cdot_X \bmod N)$ lies in the kernel of ∇.

4.4 Let Diff_S be the algebra of differential operators on A relative to $\mathbf{Z}$ and let Diff'_S be the subalgebra of Diff_S generated by the derivations of A (cf. [7] (16.11)). The Gauss-Manin connection defines a Lie algebra homomorphism $\nabla : \mathrm{Der}A \to \mathrm{End}_{\mathbf{Z}}\,(\mathbf{H}^*(X, \Omega^\cdot_{X/S}))$ so that $\nabla(D)$ is the composite of ∇ with $D \otimes 1$. This Lie algebra homomorphism extends to an algebra homomorphism

$$\nabla : \mathrm{Diff}'_S \to \mathrm{End}_{\mathbf{Z}}\,(\mathbf{H}^*(X, \Omega^\cdot_{X/S})).$$

4.5 Fix a positive integer N. Because of the relation $dF_N = NF_N d$ one can extend the homomorphism $F_N : \underline{W\mathcal{O}}_X \to \underline{W\mathcal{O}}_X \bmod N$ to a homomorphism of complexes

$$F'_N : \underline{W\mathcal{O}}_X \to \underline{W\Omega}^\cdot_X \bmod N$$

where $\underline{W\mathcal{O}}_X$ is viewed as a complex concentrated in degree 0. This leads to a homomorphism

$$F^*_N : \mathrm{H}^m(X, \underline{W\mathcal{O}}_X) \to \mathbf{H}^m(X, \underline{W\Omega}^\cdot_X \bmod N)$$

for every $m \geq 0$. One has the following commutative diagram

$$\begin{array}{ccc}
\mathrm{H}^m(X, \underline{W\mathcal{O}_X}) & \xrightarrow{F_N^*} & \mathbf{H}^m(X, \underline{W\Omega}_X^{\cdot} \bmod N) \\
F_N \downarrow & & \downarrow \tau_N \\
\mathrm{H}^m(X, \underline{W\mathcal{O}_X}) & & \mathbf{H}^m(X, \Omega_{X/S}^{\cdot}) \bmod N \\
\pi \downarrow & & \downarrow \beta \\
\mathrm{H}^m(X, \mathcal{O}_X) & \longrightarrow & \mathrm{H}^m(X, \mathcal{O}_X) \bmod N
\end{array}$$

where τ_N is induced by $\underline{W\Omega}_X^{\cdot} \bmod N \to \Omega_X^{\cdot} \bmod N \to \Omega_{X/S}^{\cdot} \bmod N$; notice that there is no $i!$-torsion in Ω_X^i and $\Omega_{X/S}^i$ because X is smooth over a subring of $\mathbf{Q}$.

4.6 Theorem *Let $f : X \to S$ be as in (4.1). Fix an integer $m \geq 0$. Take a basis $\{\omega_1, \ldots, \omega_h\}$ of $\mathrm{H}^m(X, \mathcal{O}_X)$. Let $\{\check{\omega}_1, \ldots, \check{\omega}_h\}$ be the dual basis of $\mathrm{H}^{r-m}(X, \Omega_{X/S}^r)$. Take $\xi \in \mathrm{H}^m(X, \underline{W\mathcal{O}_X})$ and define for every positive integer N $B_{N,1}, \ldots, B_{N,h} \in A$ by*

$$\pi F_N \xi = \sum_{j=1}^{h} B_{N,j} \, \omega_j$$

Suppose $P_1, \ldots, P_h \in \mathrm{Diff}'_S$ are such that

$$\nabla(P_1)\check{\omega}_1 + \cdots + \nabla(P_h)\check{\omega}_h = 0 \qquad \text{in} \quad \mathbf{H}^{2r-m}(X, \Omega_{X/S}^{\cdot}), \tag{2}$$

then

$$P_1 \, B_{N,1} + \cdots + P_h \, B_{N,h} \equiv 0 \bmod N.$$

for all $N \in \mathbf{N}$.

Proof From (4.5) one deduces for all j

$$\langle \tau_N F_N^* \xi, \check{\omega}_j \rangle \equiv B_{N,j} \bmod N.$$

The map τ_N factors via $\mathbf{H}^m(X, \Omega_X^{\cdot} \bmod N)$. Therefore the image of τ_N in $\mathbf{H}^m(X, \Omega_{X/S}^{\cdot} \bmod N)$ is contained in the kernel of the Gauss-Manin connection. So for every derivation D of A we have

$$\nabla(D)(\tau_N F_N^* \xi) = 0 \qquad \text{in} \quad \mathbf{H}^m(X, \Omega_{X/S}^{\cdot}) \bmod N.$$

In view of the compatibility of $\langle , \rangle$ and ∇ ([14] th.1) we find for all $D \in \mathrm{Der} A$ and all j

$$D B_{N,j} \equiv \langle \tau_N F_N^* \xi, \nabla(D)\check{\omega}_j \rangle \bmod N.$$

The theorem is now obvious. □

4.7 In [18](2.6) it is shown that the hypotheses in (4.1) imply that the map $\pi : \mathrm{H}^m(X, \underline{W\mathcal{O}}_X) \to \mathrm{H}^m(X, \mathcal{O}_X)$ is surjective. So there are elements $\tilde{\omega}_1, \ldots, \tilde{\omega}_h \in \mathrm{H}^m(X, \underline{W\mathcal{O}}_X)$ such that $\pi \tilde{\omega}_i = \omega_i$ for $i = 1, \ldots, h$. Define for $N \in \mathbf{N}$ the $h \times h$-matrix B_N with entries in A by

$$\pi F_N \underline{\tilde{\omega}} = B_N \underline{\omega}$$

where $\underline{\omega}$ resp. $\underline{\tilde{\omega}}$ is the column vector with components $\omega_1, \ldots, \omega_h$ resp. $\tilde{\omega}_1, \ldots, \tilde{\omega}_h$. For a prime number p the matrix $B_p \bmod p$ is known as the *Hasse-Witt matrix* of $X \otimes \mathbf{F}_p$ in degree m (cf. [11] p.27).

Corollary *The congruence differential equations in theorem* (4.6) *are valid for the rows of the matrices* B_N.

4.8 *Example* The Gauss-Manin connection makes $\mathbf{H}^{2r-m}(X, \Omega^{\cdot}_{X/S})$ a module over the algebra Diff'_S (a so-called $\mathcal{D}$-module [17]). The full set of differential equations (2) (or a generating subset thereof) gives a presentation for the sub-Diff'_S-module generated by $\mathrm{H}^{r-m}(X, \Omega^r_{X/S})$. In practice in explicit examples one finds these differential equations as Picard-Fuchs equations for the periods of regular differential forms.

For explicit examples based on families of curves of the form

$$y^n = x^a(x-1)^b(x-\lambda)^c$$

with $n, a, b, c \in \mathbf{N}$, $(n, a, b, c) = 1$, $a, b, c < n$ and connected with hypergeometric differential equations we refer to section 5 of [18]; there one also finds a full detail example illustrating (4.7).

Further explicit examples of Picard-Fuchs equations can be found in ([6] p. 73-76) for the 1-parameter family of elliptic curves

$$X^3 + Y^3 + Z^3 - 3\lambda XYZ = 0$$

and for the 1-parameter family of K3 surfaces

$$W^4 + X^4 + Y^4 + Z^4 - 4\lambda WXYZ = 0$$

and in [16] for the 2-parameter family of K3 surfaces

$$w^2 = xy(1-x)(1-y)(1-\lambda x - \mu y).$$

For these examples the matrices B_N (see (4.7)) can be calculated with the method of ([18] (5.6)).

5 Reconstruction of the unit root crystal.

In this section we prove theorem (5.6). This theorem shows great similarities with the main theorem of [13]. The two theorems seem related by a kind of Hodge symmetry. The actual congruences in our theorem look however weaker than the congruences in Katz's theorem. I do not yet understand this phenomenon.

5.1 We keep the situation and assumptions of (4.1). Fix an integer $m \geq 0$ and a prime number $p > 2$. We assume condition HW(m) of [13]:

hypothesis: For every point $\operatorname{Spec} k \longrightarrow S$ with k a perfect field of characteristic p the Frobenius endomorphism F_p on $\mathrm{H}^m(X \otimes k, \mathcal{O}_{X\otimes k})$ is bijective.

Now fix a basis $\omega_1, \ldots, \omega_h$ of $\mathrm{H}^m(X, \mathcal{O}_X)$. Take elements $\tilde{\omega}_1, \ldots, \tilde{\omega}_h$ in $\mathrm{H}^m(X, \underline{\mathcal{W}\mathcal{O}}_X)$ such that $\pi\tilde{\omega}_i = \omega_i$ for $i = 1, \ldots, h$ and define the matrices B_N as in (4.7). Then by [18] (4.2) the above hypothesis is equivalent with

hypothesis: The Hasse-Witt matrix $B_p \bmod p$ is invertible over the ring A/pA.

5.2 Set

$$\begin{array}{lll} A_n = A/p^n A, & S_n = \operatorname{Spec} A_n, & X_n = X \otimes A_n, \\ A_\infty = \varprojlim_n A_n, & S_\infty = \operatorname{Spec} A_\infty, & X_\infty = X \otimes A_\infty. \end{array}$$

Since the ring A_∞ is formally smooth over $\mathbf{Z}_p$ it carries an endomorphism σ such that for all $a \in A_\infty$

$$\sigma(a) \equiv a^p \bmod pA_\infty.$$

In general there are many endomorphisms with this property. Given one choice for σ there is a unique homomorphism of rings

$$\lambda : A_\infty \longrightarrow W(A_\infty)$$

into the ring $W(A_\infty)$ of p-typical Witt vectors over A_∞, such that $\pi\, F_p^n\, \lambda = \sigma^n$ for all $n \in \mathrm{N}$; in particular $\pi\lambda = id$ [8](17.6.9).

In the sequel we will often write a^σ instead of $\sigma(a)$. For a matrix $M = (m_{ij})$ with entries in A_∞ we set $M^{\sigma^n} = (m_{ij}^{\sigma^n})$, $\lambda(M) = (\lambda(m_{ij}))$.

5.3 In [18] theorem (3.4) it is shown that under the hypotheses of (5.1) there exist an invertible $h \times h$-matrix H with entries in A_∞ and elements $\hat{\omega}_1, \ldots, \hat{\omega}_h$ in $\mathrm{H}^m(X_\infty, \underline{W\mathcal{O}}_{X_\infty})$ such that

$$\begin{aligned}
&B_{p^{n+1}} \equiv B_{p^n}^\sigma\, H \bmod p^{n+1} \qquad \text{for all} \quad n \geq 0,\\
&\pi\hat{\omega}_i = \omega_i \quad \text{ in } \ \mathrm{H}^m(X_\infty, \mathcal{O}_{X_\infty}) = \mathrm{H}^m(X, \mathcal{O}_X) \otimes A_\infty,\\
&F_p\, \underline{\hat{\omega}} = \lambda(H)\, \underline{\hat{\omega}},
\end{aligned}$$

where $\underline{\hat{\omega}}$ is the column vector with components $\hat{\omega}_1, \ldots, \hat{\omega}_h$.

5.4 We apply (4.5) with X_n instead of X and with $N = p^n$. This provides homomorphims

$$\begin{aligned}
\tau_N\, F_N^* \ &: \ \mathrm{H}^m(X_n, \underline{W\mathcal{O}}_{X_n}) \ \to \ \mathbf{H}^m(X, \Omega^\cdot_{X/S}) \otimes_A A_n\\
\psi_N\, F_N^* \ &: \ \mathrm{H}^m(X_n, \underline{W\mathcal{O}}_{X_n}) \ \to \ \mathbf{H}^m(X_n, \underline{W\Omega}^\cdot_{X_n/S_n} \bmod N)
\end{aligned}$$

where ψ_N is induced by $\underline{W\Omega}^\cdot_{X_n} \bmod N \to \underline{W\Omega}^\cdot_{X_n/S_n} \bmod N$. Writing $\hat{\omega}_i$ also for the image of $\hat{\omega}_i$ in $\mathrm{H}^m(X_n, \underline{W\mathcal{O}}_{X_n})$ we get

$$\begin{aligned}
\tau_N\, F_N^*\, \hat{\omega}_1, \ldots, \tau_N\, F_N^*\, \hat{\omega}_h \ &\in \ \mathbf{H}^m(X, \Omega^\cdot_{X/S}) \otimes_A A_n,\\
\psi_N\, F_N^*\, \hat{\omega}_1, \ldots, \psi_N\, F_N^*\, \hat{\omega}_h \ &\in \ \mathbf{H}^m(X_n, \underline{W\Omega}^\cdot_{X_n/S_n} \bmod N).
\end{aligned}$$

Using (4.5) and (5.3) we compute

$$\tau_{p^{n+1}}\, F_{p^{n+1}}^*\, \underline{\hat{\omega}} \bmod p^n = \tau_{p^n}\, F_{p^n}^*\, F_p\, \underline{\hat{\omega}} = \tau_{p^n}\, F_{p^n}^*\, \lambda(H)\, \underline{\hat{\omega}} = H^{\sigma^n}\, \tau_{p^n}\, F_{p^n}^*\, \underline{\hat{\omega}}$$

in $\mathbf{H}^m(X, \Omega^\cdot_{X/S}) \otimes A_n$. From this computation one obtains

$$\begin{aligned}
&(H^{\sigma^n}\, H^{\sigma^{n-1}} \cdots H^\sigma\, H)^{-1}\, \tau_{p^{n+1}}\, F_{p^{n+1}}^*\, \underline{\hat{\omega}} \equiv\\
&(H^{\sigma^{n-1}}\, H^{\sigma^{n-2}} \cdots H^\sigma\, H)^{-1}\, \tau_{p^n}\, F_{p^n}^*\, \underline{\hat{\omega}} \bmod p^n
\end{aligned}$$

for all $n \geq 1$. So in $\mathbf{H}^m(X, \Omega^\cdot_{X/S}) \otimes A_\infty = \varprojlim_n \mathbf{H}^m(X, \Omega^\cdot_{X/S}) \otimes A_n$ there exist elements $\varpi_1, \ldots, \varpi_h$ such that

$$\underline{\varpi} \bmod p^n = (H^{\sigma^{n-1}}\, H^{\sigma^{n-2}} \cdots H^\sigma\, H)^{-1}\, \tau_{p^n}\, F_{p^n}^*\, \underline{\hat{\omega}}$$

in $\mathbf{H}^m(X, \Omega^\cdot_{X/S}) \otimes A_n$.

Define

$$\mathbf{H}^m(X, \underline{W\Omega}^\cdot_{X/S})_\infty = \varprojlim_n \mathbf{H}^m(X_n, \underline{W\Omega}^\cdot_{X_n/S_n} \bmod p^n)$$

With a simple computation as above one sees that there exist elements $\lambda(\varpi_1), \ldots, \lambda(\varpi_h)$ in $\mathbf{H}^m(X, \underline{W\Omega}^{\cdot}_{X/S})_\infty$ such that

$$\underline{\lambda(\varpi)} \mapsto \lambda((H^{\sigma^{n-1}} H^{\sigma^{n-2}} \cdots H^{\sigma} H)^{-1}) \psi_{p^n} F^*_{p^n} \underline{\hat{\omega}} \tag{3}$$

in $\mathbf{H}^m(X_n, \underline{W\Omega}^{\cdot}_{X_n/S_n} \bmod p^n)$.

5.5 $\mathbf{H}^m(X, \underline{W\Omega}^{\cdot}_{X/S})_\infty$ is a module over the ring $W(A_\infty)$. Via the homomorphism $\lambda : A_\infty \to W(A_\infty)$ it becomes also an A_∞-module. We set:

$\mathcal{U}$:= the sub-A_∞-module of $\mathbf{H}^m(X, \Omega^{\cdot}_{X/S}) \otimes_A A_\infty$ generated by $\varpi_1, \ldots, \varpi_h$

$\lambda(\mathcal{U})$:= the sub-A_∞-module of $\mathbf{H}^m(X, \underline{W\Omega}^{\cdot}_{X/S})_\infty$ generated by $\lambda(\varpi_1), \ldots, \lambda(\varpi_h)$

5.6 Theorem
(a) *The homomorphism* $\pi : \mathbf{H}^m(X, \underline{W\Omega}^{\cdot}_{X/S})_\infty \to \mathbf{H}^m(X, \Omega^{\cdot}_{X/S}) \otimes_A A_\infty$ *restricts to an isomorphism of* A_∞*-modules*

$$\lambda(\mathcal{U}) \simeq \mathcal{U} \qquad \text{with}$$

$$\pi\lambda(\varpi_i) = \varpi_i \quad \text{for } i = 1, \ldots, h.$$

(b) *The homomorphism* $\beta : \mathbf{H}^m(X, \Omega^{\cdot}_{X/S}) \otimes_A A_\infty \to \mathrm{H}^m(X, \mathcal{O}_X) \otimes_A A_\infty$ *(see* 4.2)) *restricts to an isomorphism of* A_∞*-modules*

$$\mathcal{U} \simeq \mathrm{H}^m(X, \mathcal{O}_X) \otimes_A A_\infty \qquad \text{with}$$

$$\beta\, \omega_i = \omega_i \quad \text{for } i = 1, \ldots, h.$$

Consequently

$$\mathbf{H}^m(X, \Omega^{\cdot}_{X/S}) \otimes_A A_\infty = \mathcal{U} \oplus \mathrm{Fil}^1_{\mathrm{Hodge}} \mathbf{H}^m(X, \Omega^{\cdot}_{X/S}) \otimes_A A_\infty.$$

(c) *The Frobenius endomorphism* F_p *on* $\mathbf{H}^m(X, \underline{W\Omega}^{\cdot}_{X/S})_\infty$ *stabilizes* $\lambda(\mathcal{U})$. *The matrix of* F_p^k *on* $\lambda(\mathcal{U})$ *with respect to the basis* $\lambda(\varpi_1), \ldots, \lambda(\varpi_h)$ *satisfies the congruences*

$$\mathrm{matrix}[F_p^k] \equiv \lambda[B_{p^n}^{-\sigma^k} B_{p^{n+k}}] \bmod p^{n+1}$$

for every $n \geq 0$.
(d) *The Gauss-Manin connection on* $\mathbf{H}^m(X, \Omega^{\cdot}_{X/S}) \otimes_A A_\infty$ *stabilizes* $\mathcal{U}$ *i.e.*

$\nabla \mathcal{U} \subset \Omega^1_S \otimes \mathcal{U}$. *If D is a derivation of A the matrix of $\nabla(D)$ on $\mathcal{U}$ with respect to the basis $\varpi_1, \ldots, \varpi_h$ satisfies the congruences*

$$\text{matrix}[\nabla(D)] \equiv -B^{-1}_{p2n}\, D(B_{p2n}) \bmod p^n$$

for every $n \geq 0$.

Proof The results in (a) and (b) are immediate consequences of the constructions in (5.4) and (4.5) and of the formula

$$\pi F_{p^n}\, \underline{\hat{\omega}} = (H^{\sigma^{n-1}}\, H^{\sigma^{n-2}} \cdots H^{\sigma}\, H)\, \underline{\omega}$$

For (c) and (d) one should first observe

$$H^{\sigma^{k-1}}\, H^{\sigma^{k-2}} \cdots H^{\sigma}\, H \equiv B^{-\sigma^k}_{p^n}\, B_{p^{n+k}} \bmod p^{n+1}$$

This together with formula (3) proves the result in (c).
One checks by induction that for every $x \in A_\infty$ there exist $x_i \in A_\infty (i \geq 0)$ such that for every $n \geq 0$

$$x^{\sigma^n} = \sum_{i=0}^{n} p^i\, x_i^{p^{n-i}}$$

Consequently $D(x^{\sigma^n}) \equiv 0 \bmod p^n$. Using that $\tau_{p^n}\, F^*_{p^n} \underline{\hat{\omega}}$ lies in the kernel of Gauss-Manin one now computes

$$\begin{aligned} \nabla(D)(\underline{\varpi}) &\equiv [D((B^{-\sigma^n}_{p^n}\, B_{p2n})^{-1})][B^{-\sigma^n}_{p^n}\, B_{p2n}]\, \underline{\varpi} \\ &\equiv [D(B^{-1}_{p2n})]\, B_{p2n}\, \underline{\varpi} \qquad \bmod p^n \end{aligned}$$

This proves (d). □

5.7 *Remark* From (c) one sees that $\underline{F}_p$ is an automorphism of $\lambda(\mathcal{U})$. Via the isomorphism in (a) it gives a Frobenius automorphism on $\mathcal{U}$. Thus $\mathcal{U}$ becomes a unit root crystal. Since the rank of $\mathcal{U}$ is $h = \text{rank } \mathrm{H}^m(X, \mathcal{O}_X)$ which is the maximal rank for a unit root sub-crystal of $\mathbf{H}^m(X, \Omega^{\cdot}_{X/S}) \otimes A_\infty$, $\mathcal{U}$ is *the unit root sub-crystal* of $\mathbf{H}^m(X, \Omega^{\cdot}_{X/S}) \otimes A_\infty$ (cf. [13] p.249).

5.8 The canonical filtration on a sheaf complex $C^{\cdot}$ consists of the subcomplexes $t_{\leq i} C^{\cdot}$ for $i \in \mathbf{Z}$ defined by $(t_{\leq i} C^{\cdot})^j = C^j$ for $j < i$, $= 0$ for $j > i$,

$= \ker(d : C^j \to C^{j+1})$ for $j = i$. It induces on the hypercohomlogy $\mathbf{H}^m(C^\cdot)$ the increasing filtration

$$\mathrm{Fil}^i_{\mathrm{con}}\mathbf{H}^m(C^\cdot) := \mathrm{image}(\mathbf{H}^m(t_{\leq i}C^\cdot) \to \mathbf{H}^m(C^\cdot)), \qquad i \in \mathbf{Z}.$$

This gives in particular the *conjugate filtrations* on $\mathbf{H}^m(X, \underline{\mathcal{W}\Omega}^\cdot_X \bmod N)$, $\mathbf{H}^m(X, \underline{\mathcal{W}\Omega}^\cdot_{X/S} \bmod N)$, $\mathbf{H}^m(X_s, \mathcal{W}\Omega^\cdot_{X_s} \bmod p^n)$, $\mathbf{H}^m(X, \Omega^\cdot_{X/S} \bmod N)$ (the terminology conjugate filtration is adopted from [11,10]). The homomorphisms

$$\begin{array}{lcl} \mathbf{H}^m(X, \underline{\mathcal{W}\Omega}^\cdot_X \bmod N) & \to & \mathbf{H}^m(X, \underline{\mathcal{W}\Omega}^\cdot_{X/S} \bmod N) \\ \mathbf{H}^m(X, \underline{\mathcal{W}\Omega}^\cdot_{X/S} \bmod N) & \to & \mathbf{H}^m(X, \Omega^\cdot_{X/S} \bmod N) \\ \mathbf{H}^m(X, \underline{\mathcal{W}\Omega}^\cdot_{X/S} \bmod N) & \to & \mathbf{H}^m(X_s, \mathcal{W}\Omega^\cdot_{X_s} \bmod N) \end{array}$$

are compatible with the conjugate filtrations on their source and target.

5.9 Let s be a closed point of S with perfect residue field $k(s)$ of characteristic p. By ([9] p.577 (3.17.3)) the canonical homomorphism

$$\mathcal{W}\Omega^\cdot_{X_s} \bmod p^n \to \mathcal{W}_n\Omega^\cdot_{X_s}$$

onto the De Rham-Witt complex of level n is a quasi-isomorphism. Thus we get an isomorphism

$$\mathbf{H}^m(X_s, \mathcal{W}\Omega^\cdot_{X_s} \bmod p^n) \simeq \mathbf{H}^m(X_s, \mathcal{W}_n\Omega^\cdot_{X_s})$$

and a specialization homomorphism

$$\mathbf{H}^m(X_n, \underline{\mathcal{W}\Omega}^\cdot_{X_n/S_n} \bmod p^n) \to \mathbf{H}^m(X_s, \mathcal{W}_n\Omega^\cdot_{X_s}). \tag{4}$$

Moreover from ([9] II(1.4), (2.8)) one knows

$$\mathrm{H}^m_{\mathrm{crys}}(X_s) \simeq \mathbf{H}^m(X_s, \mathcal{W}\Omega^\cdot_{X_s}) \simeq \varprojlim_n \mathbf{H}^m(X_s, \mathcal{W}_n\Omega^\cdot_{X_s}).$$

Thus the homomorphisms in (4) give in the limit

$$\mathbf{H}^m(X, \underline{\mathcal{W}\Omega}^\cdot_{X/S})_\infty \to \mathrm{H}^m_{\mathrm{crys}}(X_s). \tag{5}$$

The homomorphism $A \to k(s)$ corresponding with the point $s \in S$ and the homomorphism $\lambda : A_\infty \to W(A_\infty)$ lead to the composite map $A \to A_\infty \to W(A_\infty) \to W(k(s))$. From the basic comparison theorem of crystalline and De Rham cohomology ([1] (7.26.3)) and from the hypotheses in (4.1) one gets isomorphisms

$$\begin{aligned} \mathrm{H}^m_{\mathrm{crys}}(X_s) &\simeq \mathbf{H}^m(X \otimes W(k(s)), \Omega^{\cdot}_{X\otimes W(k(s))/W(k(s))}) \\ &\simeq \mathbf{H}^m(X, \Omega^{\cdot}_{X/S}) \otimes_A W(k(s)) \end{aligned}$$

So $\mathrm{H}^m_{\mathrm{crys}}(X_s)$ is a free $W(k(s))$-module.

The conjugate filtration on the finite levels $\mathbf{H}^m(X_s, \mathcal{W}_n\Omega^{\cdot}_{X_s})$ induces on the limit $\mathrm{H}^m_{\mathrm{crys}}(X_s)$ the conjugate filtration $\{\mathrm{Fil}^i_{\mathrm{con}}\mathrm{H}^m_{\mathrm{crys}}(X_s)\}_{i\geq 0}$ (see [10]). One of the main results in [10] describes $\mathrm{Fil}^i_{\mathrm{con}}\mathrm{H}^m_{\mathrm{crys}}(X_s) \otimes \mathbf{Q}$ as precisely that part of $\mathrm{H}^m_{\mathrm{crys}}(X_s) \otimes \mathbf{Q}$ where Frobenius $\underline{F}_p$ acts with slopes $\leq i$. In particular $\mathrm{Fil}^0_{\mathrm{con}}\mathrm{H}^m_{\mathrm{crys}}(X_s)$ is the *unit root part* of $\mathrm{H}^m_{\mathrm{crys}}(X_s)$ ([10] III(6.8)). Since $\mathrm{H}^m_{\mathrm{crys}}(X_s)$ is a free $W(k(s))$-module, $\mathrm{Fil}^0_{\mathrm{con}}\mathrm{H}^m_{\mathrm{crys}}(X_s)$ is also a free $W(k(s))$-module. By the theory of the *conjugate spectral sequence* [10] its rank is at most $h = \dim_{k(s)} \mathrm{H}^m(X_s, \mathcal{O}_{X_s})$.

The conjugate filtration on the finite levels induces the conjugate filtration on the limit $\mathbf{H}^m(X, \underline{\mathcal{W}\Omega}^{\cdot}_{X/S})_\infty$. The specialization homomorphism (5) is compatible with the conjugate filtrations. It is clear from the construction in (5.4) and (5.5) that $\lambda(\mathcal{U})$ is contained in $\mathrm{Fil}^0_{\mathrm{con}}\mathbf{H}^m(X, \underline{\mathcal{W}\Omega}^{\cdot}_{X/S})_\infty$. So by (5) it is mapped into $\mathrm{Fil}^0_{\mathrm{con}}\mathrm{H}^m_{\mathrm{crys}}(X_s)$.

We compose (5) with the projection $\mathrm{H}^m_{\mathrm{crys}}(X_s) \to \mathrm{H}^m(X_s, \mathcal{O}_{X_s})$. One easily checks that the composite map $\lambda(\mathcal{U}) \to \mathrm{H}^m(X_s, \mathcal{O}_{X_s})$ sends $\lambda(\varpi_i)$ to $\omega_i(s)$,= the image of ω_i under the map $\mathrm{H}^m(X, \mathcal{O}_X) \to \mathrm{H}^m(X_s, \mathcal{O}_{X_s})$. Because the $W(k(s))$-rank of $\mathrm{Fil}^0_{\mathrm{con}}\mathrm{H}^m_{\mathrm{crys}}(X_s)$ is at most h and because $\{\omega_1(s), \ldots, \omega_h(s)\}$ is a $k(s)$-basis of $\mathrm{H}^m(X_s, \mathcal{O}_{X_s})$ we conclude:

5.10 Theorem *Let s be a closed point of S with perfect residue field of characteristic p. Then the specialization homomorphism* (5) *restricts to a surjection*

$$\lambda(\mathcal{U}) \to \mathrm{Fil}^0_{\mathrm{con}}\mathrm{H}^m_{\mathrm{crys}}(X_s). \qquad \square$$

5.11 *Remark* The conjugate filtration on $\mathbf{H}^m(X, \Omega^{\cdot}_{X/S}) \otimes_A A_n$ for $n \geq 0$ induces the conjugate filtration on $\mathbf{H}^m(X, \Omega^{\cdot}_{X/S}) \otimes_A A_\infty$. Clearly $\mathcal{U}$ is contained in $\mathrm{Fil}^0_{\mathrm{con}}\mathbf{H}^m(X, \Omega^{\cdot}_{X/S}) \otimes_A A_\infty$. One may hope that this inclusion is in fact an equality (cf. [5] p.97).

References

[1] Berthelot, P., Ogus, A. *Notes on crystalline cohomology.* Princeton: Princeton University Press 1978

[2] Bloch, S. *Algebraic K-theory and crystalline cohomology.* Publ. Math. IHES 47 (1977) 187-268

[3] Cartier, P. *Groupes formels associés aux anneaux de Witt généralisés.* resp. *Modules associés à un groupe formel commutatif. Courbes typiques.* C.R. Acad. Sc. Paris 265 (1967) 49-52 resp. 129-132

[4] Deligne, P. *Théorème de Lefschetz et critères de dégénérescence de suites spectrales.* Publ. Math. IHES 35 (1968) 259-278

[5] Deligne, P. *Cristaux ordinaires et coordonnées canoniques.* in: Surfaces algébriques. Lectures Notes in Math 868, Springer Verlag 1981

[6] Dwork, B. *p-Adic cycles.* Publ. Math. IHES 37 (1969) 27-116

[7] Grothendieck, A., Dieudonné, J. *Éléments de géométrie algébrique IV, quatrième partie.* Publ. Math. IHES 32 (1967)

[8] Hazewinkel, M. *Formal groups and applications.* New York: Academic Press 1978

[9] Illusie, L. *Complexe de De Rham-Witt et cohomologie cristalline.* Ann. scient. Éc. Norm. Sup. 12 (1979) 501-661

[10] Illusie, L., Raynaud, M. *Les suites spectrales associées au complexe de De Rham-Witt* Publ. Math. IHES 57 (1983) 73-212

[11] Katz, N. *Algebraic solutions of differential equations (p-curvature and the Hodge filtration).* Inv. Math. 18 (1972) 1-118

[12] Katz, N. *Expansion-coefficients as approximate solution of differential equations.* Astérisque 119-120 (1984) 183-189

[13] Katz, N. *Internal reconstruction of unit-root F-crystals via expansion-coefficients.* Ann. scient. Éc. Norm. Sup. 18 (1985) 245-268

[14] Katz, N., Oda,T. *On the differentiation of De Rham cohomology classes with respect to parameters.* J. Math. Kyoto Univ. 8 (1968) 199-213

[15] Manin, Ju. *The Hasse-Witt matrix of an algebraic curve.* AMS Translations (2) 45 (1965) 245-264

[16] Nishiyama, S. *Appell's hypergeometric function F_2 and periods of certain elliptic K3 surfaces.* Tokyo J. Math. 10 (1987) 33-67

[17] Pham, F. *Singularités des systèmes différentiels de Gauss-Manin.* Progress in Math. 2, Boston: Birkhauser 1979

[18] Stienstra, J., van der Put, M., van der Marel, L. *On the variation of p-adic periods.* these proceedings

[19] Stienstra, J. *Cartier-Dieudonné theory for Chow groups.* J. f. d. reine u. angew. Math. 355 (1985) 1-66

[20] correction to [19] same journal vol. 362 (1985) 218-220

Arithmetic discriminants and quadratic points on curves

PAUL VOJTA*

Introduction

Let C be a curve defined over a number field k. Let $h(P)$ denote the height of an algebraic point $P \in C(\bar{k})$ relative to some fixed divisor of degree 1. For a number field F, let

$$d(F) = \frac{\log |D_{F/\mathbf{Q}}|}{[F : \mathbf{Q}]},$$

and for an algebraic point P on C, let

$$d(P) = d(k(P)).$$

Then we have

Conjecture ([**V 1**], 5.5.0.1). *Fix $\epsilon > 0$. Then for all points $P \in C(\bar{k})$ with $[k(P) : k]$ bounded by some fixed constant, we have*

$$(0.1) \qquad (2g-2)h(P) \le (1+\epsilon)d(P) + O(1).$$

Recently, the following partial result has been obtained.

Theorem ([**V 3**], 7.1). *Fix $s \in \mathbf{N}$ and $\epsilon > 0$. Then for all points $P \in C(k)$ with $[k(P) : k] \le s$,*

$$(0.2) \qquad (2g-2+2s-2s\sqrt{g})h(P) \le (1+\epsilon)d_a(P) + O(1).$$

---

*Supported by the Miller Institute for Basic Research in Science

Further refinements in the method are expected to replace (0.2) by

$$(2g-2)h(P) \leq (1+\epsilon)d_{\mathrm{a}}(P) + O(1). \tag{0.3}$$

The quantity $d_{\mathrm{a}}(P)$ is called the **arithmetic discriminant** of P; it will be defined in Section 3, below. One property is that $d_{\mathrm{a}}(P) \geq d(P)$, so (0.3) is weaker than (0.1); in particular it does not imply any form of the *abc* conjecture.

The purpose of this paper is twofold. First, we want to determine the properties of $d_{\mathrm{a}}(P)$. The second purpose is to prove the following corollary of (0.2).

Theorem A. *Let $f: C \to \mathbf{P}^1$ be a dominant morphism, let $s \in \mathbf{N}$, and let g be the genus of C. Assume also that*

$$g - 1 + s - s\sqrt{g} > (\deg f)(s-1). \tag{0.4}$$

Then the set

$$\{P \in C(\bar{k}) \mid [k(P):k] \leq s \text{ and } k(f(P)) = k(P)\}$$

is finite.

Corollary. *If $g \geq 18$ then there exists a dominant morphism $f: C \to \mathbf{P}^1$ such that the set*

$$\{P \in C(\bar{k}) \mid [k(P):k] \leq 2 \text{ and } k(f(P)) = k(P)\}$$

is finite.

Indeed, by ([**H**], 5.5.1), there exists a dominant morphism $f: C \to \mathbf{P}^1$ of degree $\leq [(g+3)/2]$.

Corollary. *If C is hyperelliptic and $g \geq 6$, then all but finitely many points on C defined over a quadratic extension of k are those obtained from k-rational points on $\mathbf{P}^1$ by pulling back via the canonical morphism $C \to \mathbf{P}^1$ of degree two.*

If (0.2) is replaced by (0.3), then (0.4) can be sharpened to

$$g - 1 > (\deg f)(s-1).$$

The corollaries would then hold for $g \geq 6$ and $g \geq 4$, respectively.

A weaker form of the second corollary (giving a growth result for the heights of quadratic points instead of a finiteness result, but with $g \geq 5$) has already been proved by Silverman [**Sil**]. For related work on quadratic points, see Hindry [**Hi**]. We also note that Faltings has recently proved the second corollary with $g \geq 4$ [**F**].

The outline of this paper is as follows. The first section gives notation which is used throughout the rest of the paper. Section 2 describes some properties of arithmetic torsion sheaves, for which I do not know a reference. Section 3 gives a (hopefully) complete discussion of the elementary properties of $d_{\mathbf{a}}$. After these two sections, the proof of Theorem A in Section 4 is very brief.

I would like to thank Noam Elkies for suggesting the ideas leading to Theorem A.

1. Notation

The notation of this paper is close to, but not identical to, the notation of ([**V 3**], 0.5). In particular, we will be dealing with arithmetical objects; *i. e.*, objects defined in the context of algebraic geometry over $\operatorname{Spec}\mathbf{Z}$ (called the finite part), together with certain analytic data on the generic fibre (called the archimedean part). Each object will be denoted with a simple symbol, and the corresponding finite part will be written with a subscript "fin." The notation for the archimedean part will vary. Note that an object and its finite part will be regarded as distinct objects.

For the purposes of this paper, an **arithmetic variety** $\mathcal{X}$ consists of the following.

(a). The finite part is a reduced scheme $\mathcal{X}_{\text{fin}}$ which is projective and flat over $\operatorname{Spec}\mathbf{Z}$ and whose generic fibre is smooth. Write $\mathcal{X}_\infty = \mathcal{X}_{\text{fin}} \times_{\operatorname{Spec}\mathbf{Z}} \operatorname{Spec}\mathbf{C}$.

(b). The arithmetic part of $\mathcal{X}$ consists of a smooth function

$$\Lambda_{\mathcal{X}} = \Lambda\colon \mathcal{X}_\infty \times \mathcal{X}_\infty \to \mathbf{R}_{\geq 0},$$

such that $\Lambda(P_1, P_2) = 0$ if and only if $P_1 = P_2$,

$$\Lambda(P_1, P_2) = \Lambda(P_2, P_1),$$

and

$$\Lambda(P_1, P_2) \gg\ll |z_1(P_1) - z_1(P_2)|^2 + \cdots + |z_n(P_1) - z_n(P_2)|^2$$

in a neighborhood of the diagonal, where $z_1, \ldots, z_n$ are local coordinates on some open subset of $\mathcal{X}_\infty$. Thus Λ will be called a **distance function** (although it actually measures the square of the distance). We also require that the form

$$\omega := d_1 d_1^c \Lambda(P, P)$$

be a Kähler form on $\mathcal{X}_\infty$ (here d_1 and d_1^c apply only to the first coordinate). Moreover, let

$$\lambda(P,Q) = -\log \Lambda(P,Q), \qquad P \neq Q.$$

In general, at the archimedean places, it is more convenient to refer to squares of things (and their logarithms).

A morphism f of arithmetic varieties is a morphism f_{fin} of their finite parts. Arithmetic curves and arithmetic surfaces are arithmetic varieties of relative dimension zero and one, respectively.

This differs from the definition in [**V 3**] because we provide the distance function on $\mathcal{X}_\infty$, which then provides the Kähler structure, instead of giving the Kähler structure directly. This is useful for applying the adjunction formula to a subscheme of an arithmetic surface.

For a number field k, let R be its ring of integers, and let B be the arithmetic scheme with $B_{\text{fin}} = \operatorname{Spec} R$ and $\Lambda_B(\sigma,\tau) = 0$ if $\sigma = \tau$ and 1 otherwise, $\sigma, \tau \in B_\infty$. Note that $B_\infty = \{\sigma : k \hookrightarrow \mathbf{C}\}$. We define the **arithmetic curve corresponding to** R as the arithmetic curve obtained by using this choice of Λ.

An arithmetic variety $\mathcal{X}$ over B is an arithmetic variety $\mathcal{X}$, together with a morphism $\mathcal{X} \to B$. For $\sigma : k \hookrightarrow \mathbf{C}$, let $\mathcal{X}_\sigma = \mathcal{X}_{\text{fin}} \times_\sigma \mathbf{C}$, so that $\mathcal{X}_\infty = \coprod_{\sigma \in B_\infty} \mathcal{X}_\sigma$. Also, we may refer to fibres of π_{fin} as (non-archimedean) fibres of π. More generally, one should regard an arithmetic scheme as a scheme with an additional fibre over the archimedean absolute value of $\mathbf{Q}$. Therefore we inherit the notions of local rings and (non-archimedean and generic) fibres from $\mathcal{X}_{\text{fin}}$.

One might think of $P \in \mathcal{X}_\infty$ as a closed point on an archimedean fibre of $\mathcal{X}$. In particular, when writing $S \subseteq B$, we will mean a subset of $B_{\text{fin}} \cup B_\infty$ (but always excluding the generic point). Nothing here requires S to be finite, but in applications it will be.

For example, arithmetic surfaces can be obtained via Arakelov theory, as follows. Given a smooth connected projective curve C/k, let X_{fin} be a regular model for C over $\operatorname{Spec} R$, and for each $\sigma \in B_\infty$ let $\Lambda_{X,\sigma}$ be an Arakelov Green's function G_σ associated to a measure μ on X_σ. We let $\Lambda = 1$ on $X_\sigma \times X_\tau$ for $\sigma \neq \tau$. Then the Kähler structure on X_∞ will be obtained from the μ-admissible metric on the tangent bundle, rather than from μ itself. (Note: this is different from [**G-S**]. The reason for this change is that it makes it easy to take the product of two arithmetic schemes.)

2. Arithmetic torsion sheaves

Basically, an arithmetic torsion sheaf is a torsion sheaf, together with certain analytic data at the archimedean places.

Definition 2.1. Let $\mathcal{X}$ be an arithmetic variety of dimension ≤ 2. An **arithmetic torsion sheaf** $\mathcal{L}$ on $\mathcal{X}$ consists of:

(a). a coherent torsion sheaf $\mathcal{L}_{\text{fin}}$ on $\mathcal{X}_{\text{fin}}$; and

(b). a smooth function $\tau_{\mathcal{L}}: (\mathcal{X}_\infty \setminus |\mathcal{L}_\infty|) \to \mathbf{R}$ (where $|\mathcal{L}_\infty|$ denotes the support of $\mathcal{L}_\infty := \mathcal{L}_{\text{fin}}|\mathcal{X}_\infty$), satisfying the growth condition that, for all fixed $P \in \mathcal{X}_\infty$, the function

$$\tau_{\mathcal{L}}(Q) - \text{length}(\mathcal{L}_P)\lambda(Q,P), \qquad Q \neq P \tag{2.1.1}$$

extends to a continuous function of Q in some neighborhood of P.

The growth condition (b) is vacuous if $\dim \mathcal{X} = 1$; if $\dim \mathcal{X} = 2$, it is equivalent to saying that $\tau_{\mathcal{L}}$ is a **Weil function** for the divisor

$$\sum_{P \in \mathcal{X}_\infty} \text{length}(\mathcal{L}_P) \cdot [P]$$

on $\mathcal{X}_\infty$ (up to changes in normalization conventions). See ([**L 2**], Chapter 10, Section 2) or ([**V 1**], Section 1.3). Also, if $\mathcal{X}$ is an arithmetic scheme over B and $\sigma \in B_\infty$, let $\mathcal{L}_\sigma = \mathcal{L}_\infty|\mathcal{X}_\sigma$ and $\tau_{\mathcal{L},\sigma} = \tau_{\mathcal{L}}|\mathcal{X}_\sigma$. Moreover, if $P \in \mathcal{X}_\infty$, then let $\mathcal{L}_P$ denote the stalk of $\mathcal{L}_\infty$ at P.

Definition 2.2. Let $\mathcal{L}$ be an arithmetic torsion sheaf on an arithmetic curve B. Then its **degree** is the real number

$$\deg \mathcal{L} := \log \#M + \frac{1}{2} \sum_{\sigma \in B_\infty} \tau_{\mathcal{L}}(\sigma),$$

where M is the module over R for which $\mathcal{L}_{\text{fin}} = \widetilde{M}$. If $S \subseteq B$ then we write

$$\deg_S \mathcal{L} := \log \#M_S + \frac{1}{2} \sum_{\sigma \notin S} \tau_{\mathcal{L}}(\sigma)$$

where M_S denotes the localization of M away from non-archimedean primes in S. Note that $\deg_S \mathcal{L} = \deg \mathcal{L}$ if $S = \emptyset$. Also, if B is a finite cover of B_0 and $S_0 \subseteq B_0$, then let

$$\deg_{S_0} \mathcal{L} = \deg_S \mathcal{L},$$

where S is the subset of B lying over S_0.

Definition 2.3. An **arithmetic divisor** D on $\mathcal{X}$ is a divisor D_{fin} on $\mathcal{X}_{\text{fin}}$, together with a smooth function $g_D: \mathcal{X}_\infty \setminus |D_\infty| \to \mathbf{R}$, such that for all open sets $U \subseteq \mathcal{X}_\infty$ on which D_∞ is locally represented by a function f, the function

$$g_D(P) + \log |f(P)|^2, \qquad P \notin |D_\infty|$$

extends to a continuous function of P on all of U. Again, we write D_σ for $D_\infty|\mathcal{X}_\sigma$ and $g_{D,\sigma} = g_D|\mathcal{X}_\sigma$. If $\mathcal{X}_{\text{fin}}$ is not regular in codimension one, then we write **arithmetic Cartier divisor** or **arithmetic Weil divisor** to indicate whether D_{fin} is a Cartier or Weil divisor.

There exists an intersection theory of arithmetic divisors, as well as a correspondence between arithmetic divisors modulo linear equivalence and metrized invertible sheaves. For details, see ([**V 3**], 0.5).

Also, if E is an arithmetic curve lying on an arithmetic surface X, then we can let E be an arithmetic divisor by setting

$$g_E(P) = \sum_i n_i \lambda_X(P, Q_i)$$

where $E_\infty = \sum_i n_i Q_i$.

Definition 2.4. Let $\mathcal{L}$ be an arithmetic torsion sheaf on $\mathcal{X}$. Then the **arithmetic divisor** D **corresponding to** $\mathcal{L}$ consists of

(a). the divisor

$$D_{\text{fin}} := \sum \text{length}(\mathcal{L}_P) \cdot P,$$

where the sum is over all points in $\mathcal{X}_{\text{fin}}$ of codimension 1; and

(b). the function

$$g_D = \tau_{\mathcal{L}}.$$

One can similarly define $\deg D$ and $\deg_S D$ for an arithmetic divisor D on an arithmetic curve B, *e. g.*,

$$\deg D = \sum_{v \in B_{\text{fin}}} \text{ord}_v D \cdot \log \#k(v) + \frac{1}{2} \sum_\sigma g_D(\sigma).$$

For D corresponding to $\mathcal{L}$, we also have

$$\deg_S D = \deg_S \mathcal{L}$$

for all $S \subseteq B$.

Definition 2.5. Let $f\colon X \to Y$ be a generically finite morphism of arithmetic varieties of dimension ≤ 2. Then let the sheaf $\Omega_{X/Y}$ be the arithmetic torsion sheaf for which

$$(\Omega_{X/Y})_{\mathrm{fin}} = \Omega_{X_{\mathrm{fin}}/Y_{\mathrm{fin}}}$$

and $\tau_{\Omega_{X/Y}} = \tau_{X/Y}$, where

$$\tau_{X/Y}(P) := \sum_{\substack{Q \in f^{-1}(f(P)) \\ Q \ne P}} \lambda_X(P,Q)$$

for points $P \in X_\infty$ not lying in the support of the ramification divisor of the morphism $f_\infty\colon X_\infty \to Y_\infty$ obtained from f_{fin} via base change.

Proposition 2.6. *If $f\colon X \to Y$ is as above, then $\Omega_{X/Y}$ is an arithmetic torsion sheaf.*

Proof. Almost by definition, we know that $(\Omega_{X/Y})_{\mathrm{fin}}$ is a torsion sheaf. What we need to check is that $\tau_{X/Y}$ is a Weil function for the corresponding ramification divisor. The case of arithmetic curves is trivial, so we need only worry about arithmetic surfaces. Clearly $\tau_{X/Y}$ is continuous outside of the ramification divisor; it remains only to check the growth near points of ramification. At such points, we can choose local coordinates on X_∞ and Y_∞ such that f_∞ is of the form

$$f_\infty(z) = z^e. \tag{2.6.1}$$

Then $f^{-1}(f(z))$ consists of $\{\zeta_e^i z, 0 \le i < e\}$ (where ζ_e is a primitive e^{th} root of unity), together with points outside the neighborhood in which (2.6.1) is valid. Thus, for $z \ne 0$,

$$\tau_{X/Y}(z) + (e-1)\log|z|^2$$

can be extended to a continuous function in a neighborhood of $z = 0$. Since the ramification divisor is locally $(e-1)\cdot[0]$, $\tau_{X/Y}$ has the desired growth. Q.E.D.

Definition 2.7. For $f\colon X \to Y$ as above, the **arithmetic ramification divisor** $R_{X/Y}$ of X over Y is the arithmetic divisor associated to the arithmetic torsion sheaf $\Omega_{X/Y}$.

It is not at all clear how one might formulate a definition of $\Omega_{X/Y}$ in higher dimensions. For example, for the function $\tau_{X/Y}$ one might take the function

$$P \mapsto \sum_{\substack{Q \in f^{-1}(f(P)) \\ Q \ne P}} \Lambda(P,Q)$$

But this does not have the desired growth: consider, for example, a function given locally by

$$(z_1, z_2) \mapsto (z_1^2, z_2^2).$$

Proposition 2.8. *Let $\mathcal{L}$ be an arithmetic torsion sheaf on a regular arithmetic surface $\pi: X \to B$, and let D be the corresponding arithmetic divisor. Then for all algebraic points $i: B' \to X$ not contained in the support of $\mathcal{L}$ and all $S \subseteq B$,*

$$|\deg_S i^*\mathcal{L} - deg_S i^* D| \leq O(\deg B'),$$

where the implicit constant depends only on $\mathcal{L}$.

Proof. Note first of all that the archimedean components of the two degrees coincide exactly. Therefore, we are reduced to the corresponding question for the corresponding finite objects $\mathcal{L}_{\text{fin}}$ and D_{fin}. Therefore, for the remainder of this proof, we will work only with finite objects, but will omit the subscript "fin." Thus, it suffices to prove that for all closed points $b \in B$ there exists a constant $c(\mathcal{L}, b)$ with the following two properties.

(a). For any $i: B' \to X$ as above and any closed point $b' \in B'$, let $e(b')$ denote the ramification index of b' over $\pi(i(b'))$ and let $m_{b'}(D)$ denote the multiplicity of b' in i^*D. Then

$$|\text{length}(i^*\mathcal{L})_{b'} - m_{b'}(D)| \leq e(b')c(\mathcal{L}, i(b')).$$

(Note that for all $b \in B$, $\sum e(b')[K(b') : K(b)] = \deg(\pi \circ i)$, where the sum is over $b' \in (\pi \circ i)^{-1}(b)$.)

(b). For fixed $\mathcal{L}$, $c(\mathcal{L}, b) = 0$ for all but finitely many $b \in B$.

This fact, in turn, will be proved in two cases, as follows.

Case 1. $\mathcal{L}$ is supported only on closed points of X. In that case $\mathcal{L}$ has finite length, and we can prove the result by induction on the length. If the length is zero, then $\mathcal{L} = 0$, and the result is trivial. Otherwise, we have an exact sequence

$$0 \to \mathcal{L}' \to \mathcal{L} \to \mathcal{O}_X/P \to 0,$$

where P is a prime ideal in $\mathcal{O}_X$ of height 2, and the length of $\mathcal{L}'$ is one less than the length of $\mathcal{L}$. Since P is a closed point on X, it has a regular sequence (h_1, h_2), with corresponding Koszul complex

$$0 \to \mathcal{O}_X \xrightarrow{h_1 \oplus h_2} \mathcal{O}_X \oplus \mathcal{O}_X \xrightarrow{h_2 \,\text{pr}_1 - h_1 \,\text{pr}_2} \mathcal{O}_X \to \mathcal{O}_X/P \to 0.$$

For any $b' \in B'$, let v denote the valuation on $\mathcal{O}_{B',b'}$. Then from the above Koszul complex, it follows that

$$\begin{aligned} \operatorname{length} \operatorname{Tor}_1^{\mathcal{O}_X}(\mathcal{O}_X/P, \mathcal{O}_{B'})_{b'} &= \operatorname{length}(\mathcal{O}_X/P \otimes_{\mathcal{O}_X} \mathcal{O}_{B'})_{b'} \\ &= \min(v(h_1), v(h_2)). \end{aligned}$$

Thus

$$\begin{aligned} |\operatorname{length}(i^*\mathcal{L})_{b'} - \operatorname{length}(i^*\mathcal{L}')_{b'}| &\leq \min(v(h_1), v(h_2)) \\ &\leq e(b'). \end{aligned}$$

It then follows that we can take $c(\mathcal{L}, b) = c(\mathcal{L}', b)$ for all $b \neq \pi(P)$, and

$$c(\mathcal{L}, \pi(P)) = c(\mathcal{L}', \pi(P)) + 1.$$

Case 2. General case. This case will be proved by induction on the effective divisor D. If $D = 0$, then the result follows by Case 1. Otherwise, let P be a prime of $\mathcal{O}_X$ corresponding to some component occurring in D. We then have a short exact sequence

$$0 \to \mathcal{L}' \to \mathcal{L} \to \mathcal{M} \to 0,$$

where the divisor D' associated to $\mathcal{L}'$ is strictly less than D, and where $\mathcal{M}$ is an invertible sheaf supported on the divisor corresponding to the prime P. Locally, identifying $\mathcal{M}$ with $\mathcal{O}_X/P$, we have the free resolution

$$0 \to \mathcal{O}_X \xrightarrow{g} \mathcal{O}_X \to \mathcal{M} \to 0$$

of $\mathcal{M}$ (where g locally defines the divisor P). From this it is possible to show that

$$\operatorname{Tor}_1^{\mathcal{O}_X}(\mathcal{M}, \mathcal{O}_{B'}) = 0$$

(where $\mathcal{O}_{B'}$ is regarded as an $\mathcal{O}_X$-module via i). We then have the exact sequence

$$0 \to \mathcal{L}' \otimes_{\mathcal{O}_X} \mathcal{O}_{B'} \to \mathcal{L} \otimes_{\mathcal{O}_X} \mathcal{O}_{B'} \to \mathcal{M} \otimes_{\mathcal{O}_X} \mathcal{O}_{B'} \to 0,$$

and consequently for any $b' \in B'$,

$$\begin{aligned} \operatorname{length}(i^*\mathcal{L})_{b'} &= \operatorname{length}(i^*\mathcal{L}')_{b'} + \operatorname{length}(i^*\mathcal{M})_{b'} \\ &= \operatorname{length}(i^*\mathcal{L}')_{b'} + m_{b'}(P). \end{aligned}$$

Thus we can let $c(\mathcal{L}, b) = c(\mathcal{L}', b)$ for all $b \in B$. Q.E.D.

3. Arithmetic discriminants

In this section we define arithmetic discriminants and determine some of their properties.

Throughout this section, let $P \in X(\bar{k})$ be an algebraic point on a regular arithmetic surface X; let $F = k(P)$. Let B' be the arithmetic curve corresponding to $\operatorname{Spec} \mathcal{O}_F$, *and let $i\colon B' \to E_P$ be the prime horizontal divisor corresponding to P.*

Definition 3.1. The **arithmetic discriminant** $d_{\mathrm{a}}(P)$ **of** P **on** X is defined by the formula

$$d_{\mathrm{a}}(P) = \frac{\deg i^* \Omega_{E_P/B}}{[F : \mathbf{Q}]}.$$

Likewise, for $S \subseteq B'$ or $S \subseteq B$, let

$$d_{\mathrm{a},S}(P) = \frac{\deg_S i^* \Omega_{E_P/B}}{[F : \mathbf{Q}]}.$$

Note that this definition depends not only on the field of definition of P, but also on its embedding in X (more precisely, its image E_P in X, together with certain data at the archimedean places).

The following two propositions give alternate characterizations of d_{a}.

Proposition 3.2. *Let $S = \emptyset$ and let $K_{X/B}$ be a divisor corresponding to $\omega_{X/B}$. Then we have*

$$d_{\mathrm{a}}(P) = \frac{(E_P \,.\, E_P + K_{X/B})}{[F : \mathbf{Q}]}. \tag{3.2.1}$$

Proof. For convenience write $E = E_P$. We have $E_{\mathrm{fin}} = \operatorname{Spec} A$ for some ring A of algebraic integers, finite over $\mathbf{Z}$. For an element a of the quotient field F of A, we have the trace $\operatorname{Tr}_{F/k}(a) \in k$; let

$$W_{A/R} = \{b \in F \mid \operatorname{Tr}_{F/k}(bA) \subseteq \mathcal{O}_k\}.$$

This is a fractional ideal of A containing A; then let

$$\mathcal{D}_{A/R,\mathrm{D}} = W_{A/R}^{-1} \subseteq A.$$

This is the **Dedekind different.** On the other hand, embed X_{fin} into a smooth scheme Y over $\operatorname{Spec} R$ (*e. g.* $\mathbf{P}^m_{\operatorname{Spec} R}$ for some $m \in \mathbf{Z}$, since $X_{\mathrm{fin}}/\operatorname{Spec} R$ is projective). Since X_{fin} and Y are regular, X_{fin} is a locally complete intersection in Y, and therefore so is E_{fin}. Thus, letting $\mathcal{I}$ denote the ideal of E_{fin} in Y, we have the following short exact sequence of sheaves on E_{fin}:

$$(3.2.2) \qquad 0 \to \mathcal{I}/\mathcal{I}^2 \to \Omega_{Y/\operatorname{Spec} R} \mid E_{\text{fin}} \to \Omega_{A/R} \to 0.$$

This sequence is exact on the left because both terms are locally free of the same rank. The determinant of the injection defines an ideal $\mathcal{D}_{A/R,\mathrm{K}}$ of A, called the **Kähler different.** Let $i\colon E' \to E$ denote the normalization of E, and let $E' = \operatorname{Spec} A'$. Then, for all closed points $x \in E'$,

$$\begin{aligned} \#(\Omega_{E/B})_x &= (A'_x : (i^*\mathcal{D}_{A/R,\mathrm{K}})_x) \\ &= (A'_x : (i^*\mathcal{D}_{A/R,\mathrm{D}})_x). \end{aligned}$$

Indeed, the first identity follows immediately from (3.2.2); the second, from ([**K**], 10.17). But, by the adjunction formula ([**L 3**], Chapter IV, 5.3), we have

$$\begin{aligned} (E \,.\, E + K_{X/B}) &= \log(A : \mathcal{D}_{A/R,\mathrm{D}}) + \frac{1}{2} \sum_{\sigma} \sum_{\substack{P,Q \in E_\sigma \\ P \neq Q}} \lambda_\sigma(P,Q) \\ &= [F : \mathbb{Q}] d_{\mathrm{a}}(P). \end{aligned}$$

Q.E.D.

The arithmetic discriminant was initially defined in [**V 2**] or [**V 3**] by means of (3.2.1); this was changed to the current definition in order to more easily define $d_{\mathrm{a},S}$. From (3.2.1) it is clear that in the function field case, the analogue of d_{a} is a function of the arithmetic genus $p_{\mathrm{a}}(E_P)$:

$$d_{\mathrm{a}}(P) = \frac{2p_{\mathrm{a}}(E_P) - 2}{[k(P) : k]} - (2g(B) - 2).$$

Compare this with the discriminant term defined in ([**V 1**], 5.1.2):

$$\begin{aligned} d(P) &= \frac{\log |D_{k(P)/\mathbb{Q}}|}{[k(P) : \mathbb{Q}]} \\ &= \frac{\deg \Omega_{B'/B}}{[k(P) : \mathbb{Q}]} \end{aligned}$$

(recall that B' is the normalization of E_P). In the function field case one has

$$d(P) = \frac{2g(B') - 2}{[k(P) : k]} - (2g(B) - 2).$$

Thus, in the function field case, the difference between $d_{\mathrm{a}}(P)$ and $d(P)$ is related to the difference between the arithmetic and geometric genera of a curve. The word "arithmetic" in the term "arithmetic discriminant" refers to this fact.

As in the function field case, the difference between $d_{\mathrm{a}}(P)$ and $d(P)$ can be written as a sum of local terms related to the singularities of E_P. This also holds in the number field case, but first we need a definition: for $S \subseteq B$, let

$$d_S(P) = \frac{\deg_S \Omega_{B'/B}}{[F : \mathbb{Q}]}.$$

Proposition 3.3. *Let* $S \supseteq B_\infty$. *Then*

$$d_{\mathrm{a},S}(P) = d_S(P) + \frac{2}{[F : \mathbb{Q}]} \sum_{\substack{x \in E_P \\ \pi(x) \notin S}} \delta_x \tag{3.3.1}$$

where

$$\delta_x = \log \#\Big(\Big(\bigoplus_{y \in i^{-1}(x)} \mathcal{O}_{B',y}\Big) \Big/ \mathcal{O}_{E,x}\Big).$$

Remark. In the function field case, this fact is due originally to Rosenlicht—see ([**S**], Chapter IV) for references. The proof used a Riemann-Roch type of argument. The following argument is more direct, but has its roots in the relative Riemann-Roch theorem for X/B.

Since the Riemann-Roch theorem holds for projective locally complete intersection morphisms, it is likely that Proposition 3.3 holds for any locally complete intersection E (not necessarily contained in an arithmetic surface).

Proof. Let E'_{fin} be the normalization of E_{fin} at places outside of S, and an isomorphism over a neighborhood of S. Write $E_{\text{fin}} = \operatorname{Spec} A$ and $E'_{\text{fin}} = \operatorname{Spec} A'$; then we have

$$\begin{aligned}(\Omega_{A/R} \otimes_A A' : \Omega_{A'/R}) &= (\mathcal{D}_{A'/R,\mathrm{D}} : \mathcal{D}_{A/R,\mathrm{D}} \otimes_A A') \\ &= (D_{A'/R} : D_{A/R}) \\ &= (A' : A)^2.\end{aligned}$$

The first step follows as in the proof of (3.2). The second step (which is an index of two ideals on R) is proved by the methods of ([**L 1**], Proposition 14, Chapter III, §3). The final step is an easy consequence of the definition of the discriminant. But this equation is merely a restatement of (3.3.1), and therefore we are done.

See also ([**L 3**], Chapter 5, Theorem 1.1) for an approach closer to Riemann-Roch. Q.E.D.

Lemma 3.4 (Elementary properties of d_{a}). *Let $S \subseteq B$. Then the following hold, in which the constants $O(1)$ are independent of P but otherwise depend on all relevant data.*

(a). *If $S' \supseteq S$ is another such subset, then*

$$d_{\mathrm{a},S'}(P) \leq d_{\mathrm{a},S}(P) + O([F:\mathbf{Q}]).$$

(b). *If X' is another model birational to X, and if P' denotes the point in $X'(\bar{k})$ corresponding to $P \in X(\bar{k})$, then*

$$d_{\mathrm{a},S}(P') = d_{\mathrm{a},S}(P) + O([F:\mathbf{Q}]).$$

(c). *If X' is the model obtained from X by making a base change and desingularizing, if P' is as defined as in (b), and if the base change is linearly disjoint from F, then*

$$d_{\mathrm{a},S}(P') = d_{\mathrm{a},S}(P) + O([F:\mathbf{Q}]).$$

(d). *If $S = \emptyset$ and $X = \mathbf{P}^1_B$, then*

$$d_{\mathrm{a}}(P) = (2[F:k] - 2)h(P) + O(1).$$

(e). *If $f\colon X \to Y$ is a morphism of arithmetic surfaces over B, and if $f \mid E_P$ is generically injective, then*

$$d_{\mathrm{a},S}(P) \leq d_{\mathrm{a},S}(f(P)).$$

Proof. Part (a) is trivial from the definitions. Note that the constant in this case is

$$-\frac{1}{2}\sum_{\sigma} \min_{X_\sigma \times X_\sigma} \lambda_\sigma,$$

taking the sum over those $\sigma \in B_\infty$ lying in $S' \setminus S$.

(b). Two functions λ_X can only differ by a bounded amount, so the result is trivial for changes at the archimedean places. This leaves the finite part.

By the Factorization Theorem for arithmetic surfaces ([C], Theorem 2.1 and Proposition 2.2), a change of model is equivalent to a succession of blowings-up and blowings-down. Therefore it will suffice to prove the result for a single blowing-up at a point x.

We will show that, if X' is the model obtained from X by blowing up at the point x, and if P' is the point on $X'(\bar{k})$ corresponding to P on $X(\bar{k})$, then

$$d_{\mathrm{a},S}(P) - \frac{[F:k]}{[k:\mathbf{Q}]} \log q_{\pi(x)} \leq d_{\mathrm{a},S}(P') \leq d_{\mathrm{a},S}(P).$$

First, note that the definition of d_{a} is unchanged by base change (unramified at places of bad reduction), so we may assume that x is rational over the residue field of $\pi(x)$. Let $p\colon X' \to X$ be the blowing-up morphism, with exceptional divisor G. Then $E_{P'} = p^* E_P - nG$, where n is the multiplicity of E_P at x. Also, we may take $K_{X'/B} = p^* K_{X/B} + G$; the claim then follows from Proposition 3.2, the inequality $n \leq [F:k]$, and the fact that

$$d_{\mathrm{a},S}(P') - d_{\mathrm{a},S}(P) = \begin{cases} 0 & \text{if } \pi(x) \in S; \\ d_{\mathrm{a}}(P') - d_{\mathrm{a}}(P) & \text{if } \pi(x) \notin S. \end{cases}$$

Part (c) is proved in a similar manner. Let $f\colon X' \to X$ be the composition of the base change and desingularization, and let $G_1, \ldots G_n$ be the exceptional divisors. Then

$$K_{X'/B'} \sim p^* K_{X/B} + \sum c_i G_i$$

for some constants c_i, and

$$E_{P'} \sim p^* E_P - \sum n_i G_i.$$

Since the c_i are independent of P and the n_i are determined by linear equations involving $(E_{P'} \,.\, G_i)$, we have $n_i = O([F:\mathbf{Q}])$, which implies the result as before.

(d). The group of Arakelov divisors on $X = \mathbf{P}^1_B$ is generated by a fixed horizontal divisor H (say, corresponding to the point at infinity) and by a fixed vertical divisor V. We define

$$h(P) = \frac{(E_P \,.\, H)}{[F:\mathbf{Q}]}$$

and therefore

$$E_P \sim [F:k](H + h(P)V),$$

assuming that V is normalized so that $(H \,.\, V) = [k:\mathbf{Q}]$. We also have $K_{X/B} = -2H$. Then the result follows immediately from Proposition 3.2.

Finally, part (e) is immediate from Proposition 3.3. Q.E.D.

This section concludes with a result of Chevalley-Weil type, analogous to ([**V 1**], 5.1.6). First, for $S \subseteq B$ and for an arithmetic divisor D on X, let

$$N_S(D,P) = \frac{1}{[F:\mathbb{Q}]} \sum_{v \notin S} (D \,.\, E_P)_v. \tag{3.5}$$

Proposition 3.6. *Let $f\colon X \to Y$ be a generically finite surjective map of arithmetic surfaces, with ramification divisor R. Let $S \subseteq B$. Then for all algebraic points P on X not lying in the support of R,*

$$d_{\mathrm{a},S}(P) \le d_{\mathrm{a},S}(f(P)) + N_S(R,P) + O(1).$$

Proof. First assume that v is non-archimedean. For this part of the proof, we work with the finite parts, but omit the subscripts "fin." The exact sequence

$$f^*\Omega_{Y/B} \to \Omega_{X/B} \to \Omega_{X/Y} \to 0$$

restricts to $E := E_{f(P)}$, giving

$$f^*\Omega_{E/B} \to \Omega_{E''/B} \to \Omega_{X/Y} \mid E'' \to 0,$$

where $E'' = f^{-1}(E)$. Also, letting $E' = E_P$, we have

$$\Omega_{E''/B} \mid E' \to \Omega_{E'/B} \to 0.$$

Let $i\colon B' \to E'$ be the normalization of E'. Then for all $b' \in (\pi \circ i)^{-1}(v)$,

$$\operatorname{length}(i^*\Omega_{E'/B})_{b'} \le \operatorname{length}((i \circ f)^*\Omega_{E/B})_{b'} + \operatorname{length}(i^*\Omega_{X/Y})_{b'}.$$

For archimedean v (corresponding to $\sigma\colon k \hookrightarrow \mathbb{C}$), let $b' \in E'_\sigma$; then

$$\begin{aligned}
\tau_\sigma(b') &= \sum_{\substack{b'' \in E'_\sigma \\ b'' \ne b'}} \lambda_{X,\sigma}(b',b'') \\
&\le \sum_{\substack{b'' \in E''_\sigma \\ b'' \ne b'}} \lambda_{X,\sigma}(b',b'') + O(1) \\
&\le \sum_{\substack{b \in E_\sigma \\ b \ne f(b')}} \sum_{\substack{b'' \in E''_\sigma \\ f(b'')=b}} \lambda_{X,\sigma}(b',b'') + \sum_{\substack{b'' \in E''_\sigma \\ f(b'')=f(b') \\ b'' \ne b'}} \lambda_{X,\sigma}(b',b'') + O(1) \\
&\le \sum_{\substack{b \in E_\sigma \\ b \ne f(b')}} \lambda_{Y,\sigma}(f(b'),b) + \sum_{\substack{b'' \in E''_\sigma \\ f(b'')=f(b') \\ b'' \ne b'}} \lambda_{X,\sigma}(b',b'') + O(1) \\
&\le \tau_{f^*\Omega_{E/B},\sigma}(b') + \tau_{X/Y,\sigma}(b') + O(1).
\end{aligned}$$

Indeed, the first and last steps follow by definitions, the second step holds since $\lambda_{X,\sigma} \geq -O(1)$, and the third step is just rearranging terms. The penultimate step follows from the inequality

$$\sum_{f(b'')=b} \lambda_{X,\sigma}(b', b'') = \lambda_{Y,\sigma}(f(b'), b) + O(1).$$

This in turn is easy to prove locally using coordinate patches, and then by compactness it holds globally.

Thus, it follows that

$$\deg_S \Omega_{E'/B} \leq \deg_S \left(f^* \Omega_{E/B} \mid E'\right) + \deg_S \left(\Omega_{X/Y} \mid E'\right).$$

Dividing by $[k(P) : \mathbf{Q}]$ and applying Proposition 2.8 (together with various definitions) then gives

$$d_{\mathrm{a},S}(P) \leq d_{\mathrm{a},S}(f(P)) + N_S(P, R) + O(1). \qquad \text{Q.E.D.}$$

4. Proof of Theorem A

This proof is immediate from the properties of d_{a} proved in the previous section. We continue using the notation of Section 3.

Proof. First we note that the conclusion is not weakened after a base change, therefore we may assume that C has semi-stable reduction over k. Fix a horizontal divisor H on X and let $h(P) = (E_P \,.\, H)/(\deg H)[k : \mathbf{Q}]$. Then for all algebraic points P on X of degree $\leq s$ for which $k(f(P)) = k(P)$, we have

$$\begin{aligned}(2g - 2 + 2s - 2s\sqrt{g} - \epsilon)h(P) &\leq d_{\mathrm{a}}(P) + O(1)\\ &\leq d_{\mathrm{a}}(f(P)) + O(1)\\ &\leq (2s - 2)h(f(P)) + O(1)\\ &\leq (\deg f)(2s - 2 + \epsilon)h(P) + O(1).\end{aligned}$$

The first step follows from ([**V 3**], 7.1) and Proposition 3.2. The second and third steps are consequences of Lemma 3.4 (e) and (d), respectively. Finally, the last step is immediate from the properties of heights relative to numerically equivalent divisors (see, *e. g.*, ([**V 1**], 1.2.9d)). Combining this with the assumption

$$g - 1 + s - s\sqrt{g} > (\deg f)(s - 1)$$

gives a bound for $h(P)$. Hence the finiteness statement follows. Q.E.D.

[C] T. Chinburg, Minimal models for curves over Dedekind rings. In: G. Cornell and J. H. Silverman, eds., *Arithmetic geometry*, Springer-Verlag, Berlin-Heidelberg-New York, 1986, 309–326.

[F] G. Faltings, Diophantine approximation on abelian varieties. Preprint, 1989.

[F-L] W. Fulton and S. Lang, *Riemann-Roch algebra* (Grundlehren der mathematischen Wissenschaften 277), Springer-Verlag, New York-Berlin-Heidelberg-Tokyo, 1985.

[G-S] H. Gillet and C. Soulé, Arithmetic intersection theory. Preprint IHES, 1988.

[H] R. Hartshorne, *Algebraic geometry* (Graduate texts in mathematics 52), Springer-Verlag, Berlin-Heidelberg-New York, 1977.

[Hi] M. Hindry, Points quadratiques sur les courbes. *C. R. Acad. Sci. Paris,* **305** (1987) 219–221.

[K] E. Kunz, *Kähler differentials,* Vieweg Verlag, Braunschweig/ Wiesbaden, 1986.

[L 1] S. Lang, *Algebraic number theory,* Addison-Wesley, Reading, Mass., 1970; reprinted by Springer-Verlag, Berlin-Heidelberg-New York, 1986.

[L 2] ———, *Fundamentals of diophantine geometry,* Springer-Verlag, Berlin-Heidelberg-New York, 1983.

[L 3] ———, *Introduction to Arakelov theory,* Springer-Verlag, Berlin-Heidelberg-New York, 1988.

[S] J.-P. Serre, *Groupes algébriques et corps de classes,* Hermann, Paris, 1959.

[Sil] J. H. Silverman, Rational points on symmetric products of a curve. Preprint, 1989.

[V 1] P. Vojta, *Diophantine approximations and value distribution theory.* (Lect. notes math., vol. 1239), Springer-Verlag, Berlin-Heidelberg-New York, 1987.

[V 2] ———, Mordell's conjecture over function fields. *Invent. Math.,* **98** (1989) 115–138.

[V 3] ———, Siegel's theorem in the compact case. *Ann. Math.*, to appear.

THE BIRCH-SWINNERTON-DYER CONJECTURE FROM A NAIVE POINT OF VIEW

Don Zagier

Throughout this paper, E will denote an elliptic curve defined over $\mathbb{Q}$ which we suppose given in $\mathbb{P}^2$ by an equation

$$f(x,y,z)=0 \qquad (f\in\mathbb{Z}[x,y,z] \text{ homogeneous of degree 3}) \tag{1}$$

with f of minimal discriminant Δ. If R is any ring with unit, then $E(R)$ denotes the set of solutions of $f=0$ in $\mathbb{P}^2(R)=\{(x,y,z)\in R^3 : xR+yR+zR=R\}/R^\times$. (In particular, $E(\mathbb{Z})$ is the same as the Mordell-Weil group $E(\mathbb{Q})$ and not, as sometimes in the literature, the finite set of integral points in the affine model $f(x,y,1)=0$ of E over $\mathbb{Z}$.) The L-series of E is the Dirichlet series given by

$$L(E,s)=\prod_p \frac{1}{1-a(p)p^{-s}+\varepsilon(p)p^{1-2s}}=\sum_{n=1}^{\infty}\frac{a(n)}{n^s} \qquad (\mathrm{Re}(s)>\frac{3}{2}), \tag{2}$$

where the product is over all primes, $a(p)=p+1-|E(\mathbb{Z}/p\mathbb{Z})|$ ($|\cdot|$ denotes cardinality) and $\varepsilon(p)=1$ or 0 depending whether $p\nmid\Delta$ or $p|\Delta$. The **Birch–Swinnerton-Dyer conjecture** consists of two statements:

(A) $L(E,s)$ continues meromorphically to $s=1$ and has a zero there of order exactly $r=\mathrm{rank}_{\mathbb{Z}}E(\mathbb{Q})$ (for $r=0$ this means that $L(E,s)$ is holomorphic and non-zero at $s=1$).

(B) Assuming this, define λ by $L(E,s)\sim\lambda\,(s-1)^r \quad (s\to 1)$. Then

$$\lambda=\frac{c\,R\,\Omega}{T^2}\,|Ш|, \tag{3}$$

where R is the determinant of the Néron-Tate height pairing with respect to a basis of the free part of $E(\mathbb{Q})$, Ω is the real period ($=$ integral over $E(\mathbb{R})$ of a Néron differential of $E/\mathbb{Z}$), $c=\prod_{p|\Delta}c_p$ where $c_p=[E(\mathbb{Z}_p):E^0(\mathbb{Z}_p)]$

($E^0(\mathbb{Z}_p)$ is the set of points reducing to non-singular points of $E(\mathbb{Z}/p\mathbb{Z})$), and Ш is the conjecturally finite Tate-Shafarevich group of E. (Each of these quantities will be discussed in more detail later.)

The Birch–Swinnerton-Dyer conjecture is very famous and has been given various formulations in more learned language, most strikingly one in terms of Tamagawa numbers in a beautiful paper by Spencer Bloch (*Invent. math.* **58** (1980), 65-76). Our purpose in this note is to go the other way, replacing as many as possible of the invariants entering into (3) by numbers that can be defined without any theoretical knowledge (in particular, without knowing what the height pairing is, how the groups $E(\mathbb{Z}_p)$ and $E^0(\mathbb{Z}_p)$ look, or even that there is a group structure on $E(R)$). More precisely, we will try to define these invariants in a purely Diophantine way, merely by *counting* solutions of (1) in $\mathbb{Q}$ and in $\mathbb{Z}/n\mathbb{Z}$. The motivation for doing this, besides the pleasure in finding a formulation of the Birch–Swinnerton-Dyer conjecture which can be explained easily to a non-specialist, comes from the analogy with the situation of zeta-functions of quadratic fields, where by defining everything in terms of counting solutios of equations one can get an easy proof of the Dirichlet class number formula. This analogy is described in §1. In §2 we interpret the numbers r and RT^{-2} (and, very vaguely, $RT^{-2}\Omega$) in terms of counting rational solutions of (1). In §3 we express the quantity $c^{-1}\lambda$ (assuming **(A)**) in terms of the number of solutions of (1) in $\mathbb{Z}/n\mathbb{Z}$ for all n. Finally, in §4 we make some remarks about the interpretation of Ш in terms of solutions of (1) and related equations over various other rings.

1. The prototype: zeta functions of quadratic fields

There is a well-known analogy between the Birch–Swinnerton-Dyer conjecture and Dirichlet's class number formula for number fields in which $\sqrt{R}$ corresponds to the regulator, T to the number of roots of unity, and Ш to the class group of the field. Like the Birch–Swinnerton-Dyer conjecture, Dirichlet's formula can be stated in a much more abstract language than the original formulation. However, it can also be stated—and, at least for quadratic fields, proved—in an entirely elementary way purely in terms of counting solutions of equations. In this section we describe how this goes, since it is the model for what we would like to do for equation (3).

Let D be an integer and consider the set of all binary quadratic forms $Q(u,v) = au^2 + buv + cv^2$ with a, b, $c \in \mathbb{Z}$ and discriminant $b^2 - 4ac = D$. The group $SL(2,\mathbb{Z})$ acts on such forms; we denote by $[Q]$ and by $\mathrm{Aut}(Q)$ the orbit and isotropy group, respectively, of Q under this action. If n is a natural number, then $\mathrm{Aut}(Q)$ acts on the set of representations of n by Q (= pairs $(u,v) \in \mathbb{Z}^2$ with $Q(u,v) = n$). We denote by $r_Q(n)$ the number of inequivalent representations under this action (this can be an integer

≥ 0 or, *a priori*, ∞) and by $r_Q^*(n)$ the number of inequivalent primitive representations (representations with u and v coprime). Then the basic identity is

$$\sum_{[Q]} r_Q^*(n) = |\{b \pmod{2n} : b^2 \equiv D \pmod{4n}\}| \tag{4}$$

(in particular , $r_Q^*(n)$ is finite, and in fact bounded by n).

We will prove (4) by a very simple counting argument which works in a uniform way for all D, positive, negative, or zero. First, however, we show how it implies the analogue of the Birch–Swinnerton-Dyer formula (3) in the case that D is the discriminant of a quadratic field (i.e. $D \neq 1$ is square-free and congruent to 1 modulo 4 or $D/4$ is square-free and congruent to 2 or 3 modulo 4). The right-hand side of (4) is clearly a multiplicative function of n. If $(\frac{D}{p}) = 1$ (or $p = 2$, $D \equiv 1 \pmod 8$), it is 2 for all $n = p^\nu > 1$; if $(\frac{D}{p}) = -1$ (or $p = 2$, $D \equiv 5 \pmod 8$), it is 0 for all $n = p^\nu > 1$; and if $p|D$ it is 1 for $n = p$ and 0 for $n = p^\nu$, $\nu > 1$. Using this and the obvious identity $r_Q(n) = \sum_{e^2|n} r_Q^*(n/e^2)$, we easily deduce from (4) the equivalent formula

$$\sum_{[Q]} r_Q(n) = \sum_{d|n} \chi_D(d), \tag{5}$$

where χ_D is defined as the totally multiplicative function on $\mathbb{N}$ with $\chi_D(p) = (\frac{D}{p})$ for p prime. By quadratic reciprocity one knows that χ_D is a periodic function of average 0; it follows that the series $L(1, \chi_D) = \sum \chi_D(n)\, n^{-1}$ converges. Taking the average $(= \lim\limits_{N\to\infty} \frac{1}{N} \sum_{n=1}^{N} (\cdots))$ of both sides of (5), we obtain

$$\sum_{[Q]} \langle r_Q \rangle = L(1, \chi_D). \tag{6}$$

But it is easily seen by counting lattice points in a sector of an ellipse or hyperbola that the average value $\langle r_Q \rangle$ of r_Q has a value $\kappa(D)$ independent of Q, given by $\dfrac{2\log\varepsilon}{\sqrt{D}}$ if $D > 0$ (where $\varepsilon = \frac{1}{2}(t + u\sqrt{D})$ for the smallest positive integer solution of Pell's equation $t^2 - Du^2 = \pm 4$ if such a solution exists, $\kappa(D) = \infty$ otherwise) and by $\dfrac{2\pi}{w\sqrt{-D}}$ if $D < 0$ (where $w = 2$, 4 or 6 is the number of integral solutions of $t^2 - Du^2 = 4$). Thus (6) says that $L(1,\chi_D) = h(D)\kappa(D)$, where $h(D)$ is the (possibly infinite) number of equivalence classes $[Q]$ of discriminant D. Since $L(1,\chi_D)$ is *a priori* known to be finite (but might be zero), while $h(D)$ and $\kappa(D)$ are *a priori* known to be non-zero (but might be infinite), this gives in one blow the finiteness of the class number $h(D)$, the existence of a solution of Pell's equation for non-square $D > 0$, the non-vanishing of $L(1,\chi_D)$, and the Dirichlet class number formula.

And now to the proof of (4), which is considerably shorter than the discussion of what to do with it. We use the following general principle: if a group G acts on two sets X and Y, and $S \subset X \times Y$ is a subset invariant under the diagonal action of G, then

$$\sum_{x\in X/G} |S_x/G_x| = \sum_{y\in Y/G} |S_y/G_y|, \tag{7}$$

where $S_x = \{y \in Y : (x,y) \in S\}$, $G_x = \{g \in G : gx = x\}$ and similarly for S_y and G_y. The proof of (7) is obvious: just count the number of orbits of G in S in the two possible orders. We apply it to the case where G is $SL(2,\mathbb{Z})$, X the set of quadratic forms Q of discriminant D, Y the set of pairs of coprime integers (u,v), and S the set of pairs $(Q,(u,v)) \in X \times Y$ with $Q(x,y) = n$. Then the left-hand side of (7) is equal to the left-hand side of (4) by definition. On the other hand, Y/G consists of a single element which we can represent by the point $y = (1,0)$. For this choice, S_y is the set of quadratic forms $nu^2 + buv + \frac{b^2-D}{4n}v^2$ with $b^2 \equiv D \pmod{4n}$ and $G_y = \{\left(\begin{smallmatrix}1 & r\\ 0 & 1\end{smallmatrix}\right) : r \in \mathbb{Z}\}$ acts on this by $b \mapsto b + 2rn$. The result follows.

In summary, the identity (4), which can be written in the equivalent form

$$\zeta(2s)^{-1} \sum_{[Q]} \sum_{\substack{(u,v)\in\mathbb{Z}^2/\mathrm{Aut}(Q)\\ Q(u,v)>0}} \frac{1}{Q(u,v)^s} = \sum_{n=1}^{\infty} \frac{|\{b\,(2n) \,:\, b^2 \equiv D\,(4n)\}|}{n^s}, \tag{8}$$

has a simple combinatorial interpretation in terms of counting orbits under a certain group action on the set of integral solutions of a certain Diophantine equation, and gives the Dirichlet class number formula and other arithmetical information by looking at the asymptotic behavior near $s = 1$. If we could interpret the Birch–Swinnerton-Dyer conjecture in the same way, we would obtain a proof of it. This of course we cannot do, but we will at least be able to write some of the invariants it involves in terms of the asymptotics of two Dirichlet series analogous to those occurring in equation (8).

2. The global Diophantine invariants r and RT^{-2} (and Ω)

Let $|\ |$ be any continuous norm on $\mathbb{R}^3$, e.g. $|(x,y,z)| = (x^2+y^2+z^2)^{1/2}$ or $|x|+|y|+|z|$ or $\max\{|x|,|y|,|z|\}$. We can consider $|\ |$ as defined on $\mathbb{P}^2(\mathbb{Q})$ or $E(\mathbb{Q})$ by identifying these sets with $\mathbb{P}^2(\mathbb{Z}) = \mathbb{Z}^3/\{\pm 1\} \subset \mathbb{R}^3/\{\pm 1\}$ and $E(\mathbb{Z})$, respectively. Now let

$$\mathcal{N}(B) = \{P \in E(\mathbb{Q}) \ :\ |P| \leq B\}$$

be the number of solutions of (1) in coprime integers (x,y,z) of norm $\leq B$, two solutions differing only in sign being counted as one. Then we have

Proposition. *The asymptotic growth of $\mathcal{N}(B)$ as $B \to \infty$ is given by*

$$\mathcal{N}(B) \sim \frac{\pi^{r/2}}{(r/2)!}\,\frac{T}{\sqrt{R}}\,(\log B)^{r/2} \qquad (B \to \infty). \tag{9}$$

Here $(r/2)!$ is to be interpreted as $\Gamma(1+\frac{r}{2})$ if r is odd. Note that the assertion of (9) is independent of the choice of norm because any two norms are bounded by multiples of one another and $(\log B + \mathrm{O}(1))^{r/2} \sim (\log B)^{r/2}$.

Proof: We recall how the Néron-Tate height pairing is defined: one shows that there exists a positive-definite quadratic form h on $E(\mathbb{Q})/(\text{torsion})$ such that

$$h(P) = \log|P| + \mathrm{O}(1) \qquad \text{for all } P \in E(\mathbb{Q}); \tag{10}$$

the height pairing is the associated non-degenerate bilinear form. Thus we can consider $E(\mathbb{Q})/(\text{torsion}) \approx \mathbb{Z}^r$ as a lattice Λ in the vector space $E(\mathbb{Q}) \otimes \mathbb{R} \approx \mathbb{R}^r$ with the metric defined by h. The number of points in Λ with $h \leq H$ for H large is asymptotically equal to $\frac{1}{\sqrt{R}}$ (the volume of $\mathbb{R}^r/\Lambda$) times $\frac{(\pi H)^{r/2}}{(r/2)!}$ (the volume of an r-dimensional sphere of radius $\sqrt{H}$), and the number of $P \in E(\mathbb{Q})$ with $h(P) \leq H$ is T times this. The result now follows from equation (10).

Notice that (9) implies that the zeta function

$$Z(E,s) = \sum_{P \in E(\mathbb{Q})} \frac{1}{|P|^s}$$

converges for all s with $\mathrm{Re}(s) > 0$ and satisfies

$$Z(E,s) \sim \frac{T}{\sqrt{R}}\,(s/\pi)^{-r/2} \qquad (s \searrow 0) \tag{11}$$

(to see this, write $Z(E,s)$ as $s\int_0^\infty \mathcal{N}(B)\,B^{-s-1}\,dB$ and use the estimate on $\mathcal{N}(B)$). Either (9) or (11) defines both r and R/T^2 in an elementary way, without reference to the group structure or the Néron-Tate height pairing on $E(\mathbb{Q})$.

Finally, we make a stab at bringing Ω into the picture. The presence of the factor $\sqrt{R}$ in (11), as well as the fact that $Z(E,s)$ has a pole of half-integral order at 0 if r is odd, suggests looking at $Z(E,s)^2$, which is the sum of $|P|^{-s}|Q|^{-s}$ over all pairs $(P,Q)\in E(\mathbb{Q})^2$. We could look instead at the subsum over pairs of points (P,Q) which are close in the real topoogy, say $|z(P-Q)/x(P-Q)|<\epsilon$ in the standard Tate-Weierstrass form $y^2z+a_1xyz+a_3yz^2=x^3+a_2x^2z+a_4xz^2+a_6z^3$. These points, at least if $E(\mathbb{Q})$ is infinite, will constitute asymptotically a proportion ν of the set of all pairs, where ν is the ratio of the length of $\{(x:y:z)\in E(\mathbb{R}):|z/x|<\epsilon\}$ to that of all of $E(\mathbb{R})$. The first of these lengths is given by an incomplete elliptic integral which is asymptotic to $2\sqrt{\epsilon}$ as $\epsilon\to 0$, while the second, a complete elliptic integral, equals Ω. Thus the two-point zeta function in question looks like $\frac{2\sqrt{\epsilon}}{\Omega}Z(E,s)^2$, and its leading terms at 0 like $2\sqrt{\epsilon}\,\frac{T^2}{\Omega R}\,(\pi/s)^r$. Obviously this is very vague, but at least it suggests that the natural combination of $T/\sqrt{R}$ and Ω is $T^2/\Omega R$, the quantity occurring in the Birch–Swinnerton-Dyer conjecture.

3. The local Diophantine invariants $L(E,s)$ and c

The L-function of E defined by (2), while natural from several points of view (in particular, the conjectured holomorphic continuation and functional equation), has several defects from a purely Diophantine point of view:

(i) The numbers $a(p)$ are defined in terms of counting solutions, but the coefficients $a(n)$ for n composite (say $n=p^2$ or $n=pq$) can only be defined by using the recursion given by the Euler product (2); we do not know how to compute $a(n)$ directly in terms of numbers of solutions of equations, or without knowing the prime factorization of n.

(ii) The function $L(E,s)$ ought to summarize all the local data about E which is relevant for the Birch–Swinnerton-Dyer conjecture, but fails to do so: at primes of bad reduction the Euler factor of $L(E,s)$ contains too little information and we are obliged to include the number c_p as an extra ("fudge") factor in (3).

(iii) $L(E,s)$ conjecturally has a zero of order r at $s=1$, but a pole would be much more convenient, since the presence of a pole in the analytic continuation of a Dirichlet series makes itself felt in the asymptotics of its coefficients, while a zero is not visible in this way.

In this section we will attempt to remedy or partially remedy these defects by introducing a new Dirichlet series which is related to, but not expressible in terms of, $L(E,s)$. Actually, we shall define three such Dirichlet series, but we give only the most natural one here, and mention the others as variants at the end of the section. We denote by $N(n)$ the cardinality of the finite set $E(\mathbb{Z}/n\mathbb{Z})$ and set

$$D(E,s) = \sum_{n=1}^{\infty} N(n)\, n^{-s}.$$

The series converges absolutely for $\mathrm{Re}(s) > 2$. We will prove:

PROPOSITION. *The product $D(E,s)\,L(E,s)$ extends meromorphically to the half-plane $\mathrm{Re}(s) > \frac{5}{6}$, is holomorphic for $\mathrm{Re}(s) \geq 1$ except for a simple pole at $s = 2$, and has the value $-c$ at $s = 1$.*

COROLLARY. *Part* **(A)** *of the Birch–Swinnerton-Dyer conjecture is equivalent to*

(A′) *$D(E,s)$ continues meromorphically to $s = 1$ and has a pole there of order exactly r (for $r = 0$ this means that $D(E,s)$ is holomorphic and non-zero at $s = 1$.)*

If this holds, then $D(E,s) \sim -\lambda^{-1}c\,(s-1)^{-r}$ as $s \to 1$.

Thus $D(E,s)$ remedies the problems (i) ($N(n)$ is defined directly by counting solutions of (1) modulo n, without reference to the prime factorization of n) and (ii) (the leading term of $D(E,s)$ at $s = 1$ involves the same combination of λ and c as occurs in the Birch–Swinnerton-Dyer conjecture, so $D(E,s)$ encodes all the interesting local information). It also partially solves (iii), since "rth order zero" has been replaced by "rth order pole" in going from **(A)** to **(A′)**. However, if $E(\mathbb{Q})$ is finite, then there is no pole, and even if $r > 0$ the effect of the pole at $s = 2$ (to say nothing of possible other poles in $\mathrm{Re}(s) > 1$ if the Riemann hypothesis is false for $L(E,s)$) will dominate the asymptotics of the coefficients $N(n)$ and tend to swamp the contribution of the pole at $s = 1$, making it hard to "see" the numbers r and $\lambda^{-1}c$ (the pole at $s = 2$ will contribute a term Ax^2 to $\sum_{n \leq x} N(n)$, while the pole at $s = 1$—if r is positive—will give a smaller order contribution $-\lambda^{-1}cx(\log x)^{r-1}/(r-1)!$). Also on the negative side, of course, is that $D(E,s)$ does not even conjecturally have an analytic continuation to all s or satisfy a functional equation.

To prove the proposition, we observe that $N(n)$ is clearly multiplicative and hence that $D(E,s)$ has an Euler product

$$D(E,s) = \prod_p D_p(p^{-s}), \qquad D_p(x) = \sum_{\nu=0}^{\infty} N(p^\nu)x^\nu.$$

We look at the Euler factors separately, starting with the case of good reduction, $p \nmid \Delta$. By definition, $N(p) = |E(\mathbb{Z}/p\mathbb{Z})| = p + 1 - a(p)$. Since the points of $E(\mathbb{Z}/p\mathbb{Z})$ are non-singular (i.e., some partial derivative of f is non-zero mod p), Hensel's lemma implies that they each lift to exactly $p^{2(\nu-1)}$ solutions of $f = 0$ in $(\mathbb{Z}/p^\nu\mathbb{Z})^3$ for $\nu \geq 1$. But the order of $(\mathbb{Z}/p^\nu\mathbb{Z})^\times$ is $p^{\nu-1}$ times that of $(\mathbb{Z}/p\mathbb{Z})^\times$, so this gives $N(p^\nu) = p^{\nu-1}N(p)$ for the number of solutions in $\mathbb{P}^2(\mathbb{Z}/p^\nu\mathbb{Z})$. Hence

$$D_p(x) = 1 + \sum_{\nu=1}^{\infty} (p + 1 - a(p))\, p^{\nu-1}x^\nu = \frac{1 - (a(p) - 1)\, x}{1 - px}.$$

It follows that the Euler factor of $D(E, s)/\zeta(s-1)$ at p is $1 - \bigl(a(p) - 1\bigr)\, p^{-s}$, which for $s = 1$ takes on the value $\dfrac{N(p)}{p}$, the reciprocal of the value of the corresponding Euler factor of $L(E, s)$. Thus the Euler product of $D(E, s)L(E, s)/\zeta(s-1)$ converges (indeed, terminates) at $s = 1$. Nevertheless, this is not the right combination to look at, since it diverges for s near 1 (its pth Euler factor for $p \nmid \Delta$ equals $1 + p^{-s}(1 + \mathrm{o}(1))$ for $\mathrm{Re}(s) > 1$). Instead, we set

$$\psi_p(x) \stackrel{\text{def}}{=} \frac{(1 - px)(1 - x)}{1 - px^2} D_p(x) L_p(x)$$

(where $L_p(p^{-s})$ is the pth Euler factor of $L(E, s)$) and find for $p \nmid \Delta$

$$\psi_p(x) = \frac{(1 - x)\bigl(1 - (a(p) - 1)x\bigr)}{\bigl(1 - px^2\bigr)\bigl(1 - a(p)x + px^2\bigr)} = 1 + \frac{x^2\bigl(1 - px\bigr)\bigl(a(p) - 1 + px\bigr)}{\bigl(1 - px^2\bigr)\bigl(1 - a(p)x + px^2\bigr)}.$$

This equals 1 at $x = p^{-1}$ and $1 + \mathrm{O}(p^{3/2}x^3)$ for $x = \mathrm{o}(p^{-1/2})$ (since $a(p) = \mathrm{O}(p^{1/2})$), so the product $\prod_{p \nmid \Delta} \psi_p(p^{-s})$ converges absolutely in $\mathrm{Re}(s) > \dfrac{5}{6}$ and equals 1 at $s = 1$.

We will show that $\psi_p(x)$ for $p|\Delta$ is a rational function having no poles in $|x| \leq \dfrac{1}{p}$ and satisfying $\psi_p\bigl(\dfrac{1}{p}\bigr) = c_p$. It follows that $\psi(s) = \prod_p \psi_p(p^{-s})$ extends to a non-zero holomorphic function in $\mathrm{Re}(s) > \dfrac{5}{6}$ with $\psi(1) = c$. This will prove the proposition, since $D(E, s)L(E, s) = \dfrac{\zeta(s)\zeta(s-1)}{\zeta(2s-1)}\psi(s)$ and $\dfrac{\zeta(s)\zeta(s-1)}{\zeta(2s-1)}$ is a meromorphic function with a simple pole at $s = 2$ and no other pole in $\mathrm{Re}(s) \geq 1$, equal to -1 at $s = 1$. So let p be a prime of bad reduction. If $E^0(\mathbb{Z}/p^\nu\mathbb{Z})$ denotes the set of points which reduce to the non-singular part of $E(\mathbb{Z}/p\mathbb{Z})$, then $|E^0(\mathbb{Z}/p\mathbb{Z})| = |E(\mathbb{Z}/p\mathbb{Z})| - 1 = p - a(p)$ and $|E^0(\mathbb{Z}/p^\nu\mathbb{Z})| = p^{\nu-1}|E^0(\mathbb{Z}/p\mathbb{Z})|$ by Hensel's lemma as before. But $|E(\mathbb{Z}/p^\nu\mathbb{Z})| \sim c_p|E^0(\mathbb{Z}/p^\nu\mathbb{Z})|$ for $\nu \to \infty$ by the definition of c_p, so $N(p^\nu) = c_p\big(p - a(p)\big)p^{\nu-1} + o(p^\nu)$. This shows that $D_p(x)$, which is a rational function by general principles, has a simple pole of principal part $c_p \dfrac{1 - a(p)/p}{1 - px}$ at $x = \dfrac{1}{p}$ and no other poles in $|x| \le \dfrac{1}{p}$. Since $L_p(x) = \dfrac{1}{1 - a(p)x}$ for $p|\Delta$,the assertions about $\psi_p(x)$ follow. This completes the proof of the proposition.

Remark. The function $D(E,s)$ cannot be expressed as a quotient of finite products of functions $\zeta(ns - m)$ and $L(E, ns - m)$ ($n \in \mathbb{N}$, $m \in \mathbb{Z}$)), because the rational function $1 - (a-1)x$ cannot be factored into functions of the form $1 - p^m x^n$ and $1 - ap^m x^n + p^{2m+1}x^{2n}$. On the other hand, we could contine the partial factorization further and get the meromorphic continuation of $D(E,s)$ (assuming that of $L(E,s)$) into a bigger half-plane. For instance, the identity (for $p \nmid \Delta$)

$$\frac{(1-x)(1-px)}{(1-px^2)(1-a(p)x+px^2)(1-pa(p)x^3+p^3x^6)} D_p(x)$$
$$= 1 + \frac{a(p)x^2(1-p^4x^7) - x^2(1-p^5x^8) + (p^2 - pa(p)^2)x^4(1-px^2)}{(1-px^2)(1-a(p)x+px^2)(1-pa(p)x^3+p^3x^6)}$$

gives the meromorphic continuation of $D(E,s)$ to the half-plane $\Re(s) > \dfrac{3}{4}$ (assuming that $L(E,s)$ is known to be meromorphic), because the expression on the right is $1 + O(p^{-1-2\epsilon})$ for $x = O(p^{-3/4-\epsilon})$. However, it seems unlikely that $D(E,s)$ continues meromorphically to the whole plane, and anyway we are mainly interested in the point $s = 1$.

Finally, we define the two variants of $D(E,s)$ mentioned at the beginning of the section. The definition of $N(n)$ can be written

$$N(n) = |\{(x,y,z) \in (\mathbb{Z}/n\mathbb{Z})_0^3 \,:\, f(x,y,z) \equiv 0 \pmod{n}\}/(\mathbb{Z}/n\mathbb{Z})^\times|$$

where $(\mathbb{Z}/n\mathbb{Z})_0^3$ denotes $\{(x,y,z) \in (\mathbb{Z}/n\mathbb{Z})^3 \,:\, \gcd(x,y,z,n) = 1\}$. The new Dirichlet series are $D_j(E,s) = \sum N_j(n)\,n^{-s}$ $(j = 1, 2)$, where

$$N_1(n) = |\{(x,y,z) \in (\mathbb{Z}/n\mathbb{Z})_0^3 \,:\, f(x,y,z) \equiv 0 \pmod{n}\}|,$$
$$N_2(n) = |\{(x,y,z) \in (\mathbb{Z}/n\mathbb{Z})^3 \,:\, f(x,y,z) \equiv 0 \pmod{n}\}|,$$

i.e., we count all coprime, or all, solutions modulo n, rather than only non-proportional ones. Since $(\mathbb{Z}/n\mathbb{Z})^\times$ acts freely on $(\mathbb{Z}/n\mathbb{Z})_0^3$, $N_1(n) =$

$N(n)\varphi(n)$, where $\varphi(n) = |(\mathbb{Z}/n\mathbb{Z})^\times|$ is Euler's function. Hence for $p \nmid \Delta$ we have

$$\begin{aligned} D_{1,p}(x) \stackrel{\text{def}}{=} \sum_{\nu=0}^{\infty} N_1(p^\nu)\, x^\nu &= 1 + \sum_{\nu=1}^{\infty} (p+1-a(p))\,(p-1)\,p^{2\nu-2}\,x^\nu \\ &= \frac{1 - \big(a(p)(p-1)+1\big)x}{1 - p^2 x} \\ &= \frac{(1-p^3x^2)\,(1-a(p)px+p^3x^2)}{1-p^2x}\psi_{1,p}(x) \end{aligned}$$

with

$$\psi_{1,p}(x) = 1 + x\frac{a(p)(1-p^4x^2) - (1-p^6x^3)}{(1-p^3x^2)\,(1-a(p)px+p^3x^2)},$$

which is equal to 1 at $x = p^{-2}$ and to $1 + O(p^{9/2}x^3)$ for $x = o(p^{-3/2})$. For $p|\Delta$ we define $\psi_{1,p}(x) = (1-p^2x)L_p(px)D_{1,p}(x)/(1-p^3x^2)$; an argument like the one before shows that this is a rational function which is holomorphic in $|x| \le p^{-2}$ and equal to c_p at $x = p^{-2}$. Thus

$$D_1(s) = \frac{\zeta(s-2)}{\zeta(2s-3)L(E,s-1)} \prod_p \psi_{1,p}(p^{-s})$$

has a meromorphic continuation to $\mathrm{Re}(s) > \dfrac{11}{6}$ with a simple pole at $s = 3$ and a leading term $-\lambda^{-1}c\,(s-2)^{1-r}$ at $s = 2$. This is even worse than for $D(s)$ because now we get an actual pole only if the rank r is at least 2. For $D_2(s)$ the calculations are messier since $(\mathbb{Z}/n\mathbb{Z})^\times$ does not act freely on $(\mathbb{Z}/n\mathbb{Z})^3$ and we have to take into account the various isotropy groups. We find for $p \nmid \Delta$

$$N_2(p^\nu) = 1 + \sum_{\frac{\nu}{3} \le \mu < \nu} p^{3(\nu-\mu-1)}(p^3-1) + \sum_{0 \le \mu < \frac{\nu}{3}} p^{3\nu-2}(p-1)(p+1-a(p))$$

(to get this, count the number of solutions with $\gcd(x,y,z,p^\nu) = p^\mu$ for each μ, $0 \le \mu \le \nu$),

$$\begin{aligned} D_{2,p}(x) &= \frac{1+x+p^3x^2}{1-p^6x^3} + \frac{(p+1-a(p))(p-1)x}{(1-p^2x)(1-p^6x^3)} \\ &= \frac{(1-p^5x^3)(1-p\,a(p)x+p^3x^2)}{(1-p^2x)(1-p^6x^3)}\left(1 + x\frac{(a(p)-p^2x)(1-p^6x^3)}{(1-p^5x^3)(1-pa(p)x+p^3x^2)}\right), \end{aligned}$$

and hence (treating the primes $p|\Delta$ as before) finally

$$D_2(s) = \frac{\zeta(s-2)\zeta(3s-6)}{\zeta(3s-2)L(E,s-1)}\,\psi_2(s)$$

with $\psi_2(s)$ holomorphic and non-zero in $\mathrm{Re}(s) > \frac{15}{8}$ and $\psi_2(2) = c$. Thus $D_2(s)$ has a meromorphic continuation to $\mathrm{Re}(s) > \frac{15}{8}$ with a simple pole at $s = 3$ and leading term $\frac{3c}{4\lambda}(s-2)^{1-r}$ at $s = 2$.

4. The Tate-Shafarevich group

Since the conjecturally finite order of the Tate-Shafarevich group Ш of $E/\mathbb{Q}$ enters into the Birch–Swinnerton-Dyer formula, we would like to relate it to Diophantine properties of E, i.e., to find a connection between the groups $E(\mathbb{Q})$ and Ш. The standard descent sequence

$$0 \longrightarrow E(\mathbb{Q})/mE(\mathbb{Q}) \longrightarrow \mathrm{Sel}_m \longrightarrow Ш[m] \longrightarrow 0 \tag{12}$$

relates the m-torsion in Ш to the cokernel of $E(\mathbb{Q}) \xrightarrow{\cdot m} E(\mathbb{Q})$ for each natural number m, and it is reasonable to ask whether this sequence "lifts" to an exact sequence independent of m. One could try to construct a sequence $0 \to E(\mathbb{Q}) \to \mathrm{Sel} \to Ш \to 0$ for some group "Sel", but a little thought show that this cannot be done in a natural way and anyway would not induce a sequence as in (12). Instead, we should look for a *four-term* sequence

$$0 \longrightarrow E(\mathbb{Q}) \longrightarrow \mathcal{E} \longrightarrow \mathcal{S} \longrightarrow Ш \longrightarrow 0. \tag{13}$$

This is motivated by two considerations:

1. An exact sequence of abelian groups

$$0 \longrightarrow A \xrightarrow{i} B \xrightarrow{f} C \xrightarrow{p} D \longrightarrow 0$$

induces a short exact sequence

$$0 \longrightarrow A/mA \longrightarrow S_m \longrightarrow D[m] \longrightarrow 0$$

for every natural number m, where

$$S_m = \frac{\{(b,c) \in B \times C \,:\, f(b) = mc\}}{\{(mb, f(b)) \,:\, b \in B\}} \tag{14}$$

and the maps $A/mA \to S_m$ and $S_m \to D[m]$ are induced by $a \mapsto (i(a), 0)$ and $(b,c) \mapsto p(c) + mD$, respectively. (The proof is an easy diagram chase.)

2. The analogy between the Birch–Swinnerton-Dyer conjecture for elliptic curves and the Dirichlet class number formula for number fields K makes $E(\mathbb{Q})$ correspond to the unit group U_K and Ш to the class group $\mathcal{C}_K$; these are related by a four-term exact sequence

$$0 \longrightarrow U_K \longrightarrow K^\times \xrightarrow{f} \mathcal{I}_K \longrightarrow \mathcal{C}_K \longrightarrow 0,$$

where $\mathcal{I}_K$ is the group of fractional ideals of K and f the map associating to each number in K the ideal it generates.

To look for an exact sequence as in (13) we must first choose a good definition of Ш. The cohomological definition of Ш can be translated in a well-known way into a definition closer to the Diophantine properties of $E/\mathbb{Q}$: Ш is the set of isomorphism classes of curves C of genus 1, defined over $\mathbb{Q}$, having a point over $\mathbb{Q}_v$ for every place v of $\mathbb{Q}$, and equipped with an action of E making them into principal homogeneous spaces over E (or equivalently, with an isomorphism defined over $\mathbb{Q}$ between the Jacobian of C and E). The isomorphisms defining the classes are required to be defined over $\mathbb{Q}$ and compatible with the E-actions (or with the isomorphisms $\mathrm{Jac(C)} \xrightarrow{\sim} E$), but not with the choices of $\mathbb{Q}_v$-rational points. The neutral element is the class of E and the sum of classes $[C']$ and $[C'']$ is the class of the curve $C = (C' \times C'')/E$, where E acts on $C' \times C''$ by $(c', c'') + e = (c' + e, c'' - e)$ and on C by $[c', c''] + e = [c' + e, c'']$.

Now let $\mathcal{E} = E(\overline{\mathbb{Q}})$, the set of $\overline{\mathbb{Q}}$-rational points on E, and $\mathcal{S}$ be the set of equivalence classes of pairs (C, P) with C as above and P a point of $C(\overline{\mathbb{Q}})$, the equivalence being given by $[C, P] = [C', P']$ if there is an E-equivariant isomorphism $C \to C'$ defined over $\mathbb{Q}$ and mapping P to P'. Both sets have natural group laws and there are obvious homomorphisms $E(\mathbb{Q}) \hookrightarrow \mathcal{E}$, $\mathcal{S} \twoheadrightarrow$ Ш and $\mathcal{E} \to \mathcal{S}$, the last sending $P \in E(\overline{\mathbb{Q}})$ to the class of (E, P). It is not hard to see that with these definitions the sequence (13) is exact.

This gives us one way to realize (13), but it is not completely satisfactory because the introduction of $\overline{\mathbb{Q}}$, suggested by the original definition of Ш in terms of Galois cohomology, takes us further away from the Diophantine properties of E. However, we could have taken instead of $\overline{\mathbb{Q}}$ any ring R satisfying

(i) R has a unit element 1 and the map $\mathbb{Z} \to \mathbb{Z} \cdot 1_R \subset R$ is injective, and

(ii) every curve C as in the definition of Ш has an R-rational point.

Then taking $\mathcal{E} = E(R)$ and $\mathcal{S}$ to be the set of isomorphism classes of (C, P) with $P \in C(R)$, we get a four-term exact sequnce (13) as before; moreover, the group S_m defined by (14) (with $B = \mathcal{E}$, $C = \mathcal{S}$) is independent up to canonical isomorphism of the choice of R and is isomorphic to Sel_m. Possible choices for R (besides $\overline{\mathbb{Q}}$) would be

—the real numbers $\mathbb{R}$,

—the complex numbers $\mathbb{C}$,

—the p-adic integers $\mathbb{Z}_p$ (or equivalently, the p-adic numbers $\mathbb{Q}_p$) for any prime p,

—the group $\hat{\mathbb{Z}} = \varprojlim(\mathbb{Z}/n\mathbb{Z}) = \prod_p \mathbb{Z}_p$ (or equivalently, the finite adeles $\mathbb{A}^f = \hat{\mathbb{Z}} \otimes \mathbb{Q}$),

—any product of these, e.g. the full adele ring $\mathbb{A} = \mathbb{A}^f \times \mathbb{R}$.

Choosing R to be $\hat{\mathbb{Z}}$ or $\mathbb{A}$ seems to be the best choice for our purpose of relating Ш to the local and global Diophantine properties of E.

5. Summary

We can summarize the contents of the paper as folows: Let

$$Z(E,s) = \sum_{\substack{(x,y,z)\in\mathbb{Z}^3/\{\pm1\} \\ \gcd(x,y,z)=1 \\ f(x,y,z)=0}} \frac{1}{(x^2+y^2+z^2)^{s/2}} \qquad (\mathrm{Re}(s) > 0),$$

$$D(E,s) = \sum_{n=1}^{\infty} \Bigg(\sum_{\substack{(x,y,z)\in(\mathbb{Z}/n\mathbb{Z})^3/(\mathbb{Z}/n\mathbb{Z})^\times \\ \gcd(x,y,z,n)=1 \\ f(x,y,z)\equiv 0 \pmod{n}}} 1 \Bigg) \frac{1}{n^s} \qquad (\mathrm{Re}(s) > 2),$$

and define numbers $\kappa \in \mathbb{R}_{>0}$ and $r \in \mathbb{Z}_{\geq 0}$ by

$$\Omega^{-1} Z(E,s)^2 \sim \kappa(\pi/s)^r \qquad (s \to 0).$$

Then the Birch–Swinnerton-Dyer conjecture is equivalent to the statement that $D(E,s)$ continues meromorphically to $s = 1$ and satisfies

$$|\text{Ш}|\, D(E,s) \sim -\frac{\kappa}{(s-1)^r} \qquad (s \to 1),$$

and the number $|\text{Ш}|$ occurring in this formula has an interpretation in terms of points on E and on principal homogeneous spaces over E over the ring $\varprojlim(\mathbb{Z}/n\mathbb{Z})$.

POLYLOGARITHMS, DEDEKIND ZETA FUNCTIONS, AND THE ALGEBRAIC K-THEORY OF FIELDS

Don Zagier

Table of Contents

§1. Introduction. The Dedekind zeta function $\zeta_F(s)$ of an algebraic number field F is the most important invariant of F. Its Euler product tells how the unramified primes of $\mathbb{Q}$ split in F. Information about the ramified primes and about the behavior at infinity is contained in three integers Δ, n_+ and n_-: the first is the absolute value of the discriminant of F and is a positive integer whose prime divisors are precisely the primes ramifying in F, and the other two ($= r_1 + r_2$ and r_2 in the standard notation) give the dimensions of the $(+1)$- and (-1)-eigenspaces of complex conjugation on $F \otimes_{\mathbb{Q}} \mathbb{R}$ $(= \mathbb{R}^{r_1} \times \mathbb{C}^{r_2})$. These invariants are in turn determined by $\zeta_F(s)$

via its functional equation

$$\zeta_F^*(s) := \Delta^{s/2}\left(\pi^{-s/2}\Gamma\left(\frac{s}{2}\right)\right)^{n_+}\left(\pi^{-s/2}\Gamma\left(\frac{s+1}{2}\right)\right)^{n_-}\zeta_F(s) = \zeta_F^*(1-s). \quad (1)$$

In the case $F = \mathbb{Q}$, the function $\zeta_F(s) = \zeta(s) = \sum_{n=1}^{\infty} n^{-s}$ was first studied by Euler, who showed that its values at negative odd integers are rational and that its values at positive even integers are rational multiples of powers of π: $\zeta(2) = \pi^2/6$, $\zeta(4) = \pi^4/90$, etc. More generally, if F is totally real ($n_- = 0$), then the Klingen-Siegel theorem says that the values of $\zeta_F(s)$ at negative odd integers are again rational numbers, or equivalently (by (1)), that the values at positive values of s are rational multiples of $\pi^{n+s}/\sqrt{\Delta}$. If F is not totally real, then one does not expect rationality results of this form. However, the Dirichlet class number formula says that the value of $\zeta_F(s)/\zeta_{\mathbb{Q}}(s)$ at $s = 1$ equals $\pi^{n_-}/\sqrt{\Delta}$ times a rational linear combination of $(n_+ - 1)$-fold products of logarithms of numbers in F (we cannot ask directly for the value of $\zeta_F(s)$ at $s = 1$ since there is a simple pole there), while the result proved in [17] by 3-dimensional hyperbolic geometry is that the value of $\zeta_F(s)$ at $s = 2$ equals $\pi^{2n_+}/\sqrt{\Delta}$ times a rational linear combination of n_--fold products of *dilogarithms* of algebraic numbers (in F or an extension of F of degree at most 4). These two cases suggest the conjecture that *the value of $\zeta_F(m)$ for an arbitrary number field F and integer $m > 1$ can be expressed in terms of the polylogarithm function $Li_k(x)$*, defined by

$$Li_k(x) = \sum_{n=1}^{\infty} \frac{x^n}{n^k} \qquad (k \in \mathbb{N}, \quad |x| \leq 1) \qquad (2)$$

(and by analytic continuation if $|x| > 1$). This conjecture was stated and a few numerical examples supporting it were given in [19], but a precise formulation was not given. It is the purpose of this paper to provide such a formulation and to describe the evidence supporting it. The conjecture will take the form that $\zeta_F(m)$ is equal up to a rational factor to $\pi^{mn_\pm}/\sqrt{\Delta}$ times the determinant of an $n_\mp \times n_\mp$-matrix whose entries are $\mathbb{Z}$-linear combinations of numbers $P_m(x)$ with $x \in F$, where the sign $\pm$ is determined by $(-1)^m = \pm 1$ and where $P_m : \mathbb{C} \to \mathbb{R}$ is a certain combination of the polylogarithm functions $Li_k(x)$, $k \leq m$ (the detailed formulation will say precisely which matrices one can take). The statement can be generalized to include values of Artin L-functions at $s = m$ and in this form reduces to a weakened form of Stark's conjectures for $m = 1$.

Finally, we must bring in the third ingredient of the title, algebraic K-theory. For each integer $i \geq 1$ we have the groups $K_i(F)$ and $K_i(\mathcal{O}_F)$, the

ith algebraic K-groups of F and of its ring of integers. For $i = 1$ we have $K_1(F) = F^\times$ and $K_1(\mathcal{O}_F) = \mathcal{O}_F^\times \approx \mathbb{Z}^{n_+-1}$ (here and in the sequel, we will use $A \approx B$ for two abelian groups A and B to mean that there is a map $A \to B$ with finite kernel and cokernel). For $i > 1$, it was shown by Borel [4] that

$$K_i(F) \approx K_i(\mathcal{O}_F) \approx \begin{cases} 0 & \text{if } i \text{ is even,} \\ \mathbb{Z}^{n_\mp} & \text{if } i = 2m-1 \text{ is odd, } (-1)^m = \pm 1 \end{cases} \tag{3}$$

and that in the latter case there is a natural map (the regulator mapping) of $K_i(F)$ into $\mathbb{R}^{n_\mp}$ whose image is a lattice with covolume a rational multiple of $\zeta_F(m)/\pi^{mn_\pm}\sqrt{\Delta}$. (Note that, by (1), the rank $\rho = n_\mp$ of $K_{2m-1}(F)$ equals the order of vanishing of $\zeta_F(s)$ at $s = 1-m$ and that $\zeta_F(m)/\pi^{mn_\pm}\sqrt{\Delta}$ is a rational multiple of $\lim_{s\to 1-m}\zeta_F(s)/(s-1+m)^\rho$.) From this point of view, our conjecture says that the algebraic K-group $K_{2m-1}(F)$ is a subquotient of the free abelian group on $F^\times$, the regulator mapping being given by the polylogarithm function P_m evaluated on the different embeddings of F into $\mathbb{C}$. (There are $n_\mp$ essentially different such embeddings as far as the function P_m is concerned, since $P_m(\overline{x}) = \mp P_m(x)$ for $x \in \mathbb{C}$.) This will be discussed in more detail in the body of the paper.

The paper contains a rather large number of numerical examples (mostly for $m = 3$) motivating and substantiating the various forms of the main conjecture. The reader who is willing to take the motivation on faith can skip straight to §§7–8, where the general formulation of the conjecture is given, and to the further examples and discussion in the following two sections.

In the final section of the paper we will also describe briefly the progress which has been made on our conjecture since the Texel conference: Deligne and Beilinson have reformulated and refined it using ideas from motivic cohomology, Goncharov has proved (most of) the case $m = 3$ and in particular has shown that $\zeta_F(3)$ for any number field F can be expressed in terms of the modified trilogarithm function P_3, and Beilinson has constructed a map from the subquotient of the free abelian group on $F^\times$ specified in the conjecture to the group $K_{2m-1}(F)$ such that the polylogarithm corresponds to the Borel regulator.

I would like to express my thanks to Christophe Soulé, Spencer Bloch, Herbert Gangl and Alexander Goncharov for useful discussions about the material in this paper, and especially to Pierre Deligne and Sasha Beilinson for trying to explain to me the mysteries of algebraic K-theory and the philosophy of motivic cohomology and for helping me formulate a reasonably precise version of the conjectural relationship between zeta values, polylogarithms, and K-groups.

§2. The dilogarithm, hyperbolic geometry, and the Bloch group. For $m = 2$ the *Bloch-Wigner function* is the modified dilogarithm function defined by

$$\begin{aligned} D(x) &= \Im\big(Li_2(x) + \log|x|\log(1-x)\big) \qquad && (|x| \le 1), \\ D(x) &= -D(1/x) && (|x| \ge 1). \end{aligned}$$

It is a continuous function from the Riemann sphere $\mathbb{P}^1(\mathbb{C})$ to $\mathbb{R}$ and is (real-) analytic except at the points 0, 1 and ∞, where it has the value 0. It satisfies a number of functional equations, most notably the six-fold symmetry property

$$D(x) = D\Big(\frac{x-1}{x}\Big) = D\Big(\frac{1}{1-x}\Big) = -D\Big(\frac{1}{x}\Big) = -D(1-x) = -D\Big(\frac{x}{x-1}\Big) \qquad (x \in \mathbb{P}^1(\mathbb{C})) \tag{4}$$

and the "five-term relation" of Spence and Abel

$$D(x) + D(y) + D\Big(\frac{1-x}{1-xy}\Big) + D(1-xy) + D\Big(\frac{1-y}{1-xy}\Big) = 0 \qquad (x, y \in \mathbb{P}^1(\mathbb{C})) \tag{5}$$

(which include equations (4) by specializing to $y = 0$ or ∞). These can be expressed in a more natural way by thinking of the argument of D as the cross-ratio of four points a, b, c, d on the complex line, i.e. by defining $\tilde{D}(a,b,c,d) = D\big(\frac{a-c}{a-d}\frac{b-d}{b-c}\big)$, in which case they take the form

$$\begin{aligned} &\tilde{D}(a_{\pi(1)}, a_{\pi(2)}, a_{\pi(3)}, a_{\pi(4)}) = \operatorname{sign}(\pi)\, \tilde{D}(a_1, a_2, a_3, a_4) \qquad (\pi \in \mathfrak{S}_4), \\ &\sum_{i \pmod 5} \tilde{D}(a_i, a_{i+1}, a_{i+2}, a_{i+3}) = 0 \qquad (a_i \in \mathbb{P}^1(\mathbb{C}), \quad a_{i+5} = a_i). \end{aligned} \tag{6}$$

This in turn has a geometric interpretation: if we think of $\mathbb{P}^1(\mathbb{C})$ as the boundary of hyperbolic 3-space, then $\tilde{D}(a,b,c,d)$ is the hyperbolic volume of the ideal hyperbolic tetrahedron $\Delta(a,b,c,d)$ with vertices a, b, c and d, and (6) just says that this volume is independent of the numbering of the vertices (except that the orientation changes under odd renumberings) and that the five tetrahedra $\Delta(a_i, a_{i+1}, a_{i+2}, a_{i+3})$ add up algebraically to the zero 3-cycle.

The relationship between $\zeta_F(2)$ and $D(x)$ was proved in [17] using this interpretation of $D(x)$ as a hyperbolic volume. For instance, in the case $n_- = 1$ we showed that

$$\zeta_F(2) \sim \frac{\pi^{2n_+}}{\sqrt{\Delta}} \sum_i D(x_i)$$

(here and from now on, $a \sim b$ means that $a/b \in \mathbb{Q}^\times$) for some numbers $x_i \in \overline{\mathbb{Q}}$ of degree ≤ 4 over F. This is because one can associate to F a hyperbolic 3-manifold M_F whose volume is $\sim \Delta^{\frac{1}{2}} \zeta_F(2)/\pi^{2n_+}$, and then triangulate M_F into ideal tetrahedra whose invariants (= cross-ratios of their vertices, taken with the proper orientation) x_i are at most quartic over F. If $n_- > 1$, then a similar argument shows that

$$\zeta_F(2) \sim \frac{\pi^{2n_+}}{\sqrt{\Delta}} \sum_i \prod_\sigma D(\sigma(x_i)), \tag{7}$$

where σ ranges over (extensions of) the n_- non-real embeddings of F into $\mathbb{C}$. (Note that $D(\overline{x}) = -D(x)$, so $D(x) = 0$ for x real.)

Now the invariants x_i of an ideal triangulation of a complete hyperbolic 3-manifold are not arbitrary: they must satisfy the relation

$$\sum_i [x_i] \wedge [1 - x_i] = 0 \qquad \in \Lambda^2(\mathbb{C}^\times), \tag{8}$$

where $\Lambda^2(\mathbb{C}^\times)$ denotes the second exterior power of $\mathbb{C}^\times$, thought of as a module over $\mathbb{Z}$. (This relation was noticed several years ago by Thurston and is mentioned in [9]. It also follows from Corollary 2.4 of [13], which describes the combinatorics of triangulations of hyperbolic 3-manifolds.) This suggests the following definition. For any field F, denote by $\mathcal{F}_F$ the free abelian group on $F^\times$ (i.e., the set of finite linear combinations $\sum_i n_i[x_i]$ with $n_i \in \mathbb{Z}$, $x_i \in F^\times$; we will also identify $\mathcal{F}_F$ with the quotient of the free abelian group on $\mathbb{P}^1(F)$ by the subgroup $\langle [0], [\infty] \rangle$), and define

$$\mathcal{A}(F) = \{ \sum_i n_i[x_i] \in \mathcal{F}_F \mid \sum_{x_i \ne 1} n_i[x_i] \wedge [1 - x_i] = 0 \text{ in } \Lambda^2(F^\times) \otimes_{\mathbb{Z}} \mathbb{Q} \}. \tag{9}$$

In other words, $\mathcal{A}(F)$ is the kernel of the map $\beta : \mathcal{F}_F \to \Lambda^2(F^\times) \otimes_{\mathbb{Z}} \mathbb{Q}$ defined by $\beta([x]) = [x] \wedge [1 - x]$ for $x \ne 1$, $\beta([1]) = 0$. It is easily checked that the expression

$$S_{xy} = [x] + [y] + [\frac{1-x}{1-xy}] + [1 - xy] + [\frac{1-y}{1-xy}], \tag{10}$$

corresponding to the 5-term relation (5), belongs to $\mathcal{A}(F)$ whenever it makes sense (i.e. for all x, $y \in \mathbb{P}^1(F)$ except $(x, y) = (0, \infty)$, $(\infty, 0)$, or $(1, 1)$; if x or y or xy equals 0, 1 or ∞, we use our convention $[0] = [\infty] = 0$). We set

$$\mathcal{C}(F) = \langle S_{xy} \rangle_{x,y \in \mathbb{P}^1(F),\, (x,y) \ne (0,\infty), (\infty,0), (1,1)} \tag{11}$$

and define the *Bloch group* $\mathcal{B}(F)$ as the quotient $\mathcal{A}(F)/\mathcal{C}(F)$. Taking $(x,y) = (0,0)$, $(1,\infty)$ $(x,0)$ and (x,∞), we find that the relations $[1] = 0$ and $[x] + [1-x] = [x] + [1/x] = 0$ $(\forall x \in \mathbb{P}^1(F))$ hold in $\mathcal{B}(F)$. The functional equation (5) implies that the function D can be defined on $\mathcal{B}(F)$ for any subfield $F \subseteq \mathbb{C}$ by $\sum n_i[x_i] \mapsto \sum n_i D(x_i)$. Therefore if F is a number field, $[F:\mathbb{Q}] = n_+ + n_-$ as usual, then there is a map $D_F : \mathcal{B}(F) \to \mathbb{R}^{n_-}$ defined by $\sum n_i[x_i] \mapsto \left(\sum n_i D(\sigma(x_i))\right)_\sigma$, where σ ranges over the non-real embeddings $F \hookrightarrow \mathbb{C}$ (taking one of each complex conjugate pair).

The algebraic theorem that explains—and strengthens—equation (7) is now the following: for each number field F, there is a map $\mathcal{B}(F) \to K_3(F)$ with finite kernel and cokernel whose composition with the Borel regulator mapping $K_3(F) \to \mathbb{R}^{n_-}$ is D_F. This result is due to Bloch and Suslin; cf. Suslin's ICM survey talk [15]. Together with Borel's theorem, it implies that $\mathcal{B}(F)$ has rank n_- and that (7) is true with arguments in F itself (rather than quartic over F) and with the sum of n_--fold products replaced by the $n_- \times n_-$ determinant $\det\left(D(\sigma(\xi_\nu))\right)_{\nu,\sigma}$, where $(\xi_\nu)_{1 \le \nu \le n_-}$ are linearly independent elements of $\mathcal{B}(F)$ and σ ranges over the non-conjugate non-real embeddings $F \hookrightarrow \mathbb{C}$.

A more leisurely expository account of the material in this section is given in [18].

§3. The trilogarithm and $\zeta_F(3)$. We would like to get an analogous statement for other special values of $\zeta_F(s)$, involving higher polylogarithms and some generalization of the Bloch groups. Specifically, we would like to define for each number field F groups $\mathcal{A}_m(F)$ and $\mathcal{C}_m(F)$ such that the quotient $\mathcal{B}_m(F) = \mathcal{A}_m(F)/\mathcal{C}_m(F)$ is a model for the K-group $K_{2m-1}(F)$ and the regulator map on $K_{2m-1}(F)$ corresponds to the mth polylogarithm function on $\mathcal{B}_m(F)$. By analogy with the case $m = 2$, $\mathcal{A}_m(F)$ should be defined as the subgroup of elements of $\mathcal{F}_F$ satisfying some algebraic relation, while $\mathcal{C}_m(F)$ should come from the functional equations of the polylogarithm function. To guess the right form, we look at numerical examples, starting with the case $m = 3$, to which the next several sections are devoted. As mentioned in the introduction, the reader not interested in the numerical motivation of the conjecture can skip to §6.

We first need an analogue of the Bloch-Wigner function $D(x)$. A modified polylogarithm function which is one-valued for all $x \neq 0, \infty$ was defined implicitly by Ramakrishnan [14] and explicitly in [18] (and, in a modified form which we will discuss in §7, in [16]), and discussed in some detail in [19]. It is given for $|x| \le 1$ by

$$D_m(x) = \Re_m\left(\sum_{r=0}^{m-1} \frac{(-1)^r}{r!} \log^r|x|\, Li_{m-r}(x) - \frac{(-1)^m}{2m!} \log^m|x|\right), \qquad (12)$$

where $\Re_m$ denotes $\Re$ or $\Im$ depending whether m is odd or even, and for $|x| \geq 1$ by the functional equation

$$D_m\Big(\frac{1}{x}\Big) = (-1)^{m-1} D_m(x) \qquad (x \in \mathbb{C}). \tag{13}$$

The function $D_m : \mathbb{P}^1(\mathbb{C}) \setminus \{0, \infty\} \to \mathbb{R}$ is real-analytic except at $x = 1$ and satisfies the functional equation

$$D_m(\overline{x}) = (-1)^{m-1} D_m(x) \qquad (x \in \mathbb{C}). \tag{14}$$

Thus, unlike the case $m = 2$ where D_m vanished on $\mathbb{R}$ and hence only the non-real embeddings of F played a role, D_m for m odd is non-trivial on $\mathbb{R}$ and we can start by looking at totally real fields and $m = 3$, the function $D_3(x)$ being given for x real by

$$D_3(x) = Li_3(x) - \log|x|\, Li_2(x) - \frac{1}{2}\log^2|x|\log(1-x) + \frac{1}{12}\log^3|x| \tag{15}$$

if $-1 \leq x \leq 1$ and by (13) otherwise. Since 3 is odd, the Klingen-Siegel theorem does not apply and we do not a priori know anything about $\zeta_F(3)$ even in the totally real case. For $F = \mathbb{Q}$, of course, the relation $\zeta_F(3) = \zeta(3) = Li_3(1) = D_3(1)$ is trivial, so we begin by looking at real quadratic fields $F = \mathbb{Q}(\sqrt{D})$. Here $n_+ = 2$, $n_- = 0$, so the group $\mathcal{B}_3(F)$ should be 2-dimensional. This means that we expect $\sqrt{D}\,\zeta_F(3)$ to be expressible as the determinant of a 2×2 matrix whose entries are combinations of trilogarithms. This is unsuitable for numerical work, since it is hard to guess the entries of a matrix of size bigger than 1×1 knowing only its determinant. However, we know that F is abelian over $\mathbb{Q}$ and that $\zeta_F(3)$ splits as $\zeta(3)$ times $L(3, \chi_D)$ for a certain Dirichlet series $L(s, \chi_D) = \sum \chi_D(n) n^{-s}$, and it is reasonable to expect that the Galois group of F over $\mathbb{Q}$ acts on $\mathcal{B}_3(F)$ and splits it into two one-dimensional spaces $\mathcal{B}_3(F)^+$ and $\mathcal{B}_3(F)^-$ on which the determinant of the regulator mapping is equal (or rationally proportional) to $\zeta(3)$ and $\sqrt{D}\,L(3, \chi_D)$, respectively. Thus we expect that $\sqrt{D}\,L(3, \chi_D)$ can be written as a linear combination of numbers $D_3(x) - D_3(x')$, where x' denotes the conjugate of $x \in F$ over $\mathbb{Q}$ (here we are thinking of F as embedded once and for all in $\mathbb{R}$, say by $\sqrt{D} > 0$).

To find such a formula empirically, we take a collection of "simple" numbers $x \in F$, where "simple" means that x and $1 - x$ factor into prime ideals of small norm, and search for numerical relations among $\sqrt{D}\,L(3, \chi_D)$ and the numbers $D_3(x) - D_3(x')$ using the L^3 (Lenstra-Lenstra-Lovasz) algorithm. We recall that this is an algorithm which finds short vectors in a finite-dimensional lattice; to use it for the problem of finding linear relations over $\mathbb{Z}$ among a collection of given real numbers $\alpha_1, \ldots, \alpha_r$, one applies it to the lattice $\mathbb{Z}^r$ with the length form

$$||(n_1, \ldots, n_r)||^2 = n_1^2 + \cdots + n_r^2 + M(n_1\alpha_1 + \cdots n_r\alpha_r)^2$$

with a very large constant M, in which case any short vector will correspond to an r-tuple of reasonably sized integers n_j with $\sum n_j\alpha_j$ equal to 0 or extremely small. Applying the algorithm to the fields $F = \mathbb{Q}(\sqrt{5})$ $(D = 5)$ and $F = \mathbb{Q}(\sqrt{2})$ $(D = 8)$ and to reasonably chosen collections of numbers $x \in F$, we found the relations

$$\frac{75\sqrt{5}}{32} L(3, \chi_5) \stackrel{?}{=} D_3(2-\sqrt{5}) - D_3(2+\sqrt{5}) + 3\big[D_3(\frac{1+\sqrt{5}}{2}) - D_3(\frac{1-\sqrt{5}}{2})\big],$$

$$\frac{20\sqrt{2}}{3} L(3, \chi_8) \stackrel{?}{=} D_3(4 + 2\sqrt{2}) - D_3(4 - 2\sqrt{2}) + 9\big[D_3(\sqrt{2}) - D_3(-\sqrt{2})\big]$$
$$- 9\big[D_3(2 + \sqrt{2}) - D_3(2 - \sqrt{2})\big] - 6\big[D_3(1 + \sqrt{2}) - D_3(1 - \sqrt{2})\big].$$

In both cases, we performed the search using a larger collection of numbers $x \in F$ than the ones which actually appear, and the relation given was the only one found by the computer. In both cases, too, the search was carried out "honestly" in the sense that the L^3 algorithm was first applied with a reasonably large value of M (about 10^{40}) and using a reasonably high precision for the numbers $L(3, \chi_D)$ and $D_3(x)$ (about 20 digits), and the relation found by the computer then checked independently by recalculating all of the numbers involved to a much higher precision (about 50 digits); thus despite the question marks over the equal sign, the above equalities carry a much higher level of certainty than many theorems.

As the next step, we looked for an algebraic relation analogous to (8) which is satisfied by the two linear combinations $[2-\sqrt{5}] - \cdots - 3[\frac{1+\sqrt{5}}{2}]$ and $[4 + \sqrt{2}] - \cdots + 6[1 - \sqrt{2}]$ occurring in the above relations and which is *not* satisfied by any other linear combination of the $[x] - [x']$ over which the search was performed. As a reasonable guess, we assumed that this relation would take place in $\bigotimes^3 F^\times$ and would involve only the numbers x and $1 - x$. After some effort, we found that there was a unique relation of this form which worked, namely

$$\sum_i n_i\, [x_i] \otimes [x_i] \otimes [\frac{x_i}{(1 - x_i)^2}] \equiv 0, \qquad (16)$$

where "$\equiv$" means "equal modulo torsion." This suggests the definition

$$\mathcal{A}_3^{(1)}(F) = \ker\big(\beta_3^{(1)} : \mathcal{F}_F \to (F^\times \otimes F^\times \otimes F^\times) \otimes \mathbb{Q}\big),$$
$$\beta_3^{(1)} : [x] \mapsto [x] \otimes [x] \otimes [\frac{x}{(1 - x)^2}], \qquad [1] \mapsto 0.$$

(We have written the superscript "(1)" because this will not be our final definition of $\mathcal{A}_3(F)$.) As before, we can extend D_3 from $\mathbb{C}^\times$ to $\mathcal{A}_3^{(1)}(\mathbb{C})$ by $D_3(\sum n_i[x_i]) = \sum n_i D_3[x_i]$; then for each embedding σ of a number field F into $\mathbb{C}$ we get a map D_3^σ from $\mathcal{A}_3^{(1)}(F)$ to $\mathbb{R}$ by composing $\mathcal{A}_3^{(1)}(F) \to \mathcal{A}_3^{(1)}(\mathbb{C})$ with D_3. We can now formulate:

CONJECTURE (FIRST VERSION). *Let F be a totally real field of degree n and discriminant D. Then the image of the map $\prod_{\sigma: F \hookrightarrow \mathbb{R}} D_3^\sigma : \mathcal{A}_3^{(1)}(F) \to \mathbb{R}^n$ is a lattice (= cocompact $\mathbb{Z}$-submodule) whose covolume is a rational multiple of $\sqrt{D}\,\zeta_F(3)$. In particular, $\sqrt{D}\,\zeta_F(3)$ is a rational linear combination of n-fold products $D_3(x^{(1)}) \cdots D_3(x^{(n)})$, $x \in F$.*

Of course, what we really want is that there should be a canonically defined subgroup $\mathcal{C}_3^{(1)}(F)$ of $\mathcal{A}_3^{(1)}(F)$ and an isomorphism (up to torsion) from the K-group $K_5(F)$ to the quotient group $\mathcal{A}_3^{(1)}(F)/\mathcal{C}_3^{(1)}(F)$ whose composite with $\prod D_3^\sigma$ is the Borel regulator map; however, we will wait with the formal statement of this until we give our final definition of $\mathcal{A}_3(F)$.

One nice aspect of the conjecture is that it makes non-trivial predictions even in the case $F = \mathbb{Q}$, which because of the identity $\zeta(3) = D_3(1)$ had at first sight appeared to be of no interest. For instance, let

$$X = X_{\{2,3\}} = \left\{-1,\ \frac{1}{2},\ -\frac{1}{2},\ \frac{1}{3},\ \frac{2}{3},\ -\frac{1}{3},\ \frac{1}{4},\ \frac{3}{4},\ -\frac{1}{8},\ \frac{1}{9},\ \frac{8}{9}\right\}$$

be the set of all rational numbers x with $|x| \le 1$ (since x and $1/x$ are equivalent from the point of view of D_3) such that x and $1-x$ contain only the prime numbers 2 and 3 in their numerator and denominator. The point of the restriction on the prime factors of x and $1-x$ is that it makes it easier to satisfy the conditions defining $\mathcal{A}_3^{(1)}(\mathbb{Q})$: specifically, each element $\beta_3^{(1)}([x]) = [x] \otimes [x] \otimes [x/(1-x)^2]$ ($x \in X$) is (up to 2-torsion) a linear combination of the six elements

$$\begin{gathered}
[2]\otimes[2]\otimes[2],\ \ ([2]\otimes[3]+[3]\otimes[2])\otimes[2],\ \ [3]\otimes[3]\otimes[2],\\
[2]\otimes[2]\otimes[3],\ \ ([2]\otimes[3]+[3]\otimes[2])\otimes[3],\ \ [3]\otimes[3]\otimes[3],
\end{gathered}$$

so any seven of the $[x]$ are linearly dependent modulo $\mathcal{A}_3^{(1)}(\mathbb{Q})$. The space $\mathcal{F}_X = \{\sum_{x \in X} n_x[x]\}$ must therefore have an intersection with $\mathcal{A}_3^{(1)}(\mathbb{Q})$ of rank at least 5. A brief computation shows that the rank is in fact exactly

5, a basis being given by

$$\xi_1 = [-1], \quad \xi_2 = 4[\tfrac{1}{2}] + 4[-\tfrac{1}{2}] - [\tfrac{1}{4}],$$
$$\xi_3 = 4[\tfrac{1}{3}] + 4[-\tfrac{1}{3}] - [\tfrac{1}{9}], \quad \xi_4 = 9[\tfrac{1}{2}] + 18[-\tfrac{1}{2}] - [-\tfrac{1}{8}], \tag{17}$$
$$\xi_5 = 3[\tfrac{1}{2}] + 2[\tfrac{1}{3}] - 6[\tfrac{2}{3}] - [-\tfrac{1}{3}] - 3[\tfrac{3}{4}] + [\tfrac{8}{9}].$$

Calculating independently with the L^3 algorithm to search for numerical linear relations between the values $D_3(x)$ ($x \in X$) and $\zeta(3)$, we found exactly the same 5-dimensional space of relations, the numerical values of the $D_3(\xi_j)$ being

j	1	2	3	4	5
$D_3(\xi_j)$	$-\frac{3}{4}\zeta(3)$	0	0	$-\frac{35}{8}\zeta(3)$	$-\frac{67}{24}\zeta(3)$

(18)

Finally, we can get other examples of the conjecture by returning to the two quadratic fields we started with. In $\mathbb{Q}(\sqrt{5})$ the images of the elements $2-\sqrt{5}$, $2+\sqrt{5}$, $\frac{1}{2}(1+\sqrt{5})$ and $\frac{1}{2}(1-\sqrt{5})$ under $\beta_3^{(1)}$ are (up to torsion)

$$\begin{aligned}
[\phi^{-3}] \otimes [\phi^{-3}] \otimes [\phi^{-1}/4] &= -18[\phi] \otimes [\phi] \otimes [2] - 9[\phi] \otimes [\phi] \otimes [\phi], \\
[\phi^{3}] \otimes [\phi^{3}] \otimes [\phi/4] &= -18[\phi] \otimes [\phi] \otimes [2] + 9[\phi] \otimes [\phi] \otimes [\phi], \\
[\phi] \otimes [\phi] \otimes [\phi^{3}] &= 3[\phi] \otimes [\phi] \otimes [\phi], \quad \text{and} \\
[\phi^{-1}] \otimes [\phi^{-1}] \otimes [\phi^{-3}] &= -\,3[\phi] \otimes [\phi] \otimes [\phi],
\end{aligned}$$

respectively, where $\phi = \frac{1}{2}(1+\sqrt{5})$, so the two elements

$$[2-\sqrt{5}] - [2+\sqrt{5}] + 3([\frac{1+\sqrt{5}}{2}] - [\frac{1-\sqrt{5}}{2}]), \quad [\frac{1+\sqrt{5}}{2}] + [\frac{1-\sqrt{5}}{2}]$$

belong to $\mathcal{A}_3^{(1)}(F)$. The first is Galois anti-invariant and gives the relation $D_3(\xi) \sim \sqrt{5}\,L(3,\chi_5)$ we already saw, while the second is Galois invariant and leads to the relation

$$D_3(\frac{1+\sqrt{5}}{2}) + D_3(\frac{1-\sqrt{5}}{2}) = \frac{1}{5}\zeta(3).$$

Similarly, for $\mathbb{Q}(\sqrt{2})$ we find that the 8 numbers $x = 4+2\sqrt{2}, \ldots$ occurring in our original numerical relation involve (up to sign) only two numbers $1+\sqrt{2}$ and $\sqrt{2}$ in the factorizations of x and $1-x$, so just as with $X_{\{2,3\}}$ there are 6 conditions which a linear combination of them must satisfy to belong to ker $\beta_3^{(1)}$, and consequently at least two linearly independent combinations belonging to the kernel. Doing the calculation, we find that there are exactly two, the Galois anti-invariant combination $[4+2\sqrt{2}]-[4-2\sqrt{2}]+\cdots$ we already exhibited and the Galois invariant combination

$$\begin{aligned}\xi &= [4+2\sqrt{2}] + [4-2\sqrt{2}] - 9([\sqrt{2}] + [-\sqrt{2}]) \\ &\quad - 9([2+\sqrt{2}] + [2-\sqrt{2}]) - 3([1+\sqrt{2}] + [1-\sqrt{2}]),\end{aligned}$$

for which $D_3(\xi)$ should be rationally proportional to $\zeta(3)$ and in fact equals $-\frac{125}{8}\zeta(3)$. Thus for both our quadratic fields F we have given elements in $\mathcal{A}_3^{(1)}(F)$ whose images under $\prod D_3^\sigma : \mathcal{A}_3^{(1)}(F) \to \mathbb{R}^2$ are rational multiples of $(\zeta(3), \zeta(3))$ and of $(\sqrt{D}\,L(3,\chi_D), -\sqrt{D}\,L(3,\chi_D))$ and hence which span a lattice with covolume $\sim \sqrt{D}\,\zeta_F(3)$, in accordance with the conjecture.

§4. **The trilogarithm (continued).** The examples we have just given show that the conjecture formulated in §3 is reasonable for totally real fields. However, when we pass to general number fields, then the group $\mathcal{A}_3^{(1)}(F)$ is too big and we must impose a further condition on elements of $\mathcal{F}_F$ to get elements in K-theory. This phenomenon, which is immediately seen in the numerical examples, was predicted by P. Deligne. To see what happens, consider an element $\xi = \sum n_i[x_i] \in \mathcal{F}_F$ and suppose that ξ belongs to ker $\beta_3^{(1)}$, i.e., that equation (16) holds. Let $v : F^\times \to \mathbb{Z}$ be any homomorphism (for instance, the valuation at a prime p in case $F = \mathbb{Q}$). Then applying $v \otimes \mathrm{Id} \otimes \mathrm{Id}$ to (16), we find

$$\sum_i n_i\, v(x_i)\, [x_i] \otimes [\frac{x_i}{(1-x_i)^2}] \equiv 0.$$

Applying to this the projection $F^\times \otimes F^\times \to \Lambda^2(F^\times)$, we deduce that

$$\sum_i n_i\, v(x_i)\, [x_i] \wedge [1-x_i] \equiv 0,$$

i.e., that $\sum n_i v(x_i)[x_i]$ belongs to the group $\mathcal{A}(F)$ defined in §2. If π denotes the canonical projection from $\mathcal{A}(F)$ to $\mathcal{B}(F) = \mathcal{A}(F)/\mathcal{C}(F) \approx K_3(F)$, then the extra condition we must impose is

$$\pi\Big(\sum_i n_i\, v(x_i)\, [x_i]\Big) \equiv 0. \tag{19}$$

(In the case of totally real fields, this extra condition was not necessary because $\mathcal{B}(F) \otimes \mathbb{Q} = \{0\}$.) Then we have:

CONJECTURE (SECOND VERSION). *Let F be an arbitrary number field of degree $n = n_+ + n_-$, Δ the absolute value of the discriminant of F. Define $\mathcal{A}_3^{(2)}(F)$ as the set of $\xi = \sum_i n_i\,[x_i] \in \mathcal{A}_3^{(1)}(F)$ such that (19) holds for all $v \in Hom(F^\times, \mathbb{Z})$. Then the image of the map $\prod_{\sigma: F \hookrightarrow \mathbb{R}} D_3^\sigma : \mathcal{A}_3^{(2)}(F) \to \mathbb{R}^{n_+}$, where the product is taken over all real embeddings and one of each pair of complex conjugate embeddings of F into $\mathbb{C}$, is a lattice whose covolume is a rational multiple of $\sqrt{\Delta}\,\zeta_F(3)/\pi^{3n_-}$. In particular, $\sqrt{\Delta}\,\zeta_F(3)/\pi^{3n_-}$ is a rational linear combination of n_+-fold products $\prod_\sigma D_3(\sigma(x))$, $x \in F$.*

We now give an example of the necessity of the condition (19), taking for F the non-real cubic field of largest discriminant, namely the field $F = \mathbb{Q}(\theta)$ where $\theta^3 - \theta - 1 = 0$. (We cannot take F to be an imaginary quadratic field because in that case the complex conjugation belongs to the Galois group of F over $\mathbb{Q}$, so any ξ can be decomposed as $\frac{1}{2}(\xi - \bar{\xi})$ plus $\frac{1}{2}(\xi + \bar{\xi})$; the first has a trivial image under D_3, so cannot be used for the regulator, and the second is always in the kernel of D_2, so the condition is automatically satisfied.) We have the real embedding given by $\theta^{(1)} = 1.32471\cdots$ and the two complex embeddings $\theta^{(2)} = \dfrac{\theta^{(1)}}{2}\left(\dfrac{\sqrt{-23}}{2\theta^{(1)} + 3} - 1\right)$ and $\theta^{(3)} = \overline{\theta^{(2)}}$. The zeta-function of F splits as $\zeta(s)L(s)$ where the L-series $L(s) = 1 - 2^{-s} - 3^{-s} + 6^{-s} + \cdots$ can be given either as a difference of Epstein zeta functions

$$L(s) = \frac{1}{2} \sum_{\substack{m,n \in \mathbb{Z} \\ (m,n) \neq (0,0)}} \left(\frac{1}{(m^2 + mn + 6n^2)^s} - \frac{1}{(2m^2 + mn + 3n^2)^s} \right)$$

or as the Mellin transform of the modular form $\eta(\tau)\eta(23\tau) = q - q^2 - q^3 + q^6 + \cdots$ of level 23 and weight 1. Using either representation and standard formulas, we can calculate $L(3)$ to a high accuracy:

$$L(3) = 0.84395\ 21532\ 37338\ 53825\ 22215\ 69424\ 77200\ 00563\ \cdots$$

(we have resisted the impulse to give this many digits of any of the numbers occurring up to now!). If $\xi = \sum n_i[x_i]$ is any element of $\mathcal{F}_F$, we denote by $\xi^{(1)}$, $\xi^{(2)}$, $\xi^{(3)}$ the images of ξ in $\mathcal{F}_{\mathbb{C}}$ under the various embeddings $F \hookrightarrow \mathbb{C}$; then for a good ξ we expect the numbers

$$\delta^+(\xi) = \frac{D_3(\xi^{(1)}) + 2D_3(\xi^{(2)})}{\zeta(3)}, \quad \delta^-(\xi) = \frac{D_3(\xi^{(1)}) - D_3(\xi^{(2)})}{\Lambda_3}, \tag{20}$$

where $\Lambda_3 = \dfrac{23^{5/2}}{96\pi^3} L(3)$, to be rational. Here the factor $23^2/96$ has been inserted for convenience, and the justification for the linear combinations of $D_3(\xi^{(1)})$ and $D_3(\xi^{(2)})$ comes from the Artin L-function version of our conjecture which will be mentioned later (§10C); note that we do not need $D_3(\xi^{(3)})$ because it equals $D_3(\xi^{(2)})$. Consider the six numbers $x_1, \dots, x_6 = \theta, -\theta, \theta^2, \theta^3, -\theta^4$ and θ^5. Each of them has the property that x_i and $1-x_i$ are, up to sign, powers of θ, so each $\beta_3^{(1)}(x_i)$ is a multiple of $[\theta]\otimes[\theta]\otimes[\theta]$, the multiples being 9, −5, 16, 9, −96 and −75, respectively. There are therefore 5 linearly independent elements $5[\theta]+9[-\theta]$, ... belonging to $\ker(\beta_3^{(1)})$. However, when we compute the corresponding values of $\delta^{\pm}(\xi)$ numerically, we do not find rational numbers. (And if we do not believe the splitting (20) of the expected 2-dimensional space $\mathcal{B}_3(F)$ into two 1-dimensional $\mathbb{Q}$-vector spaces, we can take two different ξ from our list and compute the 2×2 determinant giving the covolume of the lattice they generate; the answer is not a rational multiple of $23^{1/2}\zeta_F(3)/\pi^3$.) On the other hand, each $[x_i]$ belongs to $\mathcal{A}_2(F)$ (since x_i and $1-x_i$ are powers of θ and $[\theta]\wedge[\theta]=0$), so the value of $D_2(x_i^{(2)})$ for each i is a rational multiple of $23^{3/2}\zeta_F(2)/\pi^4$, the multiples being 1, −2, −2, 2, 1 and −1, respectively. Hence, taking v to be any homomorphism which is non-zero on θ, we see that the values of $v(x_i)\,\pi_2([x_i])$ are proportional to 1, −2, −4, 6, 4 and −5, respectively. Thus (19) imposes an extra condition on ξ and cuts down the space of solutions to dimension 4, a basis being given by $4[\theta]+4[-\theta]-[\theta^2]$, $3[\theta^5]+27[\theta]-2[\theta^3]$, $13[\theta^5]+30[-\theta]+125[\theta]$ and $5[-\theta^4]-130[\theta]-22[\theta^5]$. For these four elements we do indeed find that the values of $(\delta^+(\xi),\delta^-(\xi))$ belong to $\mathbb{Q}^2$ (or even, with our normalization of Λ, to $\mathbb{Z}^2$), the values being $(0,0)$, $(-3,69)$, $(13,259)$ and $(-12,-351)$, respectively.

§5. The trilogarithm (concluded). We now have a conjecture in the case $m=3$ which works for all number fields. However, it turns out that there is still a further improvement which can be made. To see this, we assume for convenience that we are back in the totally real case, so that a "good" element $\xi=\sum n_i[x_i]\in\mathcal{F}_F$ is one satisfying (16). Since $z\mapsto\log|z|$ is a homomorphism from $\mathbb{C}^\times$ to $\mathbb{R}$, it follows that

$$\sum_i n_i \log^2|x_i| \log\Bigl|\frac{x_i}{(1-x_i)^2}\Bigr| = 0.$$

Consequently we can modify the function $D_3(x)$ by adding to it any multiple of the function $\log^2|x|\log|x/(1-x)^2|$ without affecting the validity of our conjecture. The obvious choice is the function

$$\widetilde{D}_3(x) = D_3(x) - \frac{1}{12}\log^2|x|\log\Bigl|\frac{x}{(1-x)^2}\Bigr|, \tag{21}$$

because according to (15) it is given for $-1\le x\le 1$ by

$$\widetilde{D}_3(x) = Li_3(x) - \log|x|\,Li_2(x) - \frac{1}{3}\log^2|x|\,\log(1-x) \tag{22}$$

in which the term $\log^3|x|$ which blows up at $x=0$ has disappeared.

Now a miracle occurs. Replacing D_3 by $\widetilde{D}_3$ does not affect the validity of our previous conjecture, as already mentioned, because the two functions agree on $\mathcal{A}_3^{(1)}(F)$ anyway. But as a result of this change we find that there are suddenly many more combinations $\sum n_i[x_i]$ which work than before. For instance, return to the 11-element set $X = X_{\{2,3\}}$ which was already used as an example in §3. Applying the L^3 algorithm as before to find linear relations over $\mathbb{Z}$, now between the values $\widetilde{D}_3(x)$ $(x \in X)$ and $\zeta(3)$, we find a 9-dimensional space of solutions rather than a 5-dimensional space as before, a basis being given by the 5 old elements (17) together with the 4 new elements

$$\begin{aligned} &\xi_6 = \left[\tfrac{1}{2}\right], \quad \xi_7 = 2\left[\tfrac{1}{3}\right] - 4\left[\tfrac{2}{3}\right] - 2\left[-\tfrac{1}{3}\right] + \left[\tfrac{3}{4}\right], \\ &\xi_8 = 4\left[-\tfrac{1}{2}\right] + \left[-\tfrac{1}{3}\right] + \left[\tfrac{3}{4}\right], \quad \xi_9 = \left[-\tfrac{1}{2}\right] + \left[\tfrac{1}{3}\right] + \left[\tfrac{2}{3}\right], \end{aligned} \tag{23}$$

the corresponding values of $\widetilde{D}_3(\xi_j)$ being equal to $\frac{7}{8}\zeta(3)$, 0, $-\frac{5}{2}\zeta(3)$, and $\zeta(3)$, respectively.

Now, just as before, we look for an algebraic condition which picks out of the 11-dimensional space $\mathcal{F}_X$ precisely the 9-dimensional subspace spanned by $\xi_1, \ldots, \xi_9$, and again we find exactly one which works: the relation (16) must be replaced by the relation

$$\sum_i n_i\,[x_i] \otimes \Big([x_i] \wedge [1-x_i]\Big) \equiv 0, \tag{24}$$

which now takes place in $F^\times \otimes \Lambda^2(F^\times)$ and, as before, is required to hold only up to torsion. In other words, our new candidate for the numerator of the Bloch group in the case $m=3$ (for F totally real) is the group

$$\begin{aligned} \mathcal{A}_3^{(3)}(F) &= \ker\big(\beta_3 : \mathcal{F}_F \to (F^\times \otimes \Lambda^2(F^\times)) \otimes \mathbb{Q}\big), \\ \beta_3 &: [x] \mapsto [x] \otimes \big([x] \wedge [1-x]\big), \quad [1] \mapsto 0, \end{aligned}$$

and we make the

Conjecture (third version). *Let F be a totally real field of degree n and discriminant D. Then the image of the map $\prod_{\sigma : F \hookrightarrow \mathbb{R}} \widetilde{D}_3^\sigma : \mathcal{A}_3^{(3)}(F) \to \mathbb{R}^n$ is a lattice with covolume $\sim \sqrt{D}\,\zeta_F(3)$.*

If F is not totally real, of course, then we must add to the definition of $\mathcal{A}_3^{(3)}(F)$ the same supplementary condition as in §4, namely, that (19) should hold for every linear map $F^\times \to \mathbb{Z}$. In other words, if $\xi = \sum n_i[x_i]$ satisfies (24), then the element $\iota_v(\xi)$, where $\iota_v : \mathcal{F}_F \to \mathcal{F}_F$ is the map defined on generators by $[x] \mapsto v(x)\,[x]$, belongs to $\mathcal{A}(F) = \ker\bigl([x] \mapsto [x] \wedge [1-x]\bigr)$, and we want the image of ξ under the composite mapping

$$\mathcal{A}_3^{(3)}(F) \xrightarrow{\iota_v} \mathcal{A}(F) \xrightarrow{\pi} \mathcal{A}(F)/\mathcal{C}(F) = \mathcal{B}(F) \approx K_3(F)$$

to vanish for all v. This can be expressed more algebraically by saying that the image of ξ under the composite mapping

$$\mathcal{A}_3^{(3)}(F) \xrightarrow{\iota} F^\times \otimes \mathcal{A}(F) \xrightarrow{\mathrm{Id}\otimes\pi} F^\times \otimes \mathcal{B}(F)$$

vanishes, where $\iota : \mathcal{F}_F \to F^\times \otimes \mathcal{F}_F$ is the map defined on generators $[x]$ $(x \in F)$ by $[x] \mapsto [x] \otimes [x]$. Thus we have finally:

CONJECTURE. *Let F be a number field of degree $n = n_+ + n_-$, Δ the absolute value of the discriminant of F. Define $\mathcal{A}_3(F) = \ker\bigl((Id \otimes \pi) \circ \iota : \mathcal{A}_3^{(3)}(F) \to F^\times \otimes \mathcal{B}(F)\bigr)$ and $\widetilde{D}_3^F = \prod_\sigma \widetilde{D}_3^\sigma : \mathcal{A}_3(F) \to \mathbb{R}^{n_+}$ (product over the embeddings of F into $\mathbb{C}$ up to complex conjugation). Then $\widetilde{D}_3^F\bigl(\mathcal{A}_3(F)\bigr)$ is a lattice in $\mathbb{R}^{n_+}$ with covolume $\sim \sqrt{\Delta}\,\zeta_F(3)/\pi^{3n_-}$.*

We can summarize the various definitions of the last three sections and their interrelationships by the commutative diagram

$$\begin{array}{ccccccc} & D_3^F\ \swarrow & \mathcal{F}_F & \supseteq & \mathcal{A}_3^{(1)}(F) & \supseteq & \mathcal{A}_3^{(2)}(F) \\ \mathbb{R}^{n_+} & & & & \cap & & \cap \\ & \widetilde{D}_3^F\ \nwarrow & \mathcal{F}_F & \supseteq & \mathcal{A}_3^{(3)}(F) & \supseteq & \mathcal{A}_3(F) \end{array} \qquad (25)$$

where $\mathcal{A}_3^{(1)}(F)$ and $\mathcal{A}_3^{(3)}(F)$ denote the kernels of $[x] \mapsto [x]\otimes[x]\otimes[x/(1-x)^2]$ and $[x] \mapsto [x] \otimes \bigl([x] \wedge [1-x]\bigr)$, $\mathcal{A}_3(F)$ is the kernel of $[x] \mapsto [x] \otimes \pi_2\bigl([x]\bigr)$, $\mathcal{A}_3^{(2)}(F) = \mathcal{A}_3^{(1)}(F) \cap \mathcal{A}_3(F)$, and the horizontal inclusions are equalities when F is a totally real field.

We now give numerical examples of the above versions of the conjecture for the same number fields $\mathbb{Q}(\sqrt{5})$, $\mathbb{Q}(\sqrt{2})$ and $\mathbb{Q}(\theta)$ $(\theta^3 - \theta - 1 = 0)$ that were used to illustrate the first two versions. For $F = \mathbb{Q}(\sqrt{5})$ we had four elements $2 \pm \sqrt{5}$ and $(1 \pm \sqrt{5})/2$ and two linear combinations of them which satisfied (16) and hence belonged to $\mathcal{A}_3^{(1)}(F)$. For the last two of these elements x, both x and $1-x$ belong (up to torsion) to the group generated by $\phi = (1+\sqrt{5})/2$, so $[x]\otimes([x]\wedge[1-x])$ vanishes trivially. This gives a new element $[(1+\sqrt{5})/2] - [(1-\sqrt{5})/2]$. It is anti-invariant under the action of $\mathrm{Gal}(F/\mathbb{Q})$, so should map under $\widetilde{D}_3$ to a rational multiple of $\sqrt{5}\,L(3,\chi_5)$. The multiple turns out to be 25/24, giving the identities

$$\widetilde{D}_3\left(\frac{1\pm\sqrt{5}}{2}\right) = \frac{1}{10}\,\zeta(3) \pm \frac{25\sqrt{5}}{48}\,L(3,\chi_5).$$

For $F=\mathbb{Q}(\sqrt{2})$ the situation is similar. Here we had used a collection of 8 elements for which x and $1-x$ belonged to a sub $\mathbb{Z}$-module of $F^\times$ of rank 2. This led to 6 linear conditions needed to satisfy (16) and hence to 2 linear combinations ξ which mapped to zeta values under D_3. Equation (24), on the other hand, imposes only 2 conditions, since each $[x]\otimes([x]\wedge[1-x])$ is up to torsion a linear combination of the two elements $[\varepsilon]\otimes([\varepsilon]\wedge[\pi])$ and $[\pi]\otimes([\varepsilon]\wedge[\pi])$ ($\varepsilon = 1+\sqrt{2}$, $\pi=\sqrt{2}$), so we now get a 6-dimensional space of linear combinations which work for $\widetilde{D}_3$. As a basis of this space we can take the two previous elements and the four new ones

$$\begin{gathered}[\sqrt{2}]+[-\sqrt{2}],\ [2+\sqrt{2}]+[2-\sqrt{2}]+[1+\sqrt{2}]+[1-\sqrt{2}],\\ [1+\sqrt{2}]-[1-\sqrt{2}],\ \text{and}\ [2+\sqrt{2}]-[2-\sqrt{2}]+[\sqrt{2}]-[-\sqrt{2}].\end{gathered}$$

The first two are Galois invariant and map under $\widetilde{D}_3$ to $\frac{7}{32}\,\zeta(3)$ and $\frac{57}{32}\,\zeta(3)$, respectively, while the latter two are Galois anti-invariant and both map to $\frac{4\sqrt{2}}{3}\,L(3,\chi_8)$. Finally, look again at the field $F=\mathbb{Q}(\theta)$ and the six numbers $x_1=\theta,\ \ldots,\ x_6=\theta^5$ of §4. All six lie in $\mathcal{A}_3^{(3)}(F)$, since all have $\{x,1-x\}\subset\langle\pm1,\theta\rangle$, and for the same reason the requirement that $\pi_2(\iota_v(\xi))=0$ (or equivalently, that $D(\iota_v(\xi^{(2)}))=0$, where $(\cdot)^{(2)}$ is a non-real embedding of F into $\mathbb{C}$) imposes only a single linear condition. The space of solutions is therefore now 5-dimensional, rather than 4-dimensional as before; as a basis we can take the four elements we had before together with the new solution $[-\theta]+2[\theta]$, or, better, a new simpler basis $[-\theta]+2[\theta]$, $[\theta^2]+4[\theta]$, $[\theta^3]-6[\theta]$, $[-\theta^4]-4[\theta]$, $[\theta^5]+5[\theta]$ with images (0,3), (0,–12), (3,–15), (2,–13) and (1,13) under the (conjectural) mapping $(\delta^+,\delta^-):\mathcal{A}_3(F)\to\mathbb{Z}^2$ defined by equation (20). Examples of this kind can of course be multiplied indefinitely, but become numerically quite intricate even for relatively simple number fields. For instance, for the field $F=\mathbb{Q}(\sqrt{-11})$ the conjecture predicts that $\widetilde{D}_3(\xi)$ for any element ξ of $\mathcal{A}_3(F)$ should be a rational multiple of $\zeta(3)$ (because $n_+=1$, the same value as for $\mathbb{Q}$; note that $\zeta(3)\sim\sqrt{D}\,\zeta_F(3)/\pi^3$). By looking at the 26 numbers $x\in F\smallsetminus\mathbb{Q}$ for which x and $1-x$ involve only the three primes 2 and $\frac{1}{2}(1 \pm \sqrt{-11})$ (this seems to be the smallest set which works), we obtain enough numbers and few enough conditions to produce non-trivial relations $\widetilde{D}_3(\xi) \sim \zeta(3)$, but all involve several terms and fairly complicated coefficients, a typical example being

$$176\widetilde{D}_3\Big(\frac{4+\sqrt{-11}}{8}\Big) - 1320\widetilde{D}_3\Big(\frac{1+\sqrt{-11}}{2}\Big) - 3456\widetilde{D}_3\Big(\frac{-1+\sqrt{-11}}{3}\Big)$$
$$- 800\widetilde{D}_3\Big(\frac{4+\sqrt{-11}}{3}\Big) + \widetilde{D}_3\Big(\frac{1+\sqrt{-11}}{32}\Big) - 53\widetilde{D}_3\Big(\frac{31+\sqrt{-11}}{32}\Big) = \frac{1264}{3}\zeta(3).$$

If we also use the $x \in \mathbb{Q}$ for which x and $1-x$ involve only the primes 2 and 3, we get simpler relations, e.g.

$$\widetilde{D}_3\Big(\frac{5+\sqrt{-11}}{4}\Big) - 2\,\widetilde{D}_3\Big(\frac{-1+\sqrt{-11}}{3}\Big) + \widetilde{D}_3\big(\tfrac{1}{3}\big) + \tfrac{3}{2}\,\widetilde{D}_3\big(\tfrac{-1}{2}\big) = \zeta(3).$$

§6. Functional equations of the trilogarithm and the group $\mathcal{C}_3(F)$. The conjectures which we have stated are thus supported by considerable numerical evidence. However, if we compare them with the results described in §2 for the dilogarithm, then we see that they are still incomplete: we should still define a subgroup $\mathcal{C}_3(F)$ analogous to $\mathcal{C}(F)$ such that $\mathcal{B}_3(F) = \mathcal{A}_3(F)/\mathcal{C}_3(F)$ is conjecturally isomorphic (up to torsion) to the K-group $K_5(F)$, with $\widetilde{D}_3^F$ corresponding to the Borel regulator mapping. In analogy with $\mathcal{C}(F)$, which was defined (eq. (11)) as the space spanned by the 5-term relation satisfied by $D(x)$, we would expect $\mathcal{C}_3(F)$ to be spanned by the functional equations of the trilogarithm. We now come to the second remarkable property of $\widetilde{D}_3$: all of the functional equations of the trilogarithm become "clean" (i.e., contain no lower-order terms) if we use the trilogarithm function $\widetilde{D}_3$. Specifically, the classical functional equations of the trilogarithm are the easy 1-variable relations

$$I_x = [x]-[\tfrac{1}{x}], \quad D_x = [x^2]-4[x]-4[-x], \quad T_x = [x]+[1-x]+[1-\tfrac{1}{x}]-[1]$$

and the 2-variable relation

$$\begin{aligned} S^{(3)}_{xy} = &\Big[\frac{x(1-y)^2}{y(1-x)^2}\Big] + [xy] + \Big[\frac{x}{y}\Big] - 2\Big[\frac{x(1-y)}{y(1-x)}\Big] - 2\Big[\frac{x(1-y)}{x-1}\Big] \\ &- 2\Big[\frac{y(1-x)}{y-1}\Big] - 2\Big[\frac{1-y}{1-x}\Big] - 2[x] - 2[y] + 2[1], \end{aligned} \tag{26}$$

found by Spence (1809) and Kummer(1840); here x and y are free variables and by "relations" we mean that for each of the given expressions $\xi =$

$\sum n_i[x_i]$ the sum $\sum n_i Li_3(x_i)$ is a linear combination of lower order terms (here, products of logarithms of rational functions of x and y). The number of such lower-order terms is fairly large, but goes down considerably when the functional equation is expressed in terms of D_3 rather than Li_3; for instance, we have

$$D_3(I_x) = 0, \qquad D_3(T_x) = \frac{1}{12}\log|x(1-x)|\,\log\left|\frac{x}{(1-x)^2}\right|\,\log\left|\frac{x^2}{1-x}\right|$$

and

$$D_3(S^{(3)}_{xy}) = -\frac{1}{4}\log|xy|\,\log\left|\frac{x}{y}\right|\,\log\left|\frac{x(1-y)^2}{y(1-x)^2}\right|$$

(as observed in [18]), with 0, 1, and 1 products of logarithms on the right, whereas the corresponding formulas for Li_3 itself ([10], eqs. (6.7), (6.11) and (6.96)) have 2, 3, and 7 lower-order terms, respectively. When we pass to $\widetilde{D}_3$, however, there are no lower terms at all:

$$\widetilde{D}_3(I_x) = \widetilde{D}_3(D_x) = \widetilde{D}_3(T_x) = \widetilde{D}_3(S^{(3)}_{xy}) = 0 \qquad \forall x,\, y.$$

This property of $\widetilde{D}_3$ was noted by Lewin ([11], 3.2) and will be generalized in §7, where we will show that $\sum n_i \widetilde{D}_3(x_i)$ is constant for any collection of integers n_i and rational functions x_i satisfying the identity (24).

We thus have two ways to try to define the group $\mathcal{C}_3(F)$. One, by analogy with (11), is to find a sufficient set of functional equations for $\widetilde{D}_3$ analogous to the 5-term relation for the Bloch-Wigner function. All of the relations given above, as well as various other functional equations given in [10], can be obtained from the Spence-Kummer relation (26) by specialization, and we will present further evidence in a moment that this relation generates the full kernel of $\widetilde{D}_3^F : \mathcal{A}_3(F) \to \mathbb{R}^{n_+}$ for number fields. We can thus define, in full analogy with (11),

$$\mathcal{C}_3(F) = \left\langle S^{(3)}_{xy} \right\rangle_{x,y\,\in\,\mathbb{P}^1(F);\ \text{no indeterminate terms}} \tag{27}$$

(the phrase "no indeterminate terms" means that we exclude pairs like $(x,y) = (0,0)$ where one or more of the terms in $S^{(3)}_{xy}$ becomes $0/0$). A more modest definition, since we do not know that all functional equations follow from the Spence-Kummer ones, would be to define $\mathcal{C}_3(F)$ as the group spanned by all functional equations of $\widetilde{D}_3$ with the variables specialized to F. For higher polylogarithms, where we definitely do not possess generating collections of functional equations, we are of course forced to use the latter definition; we will therefore have to give a precise formulation of it for the general case in §7 and do not give a formulation here. Whichever way we define $\mathcal{C}_3$, we define the third Bloch group $\mathcal{B}_3(F)$ as the quotient $\mathcal{A}_3(F)/\mathcal{C}_3(F)$ and make the conjecture

CONJECTURE. *The map* $\widetilde{D}_3^F : \mathcal{B}_3(F) \to \mathbb{R}^{n_+}$ *for any number field* F *is an isomorphism onto a sublattice of* $\mathbb{R}^{n_+}$ *with covolume* $\sim \sqrt{\Delta}\,\zeta_F(3)/\pi^{3n_-}$.

Of course, what we really want is that there is a canonical map $K_5(F) \to \mathcal{B}_3(F)$ which is an isomorphism up to torsion and whose composite with $\widetilde{D}_3^F$ gives the Borel regulator mapping; we have not stated the conjecture in this form because our numerical evidence only concerns the image of the trilogarithm map $\widetilde{D}_3^F$ and gives no information about algebraic K-theory.

To support this conjecture, as well as to provide yet further evidence for the conjectures of §§3–5, let us look once more at the field $F = \mathbb{Q}$, this time allowing the prime factors 2, 3 and 5 in the numerators and denominators of x and $1-x$. The set $X = X_{\{2,3,5\}}$ of x having this property and with $|x| \le 1$ contains exactly 50 rational numbers, the one with the largest denominator being $\frac{3}{128}$. For $x \in X$, we write $x = \pm 2^{\alpha_2} 3^{\alpha_3} 5^{\alpha_5}$, $1 - x = \pm 2^{\beta_2} 3^{\beta_3} 5^{\beta_5}$ and set $\gamma_{qr} = \begin{vmatrix} \alpha_q & \alpha_r \\ \beta_q & \beta_r \end{vmatrix}$ for $q,\, r \in \{2,3,5\}$. The product $\alpha_p \gamma_{qr}$ gives the coefficient of $[p] \otimes ([q] \otimes [r])$ in $\beta_3([x])$. There are 9 such products (since $\gamma_{qr} = -\gamma_{rq}$), but only 8 of them are linearly independent since we have the relation

$$\alpha_2\,\gamma_{35} - \alpha_3\,\gamma_{25} + \alpha_5\,\gamma_{23} = \begin{vmatrix} \alpha_2 & \alpha_3 & \alpha_5 \\ \alpha_2 & \alpha_3 & \alpha_5 \\ \beta_2 & \beta_3 & \beta_5 \end{vmatrix} = 0.$$

Also, $\gamma_{25} \equiv \gamma_{35} \pmod 2$, as one sees by an easy argument using the Legendre symbol at 5. We therefore have 8 linearly independent integral invariants $\nu_1(x), \ldots, \nu_8(x)$ defined by

$$\begin{pmatrix} \nu_1(x) \\ \nu_2(x) \\ \nu_3(x) \\ \nu_4(x) \\ \nu_5(x) \\ \nu_6(x) \\ \nu_7(x) \\ \nu_8(x) \end{pmatrix} = \begin{pmatrix} 1 & 0 & 0 & 0 & -1 & 0 & -2 & -3 \\ 0 & 1 & 0 & 0 & 0 & 1 & 0 & -1 \\ 0 & 0 & 1 & 0 & 1 & 1 & 1 & 1 \\ 0 & 0 & 0 & 1 & -1 & -1 & -2 & -2 \\ 0 & 0 & 0 & 0 & 1 & 1 & 0 & 0 \\ 0 & 0 & 0 & 0 & 0 & 1 & 0 & -1 \\ 0 & 0 & 0 & 0 & 0 & 0 & 1 & 1 \\ 0 & 0 & 0 & 0 & 0 & 0 & 0 & 1 \end{pmatrix} \begin{pmatrix} \alpha_2\gamma_{23} \\ -\alpha_3\gamma_{23} \\ \frac{1}{2}\alpha_5(\gamma_{35} - \gamma_{25}) \\ -\alpha_5\gamma_{35} \\ -\alpha_2\gamma_{25} \\ \alpha_3\gamma_{35} \\ -\frac{1}{2}\alpha_2(\gamma_{25} + \gamma_{35}) \\ \frac{1}{2}\alpha_3(\gamma_{25} + \gamma_{35}) \end{pmatrix}.$$

(Example, left to the reader: the invariants $\nu_1(x), \ldots, \nu_8(x)$ for $x = \frac{3}{128}$ are 251, 5, −216, 288, −144, 12, −72, −9.) The triangular matrix and the order of the invariants on the right of this equation have been chosen so that $\nu_i(x_j) = \delta_{i,j}$ (Kronecker delta) for the 8 rational numbers $x_1, \ldots, x_8 = -\frac{1}{2}$, $\frac{1}{3}$, $\frac{1}{5}$, $-\frac{1}{5}$, $\frac{2}{5}$, $-\frac{2}{3}$, $\frac{1}{10}$, $\frac{1}{6}$. Hence the element $[x] - \sum_{i=1}^{8} \nu_i(x)\,[x_i]$ of $\mathcal{F}_{\mathbb{Q}}$ belongs to $\ker(\beta_3)$ for any $x \in X$, and if our conjecture is right we must have

$$\widetilde{D}_3(x) = \sum_{i=1}^{8} \nu_i(x)\, \widetilde{D}_3(x_i) + \frac{m(x)}{12}\, \zeta(3) \tag{28}$$

for some rational number $m(x)$ whose value can be calculated numerically. For instance, using the values of $\nu_i(\frac{3}{128})$ just given, one finds $m(\frac{3}{128}) = 7518$. In fact, $m(x)$ turns out to be integral for all $x \in X$ except $\frac{1}{2}$. What's more, the integrality of $m(x)$ is a theorem, not merely a numerical fact, since H. Gangl in Bonn has checked that the relation (28) *is a consequence of the Spence-Kummer relation for all 50 elements* $x \in X$. This provides rather strong evidence for the conjecture presented above that the kernel of $\widetilde{D}_3^F$ coincides with the group generated by the Spence-Kummer relation for all x, $y \in F$. Gangl has performed similar calculations for the much larger set $X = X_{\{2,3,5,7\}}$ and again all relations $\widetilde{D}_3(\xi) = \mu\,\zeta(3)$ ($\xi \in \mathcal{A}_3(\mathbb{Q}) \cap \mathcal{F}_X$, $\mu \in \mathbb{Q}$) which he checked turned out to be deducible from the Spence-Kummer relation and to have the property that μ is a multiple of $1/24$, though it need now no longer be an even multiple even if ξ does not contain $[\frac{1}{2}]$; for instance, the image of

$$4\left[\frac{1}{3}\right] - 12\left[\frac{1}{6}\right] - 3\left[\frac{1}{8}\right] + 3\left[\frac{1}{36}\right] + 2\left[-\frac{1}{2}\right] - 18\left[-\frac{1}{6}\right] + \left[-\frac{1}{48}\right] + \left[\frac{2}{9}\right]$$

under $\widetilde{D}_3$ is $\dfrac{89}{24}\zeta(3)$. In any event, it seems to be the case that the image of the map $\widetilde{D}_3 : \mathcal{A}_3(\mathbb{Q}) \to \mathbb{R}$ is the lattice $\dfrac{1}{24}\zeta(3)\,\mathbb{Z}$ and that the kernel is the group generated by the Spence-Kummer relation.

§7. Higher order polylogarithm functions and functional equations. Like $D_3(x)$, the function $D_m(x)$ defined by (12) has a $\log^m$ singularity at $x = 0$ (and hence, by the functional equation (13), also at $x = \infty$) if m is odd, but this can be removed in the same way as in the case $m = 3$ by adding a suitable multiple of $\log^{m-1}|x| \log|\frac{x}{(1-x)^2}|$, i.e. by looking at

$$\widetilde{D}_m(x) = \Re_m\left(\sum_{r=0}^{m-1} \frac{(-\log|x|)^r}{r!}\, Li_{m-r}(x) + \frac{(-\log|x|)^{m-1}}{m!} \log|1-x|\right). \tag{29}$$

This is the form of the function $D_m(x)$ given by Wojtkowiak [16], who denotes it $-\mathrm{L}(x)$. Observe that the function $\widetilde{D}_m$ vanishes at $x = 0$ and—since the addition of the $(-1)^{m-1}$-symmetric term $\frac{1}{2m!}\log^{m-1}|x|\log|\frac{x}{(1-x)^2}|$ does not affect the validity of the functional equation (13)—also at $x = \infty$. We now show that $\widetilde{D}_m$ for every m satisfies "clean" functional equations, i.e., functional equations involving no lower-order polylogarithms or ordinary logarithms.

PROPOSITION 1. *Let $\{n_i,\, x_i(t)\}$ be a collection of integers n_i and rational functions of one variable $x_i(t)$ satisfying the identity*

$$\sum_i n_i \left[x_i(t)\right]^{m-2} \otimes \left(\left[x_i(t)\right] \wedge \left[1 - x_i(t)\right]\right) = 0 \tag{30}$$

in $\left(Sym^{m-2}(\mathbb{C}(t)^\times) \bigotimes \Lambda^2(\mathbb{C}(t)^\times)\right) \bigotimes_Z \mathbb{Q}$. *Then* $\sum_i n_i\, \widetilde{D}_m(x_i(t)) = constant.$

PROOF: Using the identities

$$\frac{\partial \log|x|}{\partial x} = \frac{1}{2x}, \quad \frac{\partial Li_j(\overline{x})}{\partial x} = 0, \quad \frac{\partial Li_j(x)}{\partial x} = \begin{cases} \dfrac{1}{x} Li_{j-1}(x) & (j \ge 2) \\ \dfrac{1}{1-x} & (j = 1) \end{cases},$$

we find after a short calculation that

$$\frac{\partial \widetilde{D}_m(x)}{\partial x} = \frac{\alpha_m}{x}\, \widetilde{D}_{m-1}(x) + \beta_m \log^{m-2}|x| \left(\frac{\log|1-x|}{x} + \frac{\log|x|}{1-x}\right),$$

where α_m denotes $(-1)^m/2\sqrt{-1}$ and β_m is $\sqrt{-1}/2(m-1)!$ for m even, $(m-1)/2m!$ for m odd. Hence

$$\frac{\partial}{\partial t}\sum_i n_i\, \widetilde{D}_m(x_i(t)) = \alpha_m \sum_i n_i \frac{x_i'(t)}{x_i(t)}\, \widetilde{D}_{m-1}(x_i(t))$$

$$+ \beta_m \sum_i n_i \log^{m-2}|x_i(t)| \left(\frac{x_i'(t)}{x_i(t)} \log|1 - x_i(t)| - \frac{(1 - x_i(t))'}{1 - x_i(t)} \log|x_i(t)|\right).$$

The second sum vanishes because of (30) and because the maps $\lambda : x(t) \mapsto \log|x(t)|$ and $\lambda' : x(t) \mapsto x'(t)/x(t)$ are linear on $\mathbb{C}(t)^\times$. For the first sum, we use the fact that, by the fundamental theorem of algebra, any rational function $x(t)$ factors as $c\prod(t-\alpha)^{v_\alpha(x)}$ and hence $x'(t)/x(t) = \sum v_\alpha(x)/(t-\alpha)$, where the product and the sum extend over all $\alpha \in \mathbb{C}$ (only finitely many terms being non-trivial) and $v_\alpha(x)$ denotes the order of x at α. Hence the first sum can be written as

$$\sum_{\alpha \in \mathbb{C}} \frac{1}{t-\alpha}\left(\sum_i n_i\, v_\alpha(x_i(t))\, \widetilde{D}_{m-1}(x_i(t))\right).$$

But the formal sum $\sum n_i\, v_\alpha(x_i(t))\, [x_i(t)]$ satisfies the analogue of (30) with m replaced by $m-1$ (if $m = 2$, then this makes no sense, but then there is nothing to prove since $\widetilde{D}_1 \equiv 0$), so the expression in the inner sum is a constant by the proposition applied inductively to $\widetilde{D}_{m-1}$. Substituting $t = \alpha$, we see that this constant is zero, since whenever $v_\alpha(x_i(t))$ is not zero the function $\widetilde{D}_{m-1}$ is being evaluated at either 0 or ∞, where it vanishes. Hence $(\partial/\partial t)(\sum n_i \widetilde{D}_m(x_i(t))) = 0$ and similarly, of course, for $\partial/\partial \bar{t}$. □

REMARK: There is a similar result, slightly simpler to prove, for polylogarithms of a real variable. For $-1 \le x \le 1$ define

$$L_m(x) = \sum_{r=0}^{m-1} \frac{(-\log|x|)^r}{r!} Li_{m-r}(x) + \frac{(-\log|x|)^{m-1}}{m!} \log|1-x|, \tag{31}$$

and extend to $\mathbb{R}$ by $L_m(\frac{1}{x}) = (-1)^{m-1} L_m(x)$. This function agrees for m odd with the restriction of $\widetilde{D}_m$ to $\mathbb{R}$ (for m even, of course, $\widetilde{D}_m|\mathbb{R}$ vanishes identically). It was defined (for $x < 1$ and by a recurrence rather than a closed formula) by Lewin ([11], eq. (16)), who conjectured that it always satisfies "clean" functional equations and verified this for the functional equations found by Kummer for the polylogarithms of order ≤ 5. In our language this takes the form that $\sum n_i L_m(x_i(t))$ is piecewise constant (the pieces being the intervals between the real roots of the functions $x_j(t) - 1$ if m is even) for any collection of $n_i \in \mathbb{Z}$ and $x_i(t) \in \mathbb{R}(t)$ satisfying (30). The proof, as already remarked, is similar to that of Proposition 1, but even simpler: we now have

$$L'_m(x) = \frac{(-1)^{m-1}}{2m(m-2)!} \log^{m-2}|x| \left(\frac{\log|1-x|}{x} + \frac{\log|x|}{1-x} \right),$$

so

$$\begin{aligned}\frac{d}{dt} \sum_i n_i L_m(x_i(t)) &= \frac{(-1)^{m-1}}{m(m-2)!} \\ &\times (\underbrace{\lambda \otimes \ldots \otimes \lambda}_{m-1} \otimes \lambda')(\sum_i n_i \underbrace{x_i \otimes \ldots \otimes x_i}_{m-2} \otimes (x_i \otimes (1-x_i) - (1-x_i) \otimes x_i)) \\ &= 0\end{aligned}$$

where $\lambda : x \mapsto |x|$ and $\lambda' : x \mapsto x/x'$ as before. Note that this time the fundamental theorem of algebra, which is false over $\mathbb{R}$, was not needed since dL_m/dx, unlike $\partial \widetilde{D}_m/\partial x$, does not involve the $(m-1)$st polylogarithm. □

The analogue of Proposition 1 for the function D_m is:

PROPOSITION 2. *Let $\{n_i, x_i(t)\}$ be a collection of integers n_i and rational functions of one variable $x_i(t)$ satisfying the identity*

$$\sum_i n_i \, [x_i(t)]^{m-1} \otimes [\frac{x_i(t)}{(1-x_i(t))^2}] = 0 \tag{32}$$

in $\left(Sym^{m-1}(\mathbb{C}(t)^\times)\otimes \mathbb{C}(t)^\times\right)\underset{Z}{\otimes}\mathbb{Q}$. *Then* $\sum_i n_i\, D_m(x_i(t)) = constant.$

PROOF: Just as before, except that in the formula for $\partial\widetilde{D}_m(x)/\partial x$ the expression $\left(\frac{\log|1-x|}{x}+\frac{\log|x|}{1-x}\right)$ becomes $\log|x|\left(\frac{1}{x}+\frac{2}{1-x}\right)$, and $x'\left(\frac{1}{x}+\frac{2}{1-x}\right)=\lambda'\left(\frac{x}{(1-x)^2}\right)$. Alternatively, one can deduce this proposition directly from Proposition 1 since clearly $\sum n_i D_m(x_i)$ and $\sum n_i\widetilde{D}_m(x_i)$ agree for any $\sum n_i[x_i]$ satisfying (32). □

We end this section by discussing a different generalization of $\widetilde{D}_3$ which also satisfies "clean" functional equations, namely, the function

$$P_m(x)=\Re_m\left(\sum_{j=0}^{m}\frac{2^j\,B_j}{j!}\,(\log|x|)^j\,Li_{m-j}(x)\right), \tag{33}$$

where B_j is the jth Bernoulli number ($B_0=1$, $B_1=-\frac{1}{2}$, $B_2=\frac{1}{6}$, ...). (The letter "P" here stands for **P**olylogarithm; the "D" of $D_m(x)$ goes back to the Bloch-Wigner **D**ilogarithm and is illogical for $m>2$!) The motivation for the choice of coefficients, which at first sight looks a little arbitrary, is as follows. We want a function of the form $P_m(x)=D_m(x)+$(lower order terms) which, like D_m, is one-valued, real-analytic on $\mathbb{P}^1(\mathbb{C})\smallsetminus\{0,1,\infty\}$, and $(-1)^{m-1}$-symmetric with respect to $x\mapsto 1/x$ or $x\mapsto\overline{x}$, but which is continuous on all of $\mathbb{P}^1(\mathbb{C})$. To ensure the one-valuedness we make the Ansatz

$$P_m(x)=\sum_{j=0}^{m-1}\gamma_j\,(\log|x|)^j\,D_{m-j}(x) \tag{34}$$

with some universal coefficients $\gamma_0=1$, γ_1, ... yet to be determined. The requirement of $(-1)^{m-1}$-symmetry with respect to $x\mapsto 1/x$ or $x\mapsto\overline{x}$ is then equivalent to the requirement that $\gamma_j=0$ for j odd. On the other hand, the definition of D_m can be written

$$D_m(x)=\Re_m\left(\sum_{j=0}^{m}\frac{(-1)^j}{j!}\,(\log|x|)^j\,Li_{m-j}(x)\right),$$

with the convention $Li_0(x)\equiv-\frac{1}{2}$, so we can rewrite (34) as

$$P_m(x)=\Re_m\left(\sum_{j=0}^{m}\beta_j\,(\log|x|)^j\,Li_{m-j}(x)\right)$$

with $\{\beta_j\}$ defined by the generating function identity

$$\sum_{j=0}^{\infty}\gamma_j\,x^j\cdot e^{-x}=\sum_{j=0}^{\infty}\beta_j\,x^j. \tag{35}$$

Now $\log^j |x|\, Li_{m-j}(x)$ is $o(1)$ as $x \to 0$ for $j < m$ (since $Li_{m-j}(x) = O(x)$), and the term $j = m$ gives 0 if m is even because then $\Re_m(Li_{m-j}) = \Re_m(-\frac{1}{2}) = 0$. Consequently $P_m(x)$ is automatically continuous (with value 0) at $x = 0$ when m is even, but is finite at $x = 0$ for m odd if and only if $\beta_m = 0$. We therefore demand that $\beta_m = 0$ for all odd $m > 1$. Replacing x by $-x$ in (35) and subtracting, we deduce (since $\gamma_j = 0$ for j odd)

$$\left(\sum_{j=0}^{\infty} \gamma_j\, x^j\right) \left(e^{-x} - e^x\right) = 2\,\beta_1\, x,$$

whence

$$\beta_1 = -1\,, \quad \sum_{j=0}^{\infty} \gamma_j\, x^j = \frac{x}{\sinh x}\,, \quad \sum_{j=0}^{\infty} \beta_j\, x^j = \frac{2x}{e^{2x} - 1}\,, \quad \beta_j = \frac{2^j\, B_j}{j!}\,.$$

This explains the choice of coefficients in equation (33) and proves that the function it defines is continuous (with value 0) at $x = 0$ and hence, by the functional equation, at $x = \infty$; at $x = 1$, of course, it is continuous with value 0 for m even, $\zeta(m)$ for m odd. Furthermore, from the formula $\left(\sum \gamma_j x^j\right)^{-1} = \sum x^{2j}/(2j+1)!$ we deduce that the functions P_m and D_m are related by

$$D_m(x) = \sum_{0 \le 2j < m} \frac{\log^{2j} |x|}{(2j+1)!}\, P_{m-2j}(x); \tag{36}$$

thus $D_0(x) = P_0(x) = 0$, $D_1(x) = P_1(x) = \frac{1}{2} \log\left|\frac{x}{(1-x)^2}\right|$, $D_2(x) = P_2(x) = D(x)$, $D_3(x) = P_3(x) + \frac{1}{6} P_1(x) \log^2 |x|$, etc.

PROPOSITION 3. *The statement of Proposition 1 remains valid if $\widetilde{D}_m$ is replaced by P_m.*

PROOF: Let $\mathcal{P}_m(x) = \sum \beta_r \log^r |x| \big(Li_{m-r}(x) - (-1)^m Li_{m-r}(\overline{x})\big)$, which differs from $P_m(x)$ only by a factor of ± 2 or $\pm 2i$. Using the identities

$$-\sum_{k=1}^{r} \beta_k\, \beta_{r-k} = r\beta_r + 2\,\beta_{r-1}, \quad -\sum_{k=1}^{r} (-1)^k \beta_k \beta_{r-k} = r\beta_r \qquad (r \ge 1),$$

which are easily proved from the generating function $\sum \beta_r x^r = 2x/(e^{2x} - 1)$, we find after a short calculation that

$$\frac{\partial}{\partial x} \mathcal{P}_m(x) = \frac{\alpha_m}{x}\, \mathcal{P}_{m-1}(x) + \beta_m\, \log^{m-2} |x| \left(\frac{\log|1-x|}{x} + \frac{\log|x|}{1-x}\right),$$

and similarly for $\dfrac{\partial}{\partial \bar{x}}$. The rest of the proof is like that of Proposition 1. $\square$

Finally, we mention a different way to see the function $P_m(z)$ which was pointed out to me by Deligne and Beilinson, and which appears in the interpretation of the polylogarithm in terms of variations of mixed Hodge structures. Let $g(z)$ $(z \neq 0, 1)$ denote the $n \times n$ triangular matrix $(n > m)$ with entries $g_{rs}(z) = 0$ if $r < s$, $\dfrac{1}{(r-s)!}\left(\dfrac{\log z}{2\pi i}\right)^{r-s}$ if $0 < s \le r$, and $\dfrac{1}{(2\pi i)^r} Li_r(z)$ if $s = 0$. This matrix is well-defined if we have defined the polylogarithms using a common path from 0 to z in $\mathbb{C} \smallsetminus \{0,1\}$, and the many-valuedness of the logarithms and polylogarithms is expressed by the fact that $g(z)$ is multiplied on the right by a triangular matrix with rational entries when z moves around a loop in $\mathbb{C}^\times \smallsetminus \{1\}$. Since such a matrix is in particular real, the matrix $h(z) = \overline{g(z)}\, g(z)^{-1}$ has entries which are single-valued functions of z. We write $h(z) = e^{iP(z)}$ where $P(z)$ is a real triangular matrix. It is easy to see that the entries of $P(z)$ all vanish except for the subdiagonal entries, which equal $\dfrac{1}{\pi} \log|z|$, and the entries in the first column, and a short calculation shows that the mth entry in the first column is $\dfrac{\pm 1}{(2\pi)^m} P_m(z)$.

§8. Formulation of the main conjecture. In this section we state conjectures for the higher order polylogarithms analogous to the conjectures for the trilogarithm presented in §§3–6. We shall give two versions, analogous to the different versions for the trilogarithm; the second one seems preferable, but neither one obviously implies the other.

For any field F let $\mathcal{F}_F$ denote the free abelian group on $F^\times$, or alternatively, the quotient of the free abelian group on $\mathbb{P}^1(F) = F \cup \{\infty\}$ by the relations $[0] = [\infty] = 0$, and let $\iota : \mathcal{F}_F \to F^\times \otimes \mathcal{F}_F$ be the map defined on generators by $[x] \mapsto [x] \otimes [x]$. We define subgroups $\mathcal{A}_m^{(1)}(F) \subset \mathcal{F}_F$ $(m \ge 1)$ recursively by

$$\mathcal{A}_1^{(1)}(F) = \ker(\beta_1^{(1)}), \qquad \mathcal{A}_m^{(1)}(F) = \iota^{-1}(F^\times \otimes \mathcal{A}_{m-1}^{(1)}(F)) \quad (m \ge 2)$$

or in one step as $\ker(\beta_m^{(1)})$, where $F_{\mathbb{Q}}^\times = F^\times \otimes_{\mathbb{Z}} \mathbb{Q}$ and $\beta_m^{(1)}$ is the map $\mathcal{F}_F \to \mathrm{Sym}^{m-1}(F_{\mathbb{Q}}^\times) \otimes F_{\mathbb{Q}}^\times$ defined on generators by

$$\beta_m^{(1)}([x]) = [x]^{m-1} \otimes [\frac{x}{(1-x)^2}] \qquad (x \in \mathbb{P}^1(F) \smallsetminus \{0,1,\infty\})$$

and $\beta_m^{(1)}([x]) = 0$ for $x \in \{0,1,\infty\}$. Define $\mathcal{C}_m^{(1)}(F)$ to be the subgroup of $\mathcal{A}_m^{(1)}(F)$ generated by all images $(\phi_\alpha - \phi_\beta)(\mathcal{A}_m^{(1)}(\mathbb{Q}(t)))$ as α and β range over $\mathbb{P}^1(F)$, where $\phi_\alpha : \mathcal{F}_{\mathbb{Q}(t)} \to \mathcal{F}_F$ is the evaluation map defined on generators by $[x(t)] \mapsto [x(\alpha)]$. By virtue of Proposition 2 of the last section, $\mathcal{C}_m^{(1)}(F)$ is the group spanned by the functional equations (with arguments in $\mathbb{Q}(t)$) of the mth polylogarithm function D_m; this is like eq. (11) for $m = 2$, except that we no longer know the functional equations of D_m explicitly. In any case, it follows that if F is a subfield of $\mathbb{C}$, then the map $D_m : \mathcal{F}_F \to \mathbb{R}$ vanishes on $\mathcal{C}_m(F)$. Finally, and with apologies for the many superscripts, we define

$$\mathcal{A}_m^{(2)}(F) = \iota^{-1}(F^\times \otimes \mathcal{C}_{m-1}^{(1)}(F))$$

$$= \ker\big(\mathcal{A}_m^{(1)}(F) \xrightarrow{\iota} F^\times \otimes \mathcal{A}_{m-1}^{(1)}(F) \xrightarrow{\mathrm{Id}\otimes\pi_{m-1}^{(1)}} F^\times \otimes \mathcal{F}_F/\mathcal{C}_{m-1}^{(1)}(F)\big),$$

where $\pi_{m-1}^{(1)}$ is the obvious projection, and $\mathcal{B}_m^{(1)}(F)$, the (first version of the) mth *Bloch group*, as the quotient $\mathcal{A}_m^{(2)}(F)/\mathcal{C}_m^{(1)}(F)$.

Now suppose that F is an arbitrary field with $[F : \mathbb{Q}] = n_+ + n_-$ and $\Delta =$ absolute value of discriminant of F as usual. For each embedding σ of F into $\mathbb{C}$ we define D_m^σ as the composite of σ with the map $D_m : \mathbb{P}^1(\mathbb{C}) \to \mathbb{R}$ defined by (12), and define $D_m^F : \mathcal{F}_F \to \mathbb{R}^{n_\mp}$ (where $(-1)^m = \pm 1$) as the product of the D_m^σ over all the real and half the complex embeddings of F (one of each complex conjugate pair) if m is odd and over half the complex embeddings if m is even. By what we just said, the map D_m^F vanishes on $\mathcal{C}_m^{(1)}(F)$, so it defines a map from $\mathcal{B}_m^{(1)}(F)$ to $\mathbb{R}^{n_\mp}$.

MAIN CONJECTURE (FIRST VERSION). *D_m^F is an isomorphism from $\mathcal{B}_m^{(1)}(F)$ onto a lattice in $\mathbb{R}^{n_\mp}$ whose volume is a rational multiple of $\sqrt{\Delta}\,\zeta_F(m)/\pi^{mn_\pm}$.*

For the second version of the conjecture we modify the definitions as done in §5 for the trilogarithm. We define groups $\mathcal{A}_m^{(3)}(F) \subset \mathcal{F}_F$ recursively by

$$\mathcal{A}_2^{(3)}(F) = \ker(\beta_2), \qquad \mathcal{A}_m^{(3)}(F) = \iota^{-1}(F^\times \otimes \mathcal{A}_{m-1}^{(3)}(F)) \quad (m \geq 3)$$

or in one step as $\ker(\beta_m)$, where $\beta_m : \mathcal{F}_F \to \mathrm{Sym}^{m-2}(F_{\mathbb{Q}}^\times) \otimes \Lambda^2(F_{\mathbb{Q}}^\times)$ is the map defined on generators by

$$\beta_m([x]) = [x]^{m-2} \otimes ([x] \wedge [1-x]) \qquad (x \in \mathbb{P}^1(F) \smallsetminus \{0, 1, \infty\})$$

and $\beta_m([x]) = 0$ for $x \in \{0, 1, \infty\}$. Define $\mathcal{C}_m(F)$ to be the subgroup of $\mathcal{A}_m^{(3)}(F)$ generated by all images $(\phi_\alpha - \phi_\beta)(\mathcal{A}_m^{(3)}(\mathbb{Q}(t)))$ as α and β range over $\mathbb{P}^1(F)$, and set

$$\mathcal{A}_m(F) = \iota^{-1}(F^\times \otimes \mathcal{C}_{m-1}(F))$$

$$= \ker\big(\mathcal{A}_m^{(3)}(F) \xrightarrow{\iota} F^\times \otimes \mathcal{A}_{m-1}^{(3)}(F) \xrightarrow{\mathrm{Id}\otimes\pi_{m-1}} F^\times \otimes \mathcal{F}_F/\mathcal{C}_{m-1}(F)\big),$$

and $\mathcal{B}_m(F)$, the (better) mth *Bloch group*, as $\mathcal{A}_m(F)/\mathcal{C}_m(F)$.

Main Conjecture (second version). *$\widetilde{D}_m^F$ is an isomorphism from $\mathcal{B}_m(F)$ onto a lattice in $\mathbb{R}^{n_\mp}$ whose volume is a rational multiple of $\sqrt{\Delta}\,\zeta_F(m)/\pi^{mn_\pm}$.*

The various groups and maps we have defined are related by the commutative diagram (25) with "3" replaced everywhere by "m." In particular, the lattices given in the two conjectures must coincide if both conjectures are true, but the one in the second conjecture *a priori* could be larger. Note also that we could replace the function $\widetilde{D}_m$ by the function P_m in the second conjecture, since D_m and $\widetilde{D}_m$ clearly agree on $\mathcal{A}_m^{(1)}$, while P_m and $\widetilde{D}_m$ agree on $\mathcal{A}_m$ by virtue of (36) and the definition of $\mathcal{A}_m$.

Of course, as explained in §§1–2, what we really want is that the lattice in $\mathbb{R}^{n_\mp}$ arising in the two above versions of the conjecture coincides (up to torsion) with the lattice given by the Borel regulator mapping $K_{2m-1}(F) \to \mathbb{R}^{n_\mp}$, in which case the statement about the covolume is a consequence of Borel's theorem. Thus our final formulation is

Main Conjecture. *There is a canonical map $\mathcal{B}_m(F) \to K_{2m-1}(F)$, with finite kernel and cokernel, whose composite with the Borel regulator mapping coincides with P_m^F.*

Before turning to examples, we make two remarks about aspects of the above formulations which might have bothered the reader. First of all, it might be objected that the definition of the kernel $\mathcal{C}_m(F)$ is not really constructive, since we have no criterion to determine which elements of $\mathcal{F}_F$ can be obtained as linear combinations of specializations of functional equations of the polylogarithm: for $m \geq 6$ no functional equations at all beyond the trivial inversion and duplication relations are known, and even, say, for $m = 2$, where it is known that the 5-term relation (10) gives everything, it is not clear how to check whether a given element $\sum n_i[x_i] \in \mathcal{F}_F$ is a linear combination of elements of this form. This objection is especially serious since, even if we decided not to worry about the kernel of our hypothetical surjection $\mathcal{A}_m(F) \to K_{2m-1}(F)$, as was done in §§3–5 for the case $m = 3$, we would still need to know $\mathcal{C}_m(F)$ inductively in order to define the next group $\mathcal{A}_{m+1}(F)$. However, in practise we can decide whether a given element is in $\mathcal{C}_m(F)$ by first checking whether it is in $\mathcal{A}_m(F)$ (assuming the group $\mathcal{C}_{m-1}(F)$ known by induction) and then computing its image under the polylogarithm map. According to the conjecture, the answer must lie in a latttice in Euclidean space, and by computing many examples we can both see numerically that this holds—thus confirming the conjecture—and also recognize which elements are zero, since the vanishing of an element in a discrete subgroup of Euclidean space can be ascertained by a finite precision computation. In other words, the numerical verification proceeds by a bootstrap procedure from one level to the next, with the consistency of the entire numerical procedure providing the evidence for the correctness of the conjecture.

The other remark concerns the definition of the kernel using 1-variable functional equations only, i.e., by specializing elements of $\mathcal{A}_m(\mathbb{Q}(t))$ to values $t \in F$ rather than specializing elements of $\mathcal{A}_m(\mathbb{Q}(t_1, \dots, t_N))$ to values $t_1, \dots, t_N \in F$. This may seem strange in view of the fact that the basic functional equations used for the di- and trilogarithms, namely the 5-term relation (10) and the Kummer-Spence relation (26), involve two variables, and it is clear that no single one-variable equation can ever suffice to produce the whole kernel. However, there is no loss of generality involved, since the specialization of a functional equation $\sum n_i P_m(x_i(t_1, \dots, t_N)) = c$ to values $t_1 = \alpha_1 \in F, \dots, t_N = \alpha_N \in F$ is the same as the specialization of the one-variable functional equations $\sum n_i P_m(x_i(f_1(t), \dots, f_N(t))) = c$ to $t = \alpha_0$, where $F = \mathbb{Q}(\alpha_0)$ and $f_1, \dots, f_N \in \mathbb{Q}(t)$ are chosen such that $\alpha_i = f_i(\alpha_0)$; indeed, this shows that in the definition of $\mathcal{C}_m(F)$ we could have fixed the choice of α and β as α_0 and ∞. We have chosen the one-variable formulation because (i) it is simpler, (ii) it has an algebraic-topological flavor (compare the definitions of homology and homotopy groups in terms of bounding cycles of one dimension higher or of deformations of maps) and thus should lend itself to the comparison with K-theory, and (iii) it sidesteps the question, which we cannot handle anyway, of the existence of a universal functional equation like (10) for the higher polylogarithms, since as soon as we restrict to one-variable equations we are forced to look at infinitely many equations anyway. Of course, one can still speculate that for each m there may be a a single many-variable equation, or even a single two-variable equation like S_{xy} or $S_{xy}^{(3)}$, whose specializations to F generate the group $\mathcal{C}_m(F)$.

We end this section by giving a slightly different formulation of the algebraic structures we have described. For each m we have a filtration

$$\mathcal{F}_F = \mathcal{A}_{m,0}(F) \supset \mathcal{A}_{m,1}(F) \supset \dots \supset \mathcal{A}_{m,m-1}(F) \supset \mathcal{A}_{m,m}(F)$$

with $\mathcal{A}_{m,1}(F) = \mathcal{A}_m^{(3)}(F)$, $\mathcal{A}_{m,m-1}(F) = \mathcal{A}_m(F)$, and $\mathcal{A}_{m,m}(F) = \mathcal{C}_m(F)$ in our previous notation. The groups $\mathcal{A}_{m,i}$ for $i < m$ are defined inductively as $\iota^{-1}(F^\times \otimes \mathcal{A}_{m-1,i}(F))$, while $\mathcal{A}_{m,m}(F)$ is defined as before as the image of $\mathcal{A}_{m,1}(\mathbb{Q}(t))$ under specialization. By induction, one sees that $\mathcal{A}_{m,i}$ for $i > 1$ is the kernel of the (conjecturally surjective) map $\mathcal{A}_{m,i-1}(F) \to \mathrm{Sym}^{m-i}(F^\times) \otimes \mathcal{B}_i(F)$ induced from the map $\iota^{(m-i)} : \mathcal{F} \to \mathrm{Sym}^{m-i}(F) \otimes \mathcal{F}_F$ sending $[x]$ to $[x]^{m-i} \otimes [x]$. The definition of $\mathcal{A}_{m,1}(F) = \ker(\beta_m)$ can be written in a similar form by noting that there is an exact sequence

$$\begin{aligned} 0 \to \mathrm{Sym}^m(F^\times) &\xrightarrow{\kappa_0} \mathrm{Sym}^{m-1}(F^\times) \otimes F^\times \xrightarrow{\kappa_1} \mathrm{Sym}^{m-2}(F^\times) \otimes \Lambda^2(F^\times) \\ &\xrightarrow{\kappa_2} \mathrm{Sym}^{m-3}(F^\times) \otimes \Lambda^3(F^\times) \to \dots , \end{aligned} \tag{37}$$

where κ_i sends $[x_1]\dots[x_{m-i}]\otimes([y_1]\wedge\dots\wedge[y_i])$ to $\sum_{j=1}^{m-i}[x_1]\dots\widehat{[x_j]}\dots[x_{m-i}]\otimes$ $([x_j]\wedge[y_1]\wedge\dots\wedge[y_i])$, and that the image of β_m is in the kernel of κ_2. Thus we can think of β_m instead as a map to $\mathrm{Sym}^{m-1}(F^\times)\otimes F^\times/\mathrm{Sym}^m(F^\times)$. In other words, $\beta_m(\sum n_i[x_i])=0$ iff the element $\sum n_i[x_i]^{m-1}\otimes[1-x_i]$, which is already invariant under permutations of the first $m-1$ variables, is also invariant under permutation of the last two. Then we can summarize by saying that we have injective, and conjecturally surjective, maps

$$\mathcal{A}_{m,i-1}(F)/\mathcal{A}_{m,i}(F)\to\begin{cases}(\mathrm{Sym}^{m-1}(F^\times)\otimes F^\times)/\mathrm{Sym}^m(F^\times) & (i=1)\\ \mathrm{Sym}^{m-i}(F)\otimes\mathcal{B}_i(F^\times) & (1<i\le m)\end{cases}$$

induced by the map $\iota^{(m-i)}:\mathcal{F}_F\to\mathrm{Sym}^{m-i}(F^\times)\otimes\mathcal{F}_F$.

§9. **Examples.** We have given many examples supporting the conjecture for $m=3$ in previous sections. We now give a few examples for larger m.

A. Penta- and heptalogarithms of rational numbers. For our first examples, we look at the field $X=\mathbb{Q}$ and the 11-element set $X=X_{\{2,3\}}$ of §3. We look for non-trivial combinations ξ of elements of x belonging to the group $\mathcal{B}_5(\mathbb{Q})$ and consequently giving a rational multiple of $\zeta(5)$ under P_5; here for the first time we will see the necessity of the analogue of (19) for higher Bloch groups even when F is not totally real. For $x\in X$ we have $x,\ 1-x\in\langle\pm1,2,3\rangle$ and consequently $\beta_5(x)=Q_x([2],[3])\otimes([2]\wedge[3])$ where $Q_x(a,b)$ is a homogeneous cubic polynomial in two variables. For instance, for $x=1/4$ we have

$$[\tfrac{1}{4}]^3\otimes([\tfrac{1}{4}]\wedge[\tfrac{3}{4}])=(-2\,[2])^3\otimes((-2[2])\wedge([3]-2[2]))=16\,[2]^3\otimes([2]\wedge[3])$$

and consequently $Q_{1/4}(a,b)=16a^3$. Computing the values of $Q_x(a,b)$ for each $x\in X$ and forming linear combinations to eliminate these polynomials, we find that the kernel of β_5 on $\mathcal{F}_X$ is 7-dimensional, with generators $[-1]$, $[\frac{1}{2}]$, $[-\frac{1}{3}]-2[\frac{1}{3}]$, $[\frac{1}{4}]-16[-\frac{1}{2}]$, $[-\frac{1}{8}]-162[-\frac{1}{2}]$, $[\frac{1}{9}]-48[\frac{1}{3}]$, and $[\frac{8}{9}]\ \ 9[\frac{3}{4}]-36[\frac{2}{3}]-18[-\frac{1}{2}]-6[\frac{1}{3}]$. If $\xi=\sum n_i[x_i]$ is one of these elements, then each of the sums $\sum n_i v_2(x_i)^\mu v_3(x_i)^\nu[x_i]$ $(\mu+\nu=2)$ belongs to $\mathcal{A}_3(\mathbb{Q})$ and therefore gives a rational multiple of $\zeta(3)$ under P_3. We must then form linear combinations for which this triple of rational numbers vanishes. This gives 3 linear conditions on our 7 elements and finally a 4-dimensional solution space, with basis $[-1]$, $[\frac{1}{4}]-16[\frac{1}{2}]-16[-\frac{1}{2}]$, $[\frac{1}{9}]-16[\frac{1}{3}]-16[-\frac{1}{3}]$, and $[-\frac{1}{8}]-126[\frac{1}{2}]-162[-\frac{1}{2}]$. The first three elements are uninteresting, since they correspond to the inversion and duplication relations of the trilogarithm, but the last gives the non-trivial relation

$$P_5(-\frac{1}{8})-126\,P_5(\frac{1}{2})-162\,P_5(-\frac{1}{2})=\frac{403}{16}\,\zeta(5).$$

By looking at the larger set $X_{\{2,3,5\}}$ considered in §5, we get many more non-trivial relations of this sort, a typical one being

$$P_5(\tfrac{-9}{16}) - P_5(\tfrac{-1}{24}) + 30P_5(\tfrac{-1}{4}) - 30P_5(\tfrac{-2}{3}) - 3P_5(\tfrac{-1}{3}) - P_5(\tfrac{-1}{8}) + 22P_5(\tfrac{3}{4})$$
$$- P_5(\tfrac{8}{9}) - 24P_5(\tfrac{2}{3}) - 10P_5(\tfrac{3}{8}) + 10P_5(\tfrac{1}{6}) - 18P_5(\tfrac{1}{3}) + 76P_5(\tfrac{1}{2}) = \tfrac{935}{16}\,\zeta(5).$$

We did not have to put question marks above the equality signs in the last two identities because their correctness can be verified analytically using Kummer's functional equation. Christophe Soulé suggested that it might be worth finding at least one example for higher-order logarithms, where no functional equations beyond the inversion and duplication relations are known, just to be sure that the theory does not break down there. To find such an example for the heptalogarithm (since $\mathcal{B}_m(\mathbb{Q})$ is supposed to be 0 for m even), take X to be the set of $x \in \mathbb{Q}$ such that x has only the prime factors 2 and 3 in its numerator and denominator and $1-x$ only the factors 2, 3, 5 and 7, and (to avoid the inversion and duplication relations) also with $|x| < 1$, $x \neq$ square. There are 29 such elements, and 28 relations to be satisfied: 20 to make sure we are in $\ker(\beta_7)$, another 5 to make sure that the images under $[x] \mapsto v_2(x)^\mu v_3(x)^\nu [x]$ $(\mu+\nu=4)$ map to 0 in the 1-dimensional group $\mathcal{B}_3(F)$, and 3 more to make the images under $[x] \mapsto v_2(x)^\mu v_3(x)^\nu [x]$ $(\mu+\nu=2)$ vanish in $\mathcal{B}_5(\mathbb{Q})$. There must therefore be at least one non-trivial solution, and indeed we find exactly one, the value under P_7 being (numerically) a rational multiple of $\zeta(7)$ as expected:

$$25111753072P_7\left(\tfrac{1}{3}\right) + 27461584367P_7\left(\tfrac{-1}{3}\right) - 43524P_7\left(\tfrac{-1}{4374}\right)$$
$$-\frac{470985412}{3}P_7\left(\tfrac{-1}{8}\right) - \frac{17015061}{2}P_7\left(\tfrac{-1}{48}\right) + \cdots \overset{?}{=} \frac{1020149599795}{96}\,\zeta(7),$$

the omitted 24 coefficients being integers of between 6 and 12 digits. (This example was found in collaboration with H. Gangl.)

B. Cyclotomic fields. To get elements in the higher Bloch groups, we have to satisfy algebraic relations like $\sum n_i [x_i]^{m-2} \otimes \big([x_i] \wedge [1-x_i]\big) = 0$ together with supplementary conditions involving the images $\iota_v([x_i]) = v(x_i)[x_i]$ for all homomorphisms $v : F^\times \to \mathbb{Z}$. If all of the x_i are roots of unity, then these relations are automatically satisfied, since both $[x_i] \in F^\times_{\mathbb{Q}}$ and $v(x_i)$ vanish. Thus any combination of roots of unity defines an element of the Bloch group of any order of the field they generate.

Let $F = \mathbb{Q}(\zeta)$ be the cyclotomic field generated by a primitive Nth root of unity ζ $(N > 2)$. For any $M \neq 0$ and $m \geq 1$ we have the "distribution relations"

$$P_m(y) = M^{m-1} \sum_{x^M = y} P_m(x), \qquad (y \in \mathbb{C}) \tag{38}$$

(the special cases $M = -1$ and $M = 2$ are the inversion and duplication relations). Using them for the divisors M of N, we see that the values of $P_m(\zeta^j)$ for $(j, N) > 1$ can be expressed in terms of those with $(j, N) = 1$, so that we need only look at primitive Nth roots of unity. Also, $P_m(\zeta^j) = (-1)^{m-1} P_m(\zeta^{N-j})$, so we can restrict to j in the set $A = \{0 < j < N/2 \mid (j, N) = 1\}$. We have $|A| = \frac{1}{2}\phi(N) = n_+ = n_-$, the expected rank of the Bloch group $\mathcal{B}_m(F)$, so it is reasonable to expect that the elements $[\zeta^j]_{j\in A}$ form a basis. To see that they are linearly independent, we compute their images under the polylogarithm mapping P_m^F. The components of this map are the compositions of P_m with half the embeddings $\sigma : F \to \mathbb{C}$. We take these embeddings to be those given by $\zeta \mapsto \mathrm{e}_N(k) = e^{2\pi i k/N}$ as k runs over A. For each Dirichlet character $\chi \pmod N$ with $\chi(-1) = (-1)^{m-1}$ we let ξ_χ denote the element $\sum_{j\in A} \chi(j)[\zeta^j]$ of $\mathcal{F}_F \otimes \mathbb{C}$. Its image under the mapping P_m^F is the vector

$$P_m^F(\xi_\chi) = \Big(\sum_{j\in A} \chi(j)\, P_m(\mathrm{e}_N(jk))\Big)_{k\in A} = \lambda_\chi \left(\overline{\chi}(k)\right)_{k\in A}$$

where $\lambda_\chi = \sum_{j\in A} P_m(\mathrm{e}_N(j))$. The vectors $\left(\overline{\chi}(k)\right)_{k\in A}$ as χ varies are linearly independent (characters of a finite abelian group are linearly indendent). Therefore the vectors $P_m^F(\zeta^j)$ $(j \in A)$ are linearly independent if and only if all λ_χ are non-zero, and then span a lattice of covolume $\prod_\chi \lambda_\chi$, which we now calculate. Suppose that χ is induced from a primitive character χ_0 modulo a divisor $f = f_\chi$ of N. Then we have the following more or less standard calculation:

$$\begin{aligned}
\lambda_\chi &= \frac{1}{2} \sum_{\substack{1\le j\le N\\ (j,N)=1}} \chi_0(j)\, P_m\big(\mathrm{e}_N(j)\big) \\
&= \frac{1}{2} \sum_{1\le j\le N} \Big(\sum_{d\mid(j,N)} \mu(d)\Big)\, \chi_0(j)\, P_m\big(\mathrm{e}_N(j)\big) \qquad (\mu(d) = \text{Möbius function}) \\
&= \frac{1}{2} \sum_{d\mid N} \mu(d) \sum_{1\le r\le N/d} \chi_0(dr)\, P_m\big(\mathrm{e}_{N/d}(r)\big) \\
&= \frac{1}{2} \sum_{\substack{d\mid N\\ (d,f)=1}} \mu(d)\,\chi_0(d) \sum_{1\le t\le f} \chi_0(t) \sum_{\substack{1\le r\le N/d\\ r\equiv t \ (\mathrm{mod}\ f)}} P_m\big(\mathrm{e}_{N/d}(r)\big) \\
&= \frac{1}{2} \sum_{d\mid N/f} \mu(d)\,\chi_0(d) \sum_{1\le t\le f} \chi_0(t) \left(\frac{df}{N}\right)^{m-1} P_m\big(\mathrm{e}_f(t)\big) \qquad \text{(by eq. (38))} \\
&= \frac{1}{2\alpha} \left(\frac{f}{N}\right)^{m-1} \Big(\sum_{d\mid N/f} \mu(d)\,\chi_0(d)\, d^{m-1}\Big) \sum_{1\le t\le f} \chi_0(t) \sum_{n=1}^{\infty} \frac{\mathrm{e}_f(nt)}{n^m}
\end{aligned}$$

$$= \frac{1}{2\alpha} \left(\frac{f}{N}\right)^{m-1} \left(\sum_{d|N/f} \mu(d)\, \chi_0(d)\, d^{m-1} \right) \cdot \tau(\chi_0) \cdot L(m, \overline{\chi}_0),$$

where $\alpha = 1$ or i depending whether m is odd or even and $\tau(\chi_0) = \sum \chi_0(t) \mathbf{e}_f(t)$ is the Gauss sum associated to χ_0. Each of the factors in the last line is non-zero (the number $\tau(\chi_0)/\alpha$ equals $W_\chi \sqrt{f}$, with W_χ, the root number in the functional equation of $L(s, \chi)$, of absolute value 1), so λ_χ is non-zero for each χ and the vectors we have constructed in $\mathrm{Im}(\beta_m)$ form a basis of $\mathbb{R}^{n_\mp}$. The covolume of the lattice they generate, $\prod \lambda_\chi$, is rationally proportional to $\sqrt{\Delta}\, \zeta_F(m)/\pi^{n_\pm}$ by the formula we have just proven together with the facts that $\zeta_F(s)$ is the product of the L-series $L(s, \chi)$ over *all* Dirichlet characters χ (mod N), that $L(m, \chi)$ for χ with $\chi(-1) = (-1)^m$ is a rational multiple of $W\chi \sqrt{f_\chi}\, \pi^m L(1-m, \chi)$, that $\prod_\chi W_\chi = 1$ and $\prod_\chi f_\chi = \Delta$, and that the numbers $L(1-m, \chi)$ for the characters with $\chi(-1) = (-1)^m$ are algebraic numbers which are permuted among one another by the action of $\mathrm{Gal}(\overline{\mathbb{Q}}/\mathbb{Q})$. We therefore have:

PROPOSITION. *Let F be a cyclotomic field. Then P_m^F maps the subgroup of $\mathcal{B}_m(F)$ generated by the roots of unity in F onto a lattice in $\mathbb{R}^{n_\mp}$ whose volume is a rational multiple of $\sqrt{\Delta}\, \zeta_F(m)/\pi^{mn_\pm}$.* □

If $m = 2$, then it follows from the results quoted in §2 that the the roots of unity generate all of $\mathcal{B}_m(F)$. Beilinson has proved in general that the images of the roots of unity under the polylogarithm map lie in (and hence span) the image of the Borel regulator mapping (cf. §11).

C. "Ladders". It turns out that there are already in the existing literature many numerical examples supporting the conjectures of this paper. These are the so-called "ladder relations" of Lewin and his coworkers (see [1] and the references given there). Suppose given a number u, necessarily algebraic, which satisfies a relation of the form

$$\Lambda: \qquad \prod_{r \geq 1} (1 - u^r)^{a_r} = \pm u^N \tag{39}$$

for some integers a_r (Lewin always supposes that the r occurring are divisors of a fixed r_0 with $a_{r_0} = 1$, but this is not needed and spoils the property that the set of relations of the form (39) form a group). Then for each $m \geq 2$ the linear combination $\xi_m(\Lambda) = \sum_r \frac{a_r}{r^{m-1}} [u^r]$ is in the kernel of β_m (even of $\beta_m^{(1)}$). We thus get in a systematic way elements of $\mathcal{A}_m^{(3)}(F)$ (or even $\mathcal{A}_m^{(1)}(F)$) for every m for the field $F = \mathbb{Q}(u)$. The extra conditions defining the subgroup $\mathcal{A}_m(F)$ are also particularly easy to fulfill for the special elements $\xi_m(\Lambda)$, since the subgroup of $F^\times$ generated by their support is 1-dimensional: for each value $m' = 2, \ldots, m-1$ in succession we require that $\xi_{m'}(\Lambda)$ define the 0 element in $\mathcal{B}_{m'}(F)$, in which case $\xi_m(\Lambda)$ will define an element of the next Bloch group $\mathcal{B}_m(F)$. The number of linearly independent conditions this imposes at each stage is alternately n_- and n_+, so if these numbers are small and the number u happens to satisfy a large number of relations (39), we will be able to climb up several steps of the "ladder." Of the many examples given in the papers referred to, we describe only one family which is particularly spectacular. This is for the field $\mathbb{Q}(\theta)$ ($\theta^3 - \theta - 1 = 0$) already used by us as an example in §4 and §5. We had implicitly used the ladder concept by observing that the numbers $x = \theta,\ -\theta,\ \theta^3,\ -\theta^4$ and θ^5 all have the property that $1 - x$ belongs to the group generated by -1 and θ:

$$1-\theta = -\theta^{-4}, \quad 1+\theta = \theta^3, \quad 1-\theta^3 = -\theta, \quad 1+\theta^4 = \theta^5 \quad 1-\theta^5 = -\theta^4.$$

(one can use $1+x = (1-x^2)/(1-x)$ to put these in the form (39).) In [2], Lewin et al. study the same field and give the seven further ladder relations

$$\frac{1+\theta^6}{1+\theta^2} = \theta^3, \quad \frac{1+\theta^7}{1-\theta^7} = -\theta, \quad \frac{1+\theta^9}{1+\theta^3} = \theta^5, \quad \frac{1+\theta^{10}}{(1+\theta^2)^2} = \theta^3,$$
$$\frac{1+\theta^{14}}{(1+\theta^2)(1-\theta^7)} = -\theta^4, \quad \frac{1+\theta^{15}}{(1+\theta^3)(1+\theta^5)} = \theta^5, \quad \frac{1+\theta^{21}}{(1+\theta^3)^2(1+\theta^7)} = \theta^5.$$

This gives altogether 12 elements $[\theta], [-\theta], \ldots, [-\theta^{21}] - 3[-\theta^7] - 14[-\theta^3]$ of $\mathcal{A}(F)$. Each gives a rational multiple of $23^{3/2} L(2)/\pi^4$ under $P_2(\cdot) = D(\cdot)$ (cf. §4), so we get 11 linear combinations $\sum a_r[u^r]/r$ which give 0 in $\mathcal{B}_2(F)$ and hence lift to elements $\xi = \sum a_r[u^r]/r^2$ of $\mathcal{A}_3(F)$. For each of these the elements $\delta^+(\xi)$ and $\delta_-(\xi)$ defined by (20) are rational, and these must vanish in order to be able to lift again to $\sum a_r[u^r]/r^3 \in \mathcal{A}_4(F)$. Continuing in this way, we find that the number of "viable ladders" drops alternately by $n_- = 1$ or $n_+ = 2$ at each stage, the relevant linear combinations being obtained by computing the images of ξ under $P_m(\cdot)/\zeta(m)$ for m even and the map (20) (with 3 replaced by m) for m odd. This is exactly the structure which was found empirically by the authors of [2], except that they used only the real embedding of F and thus had to find one of the linear relations at each odd stage by searching numerically for a linear combination which gave a rational multiple of $\zeta(m)$ under P_m. (At the even stages they used Lewin's function L_m as defined in §7 rather than the function P_m, which vanishes on $\mathbb{R}$, and always found a rational multiple of $\zeta(m)$ here, too; this is not covered by our conjecture, which only says that $P_m(\xi^{(2)})$ should be a rational multiple of $\sqrt{23}\,L(m)/\pi^{2m}$ in this case.) At the last stage of the computation, for instance, one has two elements

$$\xi_1 = 6230\,[\theta^{42}] + 15778\,[\theta^{39}] + 164700\,[\theta^{28}] + \cdots - 382768243200\,[\theta]$$
$$\xi_2 = -20440\,[\theta^{42}] + 122794\,[\theta^{39}] + 29700\,[\theta^{28}] + \cdots + 816983193600\,[\theta]$$

in $\mathcal{A}_9(F)$ (their images under $P_9 = L_9$ equal $\frac{21}{1700}$ times the elements denoted A_9 and B_9, respectively, in [2], where the interested reader can find the remaining coefficients). Computing the values of $\delta^{\pm}(\xi_j)$ as defined by the analogue of (20) with 9 instead of 3, we find that the polylogarithm mapping $P_9^F : \xi \mapsto (P_9(\xi^{(1)}), P_9(\xi^{(2)})) \in \mathbb{R}^2$ maps ξ_1 and ξ_2 to $9850530391\mathbf{e}_1 - 3307\mathbf{e}_2$ and $-21953831855\mathbf{e}_1 + 7055\mathbf{e}_2$, respectively, where

$$\mathbf{e}_1 = 4\zeta(9)\,(1,1) \qquad \mathbf{e}_2 = \frac{21}{4}\,23^{19/2}\,\frac{L(9)}{\pi^9}\,(2,-1),$$

in agreement with the result found in [2] that $7055P_9(\xi_1) + 3307P_9(\xi_2)$ is a rational multiple of $\zeta(9)$. Observe that the covolume of the lattice in $\mathbb{R}^2$ spanned by $\mathbf{e}_1$ and $\mathbf{e}_2$ is a rational multiple of $\sqrt{23}\,\zeta_F(9)/\pi^9$, in accordance with the general theory.

§10. Complements. In this section we discuss some further aspects of the main conjecture.

A. Satisfying the conditions defining $\mathcal{A}_m(F)$. The group $\mathcal{A}_m(F)$ is defined by a number of multilinear conditions on x and $1-x \in F^\times$. To satisfy them, we need a large number of x with x and $1-x$ belonging to a relatively small subspace of $F^\times$. It is not clear a priori if such elements can be found when m is large, and indeed Deligne asked me in a letter of 1988 whether this could always be done. In this subsection we show that the answer is affirmative and discuss some of the numerical aspects involved.

We first count the number of conditions needed. Let S be a subset of $F^\times$ of cardinality s (we may assume without loss of generality that the elements of S are linearly independent) and consider the set

$$X = X(S) = \{x \in \mathbb{Q} \mid x,\, 1-x \in V\},$$

where $V = \langle S\rangle$ is the subspace of $F^\times \otimes_{\mathbb{Z}} \mathbb{Q}$ generated by S (earlier we used the notation X_S for a similarly defined set.) It generates a subgroup $\mathcal{F}_X$ of $\mathcal{F}_F$ of rank equal to $|X|$, the cardinality of X. We want to estimate the number of linear conditions on an element $\xi \in \mathcal{F}_X$ imposed by the requirement that it belong to $\mathcal{A}_m(F)$; if this number is (much) less than $|X|$, then we get (many) elements of $\mathcal{A}_m(F) \cap \mathcal{F}_X$.

The basic condition that an element $\xi \in \mathcal{F}_X$ must satisfy is that it belongs to the kernel of $\beta_m : \mathcal{F}_X \to \mathrm{Sym}^{m-2}(V) \otimes \Lambda^2(V)$. The number of conditions this imposes is $\dim(\mathrm{Sym}^{m-2}(V)) \dim(\Lambda^2(V)) = \binom{s+m-3}{m-2}\binom{s}{2}$. This can be improved slightly by noting that the image of β_m in fact belongs to the subspace $(\mathrm{Sym}^{m-1}(V) \otimes V)/\mathrm{Sym}^m(V)$ of $\mathrm{Sym}^{m-2}(V) \otimes \Lambda^2(V)$ (cf. the exact sequence (37)), so the actual number of conditions is $\dim(\mathrm{Sym}^{m-1}(V) \otimes V) - \dim(\mathrm{Sym}^m(V)) = (m-1)\binom{m+s-2}{m}$. Once our element ξ belongs to $\ker \beta_m = \mathcal{A}_m^{(3)}$, we must ensure that it lies in the subgroup $\mathcal{A}_m(F)$. This will be the case if it lies successively in the kernel of a sequence of maps to $\mathrm{Sym}^i(V) \otimes \mathcal{B}_{m-i}(F)$ for $i = 2, 3, \ldots$ (cf. the filtration of $\mathcal{A}_m(F)$ given at the end of §8). Since $\dim(\mathrm{Sym}^i(V)) = \binom{i+s-1}{i}$ and $\dim(\mathcal{B}_{m-i}(F)) = n_{(-1)^{m-i-1}}$, this gives finally

$$\dim(\mathcal{F}_X \cap \mathcal{A}_m(F)) \geq |X| - (m-1)\binom{s+m-2}{m} - \sum_{1 \leq i \leq m-2} \binom{i+s-1}{i} n_{(-1)^{m-i-1}} . \tag{40}$$

In any case, the number of conditions which an element of $\mathcal{F}_{X(S)}$ must satisfy in order to belong to $\mathcal{A}_m(F)$ grows polynomially with s. To be sure of obtaining elements of $\mathcal{A}_m(F)$ for each m, therefore, we need to know that we can find sets S of arbitrarily large cardinality s such that $|X(S)|$ grows more than polynomially in s. That this can be done is the content of a beautiful theorem of Erdös-Stewart-Tijdeman [7]. Since the result is not very well known and the proof a little hard to extract from [7], where it is presented as a series of lemmas which are specializations of much more general and complicated propositions stated elsewhere in the paper, we digress from our main theme to state the complete statement and proof.

THEOREM (ERDÖS-STEWART-TIJDEMAN). *For every sufficiently large integer s there is a set S of cardinality s for which the set $X(S)$ has $\geq e^{(4-o(1))\sqrt{s/\log s}}$ elements with an effective $o(1)$ constant (i.e., for each number $c < 4$ the expression $4 - o(1)$ can be replaced by c for s greater than an effectively computable number $s_0(c)$). The set S can be taken to consist only of primes if so desired.*

PROOF: Let $y \gg 0$ and $u \geq 3$ be real numbers and W the set of integers $\leq y^u$ all of whose prime factors are $\leq y$. For the cardinality of W, usually denoted $\psi(y^u, y)$, we have the well-known Canfield-Erdös-Pomerance estimate [5]

$$|W| \geq \left(y \frac{e + o(1)}{u \log u}\right)^u,$$

where the dependence of o(1) on u as $u \to \infty$ is effective. It follows that by choosing $u = (2+o(1))\sqrt{y}/\log y$ we can force $\binom{|W|}{2}/y^u \geq M = e^{(4+o(1))\sqrt{y}/\log y}$. But for each of the $\binom{|W|}{2}$ pairs $(a,c) \in W^2$ with $0 < a < c$, the difference $c-a$ is an integer $< y^u$, so by the pigeonhole principle there must be an integer $b < y^u$ which is representable as $c-a$ in $\geq M$ ways. Set $S = \{p \text{ prime}, p \leq y\} \cup \{b\}$; then the numbers b/c belong to $X(S)$ and there are at least M of them. The cardinality of S is given by $s = \pi(y) + 1 \leq (1+o(1))\, y/\log y$, so $M \geq e^{(4-o(1))\sqrt{s/\log s}}$. To get the last statement, we replace the choice $S = \{\text{ primes} \leq y\} \cup \{b\}$ by $S = \{\text{ primes} \leq y\} \cup \{\text{prime factors} > y \text{ of } b\}$, noting that the number of prime factors of b greater than y is bounded by $\dfrac{\log b}{\log y} \leq u = o(\dfrac{y}{\log y}) = o(s)$, so that the estimate given is not affected. □

Notice that in the statement and proof we do not necessarily take S to be the set consisting of the first s prime numbers. If we do take this set, however, then the cardinality of $X(S)$ for $s \leq 8$ is given by the following table, due to P. Vojta and R. Gross:

s	1	2	3	4	5	6	7	8
$\|X(S)\|$	3	21	99	375	1137	3267	8595	21891

(actually, these are lower bounds; there may be more solutions). We also mention that the maximum size M_s of $X(S)$ for any S of cardinality s is $< 1000 \cdot 50^s$ (J.-H. Evertse), so that the limit $\lim\limits_{s\to\infty} \dfrac{\log\log M_s}{\log s}$, if it exists, lies between 1/2 and 1; in [7] it is conjectured that the correct value is in fact 2/3. However, for our purposes only the lower bound on M_s is important.

Finally, we should remark that in practise it would be unwise to look for $\xi \in \mathcal{F}_F$ supported on the whole of $X(S)$, since $|X(S)|$, despite the theorem, grows far too slowly with $|S|$ for this to work well when m is at all large. It is far better to restrict the prime factors of x to a much smaller set than those of $1-x$, since $[x]$ occurs multilinearly and $[1-x]$ only linearly in the defining equations of $\mathcal{A}_m(F)$. Thus in §9A, by looking at the heptalogarithms for non-trivial x with $|x| \in \langle 2,3\rangle$ and $|1-x| \in \langle 2,3,5,7\rangle$ we obtained a system of 28 equations in 29 unknowns, well within the range of the computer. Had we tried instead to use the whole set $X(S)$ for some S, then we would have needed $|X_S| = \frac{1}{2}(|X(S)|+1)$ to be larger than $6\binom{s+5}{7} + \binom{s+1}{2} + \binom{s+3}{4}$ (compare eq. (40)); according to the above table, this first happens for $|S| = 8$ and we would have had to solve a system of $|X_{\{2,3,\dots,19\}}| = 10946$ equations in $6\binom{13}{7} + \binom{9}{2} + \binom{11}{4} = 10662$ unknowns!

B. Rational independence of polylogarithm values. We would like to make, or at least suggest, the conjecture that the only rational dependences over $\mathbb{Q}$ of values of the modified polylogarithm function P_m at arguments in $\overline{\mathbb{Q}} \subset \mathbb{C}$ are those which follow from the functional equations of P_m (including the reality condition $P_m(\overline{x}) = (-1)^{m-1}P_m(x)$), i.e., that the kernel of the map $P_m : \mathcal{F}_{\overline{\mathbb{Q}}} \to \mathbb{R}$ is the subgroup generated by $\mathcal{C}_m(\overline{\mathbb{Q}})$ and the group of elements $[x] + (-1)^m[\overline{x}]$. The corresponding statement when we look only at $\mathcal{A}_m(F) \subset \mathcal{F}_{\overline{\mathbb{Q}}}$ and use all embeddings $F \to \mathbb{C}$ is part of our main conjecture. In that case the polylogarithm mapping is supposed to map isomorphically to a discrete submodule of Euclidean space, so that the injectivity can be "seen" by a finite precision calculation. When we use only one embedding, the image will in general be dense, but the mapping should still be injective. There is no evidence for the conjecture beyond the trivial case $m = 1$. Indeed, so far as I know, there is not a single pair of rational linear combinations of values of the Bloch-Wigner function at algebraic arguments whose ratio is known to be irrational, so the map $P_2 : \mathcal{F}_{\overline{\mathbb{Q}}}/\langle 5\text{-term relation}, [x] + [\overline{x}]\rangle \to \mathbb{R}$, far from being injective, might conceivably have a 1-dimensional image over $\mathbb{Q}$! Nevertheless, the underlying algebraic structure of the polylogarithm and the Bloch groups makes the conjecture seem plausible. By virtue of the theorem explained in §9B, a special case of the conjecture would be that the numbers $P_m(\zeta)$ as ζ runs over all roots of unity in $\mathbb{C}$ are linearly independent over $\mathbb{Q}$ except for the relations coming from the distribution relation (38). For $m = 2$ this is a well-known conjecture of Milnor [12].

C. Generalization to Artin L-functions. The whole situation described in this paper should generalize in the more or less obvious way to the pieces of the Bloch group corresponding to various Artin L-series. Specifically, if N is a Galois extension of $\mathbb{Q}$ containing F and $G = \mathrm{Gal}(N/\mathbb{Q})$, $H = \mathrm{Gal}(N/F)$, then one knows that the Dedekind zeta-function of F factors as a product of Artin L-functions $L(s, \rho)$ corresponding to the irreducible components ρ of the representation of G on $\mathbb{C}^{G/H}$, and we expect the Bloch group $\mathcal{B}_m(F)$ to split up in the same way, with the matrix of the polylogarithm map P_m^F splitting up into a direct sum of matrices whose determinants are the appropriate multiples of the numbers $L(m, \rho)$. We do not give the details of the correct formulation, leaving these to the reader or referring him/her to the unpublished paper [9] of B. Gross, where they are described completely (in terms of algebraic K-theory, not of Bloch groups). Observe that the refined conjecture, unlike the statement given in §8, is completely unknown even for $m = 1$, where it reduces to (an imprecise formulation) of Stark's conjecture that $L(1, \rho)$ can always be expressed in closed form in terms of logarithms of algebraic units.

Several cases of the refined conjecture have occurred in this paper: in §3, where the group $\mathcal{B}_3(F)$ for a real quadratic field $F = \mathbb{Q}(\sqrt{D})$ split up into two one-dimensional pieces on which the trilogarithm was proportional to $\zeta(3)$ and $\sqrt{D}\,L(3,\chi_D)$, in §4, where the group $\mathcal{B}_3(F)$ of the non-abelian cubic field $\mathbb{Q}(\theta)$ split up in a similar way (cf. equation (19), noting that the coefficients (1,2) and (1,–1) correspond to the real parts of the trivial and non-trivial characters, respectively, on a group of order 3), and in §9B, where the entire mth Bloch group of a cyclotomic field split up into one-dimensional pieces corresponding to the various characters χ and with values of the polylogarithm proportional to the numbers $L(m,\chi)$. As a further illustration, note that a special case of the refined conjecture is that the value of the zeta-function of an *ideal class* of a number field F at $s = m$ should be expressible in terms of the mth polylogarithm function with algebraic arguments (in the Hilbert class field of F). As an example, take $F = \mathbb{Q}(\sqrt{-23})$. Here there are three ideal classes $\mathcal{A}_0$ (= {trivial ideals}), $\mathcal{A}_1$, and $\mathcal{A}_2 = \mathcal{A}_1^{-1}$. The zeta function of the former is the Epstein zeta-function $\sum(m^2+mn+6n^2)^{-s}$ (sum over non-zero pairs of integers (m,n), taken up to sign), while the zeta functions of the other two coincide and equal the Epstein zeta-function $\sum(2m^2+mn+3n^2)^{-s}$. Thus $\zeta_F(\mathcal{A}_0,3) + 2\zeta_F(\mathcal{A}_1,3) = \zeta_F(3) \sim \pi^3\zeta(3)/\sqrt{23}$, while $\zeta_F(\mathcal{A}_0,3) - \zeta_F(\mathcal{A}_1,3) = L(3)$, the value of the L-series of $\mathbb{Q}(\theta)$ discussed in §4. Thus the rationality of the numbers $\delta^{\pm}(\theta)$ defined by (19) says that both $\zeta_F(\mathcal{A}_0,3)$ and $\zeta_F(\mathcal{A}_1,3)$ can be expressed in terms of the trilogarithm at arguments in $F(\theta)$, the Hilbert class field of F.

A special case of the refined conjecture, far from obvious, is that if E is a Galois extension of F, then the subspace of $\mathcal{B}_m(E)$ invariant under $\mathrm{Gal}(E/F)$ is precisely $\mathcal{B}_m(F)$. The corresponding property for $K_{2m-1}(F)\otimes\mathbb{Q}$ is known.

§11. Recent developments. In this section, we describe very briefly some recent results concerning the conjecture presented in this paper.

In [8], A. Goncharov essentially proves the conjectures about the trilogarithm presented in §§3–5. Using the connection between K-theory and certain geometric configurations, he is led to a new functional equation for the trilogarithm having 22 terms and depending on 3 parameters, namely $\widetilde{D}_3(G_{\alpha_1\alpha_2\alpha_3}) = 0$ where

$$G_{\alpha_1\alpha_2\alpha_3} = \sum_{i=1}^{3}\Big([\alpha_i] + [\beta_i] - [\beta_i/\alpha_{i-1}] + [\beta_i/\alpha_{i-1}\alpha_i] + [\alpha_i\beta_{i-1}/\beta_{i+1}] + [-\beta_i/\alpha_i\beta_{i-1}] - [\alpha_i\alpha_{i-1}\beta_{i+1}/\beta_i]\Big) - 3\,[1] + [-\alpha_1\alpha_2\alpha_3],$$

where the α_i are arbitrary complex numbers and $\beta_i = 1 - \alpha_i(1 - \alpha_{i-1})$ (indices i taken modulo 3). The Spence-Kummer relation is deduced from this by specializing, say, to $(\alpha_1, \alpha_2, \alpha_3) = (1, x, (1-y)/(1-x))$. It is not known whether, conversely, Goncharov's relation can be deduced from Kummer's. Goncharov expresses the opinion that it cannot, but the evidence presented in §6 seems to suggest the contrary. In any case, Goncharov defines $\mathcal{B}_3(F)$ as the quotient of $\mathcal{A}_3(F)/\langle G_{\alpha_1\alpha_2\alpha_3}\rangle_{\alpha_1,\alpha_2,\alpha_3 \in F}$ and proves that there is a canonical map $K_5(F) \otimes \mathbb{Q} \to \mathcal{B}_3(F) \otimes \mathbb{Q}$, conjecturally an isomorphism, whose composite with $\widetilde{D}_3^F$ is the Borel regulator map.

In [6] and [3], the conjecture formulated in this paper is reinterpreted in terms of motivic cohomology and variations of Hodge structures. In [6], P. Deligne furthermore refines it by replacing the real-valued map P_m by a map, well-defined on $\mathcal{A}_m$, with values in $\mathbb{C}/\pi^m i\mathbb{Q}$. This map behaves correctly under complex conjugation, so gives for a number field F of degree n a map on $\mathcal{B}_m(F)$ with altogether n components, $n_\mp$ of which (the ones we have used) are well-defined real numbers, while the other $n_\pm$ are well-defined only modulo rational multiples, presumably with bounded denominator, of π^m. (An example of this phenomenon occurred in §9C, where the one-dimensional groups $\mathcal{A}_m(\mathbb{Q}(\theta))$ for m even could be detected using the value of Lewin's function L_m, which is well-defined modulo $\mathbb{Q}\pi^m$ on $\mathcal{B}_m$, even though $P_m^F|\mathbb{R}$ vanishes in this case.) In [3], A. Beilinson also constructs an element of a certain K-group that is related under the regulator mapping to polylogarithms. By specializing it to roots of unity he shows that their images under P_m^F lie in the Borel regulator lattice (cf. §9B).

Finally, Beilinson has just informed me that he can construct for all m and F a canonical map from $\mathcal{B}_m(F) \otimes \mathbb{Q}$ to $K_{2m-1}(F) \otimes \mathbb{Q}$ whose composite with the regulator mapping coincides with P_m^F. Thus the image of the polylogarithm map is always contained in the regulator lattice in $\mathbb{R}^{n_\mp}$, and the determinant $\det\big(P_m^F(\xi_j)\big)_j$ for any $n_\mp$ elements $\xi_j \in \mathcal{B}_m(F)$ is a rational multiple of $\sqrt{\Delta}\zeta_F(m)/\pi^m$. It is still not known in general whether the ξ_j can be chosen to make this multiple non-zero, i.e., whether P_m^F is a *surjection* from the Bloch group to the regulator lattice. This can, of course, be checked for any given number field F and value of m by a finite computation, and is true in the case that F is cyclotomic by the discussion in §9B and in the case $m = 3$ by Goncharov's theorem. A consequence of Beilinson's result and the discussion of §10A is that there are infinitely many relations over $\mathbb{Q}$ among the values of the polylogarithm function of arbitrary order at algebraic, or even rational, arguments.

References

[1] M. Abouzahra and L. Lewin, The polylogarithm in algebraic number fields, Journal of Number Theory **21** (1985), 214–244.

[2] M. Abouzahra, L. Lewin and H. Xiao, Polylogarithms in the field of omega (a root of a given cubic): functional equations and ladders, Aeq. Math. **33** (1987), 23–45,and Addendum, Aeq. Math. **35** (1988), 304.

[3] A. Beilinson, Polylogarithm and cyclotomic elements, preprint, M.I.T. 1990.

[4] A. Borel, Cohomologie de SL_n et valeurs de fonctions zêta aux points entiers, Ann. Sc. Norm. Sup. Pisa **4** (1977), 613–636.

[5] E.R. Canfield, P. Erdös and C. Pomerance, On a problem of Oppenheim concerning "Factorisatio Numerorum", J. Number Th. **17** (1983), 1–28.

[6] P. Deligne, Interprétation motivique de la conjecture de Zagier reliant polylogarithmes et régulateurs, preprint, Princeton, 1990.

[7] P. Erdös, C. Stewart and R. Tijdeman, Some diophantine equations with many solutions, Comp. Math. **66** (1988), 36–56.

[8] A.B. Goncharov, The classical threelogarithm, algebraic K-theory of fields and Dedekind zeta functions, preprint, Moscow, 1990.

[9] B. Gross, On the values of Artin L-functions, Brown, 1980.

[10] L. Lewin, *Polylogarithms and Associated Functions*, North-Holland, New York, 1981.

[11] L. Lewin, The order-independence of the polylogarithmic ladder structure—implications for a new category of functional equations, Aeq. Math. **30** (1986), 1–20.

[12] J. Milnor, Hyperbolic geometry: the first 150 years, Bull. AMS **6** (1982), 9–24.

[13] W. Neumann and D. Zagier, Volumes of hyperbolic three-manifolds, Topology **24** (1985), 307–332.

[14] D. Ramakrishnan, Analogs of the Bloch-Wigner function for higher polylogarithms, Contemp. Math. **55** (1986), 371–376.

[15] A.A. Suslin, Algebraic K-theory of fields, in "*Proceedings of the International Congress of Mathematicians 1986*," AMS, 1987, pp. 222–244.

[16] Z. Wojtkowiak, A construction of analogs of the Bloch-Wigner function, Math. Scandin. (to appear).

[17] D. Zagier, Hyperbolic manifolds and special values of Dedekind zeta functions, Inv. Math. **83** (1986), 285–302.

[18] D. Zagier, The remarkable dilogarithm, J. Math. Phys. Sci. **22** (1988), 131–145.

[19] D. Zagier, The Bloch-Wigner-Ramakrishnan polylogarithm function, Math. Annalen **286** (1990), 613–624.

Finiteness theorems for dimensions of irreducible λ-adic representations

Yuri G. Zarhin

In this paper we study absolutely irreducible integral λ-adic representations of the Galois groups of number fields. We assume that the representation satisfy the "Weil-Riemann conjecture" with weight n and prove that their dimension is bounded above by a constant, depending only on n and the rank of the corresponding λ-adic Lie algebras. As an application we obtain that the dimension of an Abelian variety is bounded above by the rank of its endomorphism ring times a certain constant, depending only on the semisimple rank of the corresponding ℓ-adic Lie algebra.

I am deeply grateful to Yu. I. Manin, J.-P. Serre and P. Deligne for useful discussions. This paper was written during my stay in Bonn and I am very happy to be able to thank the Max-Planck-Institut für Mathematik for its hospitality.

0. Preliminaries.

Let K be a number field of finite degree over the field $\mathbb{Q}$ of rational numbers, $K(a)$ an algebraic closure of K and $G(K) := \mathrm{Gal}(K(a)/K)$ the Galois group of K. If $K' \subset K(a)$ is a finite algebraic extension of K, then its Galois group $G(K') = \mathrm{Gal}(K(a)/K')$ is an open subgroup of finite index in $G(K)$.

Let E be a number field of finite degree over $\mathbb{Q}$ and let $\mathfrak{O} = \mathfrak{O}_E$ be the ring of integers of E. Let λ be a non-zero prime ideal in $\mathfrak{O}$ and $\ell = \ell(\lambda)$ be the characteristic of the finite residue field $\mathfrak{O}/\lambda$. We let E_λ be the completion of E in λ and regard E_λ as a finite algebraic extension of the field $\mathbb{Q}_\ell$ of ℓ-adic numbers.

0.1. λ-adic representations.

Recall (Serre[6]) that a λ-adic representation of $G(K)$ is a continuous homomorphism

$$\rho: G(K) \to \mathrm{Aut}(V)$$

where V is a finite-dimensional vector space over E_λ. The dimension of ρ is the dimension $\dim(V)$ of the corresponding representation space V. The kernel $\operatorname{Ker}(\rho)$ is a closed invariant subgroup of $G(K)$. We write $K(\rho)$ for the subfield of all $\operatorname{Ker}(\rho)$-invariants in $K(a)$. Clearly, $K(\rho)$ is a (possibly infinite) Galois extension of K.

To each $K' \subset K(a)$ corresponds the λ-adic representation

$$\rho' : G(K') \to \operatorname{Aut}(V)$$

which is the restriction of ρ to $G(K')$. Clearly, $\operatorname{Ker}(\rho') = \operatorname{Ker}(\rho) \cap G(K')$ and $K'(\rho')$ is the compositum $K'K(\rho)$ of K' and $K(\rho)$.

Since the group $\operatorname{Aut}(V)$ of all E_λ-linear automorphisms of V lies in the group $\operatorname{Aut}_{\mathbf{Q}_\ell}(V)$ of all $\mathbf{Q}_\ell$-linear automorphisms of V, it is clear that ρ also may be regarded as an ℓ-adic representation

$$\rho : G(K) \to \operatorname{Aut}_{\mathbf{Q}_\ell}(V)$$

of dimension $\dim_{\mathbf{Q}_\ell} V = [E_\lambda : \mathbf{Q}_\ell] \dim(V)$.

Recall that ρ is called absolutely irreducible if it is irreducible and the centralizer $\operatorname{End}_{G(K)} V$ is equal to E_λ.

Definition. ρ is called infinitesimally absolutely irreducible it is absolutely irreducible and for all finite algebraic extensions K' of K the λ-adic representations ρ' of $G(K')$ are also absolutely irreducible.

In order to justify this definition we need the notion of the ℓ-adic Lie algebra attached to λ-adic representation.

0.2. ℓ-adic Lie groups and Lie algebras.

Since $G(K)$ is a compact group, its image $\operatorname{Im}(\rho)$ is a closed compact subgroup of $\operatorname{Aut}(V)$. (Clearly, the compact group $\operatorname{Im}(\rho)$ is isomorphic to the profinite Galois group $\operatorname{Gal}(K(\rho)/K)$.) This implies that $\operatorname{Im}(\rho)$ is a compact $\mathbf{Q}_\ell$-Lie subgroup of $\operatorname{Aut}(V)$ but not necessarily an E_λ-Lie subgroup. We may define its Lie algebra $\operatorname{Lie}(\operatorname{Im}(\rho))$ which is a $\mathbf{Q}_\ell$-Lie subalgebra of $\operatorname{End}(V)$ but *not necessarily* an E_λ-Lie subalgebra. Clearly, $\operatorname{Im}(\rho')$ is an open subgroup of finite index in $\operatorname{Im}(\rho)$ and, therefore, $\operatorname{Lie}(\operatorname{Im}(\rho)) = \operatorname{Lie}(\operatorname{Im}(\rho'))$ for all finite algebraic extensions K' of K.

Now, one may easily check that ρ is infinitesimally absolutely irreducible if and only if the natural representation of $\operatorname{Lie}(\operatorname{Im}(\rho))$ in V is "absolutely irreducible", i.e., there is no non-trivial $\operatorname{Lie}(\operatorname{Im}(\rho))$-invariant E_λ-vector subspace in V and the centralizer of $\operatorname{Lie}(\operatorname{Im}(\rho))$ in $\operatorname{End}(V)$ coincides with E_λ.

Let us assume further that ρ is infinitesimally absolutely irreducible. In this case one may check that $\operatorname{Lie}(\operatorname{Im}(\rho))$ is a reductive $\mathbf{Q}_\ell$-Lie algebra and its center is a $\mathbf{Q}_\ell$-vector subspace of $E_\lambda \operatorname{id}$. Here $\operatorname{id}: V \to V$ is the identity map. Indeed, let B be a non-zero $\operatorname{Lie}(\operatorname{Im}(\rho))$-invariant $\mathbf{Q}_\ell$-vector subspace of V such that the natural representation of $\operatorname{Lie}(\operatorname{Im}(\rho))$ in B is irreducible. Clearly,

$$V = \sum eB \quad (e \in E_\lambda),$$

and the simple Lie(Im(ρ))-module eB is isomorphic to B for all $e \in E_\lambda \setminus \{0\}$. This implies that the representation of Lie(Im(ρ)) in the $\mathbf{Q}_\ell$-vector space V is isomorphic to the quotient of the direct sum of $[E_\lambda : \mathbf{Q}_\ell]$ copies of the simple Lie(Im(ρ))-module B. This implies, in turn, that the $\mathbf{Q}_\ell$-vector space V is an isotypic representation of Lie(Im(ρ)). In particular, it is semisimple and, therefore, Lie(Im(ρ)) is reductive.

Since it is more convenient to work with E_λ-Lie algebras, let us define E_λ Lie(Im(ρ)) as the E_λ-Lie subalgebra of End(V) spanned by Lie(Im(ρ)). Clearly, the natural representation of E_λ Lie(Im(ρ)) in V is faithful and absolutely irreducible. In particular, E_λ Lie(Im(ρ)) is a reductive E_λ-Lie algebra. Let us split E_λ Lie(Im(ρ)) into the direct sum

$$E_\lambda \operatorname{Lie}(\operatorname{Im}(\rho)) = \mathfrak{c} \oplus \mathfrak{g}_\rho$$

of its center $\mathfrak{c}$ and a semisimple E_λ-Lie algebra $\mathfrak{g}_\rho$. The absolute irreducibility implies that either $\mathfrak{c} = \{0\}$ or $\mathfrak{c} = E_\lambda \operatorname{id}$. In both cases the natural representation of $\mathfrak{g}_\rho$ in V is absolutely irreducible. In addition, E_λ Lie(Im(ρ)) is an algebraic E_λ-Lie subalgebra of End(V).

0.3. Ranks of semisimple Lie algebras

Let r be the rank of the semisimple E_λ-Lie algebra $\mathfrak{g}_\rho$. Clearly, r does not exceed the rank r' of the semisimple part of the reductive $\mathbf{Q}_\ell$-Lie algebra Lie(Im(ρ)). Notice, that if $r = 0$, then $\mathfrak{g}_\rho = \{0\}$ and the absolute irreducibility of the $\mathfrak{g}_\rho$-module V implies that $\dim(V) = 1$. Further, we will assume that $\mathfrak{g}_\rho \neq \{0\}$, i.e., $r > 0$. The aim of this paper is to give upper bounds for $\dim(V)$ in terms of r for certain class of λ-adic representations described in the next subsection.

0.4. Integral λ-adic representations of weight n.

Let us fix a positive integer n.

Definition. A λ-adic representation ρ is called E-integral of weight n if for all but finitely many places v of K the following conditions hold:

a) ρ is unramified at v;

b) let $\mathrm{Fr}_v \in \operatorname{Im}(\rho)$ be a Frobenius element attached to v (defined up to conjugacy [6,5]) and let $P_v(t) = \det(1 - t\,\mathrm{Fr}_v^{-1}, V)$ be its characteristic polynomial. Then all the coefficients of P_v lie in E and even in $\mathfrak{O}$.

c) (the Weil-Riemann conjecture). All (complex) reciprocal roots of P_v and their conjugate over $\mathbf{Q}$ have absolute value $q(v)^{n/2}$ where $q(v)$ is the number of elements of the residue field $k(v)$ at v.

Clearly, if ρ is E-integral of weigth n, then ρ' is also E-integral of weight n for every finite algebraic extension K' of K.

Remark. The Weil-Riemann conjecture easily implies that $\mathrm{Lie}(\mathrm{Im}(\rho))$ is not semisimple, i.e. $E_\lambda \mathrm{Lie}(\mathrm{Im}(\rho)) = E_\lambda \mathrm{id} \oplus \mathfrak{g}_\rho$. Indeed, the determinant $\det(\mathrm{Fr}_v^{-1}, V)$ of Fr_v^{-1} is an algebraic integer $\in E_\lambda^*$, which is not a root of 1, since its (any) archimedean absolute value is equal to $q(v)^{n \dim(V)/2} \neq 1$. Notice that $\det(\mathrm{Fr}_v^{-1}, V)$ is a λ-adic unit, because the image of the determinant map $\mathrm{Im}(\rho) \to E_\lambda^*$ is a compact subgroup. On the other hand, the logarithm map

$$\log\colon \mathrm{Im}(\rho) \to \mathrm{Lie}(\mathrm{Im}(\rho))$$

for the compact ℓ-adic Lie group $\mathrm{Im}(\rho)$ is also defined [1]. One may easily check that

$$\mathrm{tr}(\log u) = \log(\det(u, V)) \in E_\lambda \quad \text{for all} \quad u \in \mathrm{Im}(\rho) \subset \mathrm{Aut}(V).$$

Here $\mathrm{tr}: \mathrm{End}(V) \to E_\lambda$ is the trace map. Now, if we put $\mathrm{fr}_v = \log(\mathrm{Fr}_v^{-1}) = -\log(\mathrm{Fr}_v)$, then

$$\mathrm{tr}(\mathrm{fr}_v) = \log(\det(\mathrm{Fr}_v^{-1}, V) \neq 0,$$

i.e. $\mathrm{Lie}(\mathrm{Im}(\rho))$ contains an operator with non-zero trace. (Henniart [4] even proved that $\mathrm{Lie}(\mathrm{Im}(\rho))$ contains the scalar operators $\mathbb{Q}_\ell \,\mathrm{id}$.)

Our main result is the following assertion.

0.5. Main Theorem. *There exists an absolute constant $D = D(r, n)$, depending only on n and r, enjoying the following properties:*

Let $\rho: G(K) \to \mathrm{Aut}(V)$ be an infinitesimally absolutely irreducible E-integral λ-adic representation of weight n. If the rank of the semisimple E_λ-Lie algebra $\mathfrak{g}_\rho$ is equal to r then $\dim(V) < D(r, n)$.

Remark. For $r = 0$ one may put $D(0, n) = 1$ (see Sect. 0.3).

Corollary of Theorem 0.5. *Let $\rho: G(K) \to \mathrm{Aut}(V)$ be infinitesimally absolutely irreducible E-integral λ-adic representation of weight n. Let r' be the rank of the semisimple part of the reductive $\mathbb{Q}_\ell$-Lie algebra $\mathrm{Lie}(\mathrm{Im}(\rho))$. Then $\dim(V) < \max\{D(j, n), 0 \le j \le r'\}$.*

Indeed, one has only to recall that $r \le r'$ (Sect. 0.3) and apply Theorem 0.5.

0.6. Remark.

Let C be the algebraic closure of E_λ (=algebraic closure of $\mathbb{Q}_\ell$). Let us put

$$W := V \otimes_{E_\lambda} C, \quad \mathfrak{g} := \mathfrak{g}_\rho \otimes_{E_\lambda} C \subset \mathrm{End}_C W$$

and consider the simple module W over the semisimple C-Lie algebra $\mathfrak{g}$ of rank r. In order to prove Theorem 0.5 it suffices to prove that there exists a positive constant D', depending only on r and n, and such that the highest weight of the simple $\mathfrak{g}$-module W is a sum of no more than D' fundamental weights. Let us split $\mathfrak{g}$ into the direct sum

$$\mathfrak{g} = \oplus \mathfrak{g}_i \quad (1 \leq i \leq s)$$

of simple C-Lie algebras $\mathfrak{g}_i$. Clearly, $s \leq r$ and the rank of each $\mathfrak{g}_i$ does not exceed r. Then one may decompose W into the tensor product $W = \otimes W_i$ of simple $\mathfrak{g}_i$-modules W_i $(1 \leq i \leq s)$.

So, in order to prove Theorem 0.5 it suffices to prove that there exists a positive constant D'', depending only on n and r, and such that for all i the highest weight of the simple $\mathfrak{g}_i$-module W_i is a sum of no more than D'' fundamental weights.

0.7. Key Lemma. *Let*

$$f \in E_\lambda \operatorname{Lie}(\operatorname{Im}(\rho)) = E_\lambda \operatorname{id} \oplus \mathfrak{g}_\rho \subset \operatorname{End}(V)$$

be a regular element of the reductive E_λ-Lie algebra $E_\lambda \operatorname{Lie}(\operatorname{Im}(\rho))$. Since $\operatorname{End}(V) \subset \operatorname{End}_C(W)$, *one may view f as a C-linear operator in W. Let* $\operatorname{spec}(f) \subset C$ *be the set of all eigenvalues of $f: W \to W$. Let $\mathbb{Q}(f)$ be the $\mathbb{Q}$-vector subspace of C spanned by* $\operatorname{spec}(f)$. *Let us assume that there exists a finite set A of rational numbers and a finite set M of $\mathbb{Q}$-linear maps $\theta: \mathbb{Q}(f) \to \mathbb{Q}$, enjoying the following properties:*

1. *$\theta(\operatorname{spec}(f)) \subset A$ for all $\theta \in M$;*
2. *the map $\mathbb{Q}(f) \to \mathbb{Q}^M$, $a \mapsto \{\theta(a)\}_{\theta \in M}$ is an embedding.*

Then for all i (with $1 \leq i \leq s$) the highest weight of the simple $\mathfrak{g}_i$-module W_i is a sum of no more than $[\operatorname{card}(A) - 1]$ fundamental weights. Here $\operatorname{card}(A)$ *is the number of elements of A.*

We will prove this Lemma in Section 2.

So, in order to prove Theorem 0.5 it suffices to prove the existence of such f, A and M with A, depending *only* on r and n.

1. Proof of the Main theorem.

Our proof consists of the following steps.

Step 1. Replacing, if necessary, K by a suitable finite algebraic extension K' and ρ by ρ', we may and will assume that K enjoys the following properties:

1) K is a Galois extension of $\mathbb{Q}$;

2) K contains a subfield isomorphic to E.

Let us fix a prime number p and a place v of K, enjoying the following properties:

3) p is unramified in K, v lies above p and the residue field $k(v)$ at v coincides with the finite prime field $\mathbf{Z}/p\mathbf{Z}$;

4) ρ is unramified at v and the characteristic polynomial $P_v(t)$ of the corresponding Frobenius element Fr_v lies in $1 + t\mathfrak{O}[t]$ and satisfies the Weil-Riemann conjecture with weight n;

5) all the eigenvalues of Fr_v^{-1} are congruent to 1 modulo ℓ^2 and the ℓ-adic logarithm $\mathrm{fr}_v := \log(\mathrm{Fr}_v^{-1}) = -\log(\mathrm{Fr}_v)$ is a regular element of the reductive $\mathbf{Q}_\ell$-Lie algebra $\mathrm{Lie}(\mathrm{Im}(\rho))$.
(use Chebotarev density theorem).

The regularity condition implies that fr_v is a semisimple endomorphism of the $\mathbf{Q}_\ell$-vector space V and, therefore, is a semisimple endomorphism of the E_λ-vector space V. Clearly, fr_v is also regular in the reductive E_λ-Lie algebra $\mathrm{Lie}(\mathrm{Im}(\rho)) \otimes_{\mathbf{Q}_\ell} E_\lambda$, fr_v is also regular in $E_\lambda \mathrm{Lie}(\mathrm{Im}(\rho))$.

Step 2. Let us fix an embedding of E in K. Now we may and will assume that E is a subfield of K. Since K is a Galois extension of $\mathbf{Q}$, the condition 3 of step 1 implies that p splits completely in K. Since E is a subfield of K, p also splits completely in E.

Recall that C is an algebraic closure of E_λ. Let L be the subfield of C contained by adjunction to E of the set R of all eigenvalues of Fr_v^{-1}. Clearly, it is a finite Galois extension of E and all elements of R are algebraic integers. For each embedding of L into the field $\mathbf{C}$ of complex numbers all elements of R have absolute value $p^{n/2}$. Let us denote by Γ the multiplicative subgroup of L^* generated by R. Since all elements of R are congruent to 1 module ℓ^2, Γ does not contain roots of 1 different from 1. So, Γ is a finitely generated free abelian group. I claim that the rank $\mathrm{rk}(\Gamma)$ of Γ does not exceed $r+1$. Indeed, the ℓ-adic logarithm maps R into the set $\mathrm{spec}(\mathrm{fr}_v)$ of all eigenvalues of the C-linear $\mathrm{fr}_v : W \to W$, and, therefore, defines an isomorphism between Γ and the additive subgroup $\mathbf{Z}(\mathrm{fr}_v)$ of C, generated by $\mathrm{spec}(\mathrm{fr}_v)$. Recall that fr_v is a semisimple element of $E_\lambda \mathrm{id} \oplus \mathfrak{g}_\rho \subset C\mathrm{id}_W \oplus \mathfrak{g}$ where $\mathrm{id}_W : W \to W$ is the identity map and $\mathfrak{g}$ is the semisimple C-Lie subalgebra of $\mathrm{End}(W)$, having the rank r. Now, E. Cartan theory of modules with highest weight [2] easily implies that the additive subgroup, generated by the eigenvalues of any element of $\mathfrak{g}$, has rank $\leq r$. Since fr_v is the sum of a scalar operator and an operator from $\mathfrak{g}$, the rank of $\mathbf{Z}(\mathrm{fr}_v)$ does not exceed $r+1$.

Notice, that the Galois group $\mathrm{Gal}(L/E)$ acts naturally on Γ. This action defines an embedding

$$\mathrm{Gal}(L/E) \to \mathrm{Aut}(\Gamma) \cong \mathrm{GL}(\mathrm{rk}(\Gamma), \mathbf{Z}) \subset \mathrm{GL}(r+1, \mathbf{Z}).$$

Since $\mathrm{Gal}(L/E)$ is finite, it is isomorphic to a finite subgroup of $\mathrm{GL}(r+1, \mathbf{Z})$. Applying a well-known theorem of Minkowski we obtain that there exists a positive constant $D_1 = D_1(r)$, depending only on r, such that the order of Gal(L/E) divides D_1, i.e. the extension degree $[L : E]$ divides D_1.

Step 3. Let $\mathfrak{O}_L$ be the ring of integers in L. Conditions 3 and 4 of Step 1 imply that all elements α of R are algebraic integers in L and for each embedding of L into $\mathbf{C}$ we have

$$\alpha'\alpha = p^n.$$

Here $\alpha' = \bar{\alpha}$ is the complex conjugate of α and, of course, also an algebraic integer. This implies that if $\mathfrak{p}'$ is a prime ideal in $\mathfrak{O}_L$, not lying above p, then α is a $\mathfrak{p}'$-adic unit for all $\alpha \in R$. Notice, that $\alpha' = p^n/\alpha$ lies in L and even in $\mathfrak{O}_L$.

Let S be the set of prime ideals in $\mathfrak{O}_L$, lying above p. For each $\mathfrak{p}$ from S let

$$\mathrm{ord}_{\mathfrak{p}}: L^* \to \mathbf{Q}$$

be the discrete valuation of L attached to $\mathfrak{p}$ and normalized by the condition

$$\mathrm{ord}_{\mathfrak{p}}(p) = 1.$$

Recall that p completely splits in E. This implies that

$$\mathrm{ord}_{\mathfrak{p}}(E^*) = \mathrm{ord}_{\mathfrak{p}}(\mathbf{Q}^*) = \mathbf{Z},$$

$$n = \mathrm{ord}_{\mathfrak{p}}(p^n) = \mathrm{ord}_{\mathfrak{p}}(\alpha) + \mathrm{ord}_{\mathfrak{p}}(\alpha') \quad \text{for all} \quad \alpha \in R.$$

Since α, α' are algebraic integers, the rational numbers $\mathrm{ord}_{\mathfrak{p}}(\alpha)$, $\mathrm{ord}_{\mathfrak{p}}(\alpha')$ are non-negative, and, therefore, $0 \leq \mathrm{ord}_{\mathfrak{p}}(\alpha) \leq n$ for all $\alpha \in R$.

Since $[L : \mathbf{Q}]$ divides D_1, we have:

$$\mathrm{ord}_{\mathfrak{p}}(L^*) \subset (1/D_1)\,\mathrm{ord}_{\mathfrak{p}}(E^*) = (1/D_1)\mathbf{Z}.$$

Let us put $A := \{c \in \mathbf{Q}, 0 \leq c \leq n, D_1 c \in \mathbf{Z}\}$. Clearly, A is a finite set of rational numbers, consisting of $(D_1 n + 1)$ elements and depending only on n and r. We have

$$\mathrm{ord}_{\mathfrak{p}}(\alpha) \in A \quad \text{for all} \quad \alpha \in R, \mathfrak{p} \in S.$$

Let $\mathrm{ord}: \Gamma \to \mathbf{Q}^S$ be the homomorphism defined by the formula

$$\mathrm{ord}(\gamma) = \{\mathrm{ord}_{\mathfrak{p}}(\gamma)\}_{\mathfrak{p} \in S}.$$

Clearly, $\mathrm{ord}(R) \subset A^S \subset \mathbf{Q}^S$.

I claim that ord is an embedding. Indeed, if $\mathrm{ord}(\gamma) = 0$ for some $\gamma \in \Gamma$ then γ is a unit in L. The Weil-Riemann conjecture (assumption c) implies the equality of all archimedean valuations on the elements of Γ. Therefore, the product formula implies that $|\gamma| = 1$ for all archimedean valuations on L. This implies that γ is a root of 1. Since Γ does not contain non-trivial roots of 1, we have $\gamma = 1$.

One may extend ord by $\mathbf{Q}$-linearity to an embedding

$$\Gamma \otimes \mathbb{Q} \to \mathbb{Q}^S,$$

which we will also denote by ord.

Step 4. Let $\mathbb{Q}(\mathrm{fr}_v)$ be the $\mathbb{Q}$-vector subspace of C, spanned by $\mathrm{spec}(\mathrm{fr}_v)$. We have

$$\mathrm{spec}(\mathrm{fr}_v) \subset \mathbb{Z}(\mathrm{fr}_v) \subset \mathbb{Q}(\mathrm{fr}_v).$$

The ℓ-adic logarithm defines an isomorphism

$$\log\colon \Gamma \to \mathbb{Z}(\mathrm{fr}_v),$$

which can be extended by $\mathbb{Q}$-linearity to an isomorphism

$$\Gamma \otimes \mathbb{Q} \to \mathbb{Q}(\mathrm{fr}_v),$$

which we will also denote by log. Clearly, the $\mathbb{Q}$-vector space $\mathrm{Hom}(\mathbb{Q}(\mathrm{fr}_v), \mathbb{Q})$ is generated by the maps

$$\mathrm{ord}_{\mathfrak{p}} \log^{-1}\colon \mathbb{Q}(\mathrm{fr}_v) \to \Gamma \otimes \mathbb{Q} \to \mathbb{Q} \quad (\mathfrak{p} \in S).$$

Notice, that $\mathrm{ord}_{\mathfrak{p}} \log^{-1}(\mathrm{spec}(\mathrm{fr}_v)) = \mathrm{ord}_{\mathfrak{p}}(R) \subset A$ for all $\mathfrak{p} \in S$.

Now, the highest weight of each simple $\mathfrak{g}_i$-module W_i is the sum of no more than nD_1 fundamental weights. Indeed, one has only to apply Lemma 0.7 to the regular element $f = \mathrm{fr}_v$, the set

$$M = \{\mathrm{ord}_{\mathfrak{p}} \log^{-1}(\mathrm{spec}(\mathrm{fr}_v)) : \mathbb{Q}(\mathrm{fr}_v) \to \mathbb{Q} \mid \mathfrak{p} \in S\}$$

of homomorphisms $\mathbb{Q}(\mathrm{fr}_v) \to \mathbb{Q}$ and A.

2. Proof of Lemma 0.7.

We start the proof with the following remarks. First, we have natural embeddings $E_\lambda \mathrm{id} \otimes \mathfrak{g}_\rho \subset (E_\lambda \mathrm{id} \otimes \mathfrak{p}_\rho) \otimes_{E_\lambda} C = C\,\mathrm{id}_W \oplus \mathfrak{g} \subset \mathrm{End}_C W$. Since f is regular in the reductive E_λLie algebra $E_\lambda \mathrm{id} \otimes \mathfrak{g}_\rho$, it remains regular in the reductive C-Lie algebra $C\,\mathrm{id}_W \oplus \mathfrak{g}$. We have

$$f = c\,\mathrm{id} + \sum f_i \quad (1 \le i \le s)$$

with $c \in C$, $f_i \in \mathfrak{g}_i$. Since f is regular, all f_i are non-zero semisimple elements of $\mathfrak{g}_i$. Let $\mathrm{spec}(f_i) \subset C$ be the set of all eigenvalues of the C-linear operator $f_i\colon W_i \to W_i$ (recall that W_i is a faithful simple $\mathfrak{g}_i$-module). If $\alpha \in \mathrm{spec}(f_i)$ then we write $\mathrm{mult}_i(\alpha)$ for the multiplicity of the eigenvalue α of the operator f_i. Clearly,

$$\sum_{\alpha \in \mathrm{spec}(f_i)} \mathrm{mult}_i(\alpha) = \dim(W_i).$$

Since $\mathfrak{g}_i$ is a (semi)simple subalgebra of $\mathrm{End}(W_i)$, the trace

$$\operatorname{tr}(f_i, W_i) := \sum_{\alpha \in \operatorname{spec}(f_i)} \operatorname{mult}_i(\alpha)\alpha$$

is equal to 0. We have

$$\operatorname{spec}(f) = c + \sum_i \operatorname{spec}(f_i) = \{c + \sum_i \alpha_i \mid \alpha_i \in \operatorname{spec}(f_i), 1 \le i \le s\}.$$

Claim. *For all i there exists a $c_i \in \mathbf{Q}(f)$ such that*

$$\operatorname{spec}(f_i) \subset c_i + \operatorname{spec}(f).$$

In particular, $\operatorname{spec}(f_i) \subset \mathbf{Q}(f)$.

We will prove this Claim at the end of this Section.

Proof of Lemma 0.7 (modulo Claim). We identify $\mathfrak{g}_i$ with its image in $\operatorname{End}(W_i)$. Let $\mathbf{Q}(f_i)$ be the $\mathbf{Q}$-vector subspace of C spanned by $\operatorname{spec}(f_i)$. Clearly, $\mathbf{Q}(f_i) \subset \mathbf{Q}(f)$. To each homomorphism $\varphi: \mathbf{Q}(f_i) \to C$ corresponds a C-linear operator $f_i^{(\varphi)}: W_i \to W_i$ called a *replica* of f and defined as follows [10].

Each eigenvector $x \in W_i$ of f is also an eigenvector of $f_i^{(\varphi)}$ and $f_i^{(\varphi)}x = \varphi(\alpha)x$ if $fx = \alpha x$ ($\alpha \in \operatorname{spec}(f_i) \subset \mathbf{Q}(f_i)$).

Clearly, the set $\operatorname{spec}(f_i^{(\varphi)})$ of all the eigenvalues of $f_i^{(\varphi)}$ coincides with $\varphi(\operatorname{spec}(f_i))$.

Since $\mathfrak{g}_i$ is simple, it is an algebraic Lie subalgebra of $\operatorname{End}(W_i)$ and, therefore, contains all the replicas of its elements [10]. This implies that

$$f_i^{(\varphi)} \in \mathfrak{g}_i \subset \operatorname{End}(W_i)$$

for all φ. Clearly, $f_i^{(\varphi)}$ is a semisimple element of $\mathfrak{g}_i$.

Since $\mathbf{Q}(f_i) \subset \mathbf{Q}(f)$, one may attach to each homomorphism $\psi: \mathbf{Q}(f) \to C$ its restriction $\psi': \mathbf{Q}(f_i) \to C$ and consider the corresponding replica

$$f_i^{(\psi')} \in \mathfrak{g}_i \subset \operatorname{End}(W_i).$$

Clearly, $f_i^{(\psi')} \neq 0$ if and only if the restriction of ψ to $\mathbf{Q}(f_i)$ does not vanish identically. We have

$$\operatorname{spec}(f_i^{(\psi')}) = \psi'(\operatorname{spec}(f_i)) = \psi(\operatorname{spec}(f_i)) \subset \psi(c_i + \operatorname{spec}(f)) =$$
$$= \psi(c_i) + \psi(\operatorname{spec}(f)) = \{\psi(c_i) + \psi(\alpha), \alpha \in \operatorname{spec}(f)\}.$$

Now, let us choose a homomorphism $\theta: \mathbf{Q}(f) \to \mathbf{Q} \subset C$ such that $\theta \in M$ and the restriction of θ to $\mathbf{Q}(f_i)$ does not vanish identically. Then

$$f_i^{(\theta')} \in \mathfrak{g}_i \subset \operatorname{End}(W_i)$$

is a non-zero semisimple operator and

$$\operatorname{spec}(f_i^{(\theta')}) \subset \theta(c_i) + \theta(\operatorname{spec}(f)) \subset \theta(c_i) + A.$$

In particular, $f_i^{(\theta')}$ has, at most, $\operatorname{card}(A)$ different eigenvalues.

Let me recall that, if a linear irreducible simple Lie algebra contains a non-zero semisimple operator with exactly m different eigenvalues, then the highest weight of the corresponding irreducible representation is the sum of no more than $(m-1)$ fundamental weights ([11], Th. 2.2).

Applying this assertion to a non-zero semisimple element $f_i^{(\theta')}$ of the linear irreducible simple Lie algebra $\mathfrak{g}_i \subset \operatorname{End}(W_i)$, we obtain that the highest weight of the simple $\mathfrak{g}_i$-module W_i is the sum of no more than $[\operatorname{card}(A)-1]$ fundamental weights. Q.E.D.

Proof of Claim. First let us assume that $s=1$, i.e., $\mathfrak{g}=\mathfrak{g}_1$ is simple and $W=W_1$. Then $f_1 = f - c\,\mathrm{id}_W \in \mathfrak{g}_1$ and

$$c = \operatorname{tr}(f,W)/\dim(W)$$

where $\operatorname{tr}(f,W)$ is the trace of $f\colon W \to W$. This implies that $c \in \mathbb{Q}(f)$ and

$$\operatorname{spec}(f_1) = (-c) + \operatorname{spec}(f).$$

One has only to put $c_1 = -c$.

Now, let us assume that $s>1$. For each j let us choose an eigenvalue $\beta_j \in \operatorname{spec}(f_i)$ $(1 \le j \le s)$. Then, for each $\alpha \in \operatorname{spec}(f_i)$, we have:

$$c + \alpha + \sum_{j \ne i} \beta_j \in \operatorname{spec}(f).$$

So, if we put $c_i = -(c + \sum_{j\ne i} \beta_j)$, then $\alpha \in c_i + \operatorname{spec}(f)$, i.e.

$$\operatorname{spec}(f_i) \subset c_i + \operatorname{spec}(f).$$

One has only to check that $c_i \in \mathbb{Q}(f)$. But we have the following explicit formula (recall that the trace of f_i vanishes and the sum of multiplicities of all eigenvalues of f_i is equal to $\dim(W_i)$):

$$c_i = -\left(\sum_{\alpha \in \operatorname{spec}(f_i)} \operatorname{mult}_i(\alpha)(c + \alpha + \sum_{j \ne i} \beta_j) \right) / \dim(W_i).$$

This formule implies that c_i is a linear combination of eigenvalues $c + \alpha + \sum_{j\ne i} \beta_j$ of f with rational coefficients, i.e. $c_i \in \mathbb{Q}(f)$. Q.E.D.

3. Applications to Abelian varieties.

Let X be an Abelian variety defined over K. Let $T_\ell(X)$ be the Tate $\mathbb{Z}_\ell$-module of X and

$$V_\ell(X) = T_\ell(X) \otimes_{\mathbf{Z}_\ell} \mathbf{Q}_\ell.$$

It is well-known that $V_\ell(X)$ is a $\mathbf{Q}_\ell$-vector space of dimension $2 \dim X$. There is a natural ℓ-adic representation [6,5]

$$\rho_\ell: G(K) \to \operatorname{Aut} V_\ell(X).$$

A theorem of Faltings [3] asserts that ρ_ℓ is semisimple and the centralizer of $G(K)$ in $\operatorname{End} V_\ell(X)$ coincides with $\operatorname{End}_K X \otimes \mathbf{Q}_\ell$. Here $\operatorname{End}_K X$ is the ring of all K-endomorphisms of X. This implies that the $\mathbf{Q}_\ell$-Lie algebra $\operatorname{Lie}(\operatorname{Im}(\rho_\ell))$ is reductive, its natural representation in $V_\ell(X)$ is semisimple and the centralizer of $\operatorname{Lie}(\operatorname{Im}(\rho_\ell))$ in $\operatorname{End} V_\ell(X)$ coincides with $\operatorname{End} X \otimes \mathbf{Q}_\ell$. Here $\operatorname{End} X$ is the ring of all endomorphisms of X (over $K(a)$). Recall that the ring $\operatorname{End} X$ is a free abelian group of finite rank. We write $\operatorname{rk}(\operatorname{End} X)$ for the rank of $\operatorname{End} X$.

Let us split the reductive $\mathbf{Q}_\ell$-Lie algebra $\operatorname{Lie}(\operatorname{Im}(\rho_\ell))$ into the direct sum

$$\operatorname{Lie}(\operatorname{Im}(\rho_\ell)) = \mathfrak{c}_\ell \oplus \mathfrak{g}_\ell$$

of its center $\mathfrak{c}_\ell$ and a semisimple $\mathbf{Q}_\ell$-Lie algebra $\mathfrak{g}_\ell$. Let $r(X)$ be the rank of $\mathfrak{g}_\ell$. The results of [7] combined with Faltings' theorem imply that $r(X)$ does not depend on ℓ.

3.1 Theorem. *Let us put*

$$H = H(r(X)) = \max\{D(j,1),\ 0 \leq j \leq r(X)\}$$

where D are as in Theorem 0.5. Then

$$\dim(X) \leq H \operatorname{rk}(\operatorname{End} X)/2.$$

In particular, the dimension of X is bounded above by $\operatorname{rk}(\operatorname{End} X)$ times a certain constant, depending only on $r(X)$.

Example. If $r(X) = 0$, then X is of CM-type and $\dim X \leq \operatorname{rk}(\operatorname{End} X)/2$.

Remark. If $r(X) = 1$, then the results of [9] imply that $\dim X \leq \operatorname{rk}(\operatorname{End} X)$.

Remark. One may deduce from several conjectures [8] (e.g., the conjecture of Mumford-Tate or a conjecture of Serre [12]) that $\dim X$ does not exceed $2^{r(X)-1} \operatorname{rk}(\operatorname{End} X)$.

3.2. Proof of Theorem 3.1.

In the course of the proof we may and will assume that all endomorphisms of X are defined over K and X is absolutely simple. Then $\operatorname{End}_0 X = \operatorname{End} X \otimes \mathbf{Q}$ is a division algebra of finite dimension over $\mathbf{Q}$. Let us fix a maximal commutative $\mathbf{Q}$-subalgebra E in $\operatorname{End}_0 X$. Then E is a number field, coinciding with its centralizer in $\operatorname{End}_0 X$; the degree $[E : \mathbf{Q}]$ divides divides $\operatorname{rk}(\operatorname{End} X)$. In particular,

$$[E:\mathbb{Q}] \leq \operatorname{rk}(\operatorname{End} X).$$

In addition, $[E:\mathbb{Q}]$ divides $2\dim X$ and the natural embedding

$$E \otimes_{\mathbb{Q}} \mathbb{Q}_\ell \to \operatorname{End}_0 X \otimes_{\mathbb{Q}} \mathbb{Q}_\ell = \operatorname{End} X \otimes \mathbb{Q}_\ell \subset \operatorname{End} V_\ell(X)$$

provides $V_\ell(X)$ with the structure of a free $E \otimes_{\mathbb{Q}} \mathbb{Q}_\ell$-module of rank $2\dim X/[E:\mathbb{Q}]$ [5].

Let $\mathfrak{O}$ be the ring of integers in E. There is a natural splitting

$$E \otimes_{\mathbb{Q}} \mathbb{Q}_\ell = \oplus_\lambda E_\lambda$$

where λ runs through the set of dividing ℓ prime ideals in $\mathfrak{O}$. Clearly,

$$[E:\mathbb{Q}] = \sum [E_\lambda : \mathbb{Q}_\ell].$$

Since $V_\ell(X)$ is a free $E \otimes_{\mathbb{Q}} \mathbb{Q}_\ell$-module of rank $2\dim X/[E:\mathbb{Q}]$, there is a natural splitting

$$V_\ell(X) = \oplus V_\lambda$$

where $V_\lambda = E_\lambda V_\ell(X)$ is an E_λ-vector space of dimension $2\dim X/[E:\mathbb{Q}]$. Clearly, each V_λ is $G(K)$-invariant and ρ_ℓ is the direct sum of the corresponding λ-adic representations

$$\rho_\lambda : G(K) \to \operatorname{Aut}_{E_\lambda} V_\lambda.$$

One may easily check, using Faltings' theorem, that each ρ_λ is an absolutely irreducible (and even infinitesimally absolutely irreducible) λ-adic representation (see [9], Sect. 0.11.1).

Let us split the reductive $\mathbb{Q}_\ell$-Lie algebra $\operatorname{Lie}(\operatorname{Im}(\rho_\lambda))$ into the direct sum

$$\operatorname{Lie}(\operatorname{Im}(\rho_\lambda)) = \mathfrak{c}_\lambda \oplus \mathfrak{g}_\lambda$$

of its center $\mathfrak{c}_\lambda$ and a semisimple $\mathbb{Q}_\ell$-Lie algebra $\mathfrak{g}_\lambda$. Let r'_λ be the rank of $\mathfrak{g}_\lambda$.

Claim. $r'_\lambda \leq r(X)$.

In order to prove this inequality it suffices to construct a surjective homomorphism $\mathfrak{g}_\ell \to \mathfrak{g}_\lambda$ of semisimple $\mathbb{Q}_\ell$-Lie algebras. In turn, in order to construct such a homomorphism it suffices to construct a surjective homomorphism

$$\mathfrak{c}_\ell \oplus \mathfrak{g}_\ell = \operatorname{Lie}(\operatorname{Im}(\rho_\ell)) \to \mathfrak{c}_\lambda \oplus \mathfrak{g}_\lambda = \operatorname{Lie}(\operatorname{Im}(\rho_\lambda))$$

of reductive $\mathbb{Q}_\ell$-Lie algebras and take its restriction to $\mathfrak{g}_\ell$. But it is very easy to construct the latter homomorphism. One has only to consider the surjective homomorphism $\operatorname{Im}(\rho_\ell) \to \operatorname{Im}(\rho_\lambda)$ of $\mathbb{Q}_\ell$-Lie groups induced by the projection map $V_\ell(X) \to V_\lambda$, and take the corresponding homomorphism of $\mathbb{Q}_\ell$-Lie algebras.

It is well-known [6,5] that for all but finitely many places v of K the following conditions hold:

1) ρ is unramified at v;

2) the characteristic polynomial $\det(t\,\mathrm{id} - \mathrm{Fr}_v, V_\lambda)$ lies in $\mathfrak{O}[t]$; all its (complex) roots and their conjugate over $\mathbf{Q}$ have absolute value $q(v)^{1/2}$ (a theorem of A. Weil).

In order to obtain an E-integral λ-adic representation of weight 1 let us consider the dual E_λ-vector space

$$V_\lambda^* = \mathrm{Hom}_{E_\lambda}(V_\lambda, E_\lambda)$$

and the isomorphism

$$\tau\colon \mathrm{Aut}_{E_\lambda}(V_\lambda) \to \mathrm{Aut}_{E_\lambda}(V_\lambda^*)$$

defined by the formula $\tau(u) = (u^*)^{-1}$ where u^* is the adjoint of u.

Clearly, $\dim_{E_\lambda} V_\lambda = \dim_{E_\lambda} V_\lambda^*$.

Let us consider the dual λ-adic representation

$$\rho_\lambda^* = \tau\rho_\lambda\colon G(K) \to \mathrm{Aut}_{E_\lambda}(V_\lambda) \to \mathrm{Aut}_{E_\lambda}(V_\lambda^*).$$

Clearly, ρ_λ^* is an E-integral λ-adic representation of weight 1. One may easily check that ρ_λ^* is also infinitesimally absolutely irreducible. Notice that τ induces an isomorphism $\mathrm{Im}(\rho_\lambda) \cong \mathrm{Im}(\rho_\lambda^*)$ of $\mathbf{Q}_\ell$-Lie groups, which, in turn, induces an isomorphism

$$\mathrm{Lie}(\mathrm{Im}(\rho_\lambda)) \cong \mathrm{Lie}(\mathrm{Im}(\rho_\lambda^*))$$

of the corresponding $\mathbf{Q}_\ell$-Lie algebras. This implies that the rank of the semisimple part of the $\mathbf{Q}_\ell$-Lie algebra $\mathrm{Lie}(\mathrm{Im}(\rho_\lambda^*))$ is also equal to r'_λ and, therefore, does not exceed $r(X)$.

Applying Corollary of Theorem 0.5 to the infinitesimally absolutely irreducible E-integral λ-adic representation ρ_λ^* of weight 1 we obtain that

$$\dim_{E_\lambda} V_\lambda^* \le \max\{D(j,1), 0 \le j \le r'_\lambda\}.$$

Since $r'_\lambda \le r(X)$ and $\dim_{E_\lambda} V_\lambda^* = \dim_{E_\lambda} V_\lambda$,

$$\begin{aligned}\dim_{\mathbf{Q}} V_\lambda &= [E_\lambda : \mathbf{Q}_\ell]\dim_{E_\lambda} V_\lambda\\ &\le [E_\lambda : \mathbf{Q}_\ell]\max\{D(j,1), 0 \le j \le r(X)\}\\ &= [E_\lambda : \mathbf{Q}_\ell]H.\end{aligned}$$

Summing up over λ we obtain that

$$\begin{aligned}2\dim X = \dim_{\mathbf{Q}_\ell} V_\ell(X) = \sum \dim_{\mathbf{Q}_\ell} V_\lambda(X) \le\\ \le H\sum [E_\lambda : \mathbf{Q}_\ell] = H[E : \mathbf{Q}] \le H\,\mathrm{rk}(\mathrm{End}\,X).\end{aligned}$$

References

[1] N. Bourbaki, Groupes et algèbres de Lie, Ch. 2–3, Hermann, Paris, 1972.

[2] N. Bourbaki, Groupes et algèbres de Lie, Ch. 7–8, Hermann, Paris, 1975.

[3] G. Faltings, Endlichkeitssätze für Abelsche Varietäten über Zahlkörpern, Invent. Math., 73, 349–366 (1983); erratum, 75, 381 (1984).

[4] G. Henniart, Répresentations ℓ-adiques abéliennes, Séminaire de Théorie des Nombres, Paris, 1980–81, Progress in Math., 22, Birkhäuser, 1981, 107–126.

[5] K.A. Ribet, Galois actions on division points of abelian varieties with many real multiplications, Amer. J. of Math, 98, 751–804 (1976).

[6] J.-P. Serre, Abelian ℓ-adic representations and elliptic curves, Benjamin, New York, Amsterdam, 1968; 2nd edition, Addison-Wesley, 1989.

[7] Yu. G. Zarhin, Abelian ℓ-adic representations and Lie algebras. Rank indepence on ℓ, Invent. Math., 55, 165–176 (1979).

[8] Yu. G. Zarhin, Weights of simple Lie algebras in the cohomology of algebraic varieties, Math. USSR Izvestija, 24, 245–282 (1985).

[9] Yu. G. Zarhin, Torsion of Abelian varieties over GL (2)-extensions of number fields, Math. Ann., 284, 631-646 (1989).

[10] C. Chevalley, Théorie des groupes de Lie, Groupes algébriques, Hermann, Paris, 1961.

[11] Yu. G. Zarhin, Linear semisimple Lie algebras containing an operator with small number of eigenvalues, Arch. Math., Basel, 46, 522–532 (1986).

[12] J.-P. Wintenberger, Motifs et ramification pour les points d'ordre fini des variétés abéliennes, Séminaire de Théorie des Nombres, Paris, 1986–87, Progress in Math., 75, Birkhäuser, 1988, 453–471.

Progress in Mathematics

Edited by:

J. Oesterlé
Département de Mathématiques
Université de Paris VI
4, Place Jussieu
75230 Paris Cedex 05
France

A. Weinstein
Department of Mathematics
University of California
Berkeley, CA 94720
U.S.A.

Progress in Mathematics is a series of books intended for professional mathematicians and scientists, encompassing all areas of pure mathematics. This distinguished series, which began in 1979, includes authored monographs and edited collections of papers on important research developments as well as expositions of particular subject areas.

All books in the series are "camera-ready," that is they are photographically reproduced and printed directly from a final-edited manuscript that has been prepared by the author. Manuscripts should be no less than 100 and preferably no more than 500 pages.

Proposals should be sent directly to the editors or to: Birkhäuser Boston, 675 Massachusetts Avenue, Suite 601, Cambridge, MA 02139, U.S.A.

A complete list of titles in this series is available from the publisher.

37 BOUTET DE MONVEL/DOUADY/VERDIER. Mathématique et Physique
38 BERTIN. Séminaire de Théorie des Nombres, Paris 1981–82
39 UENO. Classification of Algebraic and Analytic Manifolds
40 TROMBI. Representation Theory of Reductive Groups
41 STANELY. Combinatorics and Commutative Algebra
42 JOUANOLOU. Théorèmes de Bertini et Applications
43 MUMFORD. Tata Lectures on Theta II
44 KAC. Infinite Dimensional Lie Algebras
45 BISMUT. Large Deviations and the Malliavin Calculus
46 SATAKE/MORITA. Automorphic Forms of Several Variables, Taniguchi Symposium, Katata, 1983
47 TATE. Les Conjectures de Stark sur les Fonctions L d'Artin en s = 0
48 FRÖHLICH. Classgroups and Hermitian Modules
49 SCHLICHTKRULL. Hyperfunctions and Harmonic Analysis on Symmetric Spaces
50 BOREL, ET AL. Intersection Cohomology
51 BERTIN/GOLDSTEIN. Séminaire de Théorie des Nombres, Paris 1982–83
52 GASQUI/GOLDSCHMIDT. Déformations Infinitésimales des Structures Conformes Plates
53 LAURENT. Théorie de la Deuxième Microlocalisation dans le Domaine Complexe
54 VERDIER/LE POTIER. Module des Fibres Stables sur les Courbes Algébriques: Notes de l'Ecole Normale Supérieure, Printemps, 1983
55 EICHLER/ZAGIER. The Theory of Jacobi Forms

56 Shiffman/Sommese. Vanishing Theorems on Complex Manifolds
57 Riesel. Prime Numbers and Computer Methods for Factorization
58 Helffer/Nourrigat. Hypoellipticité Maximale pour des Opérateurs Polynomes de Champs de Vecteurs
59 Goldstein. Séminaire de Théorie des Nombres, Paris 1983–84
60 Procesi. Geometry Today: Giornate Di Geometria, Roma. 1984
61 Ballmann/Gromov/Schroeder. Manifolds of Nonpositive Curvature
62 Guillou/Marin. A la Recherche de la Topologie Perdue
63 Goldstein. Séminaire de Théorie des Nombres, Paris 1984–85
64 Myung. Malcev-Admissible Algebras
65 Grubb. Functional Calculus of Pseudo-Differential Boundary Problems
66 Cassou-noguès/Taylor. Elliptic Functions and Rings and Integers
67 Howe. Discrete Groups in Geometry and Analysis: Papers in Honor of G.D. Mostow on His Sixtieth Birthday
68 Robert. Autour de L'Approximation Semi-Classique
69 Faraut/Harzallah. Deux Cours d'Analyse Harmonique
70 Adolphson/Conrey/Ghosh/Yager. Number Theory and Diophantine Problems: Proceedings of a Conference at Oklahoma State University
71 Goldstein. Séminaire de Théorie des Nombres, Paris 1985–1986
72 Vaisman. Symplectic Geometry and Secondary Characteristic Classes
73 Molino. Riemannian Foliations
74 Henkin/Leiterer. Andreotti–Grauert Theory by Integral Formulas
75 Goldstein. Séminaire de Théorie des Nombres, Paris 1986–87
76 Cossec/Dolgachev. Enriques Surfaces I
77 Reyssat. Quelques Aspects des Surfaces de Riemann
78 Borho/Brylinski/MacPherson. Nilpotent Orbits, Primitive Ideals, and Characteristic Classes
79 McKenzie/Valeriote. The Structure of Decidable Locally Finite Varieties
80 Kraft/Petrie/Schwarz. Topological Methods in Algebraic Transformation Groups
81 Goldstein. Séminaire de Théorie des Nombres, Paris 1987–1988
82 Duflo/Pedersen/Vergne. The Orbit Method in Representation Theory: Proceedings of a Conference Held in Copenhagen, August to September 1988
83 Ghys/de la Harpe. Sur les Groupes Hyperboliques d'après Mikhael Gromov
84 Araki/Kadison. Mappings of Operator Algebras: Proceedings of the Japan-U.S. Joint Seminar, University of Pennsylvania, Philadelphia, Pennsylvania, 1988
85 Berndt/Diamond/Halberstam/Hildebrand. Analytic Number Theory: Proceedings of a Conference in Honor of Paul T. Bateman
86 Cartier et al. The Grothendieck Festschrift, Volume One
87 Cartier et al. The Grothendieck Festschrift, Volume Two
88 Cartier et al. The Grothendieck Festschrift, Volume Three
89 van der Geer/Oort/Steenbrink. Arithmetic Algebraic Geometry
90 Srinivas. Algebraic K-Theory
91 Goldstein. Seminaire de Theorie des Nombres, Paris 1988-89
92 Connes/Duflo/Joseph/Rentschler. Operator Algebras, Unitary Representations, Enveloping Algebras, and Invariant Theory: Actues du Colloque en l'honneur de Jacques Dixmier